PHYSICAL
ELECTROCHEMISTRY

MONOGRAPHS IN ELECTROANALYTICAL CHEMISTRY AND ELECTROCHEMISTRY

consulting editor
Allen J. Bard
Department of Chemistry
University of Texas
Austin, Texas

Electrochemistry at Solid Electrodes, *Ralph N. Adams*

Electrochemical Reactions in Nonaqueous Systems, *Charles K. Mann and Karen Barnes*

Electrochemistry of Metals and Semiconductors: The Application of Solid State Science to Electrochemical Phenomena, *Ashok K. Vijh*

Modern Polarographic Methods in Analytical Chemistry, *A. M. Bond*

Laboratory Techniques in Electroanalytical Chemistry, *edited by Peter T. Kissinger and William R. Heineman*

Standard Potentials in Aqueous Solution, *edited by Allen J. Bard, Roger Parsons, and Joseph Jordan*

Physical Electrochemistry: Principles, Methods, and Applications, *edited by Israel Rubinstein*

ADDITIONAL VOLUMES IN PREPARATION

PHYSICAL ELECTROCHEMISTRY

PRINCIPLES, METHODS, AND APPLICATIONS

EDITED BY
ISRAEL RUBINSTEIN

Department of Materials and Interfaces
Weizmann Institute of Science
Rehovot, Israel

Marcel Dekker, Inc.　　　　New York•Basel•Hong Kong

Library of Congress Cataloging-in-Publication Data

Physical electrochemistry : principles, methods, and applications /
    edited by Israel Rubinstein.
        p.   cm. — (Monographs in electroanalytical chemistry and
    electrochemistry)
        Includes bibliographical references and index.
        ISBN 0-8247-9452-4 (acid-free)
        1. Electrochemistry.   I. Rubinstein, Israel
    II. Series.
    QD553.P48   1995
    541.3'7—dc20   94-47122
                                                                         CIP

The publisher offers discounts on this book when ordered in bulk quantities. For more information, write to Special Sales/Professional Marketing at the address below.

This book is printed on acid-free paper.

Marcel Dekker, Inc.
270 Madison Avenue, New York, New York 10016

Current printing (last digit):
10   9   8   7   6   5   4   3   2   1

PRINTED IN THE UNITED STATES OF AMERICA

Preface

The science of electrochemistry dates back to the late 18th century, when the first demonstrations of the relationship between chemistry and electricity were performed by Galvani and later by Volta. It was Michael Faraday who, in 1835, established electrochemistry as a quantitative science and contributed more than anyone else to the advancement of this field. The lead-acid battery, an incredible workhorse in applied electrochemistry, has already celebrated its 130th birthday. Throughout the years, electrochemistry has developed in various directions, including analytical, physical, synthetic, biological, and industrial electrochemistry.

For many years, electrochemistry has been considered a highly specialized field with relevance to certain specific scientific and technological endeavors. This included, in the early days, areas such as polarography and electroanalytical chemistry, reaching a high point with the granting of the Nobel Prize to J. Heyrovsky in 1959; electrode kinetics and organic electrochemistry in the 1950s and 1960s; energy-oriented problems in the 1970s, e.g., semiconductor electrodes and novel batteries and fuel cells; and chemically modified electrodes, primarily for electrocatalysis, in the 1980s.

This situation has changed quite dramatically in recent years, with the general trend in science towards interdisciplinary research. This has been expressed in two ways: 1) The introduction of numerous nonelectrochemical techniques to electrochemical studies, including *in situ* techniques, and 2) a notable trend of introducing electrochemical measurements to many fields of science. More than ever before, electrochemists now address problems of general scientific interest and use a large variety of other techniques, while researchers in many fields routinely resort to electrochemical measurements to obtain essential information. As a consequence, electrochemistry has become an important facet of modern science, especially in surface and materials science.

The aim of this book is to respond to the above trend by providing updated presentations of various areas of current interest to the electrochemist of broad scope, as well as to surface scientists, materials scientists, and physicists, who have already discovered or may discover that electrochemistry is relevant to their own research.

"Traditional" areas of electrochemistry (e.g., batteries, metal deposition, corrosion) and standard electrochemical techniques (e.g., polarography, voltammetry, transients, rotating disk electrode) which are well covered in existing textbooks, series, and reviews, are not directly addressed in this book. Areas related to organic, biological, and industrial electrochemistry are mentioned only indirectly, due to the stated scope of the book. As a rule, this book is concerned with *interfacial* electrochemistry, hence topics related to *solution* electrochemistry (such as ions in solution, ion transport, and electrolytic conductivity) are also not included.

For additional information, the interested reader may consult the list of selected texts on basic electrochemical concepts and techniques that appears in Chapter 1 of this book.

Israel Rubinstein

Contents

Contributors

Christian Amatore   Département de Chimie, Ecole Normale Supérieure, URA CNRS 1679, Paris, France

Allen J. Bard   Department of Chemistry and Biochemistry, The University of Texas at Austin, Austin, Texas

Eugene Y. Cao   Department of Radiation Therapy, Medical College of Ohio, Toledo, Ohio

A. F. Diaz   IBM Alamaden Research Center, San Jose, California

Fu-Ren Fan   Department of Chemistry and Biochemistry, The University of Texas at Austin, Austin, Texas

Claude Gabrielli   Physique des Liquides et Electrochimie, UPR15 CNRS, Université Pierre et Marie Curie, Paris, France

Shimshon Gottesfeld   Electronic and Electrochemical Materials and Devices, Los Alamos National Laboratory, Los Alamos, New Mexico

Gary Hodes   Department of Materials and Interfaces, Weizmann Institute of Science, Rehovot, Israel

Arthur T. Hubbard   Department of Chemistry, University of Cincinnati, Cincinnati, Ohio

Yeon-Taik Kim*   Electronic and Electrochemical Materials and Devices, Los Alamos National Laboratory, Los Alamos, New Mexico

Mario Leclerc   Department of Chemistry, University of Montreal, Montreal, Quebec, Canada

James McBreen   Department of Applied Science, Brookhaven National Laboratory, Upton, New York

Cary J. Miller   Department of Chemistry and Biochemistry, University of Maryland, College Park, Maryland

Present affiliation: Department of Chemistry, University of Alabama, Tuscaloosa, Alabama.

Michael Mirkin   Department of Chemistry, Queens College—CUNY, Flushing, New York

My T. Nguyen   Corporate Research and Development, Polychrome Corporation, Carlstadt, New Jersey

Antonio Redondo   Electronic and Electrochemical Materials and Devices, Los Alamos National Laboratory, Los Alamos, New Mexico

Israel Rubinstein   Department of Materials and Interfaces, Weizmann Institute of Science, Rehovot, Israel

M. Rudolph   Department of Chemistry, Friedrich-Schiller-Universität, Jena, Germany

Donald A. Stern   Chevron Chemical Company, Kingwood, Texas

Michael D. Ward   Department of Chemical Engineering and Materials Science, University of Minnesota, Minneapolis, Minnesota

1

Fundamentals of Physical Electrochemistry

ISRAEL RUBINSTEIN
Weizmann Institute of Science
Rehovot, Israel

I. INTRODUCTION

Electrochemistry, as the name suggests, is the field of science that deals with the relation between electrical current or potential and chemical systems. It covers quite a number of areas, including physical electrochemistry, analytical electrochemistry, electroorganic chemistry, and bioelectrochemistry, as well as applied areas such as corrosion, batteries, fuel cells, and solar energy conversion. In light of the scope of this book, this chapter is concerned only with physical electrochemistry: the basic physical principles that govern electrochemical systems.

Traditionally, physical electrochemistry has been divided into two major areas:

1. *Bulk electrochemistry* (i.e., ions in solution). This includes such topics as ion–solvent and ion–ion interactions, activity coefficients, the Debye–Hückel theory, ionic mobility and conductivity, and transport

numbers. This facet of physical electrochemistry is not addressed directly in this chapter.*

2. *Interfacial electrochemistry.* This includes topics such as the nature of the electrode–solution interphase, thermodynamics and kinetics of electrode reactions, and mass-transport effects.

This chapter deals with the latter: that is, interfacial phenomena and electrode reactions.

II. BASIC FEATURES OF ELECTRODE REACTIONS

Electrode reactions are chemical reactions in essence, and as a rule proceed according to the usual chemical principles. There are, however, certain properties that are characteristic of, or unique to, electrode reactions, and these are listed below.

1. Electrode reactions always involve electron transfer and hence are categorized as oxidation–reduction reactions.
2. Electrode reactions are heterogeneous in nature, which implies heterogeneous kinetics, reaction rates that scale with the electrode area, and a strong effect of interfacial properties on the electrode reaction.
3. The current is a simple and direct measurement of the reaction rate.
4. Changes in the reaction rate can be induced by variation of the concentrations or the temperature, much the same as in usual chemical kinetics. Unique to electrode reactions is the possibility of influencing the rate upon changing the electrode potential.
5. The passage of a current through an electrochemical cell represents a transition between electronic conductivity (in the electrodes) and ionic conductivity (in the solution). This implies that the current would be controlled not only by the solution resistance, but to a great extent by the interfacial impedances at the two electrode–solution interfaces. The latter, in turn, depend on such factors as electrode kinetics, interfacial capacitance, and mass-transport limitations.

III. ELECTRICAL ANALOG OF AN ELECTROCHEMICAL INTERPHASE

The electrode–solution interphase behaves in a manner analogous to a parallel combination of a resistor R_{ct} and a capacitor C_d, and hence can be represented by an electrical equivalent circuit (Fig. 1). R_{ct} represents the current–potential relationship associated with the charge-transfer process (see Section VI) and is

---

*The reader interested in these aspects may refer, for example, to Vol. 1 of Ref. 1.

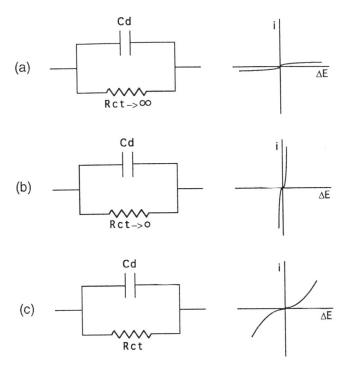

FIGURE 1 Electrical equivalent circuits for the electrode–solution interface and corresponding current–potential (i–ΔE) curves: (a) highly polarizable; (b) reversible; (c) intermediate. WE and CE are the working and counter electrode connections, respectively. The various electrical elements are described in the text.

potential dependent. C_d represents the capacitance associated with the ionic double layer (see Section V) and is also potential dependent.

Figure 1a corresponds to a situation where R_{ct} is very large. Under these conditions the potential across the interface can be changed substantially without causing any significant flow of current. This is a *polarizable interface*. Figure 1b represents the opposite situation (i.e., R_{ct} is very small). Here the potential is fixed by the solution composition (and is given by the Nernst equation; see Section VII.C). Any change in the potential causes substantial flow of current until new equilibrium concentrations are established at the interface. This is a *nonpolarizable* or *reversible* interface. Figure 1c represents an intermediate case, where R_{ct} has a certain finite value.

It should be noted that a complete electrochemical cell includes at least two interfaces (two electrodes), as shown in Figure 2a. R_s is the solution resis-

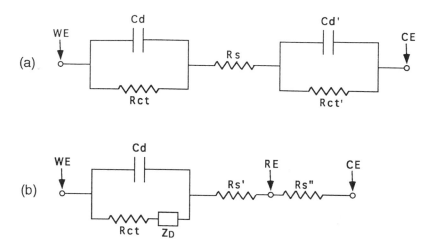

FIGURE 2 Electrical equivalent circuits for (a) a two-electrode system, and (b) a single electrode–solution interface with a mass-transport element (Z_D) and a reference connection (RE).

tance. WE and CE are the working and counter (auxiliary) electrodes, respectively. In a two-electrode configuration they are chosen such that the CE is highly reversible compared with the WE. Under these conditions changes in the applied (or measured) cell voltage can be assigned, to a good approximation, to changes in the WE polarization.

To complete the picture, Figure 2b shows a half-cell equivalent circuit which includes, in addition to the above-mentioned elements, a high-impedance reference electrode connection (RE), which serves to measure the WE potential virtually at open circuit. The RE therefore separates the compensated (R'_s) and uncompensated (R''_s) parts of the solution resistance. In addition, the circuit includes a mass-transport impedance Z_D in series with R_{ct}. This latter impedance corresponds to overpotential that is developed as a result of concentration gradients of reactants and products that are built up upon passage of the current.

IV. ELECTROCHEMICAL THERMODYNAMICS

A. Electrochemical Potentials

Chemical systems (at constant temperature T and pressure P) will proceed spontaneously in the direction that effects a decrease in the Gibbs free energy G or the chemical potential μ:

$$^{\alpha}\mu_i = \left(\frac{\partial G}{\partial n_i}\right)_{T,P,n_{j \neq i}} \tag{1}$$

where $^{\alpha}\mu_i$ is the chemical potential of species i in phase α. Chemical equilibrium is therefore characterized by the equality of chemical potentials.

If the system includes charged species, an electrochemical potential $\overline{\mu}$ can be defined as

$$\overline{\mu}_i = \mu_i + z_iF\phi = \mu_i^{\circ} + RT \ln a_i + z_iF\phi \tag{2}$$

where z_i is the charge of species i (in electron charge units), F the Faraday constant, ϕ the inner electrical potential of the phase (the *Galvani potential*), μ_i° a constant, and a_i the activity of species i. Hence the driving force in electrochemical systems is the gradient of the *electrochemical potential*. In other words, species i is at equilibrium between two phases α and β if $^{\alpha}\overline{\mu}_i = {}^{\beta}\overline{\mu}_i$.

Several points should be noted here:

1. The inner potential $^{\alpha}\phi$ of a single phase α cannot be measured. This is a direct consequence of the basic definition of electrical potential,* namely, that any test charge would interact with the medium upon crossing the phase boundary.
2. The inner potential difference across a single phase boundary, $^{\alpha}\Delta^{\beta}\phi$, is also not measurable. This results from the fact that the measurement necessarily involves *two* interfaces (e.g., two electrodes).
3. The quantity usually measured in electrochemistry is $\delta\Delta\phi$ (i.e., *changes* in the inner potential difference). Under suitable conditions, $\delta\Delta\phi$ for a single electrode–solution interface can be measured.†
4. The separation of the electrochemical potential $\overline{\mu}$ into a chemical term (μ) and an electrical term ($zF\phi$) is artificial, as any "chemical" interaction is electrical in nature. Hence the terms μ and $zF\phi$ cannot be determined independently, while $\overline{\mu}$ is measurable in principle. The separation is, however, useful in deriving various electrochemical relationships, as exemplified in the following.

Consider an electrochemical cell comprising a Zn/Zn^{2+} half-cell connected to a Br^-/Br_2 half-cell; both electrodes (Zn and Pt) are connected to a high-

---

*The electrical potential at a certain point is the coulombic energy required to bring a unit positive charge from infinity to that point.

†Such conditions are obtained when changes in the working electrode potential are measured vs. a highly reversible counterelectrode, or vs. a nonpolarized reference electrode (in the latter case, using a potentiostat).

impedance voltmeter through copper leads:

$$Cu/Zn/Zn^{2+} : H_2SO_4 : Br^-, Br_2/Pt/Cu'$$

The two half-cell reactions are

$$Zn \rightleftharpoons Zn^{2+} + 2e(Cu)$$
$$Br_2 + 2e(Pt) \rightleftharpoons 2Br^-$$

The overall cell reaction is

$$Zn + Br_2 + 2e(Pt) \rightleftharpoons Zn^{2+} + 2Br^- + 2e(Cu)$$

Equilibrium conditions can be expressed as

$$^{Zn}\overline{\mu}_{Zn} + {}^{S}\overline{\mu}_{Br_2} + 2\,{}^{Pt}\overline{\mu}_e = {}^{S}\overline{\mu}_{Zn2+} + 2\,{}^{S}\overline{\mu}_{Br-} + 2\,{}^{Cu}\overline{\mu}_e$$

The electrons are at equilibrium between the Pt electrode and the Cu' lead, (i.e., $^{Pt}\overline{\mu}_e = {}^{Cu'}\overline{\mu}_e$); hence

$$2({}^{Cu'}\overline{\mu}_e - {}^{Cu}\overline{\mu}_e) = {}^{S}\overline{\mu}_{Zn2+} + 2\,{}^{S}\overline{\mu}_{Br-} - {}^{Zn}\overline{\mu}_{Zn} - {}^{S}\overline{\mu}_{Br_2}$$

For electrons in a metal M one can write, using equation (2): $^{M}\overline{\mu}_e = \mu_e^\circ - F{}^{M}\phi$; hence

$$-2F({}^{Cu'}\phi - {}^{Cu}\phi) = {}^{S}\overline{\mu}_{Zn2+} + 2\,{}^{S}\overline{\mu}_{Br-} - {}^{Zn}\overline{\mu}_{Zn} - {}^{S}\overline{\mu}_{Br_2}$$

The term in parentheses, $\Delta\phi$, is the equilibrium (open-circuit) cell voltage ΔE_r, measured between the two Cu leads. Substituting and using equation (2), one obtains

$$-2F\,\Delta E_r = \mu_{Zn2+}^\circ + RT \ln a_{Zn2+} + 2F^S\phi + 2\mu_{Br-}^\circ$$
$$+ 2RT \ln a_{Br-} - 2F^S\phi - \mu_{Zn}^\circ - \mu_{Br_2}^\circ - RT \ln a_{Br_2}$$

or, upon gathering all the constants,

$$-2F\,\Delta E_r = \Delta G^\circ + RT \ln \frac{a_{Zn2+}\, a_{Br-}^2}{a_{Br_2}} \tag{3}$$

which is the *Nernst equation* for the cell, as shown in Section IV.B.

B. Basic Equations

Here one wishes to correlate the open-circuit cell voltage ΔE_r with the Gibbs free energy ΔG. In an electrochemical system, ΔG is defined as the reversible energy dissipated through an infinite load resistance such that the current $i \rightarrow 0$. Under these conditions,

$$|\Delta G| = q\,|\Delta E_r| = nF\,|\Delta E_r| \tag{4}$$

where q is the charge and n is the number of electrons transferred in the overall reaction.

A positive ΔE_r is defined as that corresponding to the spontaneous direction of the reaction, and accordingly,

$$\Delta G = -nF\,\Delta E_r \tag{5}$$

When the cell is under standard conditions and all the components are at unit activity, one can write

$$\Delta G° = -nF\,\Delta E_r° \tag{6}$$

Since $\Delta G = \Delta H - T\,\Delta S$, it follows that

$$\Delta S = -\left(\frac{\partial \Delta G}{\partial T}\right)_p = nF\left(\frac{\partial \Delta E r}{\partial T}\right)_p \tag{7}$$

and using standard thermodynamic relationships, one obtains readily

$$-\Delta G° = nF\,\Delta E_r° = RT\ln K_{eq} \tag{8}$$

where K_{eq} is the equilibrium constant of the overall cell reaction.

The *Nernst equation* for the electrode reaction $Ox + ne \rightleftharpoons Red$ can be derived from the relationship $\Delta G = \Delta G° + RT\ln(a_{Red}/a_{Ox})$. Using equations (5) and (6), one obtains the Nernst equation:

$$\Delta E_r = \Delta E_r° + \frac{RT}{nF}\ln\frac{a_{Ox}}{a_{Red}} \tag{9}$$

Note that substituting equation (6) in (3) produces the usual form of the Nernst equation for the cell.

C. Half-Cell Reactions and the Hydrogen Scale

An electrochemical cell is composed of two electrochemical half-cells, to which individual half-cell potentials can be assigned. The cell voltage will be the sum of the two individual half-cell potentials, which can readily be justified on the basis of equation (5) or (6). The standard way of performing such calculations is by using *reduction potentials* and a *hydrogen scale* (i.e., the half-cell potential of the standard hydrogen electrode is defined arbitrarily as zero). An example is shown below, where the electrochemical cell comprises a Fe^{3+}/Fe^{2+} half-cell connected to a standard hydrogen electrode:

$$Fe^{3+} + e \rightleftharpoons Fe^{2+} \qquad E°(Fe^{3+}/Fe^{2+})$$
$$2H^+(a = 1) + 2e \rightleftharpoons H_2(a = 1) \qquad E°(H^+/H_2) \equiv 0.000\ V$$

where $E°$ is a standard electrode potential. The measured cell voltage is 0.771 V, and the Fe^{3+}/Fe^{2+} electrode is positive.

The overall cell reaction

$$Fe^{3+} + \tfrac{1}{2}H_2(a = 1) \rightleftharpoons Fe^{2+} + H^+(a = 1)$$

$$\Delta E^\circ = E^\circ \ (Fe^{3+}/Fe^{2+}) - E^\circ \ (H^+/H_2) = E^\circ \ (Fe^{3+}/Fe^{2+}) = 0.771 \text{ V}$$

Using the relationship

$$\Delta G = \Delta G^\circ + RT \ln \frac{a_{Fe2+} \ a_{H+}}{a_{Fe3+} \ a_{H_2}^{1/2}}$$

and the unit activities of H_2 and H^+, one readily obtains

$$E_r = E^\circ \ (Fe^{3+}/Fe^{2+}) + \frac{RT}{F} \ln \frac{a_{Fe3+}}{a_{Fe2+}}$$

which is the Nernst equation for the half-cell.

D. Formal Potentials

For the general half-cell reaction $Ox + ne \rightleftharpoons Red$, the reversible (open-circuit) half-cell potential is given by the Nernst equation:

$$E_r = E^\circ + \frac{RT}{nF} \ln \frac{a_{Ox}}{a_{Red}}$$

For convenience, it is common to use concentrations rather than activities:

$$E_r = E^\circ + \frac{RT}{nF} \ln \frac{\gamma_{Ox}[Ox]}{\gamma_{Red}[Red]}$$

where γ is a corresponding activity coefficient. It follows that

$$E_r = E^{\circ\prime} + \frac{RT}{nF} \ln \frac{[Ox]}{[Red]} \qquad \text{where } E^{\circ\prime} = E^\circ + \frac{RT}{nF} \ln \frac{\gamma_{Ox}}{\gamma_{Red}}$$

Therefore, the formal potential $E^{\circ\prime}$ is approximately equal to the standard potential E° when the activity coefficients of Ox and Red are similar.*

E. Comment on Equilibrium in Electrochemical Cells

Electrochemical equilibrium should be distinguished from chemical equilibrium. An electrochemical cell at open circuit is at *electrochemical* equilibrium (i.e., the sum of all the electrochemical potentials is zero). At open circuit, the potential difference measured by the voltmeter is equal and of opposite sign to the potential drop across the membrane that separates the two half-cells.

---

*And not necessarily when the activity coefficients are close to unity, a common misconception.

The cell *is not*, however, at *chemical* equilibrium. When the circuit is closed (shortened), a current will flow instantaneously and the chemical reaction will proceed toward *chemical* equilibrium, expressed by the equilibrium constant of the reaction.

Note that at open circuit each electrode (half-cell) is at *electrochemical* equilibrium; that is, the electrochemical potential of the electrons in Ox and Red are equal, hence one can write the Nernst equation for each half-cell or for the complete cell.*

V. IONIC DOUBLE LAYER

A. Background

Models for the structure of the ionic double layer at an electrode–solution interphase have been developed to account for the electrical behavior of such systems. In particular, one is interested in the capacitive behavior of electrodes in electrolyte solutions and its variation with applied potential and electrolyte concentration.

Under equilibrium conditions, the time-averaged forces acting on a species (e.g., ion) in the bulk of an electrolyte solution are isotropic and homogeneous, and no net electric field is developed. At a phase boundary (e.g., electrode surface) the situation is quite different, and the symmetry breaks. In the interphase region the species experience anisotropic forces (i.e., different toward the electrode or toward the bulk solution) that vary with the distance from the electrode. This gives rise to net orientation of solvent dipoles and a net excess ionic charge near the phase boundary, on the solution side.

Once the electrolyte side acquires a net charge, the electrode responds by acquiring on its surface an induced charge of the same quantity and opposite sign. Thus a charge separation has occurred at the electrode–solution interface, and a potential gradient developed. In other words, the interface behaves as a capacitor.

The models described below correspond to ideally polarizable electrode–solution interfaces (Fig. 1a); that is, the electrode behaves purely capacitively in the absence of any faradaic reaction. The conclusions concerning the electrode capacitance apply, however, to the other situations as well.

B. Helmholtz Model

The Helmholtz model for the ionic double layer is a basic parallel-plate capacitor model. It assumes that while all the excess charge in the electrode is located at

---

*The Nernst equation expresses the conditions of *electrochemical* equilibrium.

the surface,* on the solution side a rigidly held layer of oppositely charged ions exists, located in a plane parallel to the electrode surface and very close to it. This plane [the *outer Helmholtz plane* (OHP)] is defined by the centers of the ions, which provide the excess charge on the solution side.

The Helmholtz double-layer capacitance per unit area, C_H, is given by (in MKS units)

$$C_H = \frac{\epsilon\epsilon^\circ}{d} \tag{10}$$

where ϵ is the dielectric constant, ϵ° the permittivity of free space (8.85×10^{-12} F/m), and d the distance of closest approach of hydrated ions, usually assumed to be about 3 Å. The great advantage of the Helmholtz model is its simplicity. It fails, however, to explain the often observed dependence of the double-layer capacitance on the potential and the solution composition.

C. Gouy–Chapman Model

The Gouy–Chapman model (the diffuse double-layer model) assumes a different distribution of the excess charge on the solution side. The electrostatic interaction between the field and the ions is countered by random thermal motion; an equilibrium is reached, represented by a Boltzmann distribution of ions of both signs (in one dimension):

$$c_i(x) = c_i^\circ \exp\left(\frac{-z_i F\phi_x}{RT}\right) \tag{11}$$

where $c_i(x)$ and c_i° are, respectively, the ionic concentrations at a distance x from the electrode surface (where the potential is ϕ_x with respect to the potential in the bulk of the solution) and in the bulk of the solution (where the potential is usually taken as zero, for convenience).

For the relationship between potential, distance, and charge density, one can use the *Poisson equation*:

$$\frac{d^2\phi_x}{dx^2} = -\frac{4\pi\rho_x}{\epsilon} \tag{12}$$

where ρ_x is the charge density per unit volume:

$$\rho_x = \sum_i Fz_i c_i(x) \tag{13}$$

---

*According to Gauss's law, excess charge in a conductor is confined entirely to the outer surface of the conductor.

Combining equations (11), (12), and (13) results in the *Poisson–Boltzmann equation*, in one dimension:

$$\frac{d^2\phi_x}{dx^2} = -\frac{4\pi F}{\epsilon} \sum_i z_i c_i^\circ \exp\left(\frac{-z_i F\phi_x}{RT}\right) \tag{14}$$

The charge on the electrode (which is equal to the total net ionic charge on the solution side), Q, can be obtained upon integration of equation (14),* to yield

$$Q = \pm \left(\frac{RT\epsilon}{2\pi}\right)^{1/2} \left\{\sum_i c_i \left[\exp\left(\frac{-z_i F\phi_0}{RT}\right) - 1\right]\right\}^{1/2} \tag{15}$$

where ϕ_0 is the potential at $x = 0$ (i.e., at the electrode surface). For a symmetrical electrolyte, where $c_+ = c_- = c$ and $|z_+| = |z_-| = z$, equation (15) becomes

$$Q = \left(\frac{2RT\epsilon}{\pi}\right)^{1/2} c^{1/2} \sinh \frac{zF\phi_0}{2RT} \tag{16}$$

The differential diffuse double-layer capacitance $C_G = (dQ/d\phi_0)$ is given by

$$C_G = \left(\frac{z^2 F^2 \epsilon}{2\pi RT}\right)^{1/2} c^{1/2} \cosh \frac{zF\phi_0}{2RT} \tag{17}$$

Equation (17) predicts a minimal capacitance at $Q = 0$ [i.e., at the *potential of zero charge* (PZC)][†]. This minimal capacitance is given by

$$C_G(\text{min}) = zF \left(\frac{\epsilon}{2\pi RT}\right)^{1/2} c^{1/2} \tag{18}$$

For small values of the potential ($\phi_0 \leq \sim 50/z$ mV), the potential decays approximately exponentially with the distance from the electrode:

$$\phi_x = \phi_0 \exp(-Kx) \tag{19}$$

where K^{-1} is the *Debye–Hückel reciprocal length*, representing the effective radius of the ionic atmosphere around a given ion in an electrolyte solution:

$$K = \left(\frac{8\pi}{\epsilon RT}\right)^{1/2} zF c^{1/2} \tag{20}$$

Hence an electrode in solution behaves as a "macroscopic ion," with the diffuse double-layer thickness being analogous to the ionic cloud surrounding an ion.

The Gouy–Chapman model is attractive in that it predicts a logical dependence of the double–layer capacitance on the applied potential and on the

---

*Using the relationship $Q = -(\epsilon/4\pi) (d\phi/dx)_{x=0}$.

[†]It follows immediately that at the PZC, $\phi_0 = O$.

ionic concentrations. However, only in very dilute electrolyte solutions is the measured capacitance in good agreement with the values calculated using equation (17), while at higher concentrations the measured capacitance is much lower* than predicted by the theory.

D. Stern Model

The Stern model overcomes the difficulties encountered with the two previous theories by combining them. Hence it extends the Gouy–Chapman model by considering the finite size of the ions and therefore a finite distance of closest approach of the ions from the electrode surface, identified as the outer Helmholtz plane (OHP; see Section IV.B). The ions located at the OHP are assumed to form a *compact double layer* (a Helmholtz layer), and the potential at this plane is denoted ϕ_2. The OHP is then taken as $x = 0$ for the Gouy–Chapman model, and thus ϕ decays with x from ϕ_2 towards ϕ_s, the potential of the bulk of the solution, which is defined arbitrarily as $\phi_s = 0$. The total potential drop at the interphase is the sum of the two contributions:

$$\phi_M - \phi_s = (\phi_M - \phi_2) + (\phi_2 - \phi_s) \tag{21}$$

where ϕ_M is the potential at the electrode surface. Differentiation, and using $\phi_s = 0$, leads to

$$\frac{d\phi_M}{dQ} = \frac{d(\phi_M - \phi_2)}{dQ} + \frac{d\phi_2}{dQ} \tag{22}$$

which, by definition, can be written as

$$\frac{1}{C_d} = \frac{1}{C_{M-2}} + \frac{1}{C_{2-S}} = \frac{1}{C_H} + \frac{1}{C_G} \tag{23}$$

Hence the measured double-layer capacitance C_d behaves as the sum of two capacitors connected effectively in series: C_H, which corresponds to the compact double layer and is given by the Helmholtz expression [equation (10)], and C_G, which corresponds to the diffuse double layer and is given by the Gouy–Chapman expression (17) for a symmetrical electrolyte.

E. Effect of Concentration

The diffuse double-layer capacitance C_G increases substantially with electrolyte concentration [equation (17)], and it is therefore clear from equation (23) that at high concentrations the diffuse double layer can be neglected and $C_d \approx C_H$.*

---

*It can be seen from equation (16) that for a certain Q, ϕ_0 (ϕ_2 in the Stern model) decreases upon increasing the electrolyte concentration and at high concentrations becomes effectively zero.

For the same reason, C_G is predominant in very dilute electrolyte solutions, where $C_d \approx C_G$.

F. Other Models

The Stern model, although quite satisfactory in many cases, shows deviations from the expected double-layer capacitance under certain conditions. Therefore, a number of modifications of the Stern model have been developed. The most prominent model that should be noted is the *Grahame modification*, which emphasizes contact adsorption of specifically adsorbed (nonhydrated) ions at a plane denoted the *inner Helmholtz plane* (IHP). This, as well as other modifications of the Stern model, are beyond the scope of this chapter.

VI. BASIC ELECTRODE KINETICS

A. One-Step, One-Electron Reaction

Consider the following one-step electrode reaction:

$$\text{Red} \underset{k_b}{\overset{k_f}{\rightleftharpoons}} \text{Ox} + \text{e}$$

The basic electrochemical kinetic equations can be derived from the *absolute reaction-rate theory*. The rate constant of a general chemical reaction, according to the simplified *activated complex theory*, is given by

$$k_f' = \frac{kT}{h} \exp\left(-\frac{\Delta G^{\circ\#}}{RT}\right) \tag{24}$$

where k_f' is the forward rate constant, k and h are the Boltzmann and Planck constants, respectively, and $\Delta G^{\circ\#}$ is the standard chemical free energy of activation of the reaction. For the electrode reaction one can write a similar equation, using a standard *electrochemical* free energy of activation:

$$k_f = \frac{kT}{h} \exp\left(-\frac{\overline{\Delta G}^{\circ\#}}{RT}\right) \tag{25}$$

The potential dependence of $\overline{\Delta G}^{\circ\#}$ is shown schematically in Figure 3, where Ox is assumed to be ionic and stabilized by the electric field. The energy curves are plotted as straight lines to simplify the geometric considerations. It is seen that the change in the free energy of activation $\delta\overline{\Delta G}^{\circ\#}$ due to the applied potential is a fraction of the change in the free energy of the reaction $\delta\overline{\Delta G}^{\circ}$. The *symmetry factor* β is defined as

$$\beta = \frac{\delta\overline{\Delta G}^{\circ\#}}{\delta\overline{\Delta G}^{\circ}} \tag{26}$$

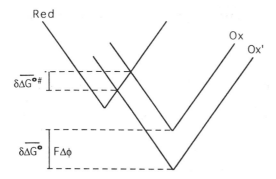

FIGURE 3   Schematic presentation of the dependence of the electrochemical energy of activation $\overline{\Delta G}^{\circ\#}$ on the applied potential $\Delta\phi$, and the relationship between changes in the energy of activation $\delta\overline{\Delta G}^{\circ\#}$ and changes in the Gibbs free energy of the reaction $\delta\overline{\Delta G}^{\circ}$.

β is usually assumed to be close to 0.5. It can be shown from simple geometry that

$$\overline{\Delta G}_f^{\circ\#} = \Delta G_f^{\circ\#} - \beta FE \tag{27a}$$

$$\overline{\Delta G}_b^{\circ\#} = \Delta G_b^{\circ\#} + (1 - \beta)FE \tag{27b}$$

In equations (27a) and (27b), $\Delta\phi$ has been replaced by the more common sign E for the potential. Equations (25) and (27a) lead to

$$k_f = \frac{kT}{h} \exp\left(-\frac{\Delta G^{\circ\#}}{RT}\right) \exp\left(\frac{\beta FE}{RT}\right) \tag{28}$$

Combining the potential independent terms, one obtains the expression for the forward rate constant k_f, and similarly for k_b:

$$k_f = k_f^{\circ} \exp\left(\frac{\beta FE}{RT}\right) \tag{29a}$$

$$k_b = k_b^{\circ} \exp\left[-\frac{(1 - \beta)FE}{RT}\right] \tag{29b}$$

E is the potential with respect to an arbitrary reference (i.e., k_f° and k_b° are dependent on the choice of reference). The partial forward and backward currents are given by*

$$i_f = FAk_fC_R \qquad i_b = FAk_bC_O \tag{30}$$

---

*If the electrode reaction involves the transfer of n electrons, equations (30) become $i_f = nFAk_fC_R$ and $i_b = nFAk_bC_O$.

where A is the electrode area and C_R and C_O are, respectively, the concentrations of Red and Ox at the electrode surface.

At *equilibrium*, $C_R = C_R^\circ$ and $C_O = C_O^\circ$ (where C_R° and C_O° are, respectively, the bulk concentrations of Red and Ox); if we choose $C_R^\circ = C_O^\circ$, then from the Nernst equation $E = E^\circ$ (the standard reduction potential of the electrode reaction). Since under these conditions the net current i is zero, it follows that $k_f C_R^\circ = k_b C_O^\circ$, which implies that $k_f = k_b$. From equations (29),

$$k_f^\circ \exp\left(\frac{\beta F E^\circ}{RT}\right) = k_b^\circ \exp\left[-\frac{(1-\beta)FE^\circ}{RT}\right] = k_s \tag{31}$$

where k_s is the *standard heterogeneous rate constant* of the electrode reaction (in cm s^{-1}). Equations (29) then become

$$k_f = k_s \exp\left[\frac{\beta F(E-E^\circ)}{RT}\right] \tag{32a}$$

$$k_b = k_s \exp\left[-\frac{(1-\beta)F(E-E^\circ)}{RT}\right] \tag{32b}$$

The current is given by*

$$i = FA(k_f C_R - k_b C_O) \tag{33}$$

or, substituting equations (32) in (33) gives

$$i = FAk_s \left\{ C_R \exp\left[\frac{\beta F(E-E^\circ)}{RT}\right] - C_O \right.$$
$$\left. \exp\left[-\frac{(1-\beta)F(E-E^\circ)}{RT}\right]\right\} \tag{34}$$

At *equilibrium* the partial forward (anodic) and backward (cathodic) currents are equal: $i_f = i_b = i_0$, where i_0 is the *exchange current*. Using the Nernst equation, $C_O^\circ/C_R^\circ = \exp[F(E_r - E^\circ)/RT]$ (substituting concentrations for activities) and the equilibrium conditions $C_R = C_R^\circ$, $C_O = C_O^\circ$, $E = E_r$, and $i = 0$, and performing some tedious algebra, one arrives at an expression for the exchange current:

$$i_0 = FAk_s C_R^{\circ(1-\beta)} C_O^{\circ\beta} \tag{35}$$

The *overpotential* η is defined as

$$\eta = E - E_r \tag{36}$$

---

*If the electrode reaction involves the transfer of n electrons, equation (33) becomes $i = nFA(k_f C_R - k_b C_O)$.

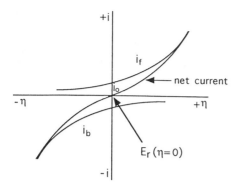

FIGURE 4　Schematic presentation of the current–overpotential relationship for an electrode reaction. i_f and i_b are, respectively, the forward (anodic) and backward (cathodic) partial currents.

Dividing equation (34) and (35), using the definition of η, one obtains

$$i = i_0 \left\{ \frac{C_R}{C_R^\circ} \exp\left(\frac{\beta F \eta}{RT}\right) - \frac{C_O}{C_O^\circ} \exp\left[-\frac{(1 - \beta)F\eta}{RT}\right] \right\} \qquad (37)$$

If mass-transport limitations can be neglected, $C_O \cong C_O^\circ$ and $C_R \cong C_R^\circ$; then

$$i = i_0 \left\{ \exp\left(\frac{\beta F \eta}{RT}\right) - \exp\left[-\frac{(1 - \beta)F\eta}{RT}\right] \right\} \qquad (38)$$

A schematic graphic presentation of equation (37) or (38) is shown in Figure 4. Equation (38) has two limiting cases:

1. *At high overpotentials* ($|\beta F\eta/RT| \gg 1$), one of the two terms (partial currents) becomes negligible. Equation (38) then becomes (for large positive or negative overpotentials, respectively)

$$i = i_0 \exp\left(\frac{\beta F \eta}{RT}\right) \qquad \text{or} \qquad i = i_0 \exp\left[-\frac{(1 - \beta)F\eta}{RT}\right] \qquad (39)$$

which are the *Tafel equations* for the electrode reaction. The *Tafel slope b* is defined as (for, e.g., the anodic reaction)

$$b = \left| \frac{\partial \eta}{\partial \log i} \right| = \frac{2.30RT}{\beta F} \qquad (40)$$

In the *Tafel region* (high $|\eta|$), the kinetic parameters can be extracted from *Tafel plots* (i.e., experimental η vs. $\log i$ plots) [equations (39)].

i_0 is obtained from the extrapolated intercept, and β (or $1 - \beta$) from the slope.* k_s can be calculated from equation (35).

2. At *low overpotentials* ($|\beta F\eta/RT| \leq 0.2$), the exponent in equation (38) can be linearized. One then obtains

$$i = i_0 \frac{F\eta}{RT} \tag{41}$$

Thus at low $|\eta|$ (near equilibrium) the current varies linearly with the overpotential (i.e., an ohmic behavior). The *charge-transfer resistance* R_{ct} near equilibrium can be defined as[†]

$$R_{ct} = \frac{\eta}{i} = \frac{1}{i_0} \frac{RT}{F} \tag{42}$$

In the linear region (low $|\eta|$), i_0 is readily obtained from the slope of experimental η vs. i plots [equation (41)].

B. Elucidation of Reaction Mechanism

1. Basic Equations

In many electrode reactions more than one electron is transferred in more than one step. Such mechanisms are usually analyzed using the following simplifying assumptions:

1. Only one electron is transferred in an elementary step.
2. One of the elementary steps is much slower than the others, and there-fore is the *rate-determining step* (RDS); the elementary processes preceding the RDS are at *quasi-equilibrium*.

The quasi-equilibrium assumption is justified by the fact that the preceding elementary steps, which are intrinsically much faster than the RDS (i.e., have a much larger i_0), must proceed at the rate dictated by the RDS, and therefore are only marginally perturbed from equilibrium. In other words, the deviation from equilibrium, expressed by the value of i/i_0, may be substantial for the RDS but negligible for the preceding elementary steps.

For multistep electrode reactions, equations (37) to (41) may be replaced by the following equations:

$$i = i_0 \left[\frac{C_R}{C_R^\circ} \exp\left(\frac{\alpha_a F\eta}{RT}\right) - \frac{C_O}{C_O^\circ} \exp\left(-\frac{\alpha_c F\eta}{RT}\right) \right] \tag{37a}$$

---

*If $\beta \approx 0.5$, $b \approx 118$ mV/decade at room temperature.
[†]At larger overpotentials R_{ct} is potential dependent, as in, e.g., equations (39).

In the absence of mass-transport limitations,

$$i = i_0 \left[\exp\left(\frac{\alpha_a F \eta}{RT}\right) - \exp\left(-\frac{\alpha_c F \eta}{RT}\right) \right] \tag{38a}$$

At high overpotentials (positive or negative η, respectively):

$$i = i_0 \exp\left(\frac{\alpha_a F \eta}{RT}\right) \quad \text{or} \quad i = i_0 \exp\left(-\frac{\alpha_c F \eta}{RT}\right) \tag{39a}$$

$$b = \left| \frac{\partial \eta}{\partial \log i} \right| = \frac{2.30 RT}{\alpha F} \tag{40a}$$

At low overpotentials,

$$i = i_0 \frac{(\alpha_a + \alpha_c) F \eta}{RT} \tag{41a}$$

where α_a and α_c are, respectively, the anodic (forward) and cathodic (backward) *transfer coefficients* of the electrode reaction. α_a and α_c are experimental parameters, determined by the reaction mechanism, as shown below.

2. Example: Hydrogen Evolution Reaction

As an example, the cathodic hydrogen evolution reaction in acid solution, $2H^+ + 2e \rightleftharpoons H_2$, was chosen. Several possible mechanisms are analyzed. For simplicity, the hydronium ion is symbolized by H^+. Consider the following mechanism:

$$H^+ + e \underset{k_{-1}}{\overset{k_1}{\rightleftharpoons}} H_{ads} \tag{43}$$

$$2H_{ads} \underset{k_{-2}}{\overset{k_2}{\rightleftharpoons}} H_2 \tag{44}$$

1. *If step (43) is the RDS*, under conditions of a high cathodic overpotential, the electrode reaction rate v (in mol/cm$^2 \cdot$s) is given by

$$v = k_1 [H^+] (1 - \theta) \exp\left[-\frac{(1 - \beta) FE}{RT} \right] \tag{45}$$

The measured cathodic current density (in A cm^{-2}) is given by $i/A = nFv$, where n is the number of electrons transferred in the overall reaction. θ is the fractional surface coverage of the intermediate H_{ads} ($0 \leq \theta \leq 1$). The term $1 - \theta$ arises from the need for an available adsorption site on the electrode surface.

At this point one has to assume a certain adsorption isotherm for H_{ads}. Here we consider only the Langmuir isotherm,* which assumes (1) a very low surface coverage by the intermediate (i.e., $1 - \theta \approx 1$), and (2) that the heat of adsorption is independent of the coverage θ, equivalent to saying that the rate constant k_1 is independent of θ. Under these conditions, it can readily be seen [using equations (39a) and (40a) and assuming that $\beta = 0.5$] that this mechanism predicts that $\alpha_c = 1 - \beta = 0.5$ and $b = 118$ mV/decade. The reaction order with respect to protons, defined as $\rho_{H+} = (\partial \log i / \partial \log c_{H+})_E$, is 1.[†]

2. *If step (44) is the RDS,* step (43) is at quasi-equilibrium, or (under Langmuir conditions)

$$k_1 [H^+] (1 - \theta) \exp\left[-\frac{(1 - \beta)FE}{RT} \right] = k_{-1} \theta \exp\left(\frac{\beta FE}{RT} \right) \quad (46)$$

$$\frac{\theta}{1 - \theta} \approx \theta = \frac{k_1}{k_{-1}} [H^+] \exp\left(-\frac{FE}{RT} \right) \quad (47)$$

The reaction rate will be given by the rate of step (44), which does not involve electron transfer:

$$v = k_2 \theta^2 = k_2 \left(\frac{k_1}{k_{-1}} \right)^2 [H^+]^2 \exp\left(-\frac{2FE}{RT} \right)$$

$$= k_b [H^+]^2 \exp\left(-\frac{2FE}{RT} \right) \quad (48)$$

Thus this mechanism predicts that $\alpha_c = 2$, $b = 30$ mV/decade, and $\rho_{H+} = 2$.

3. *If step (44) is replaced with step (44a) and we assume that (44a) is the RDS:*

$$H_{ads} + H^+ + e \underset{k_{-2}}{\overset{k_2}{\rightleftharpoons}} H_2 \quad (44a)$$

Equation (47) still holds. For the reaction rate one can write

$$v = k_2 \theta [H^+] \exp\left[-\frac{(1 - \beta)FE}{RT} \right]$$

$$= k_b' [H^+]^2 \exp\left[-\frac{(2 - \beta)FE}{RT} \right] \quad (49)$$

---

*Other adsorption isotherms, such as the Frumkin or Temkin isotherms, are beyond the scope of this chapter.

[†]This is the reaction order at a constant *potential*. One can also define the reaction order at a constant *overpotential*, $(\partial \log i / \partial \log c_{H+})_\eta$, which is related to the former by the Nernst equation and will not be discussed here.

This mechanism predicts that $\alpha_c = 2 - \beta = 1.5$, $b = 39$ mV/decade, and $\rho_{H^+} = 2$.

3. General Equation

It can be shown that under the same conditions as those discussed above, a general equation can be derived for the dependence of the current on the electrode potential:

$$i = i_0 \left\{ \exp\left[\frac{\beta(n - \overleftarrow{n} - \overrightarrow{n}) + \overrightarrow{n}}{\nu} \frac{F\eta}{RT} \right] - \exp\left[-\frac{(1 - \beta)(n - \overleftarrow{n} - \overrightarrow{n}) + \overleftarrow{n}}{\nu} \frac{F\eta}{RT} \right] \right\} \tag{50}$$

where n is the number of electrons transferred in the overall reaction; $\overrightarrow{n}$ and $\overleftarrow{n}$ are, respectively, the number of electrons transferred before and after the RDS; and ν is the *stoichiometric number* (i.e., the number of times the RDS must take place in order for the overall reaction to occur once).

It is evident from equations (50) and (38a) that

$$\alpha_a = \frac{\beta(n - \overrightarrow{n} - \overleftarrow{n}) + \overrightarrow{n}}{\nu} \qquad \alpha_c = \frac{(1 - \beta)(n - \overrightarrow{n} - \overleftarrow{n}) + \overleftarrow{n}}{\nu} \tag{51}$$

Substituting equation (51) in equation (41a) yields, for low overpotentials,

$$i = i_0 \frac{nF\eta}{\nu RT} \tag{52}$$

VII. DIFFUSION CONTROL IN ELECTRODE REACTIONS

A. Mass-Transport Limitations

Mass transport of reactants to the electrode (and of products from the electrode) may be controlled by one, or a combination, of the following processes: (a) *migration*, transport of charged species under the influence of an electric field; (b) *convection* (natural or forced), transport of solute species caused by hydrodynamic motion of the liquid; and (c) *diffusion*, transport brought about by concentration gradients. The latter is a stochastic process, the direction of which is always opposite to the concentration gradient.

It is often desirable to work under conditions of pure diffusion control. One of the major reasons is that diffusion control presents well-defined and easily reproducible conditions. As a result, electrochemists often prefer to work under diffusion-control conditions, and most of the mathematical derivations worked out for electrochemical experiments have been derived for pure diffusion

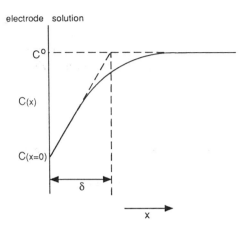

electrode   solution

FIGURE 5 Schematic presentation of the concentration profile of the reactant at an electrode–solution interphase. The diffusion layer thickness δ is shown.

control.* This is conveniently achieved as follows: (a) The effect of migration can be minimized by using a high concentration of supporting (inert) electrolyte such that the solution resistance, and hence the electric field in the solution, is kept as low as possible; and (b) forced and natural convection, respectively, can be minimized by working in quiescent solutions and at relatively short times. In what follows, only diffusion control is considered.

B.   Basic Equations for Diffusion Control

Fick's first law, in one dimension, is written

$$\text{flux} = -D \left(\frac{\partial c}{\partial x} \right)_{x=0} \tag{53}$$

where D is the *diffusion coefficient* (in $cm^2 \ s^{-1}$). The electrochemical analog of equation (53) is

$$i = nFAD \left(\frac{\partial c}{\partial x} \right)_{x=0} \tag{54}$$

Figure 5 is a schematic representation of the concentration gradient of a reactant at the electrode–solution interphase (c° is the concentration in the bulk

---

*A prominent exception is the *rotating disk electrode*, where the reaction proceeds under mixed diffusion–covection control.

of the solution). As expressed in equation (54), the current is determined by the slope of c vs. x at $x = 0$, which coincides with the electrode surface.

It is common to define the *diffusion-layer thickness* δ by a linear extrapolation of the c vs. x slope at $x = 0$, as shown in Figure 5. The current will then be given by

$$i = nFAD \frac{c^{\circ} - c(x = 0)}{\delta} \tag{55}$$

It can be seen from simple geometric considerations that for a certain δ (i.e., a certain time t; see Section VII.D), a maximal current is obtained when $c(x = 0) = 0$. According to equation (55), this limiting current i_L is given by

$$i_L = \frac{nFADc^{\circ}}{\delta} \tag{56}$$

A useful relationship that provides the extent of diffusion control is obtained upon combining equations (55) and (56):

$$\frac{c(x = 0)}{c^{\circ}} = 1 - \frac{i}{i_L} \tag{57}$$

Another useful expression, for mixed kinetic-diffusion control, is obtained upon substituting equation (57) in (37a):

$$\frac{i}{i_0} = \left(1 - \frac{i}{i_{L,a}}\right) \exp\left(\frac{\alpha_a F \eta}{RT}\right) - \left(1 - \frac{i}{i_{L,c}}\right) \exp\left(-\frac{\alpha_c F \eta}{RT}\right) \tag{58}$$

For small values of η, equation (58) can be linearized, to yield

$$\frac{i}{i_0} = \frac{i}{i_{L,c}} - \frac{i}{i_{L,a}} + \frac{(\alpha_a + \alpha_c)F\eta}{RT} \tag{59}$$

C. Current–Voltage Relationship for a Nernstian Reaction

If the electrode reaction is very fast [i.e., has a very large i_0 ($i/i_0 \rightarrow 0$)], the reaction is *totally reversible* or *Nernstian*. Under these conditions the *surface concentrations* of Red and Ox are given by the Nernst equation. In other words, the very fast charge transfer maintains the surface concentrations at the values determined by the potential according to the Nernst equation, as long as diffusion is fast enough to overcome concentration changes resulting from the net current. Hence, for a Nernstian electrode reaction (using concentrations rather than activities),

$$E = E^{\circ} + \frac{RT}{nF} \ln \frac{C_O(x = 0)}{C_R(x = 0)} \tag{60}$$

When only one form (either Ox or Red) is present in the solution, using equations (55), (60), and the Einstein–Smoluchowski relationship (see Section VII.D), one obtains, after some rearrangement,

$$E = E° - \frac{RT}{nF} \ln\left(\frac{D_O}{D_R}\right)^{1/2} + \frac{RT}{nF} \ln \frac{i_L - i}{i} \tag{61}$$

Equation (61) describes the current–potential relationship of a Nernstian electrode reaction, at a certain given time. Figure 6 is a schematic presentation of equation (61) (i.e., an i–E plot for a Nernstian reaction). Note that since i and i_L vary with time (see Section VII.D), Figure 6 refers to a situation wherein all the points on the i–E curve correspond to the same time of measurement.

For $i = i_L/2$, one obtains from equation (61),

$$E_{1/2} = E° - \frac{RT}{nF} \ln\left(\frac{D_O}{D_R}\right)^{1/2} \tag{62}$$

where $E_{1/2}$ is the *half-wave potential*, as shown in Figure 6. It follows from equation (62) that if $D_O \approx D_R$, then $E_{1/2} \approx E°$. Hence in many cases an experimentally measured $E_{1/2}$ provides a good estimate of $E°$ of the electrode reaction.[*] Substituting equation (62) in (61) yields

$$E = E_{1/2} + \frac{RT}{nF} \ln \frac{i_L - i}{i} \tag{63}$$

D. Effect of Time

Figure 7 presents linearized concentration profiles at the electrode–solution interface, under diffusion control. At low overpotentials (or very short times), a Nernstian situation prevails, with constant concentrations at the electrode surface (Fig. 7a). At higher overpotentials, or as time progresses, the reactant concentration at the surface decreases to zero, and limiting conditions (Cottrell conditions, see below) are reached (Fig. 7b).

The dependence of the diffusion-layer thickness δ on time is given by the Einstein–Smoluchowski equation, which, in this case, has the following form:

$$\delta = (\pi D t)^{1/2} \tag{64}$$

Combining equations (55) and (64), one obtains for the dependence of the diffusion-controlled current on time:

$$i = \frac{nFAD^{1/2}[c° - c(x = 0)]}{\pi^{1/2} t^{1/2}} \tag{65}$$

---

[*]$E_{1/2}$ is commonly obtained as the potential halfway between the anodic and cathodic peaks in cyclic voltammetry of reversible reactions.

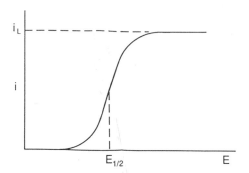

FIGURE 6 Schematic presentation of the current–voltage relationship for a Nernstian electrode reaction. The limiting current i_L and the half-wave potential $E_{1/2}$ are shown. All the points on the i–E curve correspond to the same time of measurement.

and for the limiting current:

$$i_L = \frac{nFAD^{1/2}c^\circ}{\pi^{1/2}t^{1/2}} \tag{66}$$

Equation (66), known as the *Cottrell equation*, describes the variation of the limiting current with time upon applying a constant potential.

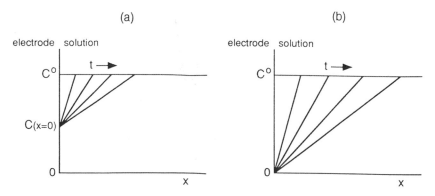

FIGURE 7 Evolution of the diffusion-layer thickness with time for (a) Nernstian conditions, where the concentrations at the electrode surface are fixed and of finite values, and (b) Cottrell conditions, where the reactant concentration at the electrode surface is zero.

VIII. CONCLUSION

An effort has been made to encapsulate the fundamental principles and equations of physical electrochemistry. Due to the size limitations of this chapter, many details and some basic issues have necessarily been left out. Hence topics such as electrocapillary thermodynamics, adsorption, electrochemical measurements and instrumentation, cell design, reference electrodes, electrochemical techniques, and others have not been dealt with. For the reader interested in more information, a list of texts is given (in chronological order) in the References.

REFERENCES

1. J. O'M. Bockris and A. K. N. Reddy, *Modern Electrochemistry*, Vols. 1 and 2, Plenum Press, New York, 1970.
2. E. Gileadi, E. Kirowa-Eisner, and J. Penciner, *Interfacial Electrochemistry*, Addison-Wesley, Reading, Mass., 1975.
3. A. J. Bard and L. R. Faulkner, *Electrochemical Methods*, Wiley, New York, 1980.
4. P. T. Kissinger and W. R. Heineman, eds., *Laboratory Techniques in Electroanalytical Chemistry*, Marcel Dekker, New York, 1984.
5. Southampton Electrochemistry Group, *Instrumental Methods in Electrochemistry*, Ellis Horwood, Chichester, West Sussex, England, 1985.
6. J. Koryta and J. Dvorák, *Principles of Electrochemistry*, Wiley, New York, 1987.
7. H. D. Abruña, ed., *Electrochemical Interfaces*, VCH Publishers, New York, 1991.
8. R. W. Murray, ed., *Molecular Design of Electrode Surfaces*, in *Techniques of Chemistry*, Vol. XXII, Wiley, New York, 1992.
9. J. O'M. Bockris and S. U. M. Khan, *Surface Electrochemistry*, Plenum Press, New York, 1993.
10. E. Gileadi, *Electrode Kinetics for Chemists, Chemical Engineers, and Materials Scientists*, VCH Publishers, New York, 1993.

2

Heterogeneous Electron Transfer Kinetics at Metallic Electrodes

CARY J. MILLER
University of Maryland
College Park, Maryland

I. INTRODUCTION

A. Motivations

The study of heterogeneous electron transfer kinetics has experienced a revitalization in recent years. This increased interest in both the experimental and theoretical aspects of electrode kinetics results from several factors. First among these factors is a general increase in the interest in homogeneous electron transfer dynamics. Current areas of interest include long-range electron transfers in biological systems [1], the influences of the static and dynamic properties of solvents on the electron transfer rate [2] and on the driving force dependence of electron transfer rates [3]. Because electron transfers occurring at electrode surfaces share many of the same characteristics of homogeneous electron transfers, electrochemical techniques can play an important role in the elucidation and validation of electron transfer theories and in the kinetic characterization of redox active molecules.

A second reason for the increased interest in heterogeneous electron transfer kinetics has been the result of recent experimental advances on several unrelated fronts that have increased the applicability and range of electrochemical kinetic measurements. These include:

1. The increasing popularity and utility of ultramicroelectrodes and their ever-decreasing dimensions [4]. Because of their increased mass transport rates and immunity to the problem of the solution resistance, micron- and submicron-sized electrodes are allowing electron transfer rates to be measured for more rapid redox reactions and in much more resistive solutions than is possible with larger electrodes.
2. The development of in situ scanning tunneling microscopes that allow one atomically to image electrode surfaces [5] and scanning electrochemical microscopy [6,7], which can be used to measure local electron transfer rates and thereby map out the reactivity of an electrode surface.
3. The revival of ultralow temperature voltammetric studies [8–10] spurred on in part by the quest for investigating the electron transfer properties of the high-temperature superconductors [11].
4. The development of compact organic layers of controlled dimension and structure via the spontaneous self-assembly of organic thiols and other amphiphiles [12]. These monolayer films can be used to control the interfacial structure of the electrode and to probe both the reactivity of redox molecules and the nature of the electronic coupling through the monolayer film.

Several of these experimental areas are addressed in greater length in other chapters in this volume.

B. Scope

In the limited format of this chapter, it would be quite impossible to present all (or even most) of the current theoretical and experimental results in electron transfer. The purpose of this chapter is to present a modern description of heterogeneous electron transfer that will be used subsequently to describe two recent experimental developments in the measurement of electron transfer rates at metal electrodes. The first involves the use of adsorbed thiol monolayers that control the reactivity of the electrode [13]. The second focuses on how the dynamic properties of the solvent can influence the measured heterogeneous electron transfer rate [2]. These two limited experimental areas were chosen because they allow a number of direct comparisons with the heterogeneous electron transfer theory.

Part of the difficulty in following the recent advances in heterogeneous electron transfer kinetics is that one cannot use the standard Butler–Volmer equation, which fits the potential dependence of the electron transfer rate with an expression containing a minimum number of parameters. The Butler–Volmer equation for the reduction rate constant at an electrode polarized to a potential, E, $k_f(E)$, is [14]

$$k_f(E) = k^\circ e^{-\alpha n \mathfrak{f}(E-E^{\circ\prime})} \tag{1}$$

where k° is the rate constant measured when the electrode is polarized to the formal potential of the redox molecule, $E^{\circ\prime}$; α is the transmission coefficient; n is the number of electrons transferred; and $\mathfrak{f} = F/RT$, where F is Faraday's constant, R the gas constant, and T the absolute temperature. From a plot of the logarithm of the measured heterogeneous rate constant versus the formal overpotential,* one obtains α and k° parameters from the slope and intercept, which characterize the electron transfer reaction. Stemming from an empirical approach, this Butler–Volmer analysis is seriously deficient in describing the molecular factors that govern the electron transfer rate. To delve more deeply into these molecular factors, one must adopt a more general (and unfortunately, a more complex) mathematical and conceptual framework for predicting and evaluating heterogeneous electron transfer data. In this chapter electron transfer rate expressions will be developed from the next-higher level of sophistication using a transition-state approach relying heavily on the Marcus theory approximations [15].

Electron transfer reactions can be mechanistically quite complex involving homogeneous as well as heterogeneous chemical steps prior to or after the actual electron transfer step. To focus exclusively on the electron transfer step, the description presented here will assume that the electron transfer step is the rate-determining heterogeneous step and that the redox reaction is not complicated by coupled homogeneous reactions. The initial description of electron transfer kinetics presented here will be quite general involving the transition-state approach. While this description will focus entirely on heterogeneous electron transfer at metallic electrodes, many of the considerations are identical for both homogeneous and heterogeneous electron transfer. Indeed, most of the theoretical and experimental work that has led to our current understanding of electron

---

*The formal overpotential is the potential difference between the applied electrode potential, E, and the formal potential of the redox molecule, $E^{\circ\prime}$. In this chapter the formal potential is used rather than the standard potential to reference electrode potentials. The formal potential of a redox couple is the equilibrium potential of a solution containing equal concentrations of the oxidized and reduced forms of the redox couple. It is different from the standard potential, which is measured with equal activities of the two redox forms. The selection of any particular voltage reference is somewhat arbitrary. Because formal potentials are much easier to measure, they are more commonly used to characterize redox molecules. The electric double layer also affects the observed heterogeneous electron transfer rate by changing the potential drop between the bulk of the metal and the redox molecule within the double layer. The electron transfer rate depends on the local potential difference between the redox molecule and the electrode surface rather than the applied potential measured between the electrode surface and the bulk solution.

transfers has been the result of inquiries into homogeneous electron transfer reactions.

A useful characterization of electron transfer reactions is whether they involve an inner-sphere or an outer-sphere mechanism [16]. In the simpler outer-sphere mechanism, the electron transfer occurs without a significant change in the redox molecule's internal structure. Qualitatively, *outer-sphere* mechanisms occur when there are no strong chemical interactions between the reactants. For a heterogeneous electron transfer, an outer-sphere mechanism requires only weak interactions between the redox molecule and the electrode surface. The electron transfer properties of such an outer-sphere redox molecule are the same at the electrode surface as in the bulk solution. In contrast, electron transfers that involve strong interactions between the redox molecule and the electrode can be characterized as *inner-sphere* electron transfers. In these cases the adsorbed redox molecule's internal structure and extent of solvation can be significantly different from that in solution. The actual reacting species can be considered to be a different molecule from its precursor in the solution. Because the structure of the reactive form of the redox species adsorbed at the electrode surface is not generally known, these inner-sphere electron transfer mechanisms are considerably more difficult to study and predict. Because of the added complications in describing inner-sphere electron transfers, most of the experimental and theoretical effort has been focused on the simpler outer-sphere mechanism.

II. THEORY

A. Conceptual Description of Electron Transfer Kinetics

An electron transfer reaction occurring at an electrode in an electrolyte solution can be thought of as occurring through a series of steps that are depicted in Figure 1. The separation of the electron transfer into a particular set of elementary steps is somewhat arbitrary but is useful in the theoretical and conceptual development of a mathematical description. Redox active molecules are transported (via diffusion and possibly via convection) to the vicinity of the electrode surface. Due to the possibility of specific chemical and electrostatic interactions with the electrode, redox molecules may be attracted or repelled from the electrode surface. These interactions can give rise to markedly different concentrations of redox species near the electrode surface compared with the bulk solution. The concentration profile in the near-electrode surface region may be a complex function of the electrode surface structure, the concentration and identity of all the species in the solution, as well as the temperature. As the redox molecules approach the electrode surface, they become increasingly electronically coupled to electronic states within the metal. The actual electron transfer step usually requires some thermal activation of the redox molecule. Through random thermal fluctuation in the redox molecule's structure and solvation shell, the redox molecule comes into resonance with the appropriate electronic states within the metal (filled electronic states for reductions, unfilled for oxidations). Electrons then tunnel between the electrode and the redox molecule with a certain probability dependent on the redox molecule's proximity and orientation to the electrode surface, the density of electronic states within the electrode, and the structure of the intervening space between the electrode and redox molecule. The product of the electron transfer is generally activated relative to its most stable internal geometry and solvation. This product then relaxes toward its equilibrium structure and solvation, moves away from the near-surface region, and is transported away from the electrode again via diffusion and/or convection.

The goal of a theoretical description of electron transfer is to obtain a mathematical equation that relates the measured electrode current on such experimental parameters as the structure and properties of the redox reactant, the electrode, solvent, electrolyte along with the solution concentrations, electrode potential, and temperature. The general approach to modeling electrode kinetics has been to consider independently each of the elementary steps listed above. Once each of the steps is summed appropriately, one obtains expressions for the electrode current which is a direct measure of the net heterogeneous electron transfer rate. The current is most commonly normalized to the unit area of the electrode and bulk solution concentration of the reactant (typically, 1 cm^2 and $1/nF$ mol/cm^3, respectively) to obtain the observed heterogeneous electron transfer rate constant k (with units of cm/s), so that

$$i(E) = nFAk(E)C^*$$ (2)

where $i(E)$ is the current measured at the electrode potential E, A is the electrode area, and C^* is the bulk concentration of the redox active reactant.

A difficulty in using the theoretical predictions obtained from these elementary steps is that the mathematical expression generally has too many parameters. The theory overdetermines the experimental results. One can obtain experimentally only a few kinetic parameters from a single voltammetric experiment, each of which is dependent on a number of elementary parameters used in the model. Indeed, the job of measuring heterogeneous electron transfer kinetics for redox active species has been one of delineating these elementary steps through careful measurement of electrode currents made as a function of electrode potential, temperature, or some structural parameter of either the electrode surface, redox molecule, or other component within the solution. Extracting accurate information about these underlying elementary steps is often a difficult but essential prerequisite to understanding heterogeneous electron transfers.

Mass Transport

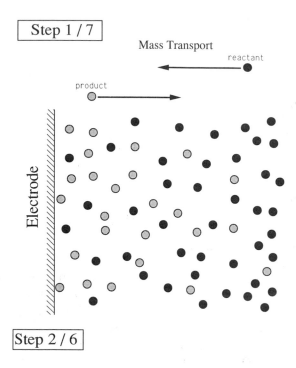

Surface Interactions Between Electrode and Redox Molecules

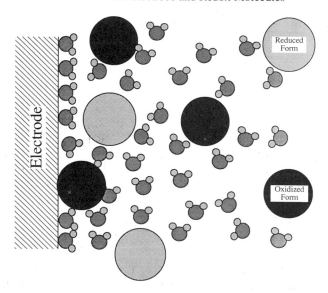

FIGURE 1 Elementary steps involved in heterogeneous electron transfer. This figure depicts schematically the seven elementary steps that occur during the course of a heterogeneous electron transfer reaction. As the first three steps are mirrored by the last three, each pair is represented by the same figure as indicated (see the text).

Thermal Activation and Relaxation

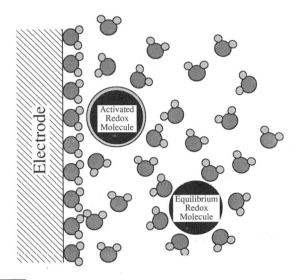

Step 4

Electron Tunneling Between Electrode and
Redox Molecule

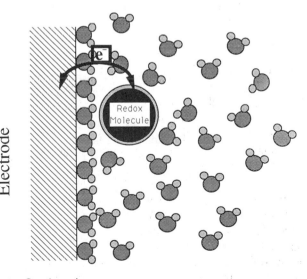

FIGURE 1   Continued

B. Elementary Steps

1. Transport

The transport of redox active molecules between the electrode surface and the bulk of solution can play a major role in determining the observed electron transfer rate. Finite rates of material transport to the electrode surface can lead to strong deviations in the concentrations of reactants and products near the electrode surface from their initial values. When the rate of material transport becomes much lower than the other elementary steps, the overall electron transfer rate become equal to this material transport rate. The concentrations of the reactants at the electrode surface adjust so that the heterogeneous electron transfer rate matches the material transport rate. This transport limited current response is often termed the diffusion-controlled or diffusion-limited rate [14].

Diffusion-controlled electrode currents are extremely common in voltammetric experiments. Because diffusion-limited currents reflect only the material transport rate within the bulk solution, no information about the other elementary steps is contained in these measured currents. This easy separation between the rate of material transport and the heterogeneous electron transfer kinetics has allowed quite detailed studies of transport phenomena and is the basis for nearly all amperometric analysis techniques. However, for experimental studies of heterogeneous electron kinetics, material transport limitations must be eliminated from the measured currents. A detailed discussion of eliminating transport limitations is beyond the scope of this chapter, but an important consequence of the finite rate of material transport is that it limits the magnitude of heterogeneous electron transfer rate that can be measured.

2. Surface Interactions

Once transported to the vicinity of the electrode surface, the redox centers may be preconcentrated or depleted from their bulk solution concentrations due to specific chemical interactions with the electrode surface, changes in the solvent properties within the near-surface region, and electrostatic interactions of charged redox centers with the electric fields present at the electrode surface. These interactions can cause the concentration profile near the electrode surface to deviate greatly from that present in the solution [17]. Figure 2 shows a concentration profile for a cationic redox molecule which has no specific interactions with the electrode surface but is attracted by a negative surface charge of the electrode double layer. The potential difference between the plane of closest approach of nonspecifically adsorbed molecules [the outer Helmholtz plane (OHP)] and the bulk solution potentials is the diffuse-layer potential. This is the potential difference which attracts cations to and repels anions from the electrode surface. The concentration of the cationic redox center at the electrode surface is higher than its solution concentration for the case shown in Figure 2, giving rise to

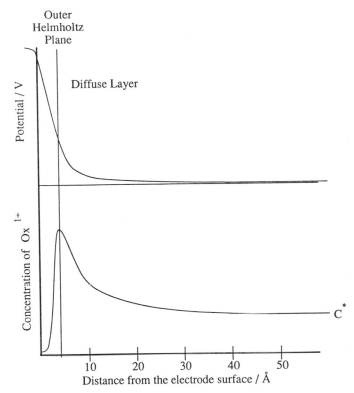

FIGURE 2 Concentration and potential profile diagrams near an electrode surface. The top portion of the figure plots the potential profile for an electrode polarized negative of the potential of zero charge (PZC). The bottom portion plots the concentration profile for a monocationic species, Ox^+.

an acceleration in the measured electron transfer rate. The perturbation of the concentration profile of a redox molecule within the electrode double layer is a large part of the double-layer effect on the observed heterogeneous electron transfer rate.

The partitioning of the redox center into the near-electrode surface region can enter the heterogeneous electron transfer kinetics in two distinct ways. If the local transport necessary for maintaining this concentration profile is faster than the electron transfer rate, the concentration of the redox center near the surface of the electrode will remain in equilibrium with the bulk concentration throughout the electrochemical experiment. The effect of specific interactions on the electron transfer rate can be accounted for using a partition coefficient κ

[18,19]. If the local transport is not fast enough to maintain the equilibrium between the bulk solution and the surface region, the heterogeneous electron transfer rate expression will contain this local transport rate. In an inner-sphere mechanism, the rate of formation of the reactive surface adsorbed form of the redox species may be rate determining. Such a limitation in the local transport rate would not be expected for an outer-sphere mechanism. In this case the preconcentration or depletion of a redox species occurs via diffusion over quite short distances (1 to 100 nm) which are fast relative to accessible heterogeneous electron transfer rates and therefore do not impede the electron transfer rate.

3. Partition Coefficient

As a redox center approaches the electrode surface, its electronic coupling with electronic states within the electrode typically increases exponentially [20]. An important consequence of this distance dependence of the electronic coupling is that the heterogeneous electron transfer rate constant for a given redox molecule is distance dependent. If one knows the concentration and rate constant profiles, one can integrate this distance-dependent rate expression and obtain the partition coefficient κ such that heterogeneous reaction i/nFA equal

$$\frac{i}{nFA} = \kappa k(0)C^* = \int_0^\infty C(x)k(x)\,dx \qquad (3)$$

where i is the electrode current, A the electrode area, and C^* the bulk concentration of the redox species. The units of the partition coefficient κ are distance (cm) and describe an equivalent reaction-layer thickness [21]. This definition allows one to cast the heterogeneous electron transfer expression into a simplified case of a constant concentration profile held at the bulk concentration of the redox species reacting with a constant rate constant $k(0)$ (now expressed as a pseudo-first order rate constant with units S^{-1}) and allowing only those redox centers within a distance κ from the electrode surface to undergo electron transfer with the electrode. Unfortunately, the concentration and rate constant profiles near the electrode surface are generally not known, so that this κ can become an adjustable parameter in the kinetic description.

4. Time Scale of the Electron Transfer

Due to the relatively small mass of electrons, the motion of electrons between the electrode and redox centers occurs more rapidly than nuclear vibrations, rotations, and translations. This difference in the time scales of motion between electrons and nuclei results in the nuclear motions being frozen on the time scale of the electron transfer.* At the instant of electron transfer, the redox center

---

*This is the Born–Oppenheimer approximation, which separates the electron dynamics from nuclear motions.

changes from one oxidation state to another, maintaining its internal and solvent structure. The internal energy of the reactants and products must be identical because there is not enough time within the actual electron transfer event for energy transfer to or from the surrounding medium.

Incorporating this conservation of energy between the reactants and products of the electron transfer reaction into a description of the electron transfer is facilitated by the definition of a useful fiction. The electron transfer is described as occurring from a one electron donor orbital on the reductant to the acceptor orbital of the oxidant. This is essentially a molecular orbital approach in which all the changes in the internal energy of the redox molecule upon addition or removal of an electron are reflected in the energy level of the one electron acceptor or donor orbital. Maintaining the internal energy of the reactants and products then becomes equivalent to matching the energy levels of the donor and acceptor orbitals involved in the electron transfer.

5. Nuclear Activation

The energy level of a donor or acceptor orbital of an individual redox molecule is a dynamic rather than a static parameter. Random thermal fluctuations in the internal energy and solvation of the redox center cause its donor and acceptor levels to vary about some characteristic energy corresponding to the redox molecule's most stable molecular and solvent structure. In other words, thermal fluctuations in the redox center's structure cause it to become alternately easier or harder to reduce (or oxidize). One can define an *instantaneous redox potential* for the molecule that is dependent on its instantaneous internal and solvent structure. For the reduction of a redox molecule, Ox, at the electrode surface under typical voltammetric conditions, the percentage of redox centers within the reaction layer that are sufficiently activated to bring their relevant electronic states in resonance with the filled metal electronic states is extremely low. The free energy of these activated redox sites relative to their equilibrium value is typically many times higher than the average thermal energy. In obtaining an expression for $k(0)$ in equation (3), one needs to determine the percentage of sufficiently activated redox sites within the reaction layer at the electrode.

6. Reaction Coordinate Diagram

Not all the thermal fluctuations in the internal structure and solvation of the redox molecule will affect its instantaneous redox potential equally. For example, if the reduction of a redox molecule localizes electron density preferentially within an antibonding atomic orbital, the elongation of that bond will reduce the electron–electron repulsion, resulting in a lower overall energy and a more positive instantaneous reduction potential. In contrast, the vibration of bonds that make only a small contribution to the acceptor orbital will not be nearly as effective in reducing the energy of the acceptor orbital. There will be a combination of vibrations, internal rotations, and solvent orientations that give rise to the largest change in the instantaneous redox potential. This combination of nuclear motions is termed the *reaction coordinate*. As the oxidized molecule moves from its equilibrium structure along this reaction coordinate in the direction of making the reduction potential more positive, its internal and solvent structure qualitatively becomes more like the equilibrium structure of the reduced form. Indeed, a more common but less general definition for the reaction coordinate is that it is the collection of nuclear modes that interconverts the structrure of the oxidized and reduced forms of the redox couple in their most stable nuclear configurations.

A graphical presentation of this reaction coordinate is shown in Figure 3. Figure 3A plots the free energy of the oxidized and reduced forms of a redox couple as a function of the displacement of the molecules' structure along the

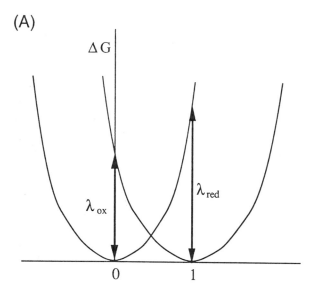

(A)

FIGURE 3 Reaction coordinate diagrams. (A) Plot of the free energy of Ox and Red species as a function of a normalized reaction coordinate parameter. The reorganization energies for the two redox molecules λ_{Ox} and λ_{Red} are shown with double arrows. (B) Same plot as in (A) except plotted as a function of the instantaneous redox potential of Ox and Red. (C) Density of electronic states representation of (A) and (B). The abscissa plots the density of electronic states within the metal and reaction layer, and the ordinate plots the electronic energy level. The shaded distributions represented filled electronic states.

(B)

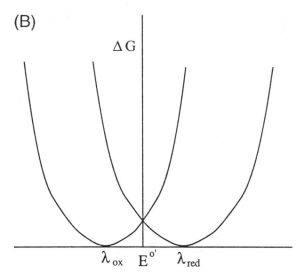

(C)

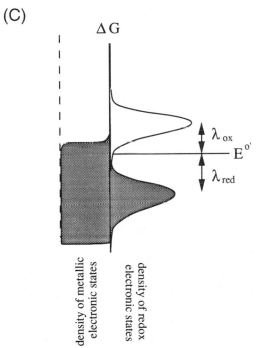

FIGURE 3 Continued

reaction coordinate. The origin of this reaction coordinate is set on the minimum for the oxidized form of the redox couple, and the abscissa is normalized to the position of the reduced form's minimum free energy. In this figure the potential of the electrode is set at the formal potential of the redox couple so that the free energies of the oxidized and reduced forms of the molecules [in equilibrium with the electrons and electron vacancies (holes) within the metal] are the same.

The reduction of Ox in its most stable nuclear configuration produces a highly activated form of Red. Graphically, this electron transfer is denoted by a vertical transition along reaction coordinate diagram. The energy required for this electron transfer is the reorganization energy, λ_{Ox}, which must be supplied by an electron that is λ_{Ox}/e more negative than the formal potential.* This is to say that the instantaneous reduction potential of Ox in its most stable nuclear state is λ_{Ox}/e more negative in energy than the formal potential of the Red/Ox couple. The oxidation of Red in its most stable nuclear configuration requires an acceptor orbital that is λ_{Red}/e more positive of the formal potential. Similarly, for any extent of activation in Figure 3A, one can determine the instantaneous redox potential of the redox molecule required for the electron transfer. The instantaneous reduction potential is simply the energy difference between the Ox and Red free-energy curves divided by an electronic charge, e.

Figure 3B shows a plot of the free energy of Ox and Red as a function of their instantaneous redox potential [22]. The minima for the Ox and Red forms are located at λ_{Ox}/e and λ_{Red}/e, respectively. The intersection of the two free-energy curves at the formal potential (0 V in this diagram) marks the potential at which the oxidized and reduced forms of the redox couple can undergo electron exchange with the same nuclear activation.

7.   Density of Electronic States Distributions

The reaction coordinate diagram shown in Figure 3B can be redrawn showing the distribution of the molecules' instantaneous redox potentials (in states/eV) [23]. In this particular density of electronic states distribution diagram, the number of oxidized and reduced redox molecules within the reaction layer are equal, so that the areas of the two distribution functions are identical. The potential of the electrode is set negative of the formal potential of the redox couple. These redox electronic states distributions can be thought of as a time-averaged distribution of a single Ox or Red molecule's instantaneous redox potential or an instantaneous snapshot of the density of one electron donor/acceptor orbitals for all Ox and Red within the reaction layer at the surface of the electrode. Math-

---

*For reorganization energies measured in electron volts, λ_{Ox}/e and λ_{Red}/e are simply in volts. As indicated in this figure, λ_{Ox} and λ_{Red} need not be identical.

ematically, this distribution function is simply the Boltzmann distribution of donor/acceptor levels given by

$$D(E) = NC_n e^{-[\Delta G(E)/kT]} \tag{4}$$

where $D(E)$ is the density of electronics states of the donor/acceptor orbitals as a function of its energy E; N the number of Ox or Red centers within the reaction layer; C_n a normalization constant, which is dependent on the exact form of the free energy versus reaction coordinate diagram; $\Delta G(E)$ the free-energy function plotted in Figure 3B; and kT the Boltzmann constant multiplied by the absolute temperature.

The peaks of these distributions occur at the instantaneous reduction potential of Ox and the instantaneous oxidation potential of Red, corresponding to their most stable nuclear configurations. The electronic orbitals available at the electrode surface can also be displayed on the same potential axis. A metallic electrode is characterized by a continuum of one-electron states that are occupied on average up to the Fermi level [24]. In contrast to the broad distribution of electronic state for the redox molecules, the distribution of filled and unfilled orbitals in a conductor is much sharper and is given by the Fermi distribution function.*

$$D_e(E) = D_m(E) \frac{1}{e^{-(E_f - E)/kT} + 1} \tag{5}$$

$$D_h(E) = D_m(E) - D_{e^-}(E) \tag{6}$$

where $D_e(E)$ is the density of filled electronic states as a function of their energy, $D_m(E)$ the density of electronic states within the metal, E_f the Fermi-level energy, and $D_h(E)$ the density of unfilled electronic states. As the electrode is polarized negative of the formal redox potential, the Fermi level would be raised in Figure 3C, increasing the number of Ox molecules which are in resonance with filled electronic states in the electrode. This increasing overlap between the filled electronic states in the metal and the "unfilled" or oxidized redox molecules within the reaction layer thickness results in an approximately proportional increase in the reduction rate as the electrode is polarized negatively.

By virtue of this distribution of filled and unfilled electronic states, the electrode behaves not as a single reactant but as a distribution oxidants and

---

*The abscissa in Figure 3B shows the energy level of the one-electron acceptor and donor levels of Ox and Red. Displacement of the redox molecule along this axis is not equivalent to an increase in the free energy of the redox molecule but only of the one-electron donor or acceptor orbital. For this reason the excursions of the acceptor and donor orbitals from the energy level of the most stable form of the redox molecule need not follow a narrower Boltzmann distribution.

reductants. A general expression for the electron transfer rate would have to sum the rates for all the filled (or unfilled) electronic states within the metal. In this summation a different activation energy, ΔG, would have to be used for each metal energy level. Luckily, the overlap between the filled electronic states in the metal and the oxidized redox species within the reaction layer is usually sharply peaked centered about the Fermi level. A large majority of the electron transfers occur from electronic states within kT of the Fermi level (within 0.026 V at 25°C). In this situation one can approximate that all the electron transfers are occurring from states at the Fermi level, with the activation of the redox molecule being appropriate for an electronic state at the Fermi level. This is equivalent to considering only a single reaction coordinate curve for both the oxidized and reduced forms of the redox couple, as shown in Figure 3B.

8.   Electronic Coupling

The resonance between the donor and acceptor orbitals of the metal surface and redox molecule is only one of the requirements for an electron transfer to take place. Once in resonance, the electron transfer depends on the extent of electronic coupling of the donor and acceptor orbitals. Above, we considered the dependence of this electronic coupling on the distance of the redox molecule from the electrode surface. A knowledge of this distance dependence was essential in determining the reaction layer thickness κ, which simplified the kinetic description to an equivalent number of molecules reacting at closest approach to the electrode surface. At closest approach to the electrode surface, the probability that a pair of acceptor and donor orbitals will exchange an electron can vary from zero to unity depending on the electronic structure and physical separation of the orbitals involved. Focusing on a single Ox molecule at closest approach to the electrode, its electronic coupling with filled states of the metal is directly proportional to the density of filled electronic states in the metal at its instantaneous reduction potential. If Ox initially is in its most stable configuration (with an instantaneous reduction potential of $-\lambda_{Ox}/e$ relative to the formal potential of the redox couple), the electronic coupling with the electrode held at the formal potential will be extremely small due to the infinitesimally small density of filled electronic states in the metal at $-\lambda_{Ox}/e$. As Ox moves along the reaction coordinate and its acceptor-level orbital moves closer to the Fermi level, the density of filled electronic states in the metal increases, increasing the probability per unit time that an electron will be transferred to Ox.*

This increasing probability of electron transfer has a significant effect on the motion of the redox molecule along the reaction coordinate. As Ox moves

---

*At instantaneous redox potentials more positive than the Fermi level, the Fermi distribution function becomes unity, so that the electronic coupling of Ox with the filled electrode states becomes relatively constant and depends on the density of electronic states distribution of the electrode material.

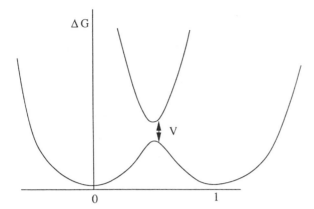

FIGURE 4 Splitting in the reaction coordinate diagram due to electron transfer with the electrode. This is the same reaction coordinate plot as that shown in Figure 3A, showing the effect of the electronic coupling between the electrode and redox couple.

toward Red along the reaction coordinate, it experiences an ever-stronger restoring force directed toward its most stable structure. Upon electron transfer, the direction of this restoring force changes toward the most stable structure of Red. For an instant there is a reduced molecule at the electrode surface that can return an electron to an unfilled metal electronic states and again become Ox. Through this electronic exchange with the metal, the exact oxidation state of the redox molecule becomes uncertain when its instantaneous redox potential approaches the Fermi level. The electron exchange couples the oxidized and reduced forms of the redox couple resulting in a splitting of the free-energy curves shown in Figure 4. The splitting in the free-energy curves gives a measure of the electronic coupling. For redox molecules with small electronic couplings at closest approach to the electrode surface ($V \ll kT$), the splitting between the two free-energy curves is small enough that a redox molecule approaching the intersection point will be easily thermally excited and continue along the upper curve rather than undergoing electron exchange. These poorly electronically coupled electron transfer reactions are termed nonadiabatic in that they avoid undergoing electron transfer via becoming thermally activated.* As the electronic coupling between the electrode and the redox molecule becomes larger (reflected in a large splitting of the free-energy curves, $V > kT$), electron transfer will

---

*This extra energy needed to overcome the barrier between the lower and upper curves is transferred to the reaction coordinate by other modes of the redox molecule and solvent.

efficiently couple the oxidized and reduced forms of the redox couple, causing the reaction to proceed along the lower free-energy curve. Such a strongly electronically coupled electron transfer is termed adiabatic because no thermal excitation is required to proceed along the lower free-energy curve. In the case of even stronger electronic coupling, the very notion of an oxidized and reduced state may come into question. When the electronic coupling is strong enough to lower the barrier between the oxidized and reduced free-energy curves to below the thermal energy, kT, electrons become delocalized between the redox molecule and the electrode surface, and the idea of a discrete electron transfer reaction ceases to be an appropriate description.

9. Transition-State Rate Expressions

With the complex motion of the redox molecule and solvent and the involvement of quantum mechanical tunneling, the task of developing a mathematical expression for the electron transfer rate may seem daunting. However, because one seeks only a measure of the electron transfer rate and not a complete description of the reacting system, one can focus on the portion of the reaction coordinate diagram, which largely determines the electron transfer rate, the transition-state region near the intersection of the oxidized and reduced free-energy curves shown in Figure 3A. A particularly simple starting point for the mathematical description can be obtained by considering only the reaction coordinate curve at the intersection point, the transition state of the electron transfer reaction [25]. In this transition-state theory, the rate of the reaction is calculated from the number of reacting molecules reaching the transition state at any particular instant in time and their rate of crossing over the transition state. The number of molecules at the transition state is the density of electronic-state distribution at the transition state free energy, ΔG. Considering only electron transfers from the Fermi level, the transition-state theory rate (in mol s^{-1}) becomes

$$R^{\mathrm{TST}} = \nu N_{\mathrm{Ox}} e^{-\Delta G/kT} \tag{7}$$

where ν is the crossing rate at the transition state and the ΔG is taken for the reduction occurring at the Fermi level. Replacing N_{Ox} with the reaction-layer thickness, this rate becomes

$$R^{\mathrm{TST}} = \nu \kappa A C_{Ox} \, e^{-\Delta G/kT} \tag{8}$$

From this one obtains an expression for the heterogeneous electron transfer rate constant as

$$k^{\mathrm{TST}} = \nu \kappa e^{-\Delta G/kT} \tag{9}$$

C. Marcus Theory of Electron Transfer

The development of this general transition-state expression has relied on only the most simple assumptions about the electron transfer mechanism. Equation (9) is therefore quite general and not particularly useful for the analysis of kinetic data. One would like to express v and ΔG in terms of more experimentally derived parameters. Expressions derived by Marcus for these parameters have formed the basis of much of the theoretical work in modern electron transfer theory [15]. Below is outlined a presentation of the Marcus theory. Refinements to this initial description will be addressed in response to experimental observations in the next sections.

1. Free Energy of Activation

The best known assumption of the Marcus theory of electron transfer is that the free energy of activation of the redox molecule can be approximated via a simple parabolic well with respect to displacement along the reaction coordinate [26]. This harmonic approximation can be viewed as the simplest polynominal approximation to the free energy vs. reaction coordinate diagram shown in Figure 3A. In a more favorable light, the parabolic approximation of the activation energy relies on each vibration involved in the activation of the molecule being describable as a harmonic oscillator and the solvent energy being quadratically dependent on the charge density of the redox molecule. These two components to the activation, the inner-shell vibrations of the redox molecule and the outer-shell motions of the surrounding solvent, are considered separately, so that

$$\Delta G_{\text{total}} = \Delta G_{\text{is}} + \Delta G_{\text{os}} \tag{10}$$

where ΔG_{total} is the total activation energy and ΔG_{is} and ΔG_{os} are the inner- and outer-shell activation energies, respectively. The inner-shell activation is obtained as the minimum energy required to change the geometry of the redox molecule to its transition-state nuclear configuration. This activation energy is related to the inner-shell reorganization energy of the redox molecule, λ_{is}, which is the energy required to interconvert the most stable structures of the oxidized and reduced forms of the redox couple while maintaining their initial oxidation states. This inner-shell reorganization energy can be calculated from the force constants of the each vibration mode by summing over all vibrations using the equation [18]

$$\lambda_{\text{is}} = \sum_{\text{all modes}} \frac{1}{2} f_i (\Delta l_i)^2 \tag{11}$$

where f_i is the force constant for the ith vibration mode and Δl_i is the difference in the equilibrium (most stable) structure of the oxidized and reduced forms of the redox molecule for the ith mode.

When the electrode potential is held at the formal potential of the redox couple as in Figure 3A, the transition-state geometry is most commonly located midway between the most stable nuclear configurations of the oxidized and reduced forms of the redox molecule. A good approximation to the transition-state geometry is therefore the average between the oxidized and reduced forms. In other words, the change in the bond lengths and angles is one-half the Δl_i's used in equation (11) to obtain the inner-shell reorganization energy. The inner-shell activation energy can therefore be approximated as one-fourth of the inner-shell reorganization energy λ_{is}.

The changes in the solvent structure between the oxidized and reduced forms of a redox couple give rise to an outer-shell component to the activation barrier. Because there are typically a large number of solvent molecules whose exact positions relative to the redox molecule are not known, calculating the solvent activation using equation (11) is not feasible. Instead, the Marcus theory approximates the solvent as a uniform dielectric medium. A particularly simple and widely used expression for the outer-shell activation energy was obtained by Marcus using a two-step charging model. Because a good description of the derivation has been given by Marcus, only a brief outline of the essential elements of the derivation will be given here [27].

The outer-shell or solvent component of the activation energy arises from the rapid charging of the redox molecule during the electron transfer event. The energy (in joules) required to charge a sphere in a uniform dielectric is given by

$$U = \frac{n^2 e^2}{8\pi\epsilon_0\epsilon r} \tag{12}$$

where n is the number of electrons transferred, e the electronic charge in coulombs, ϵ_0 the permittivity of free space (8.85×10^{-12} C V^{-1} m^{-1}), ϵ the relative dielectric constant of the medium, and r the radius of the sphere in meters [28]. On the time scale of the electron transfer, only the high-frequency electronic modes of the solvent can respond to the change in the redox molecule's charge. The initial stabilization of the charge-transfer product depends on the *optical dielectric constant*, ϵ_{op}, which is the square of the solvent's refractive index. On a longer time scale, the solvent dipoles can also rotate and the position of the solvent relative to the redox center can change, further stabilizing the electron transfer product to its equilibrium energy, which is dependent on the static dielectric constant of the solvent, ϵ_s. These static charging energies of the reduced and oxidized forms of the redox molecule are thermodynamic quantities which are reflected in the formal potential of the redox couple. The solvent activation energy ΔG_{os} is therefore the extra charging energy needed initially to change the redox state and is obtained as the difference between the optical and static charging energies:

$$\Delta G_{os} = \frac{n^2 e^2}{8\pi\epsilon_0 r} \left(\frac{1}{\epsilon_{op}} - \frac{1}{\epsilon_s} \right) \tag{13}$$

When n is set to 1, one obtains the outer-shell reorganization energy. At the transition state one need polarize the reactant by roughly half this amount ($n = \frac{1}{2}$) so that the solvent activation energy becomes

$$\Delta G_{os} = \frac{e^2}{32\pi\epsilon_0 r} \left(\frac{1}{\epsilon_{op}} - \frac{1}{\epsilon_s} \right) \tag{14}$$

For the polar solvents typically used in electrochemical experiments, $\epsilon_{op} \ll \epsilon_s$, so that the solvent activation energy depends almost solely on the optical properties of the solvent.

A further complication of this expression for the activation energy stems from the dielectric properties of the metallic electrode. When an ion is brought to a distance R of the conducting surface, the mobile electrons within the conductor redistribute in response to the ion's charge. Electrostatically, this redistribution of charge can be modeled as an equal and oppositely charged image located a distance R inside the metal surface. This image charge screens part of the redox molecule's charge, reducing the outer-shell activation. Its effects are quantitatively accounted for in the outer-shell activation as

$$\Delta G_{os} = \frac{e^2}{32\pi\epsilon_0} \left(\frac{1}{r} - \frac{1}{2R} \right) \left(\frac{1}{\epsilon_{op}} - \frac{1}{\epsilon_s} \right) \tag{15}$$

2. Frequency Factor

The frequency factor, ν, in the transition-state expression for the electron transfer rate measures the rate of reactive crossings of the transition state. It is therefore dependent on the rate of motion along the reaction coordinate and the probability of crossing over from the reactant to the product free-energy curve [25]. These two components to the frequency factor can be introduced separately such that

$$\nu = \nu_n \kappa_{el} \tag{16}$$

The ν_n term measures the frequency of passage of the reactant through the transition state. κ_{el} is the electronic transmission coefficient, which can be viewed as the percentage of these transition-state crossings that are reactive and therefore proceed along to the product free-energy curve.

The frequency of motion along the reaction coordinate depends on the frequencies of motion of the internal vibrational and solvent modes which comprise the reaction coordinate. An expression for ν_n can be obtained by taking a weighted average of the frequencies associated with the inner- and outer-shell modes [18]:

$$v_n = \sum_{\text{all modes}} \left(v_i^2 \frac{\Delta G_i}{\Delta G_{\text{total}}} \right)^{1/2} \tag{17}$$

where the v_i's and ΔG_i's refer to the frequencies and free energy of activation for each mode. Although simple in form, this equation hides considerable complexity present in the nuclear frequency factor. One can make a distinction between the inner- and outer-shell modes which are summed in this equation. The inner-shell modes are typically faster than the solvent modes (due primarily to the stronger force constants between the bonded atoms of the redox molecules relative to the charge–dipole interactions of the solvent). When the inner-shell activation energy dominates the total activation energy for the electron transfer reaction, the frequency factor is also dominated by these faster inner-shell modes. Typical values for v_n in this case may range from 10^{13} to 10^{14} s^{-1} [29]. In contrast, for redox reactions characterized by a small change in structure between the oxidized and reduced forms, the outer-shell activation dynamics can determine this nuclear frequency factor. The frequency factor then depends primarily on rotational motions of the surrounding solvent molecules, which give rise to frequency factors typically in the range 10^{11} to 10^{12} s^{-1}.

The extent of electronic coupling between the redox molecule and the electrode also influences these nuclear frequency factors by changing the shape of the free-energy barrier at the transition state. For weakly electronically coupled systems, the transition-state region appears as a sharp cusp. For these cases the rate of motion of the reactant through the transition state to the product is well approximated by equation (17). In contrast for strong electronic coupling between the redox molecule and the metal electrode, the transition-state region is flattened by the rapid electron exchange that occurs near the transition state. This flattened transition-state region results in a reduced rate of crossing across the barrier region [2].

As the electronic coupling between the metal electrode and the redox molecule becomes weaker and weaker, the rate of electron transfer can become independent of the motion of the reactant along the reaction coordinate [30]. In the limit of weak electronic coupling, each passage of the reactant through the transition state will only rarely result in an electron transfer. The probability of an electron transfer for each passage, κ_{el}, depends on both the electronic splitting energy, V, and the amount of time the reactant remains in the transition region, which is proportional to v_n^{-1}. The nuclear frequency term therefore cancels from the preexponential term in the electron-rate expression. In effect, the dynamic motion of the reactant within the transition-state region is averaged out in this extremely nonadiabatic case.

An adiabatic passage of the reactant through the transition state can also be nonreactive due to energy transfer between the reaction coordinate and other molecular and solvent modes which alter the direction of the reactant's motion along the reaction path. This second mechanism for unreactive trajectories of the reactant is closely related to the dynamic properties of the solvent. The study of such solvent dynamic effects on the κ_{el} term is discussed in greater detail in Section III.B.

3. Marcus Theory Predictions

Although some effort has been expended to use these Marcus theory expressions to predict a priori the heterogeneous electron transfer rates [31], the agreement between theory and experiments is not perfect and often relies on simplifying assumptions, particularly with respect to the more difficult input parameters needed for the calculation. Each of the four principal components of the heterogeneous electron transfer rate expression, ν_n, κ_{el}, κ, and ΔG, are interrelated in subtle ways which complicate their estimation and use in equations (9) to (17). A less demanding use of the Marcus theory expressions is for quantitatively predicting how changes in a redox molecule's structure, the solvent, the electrode potential, or the temperature affect the heterogeneous electron transfer rate. In these simpler experiments, many of the input parameters can be assumed to be unchanged in the experimental series, so that isolated features of the electron transfer kinetics can be probed. Much emphasis in the study of electron transfer reactions at electrodes has been directed toward testing the validity of the Marcus theory and toward extracting fundamental parameters that describe the kinetic facility of redox molecules [32–36]. Such experimental studies and additional theoretical work [37–42] have both supported the overall Marcus description and improved (at the expense of some simplicity) certain of the approximations used to implement equation (9).

As stated above, the Marcus theory expressions contain more parameters than can be determined in a simple voltammetric experiment. Because the Marcus theory parameters are generally seriously underdetermined in a typical voltammetric experiment, the assignment of a particular underlying parameter can be somewhat ambiguous. Perhaps the most difficult aspect of an electron transfer reaction to characterize is its adiabaticity [43]. Experimental determinations of the adiabaticity of a given redox reaction are difficult because the adiabaticity of the electron transfer affects primarily the preexponential term in the rate expression. To assign changes in the preexponential term between two different redox molecules to changes in their electronic coupling with the electrode, one must have accurate estimates of the nuclear frequency factor and partition coefficient for the two redox molecules. Such accurate estimates are not typically available, so the adiabaticity of a particular redox reaction is usually uncertain.

Heterogeneous electron transfer reactions are very commonly *assumed* to be adiabatic. Part of the motivation for this assumption is that it simplifies the kinetic analysis because the electronic transmission coefficient can be set to unity. Because the adiabaticity of an electron transfer affects many of the elementary steps in the electron transfer, incorrect assumptions as to the extent of electronic coupling between the electrode and the redox molecule can invalidate much of the subsequent analysis.

The other major factor that determines the heterogeneous electron transfer rate is the transition-state activation energy. Activation parameters for electron transfer reactions have generally been obtained via observations of the temperature dependence of the heterogeneous electron transfer rate. The slope of an Arrhenius plot [$\ln(k)$ versus $1/T$] gives a measure of the activation free energy as the slope. This analysis can be continued separating the activation free energy into its enthalpic and entropic components [43–45]. A more direct way of investigating the activation free energy of a particular electron transfer reaction is via measurements of the heterogeneous electron transfer rate dependence on the electrode potential [32]. As the electrode potential is changed from the formal potential toward the reorganization energy of a species, Ox, in the solution, the electron transfer rate increases due to the increased overlap between filled electrode states and unfilled molecular states at the Fermi-level energy. In more common language, polarization of the electrode to more negative potentials makes the electrode surface a stronger reductant.

The exact potential dependence of the heterogeneous electron transfer rate depends on the extent of electronic coupling between the electrode and redox molecule. For this discussion it is convenient to describe the electron transfer kinetics from the viewpoint of the density of electronic states. In the Marcus theory approximation, the parabolic dependence of the redox molecules free energy on the reaction coordinate produces a Gaussian density of electronic states distribution [46]. The two limiting cases for the potential dependence on the electron transfer rate, strongly adiabatic and strongly nonadiabatic, are depicted in Figure 5. For adiabatic electron transfers shown in Figure 5A, the strong electronic coupling between electrons in the electrode and the Ox states at the Fermi level results in the reduction of all Ox centers activated to the Fermi level. The depletion of Ox centers with instantaneous redox potential more positive than the Fermi level shown in Figure 5A is equivalent to the electron transfer proceeding along the lower free energy vs. reaction coordinate curve shown in Figure 3. The electron transfer rate will therefore occur primarily through electrode states near the Fermi level so that the rate of electron transfer can be approximated by:

$$k = v_n D_{Ox}(E_f)$$
                                                                                                       (18)

Expanding the D_{Ox} term, one obtains the expression that is plotted in Figure 6A:

$$k = v_n N_{Ox}(4\pi\lambda_{Ox}kT)^{-1/2}e^{-[(\lambda_{Ox}-E)^2/4\lambda_{Ox}kT]} \tag{19}$$

The second limiting case shown in Figure 5B is for a nonadiabatic electron transfer in which the electronic coupling between the metal and redox molecule is small enough so that the rate of electron transfer is slow relative to the motion of Ox along the reaction coordinate. In this strongly nonadiabatic regime, the electron transfer reaction is not fast enough to distort the density of electronic states distribution of Ox. The overlap between the electrode and redox molecule's density of electronic states distributions will be broader than for the adiabatic case, with some of the electron transfers occurring from electrode states beneath the Fermi level. In an expression for the total electron transfer rate, one must sum over all the electron energies rather than only at the Fermi level. This gives:

$$k = v_n N_{Ox}(4\pi\lambda_{Ox}kT)^{-1/2} \int_{-\infty}^{+\infty} \frac{e^{-[(\lambda_{Ox}-E)^2/4\lambda_{Ox}kT]}}{e^{(E-E_f)/kT} + 1} \, dE \tag{20}$$

Figure 6 compares these two limiting cases showing plots of the heterogeneous electron transfer rate constant plotted against the formal overpotential. These plots depict the reduction rate constant of a species, Ox, which has a reorganization energy of 1 eV and an effective preexponential factor of 10,000 cm/s. The two plots have been normalized to the same formal rate constant (0.082 cm/s). As seen in the semilogarithmic plot shown in Figure 6B, the two cases give nearly identical potential dependences on the heterogeneous electron transfer rate constant for formal overpotentials which are less than half the reorganization energy. When viewed over a wide range of electrode potentials, the logarithm of the heterogeneous rate constant displays substantial curvature. This Marcus theory prediction is in sharp contrast to the Butler–Volmer equation's linear relationship between the logarithm of the heterogeneous rate constant and the electrode potential.

The deviations between the adiabatic and nonadiabatic cases occur at potentials near the reorganization energy of the redox molecule. As the Fermi level in the electrode is polarized beyond the reorganization energy, the reduction rate for the nonadiabatic case asymptotically approaches a limiting electron transfer rate. At these high negative electrode potentials, the overlap between the filled electronic states in the metal and the unfilled redox states within the reaction layer is complete. Because of the approximately constant density of filled electronic states within the metal beneath the Fermi level,* each Ox molecule within

---

*The density of electronic states distribution for metals is often a slowly varying function of energy, so that within several tenths of a volt of the Fermi level, the density of electronic states can be approximated as being constant (see Ref. 47).

(A)

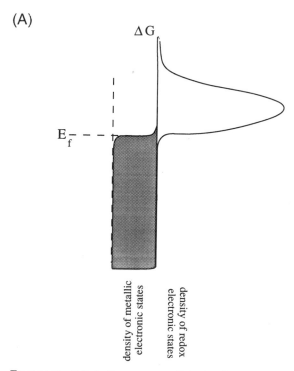

ΔG

E_f

density of metallic electronic states

density of redox electronic states

FIGURE 5 Adiabatic vs. nonadiabatic density of electronic states plots for the reduction of Ox (same figure as shown in Figure 3C for an electrode in contact with a solution containing only the oxidized form of the redox couple): (A) distortion of density of electronic states distribution for Ox caused by an adiabatic electron transfer reaction; (B) undistorted density of electronic states distribution for Ox predicted for an extremely nonadiabatic electron transfer.

the reaction layer has the same electron transfer rate, independent of its particular extent of activation. This limiting electron transfer rate depends only on the number of Ox molecules within the reaction layer and their electronic coupling to the filled states within the metal electrode.

The potential dependence of the heterogeneous electron transfer rate constant appears somewhat stranger for the adiabatic case because in it one assumes that all the electron transfers occurs via electrode states at the Fermi level. Its Gaussian shape results from the depletion of the Ox states whose instantaneous redox potentials are more negative than the Fermi level during the finite time required to polarize the electrode to a given potential. As the electrode potential is polarized negative of the reorganization energy, the number of reducible redox

(B)

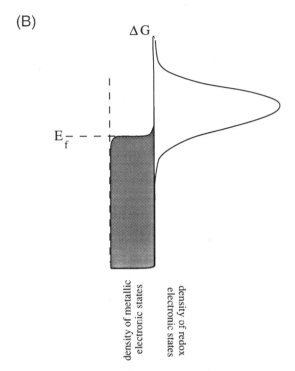

FIGURE 5 Continued

molecules decreases exponentially. An extremely short time scale for this heterogeneous electron transfer rate determination was chosen so that no exchange of the redox centers occurs between the reaction layer and the bulk solution.

One of the more controversial predictions of the Marcus theory is the existence of an *inverted* or *abnormal* reaction free-energy region in which increasing the reaction free energy results in a decrease in the electron transfer rate [15,48]. This prediction is the direct consequence of the concave shape of the free energy vs. reaction coordinate diagram, which is approximated as a parabola in the Marcus theory. The instantaneous redox potential of Ox at the minimum of the free energy vs. reaction coordinate plot is the potential at which Ox can be reduced with no activation. Reductions that occur at other potentials (either more negative or positive) require Ox to be activated. The rate of electron transfer is therefore maximal at an electrode potential equal to the reorganization potential.

This slowing of the electron transfer rate within the inverted free-energy region can be observed in Figure 6 at potentials negative of the reorganization

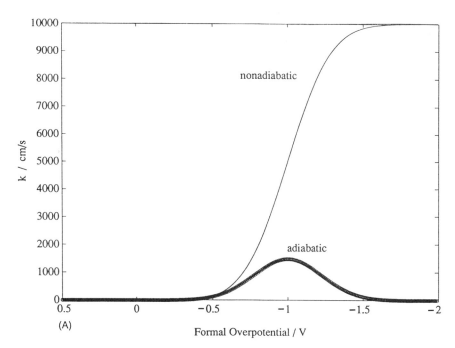

FIGURE 6 Heterogeneous electron transfer rate constant vs. formal overpotential for the adiabatic and nonadiabatic electron transfer cases: (A) linear scale; (B) semilogarithmic scale. The solid line plots the heterogeneous electron transfer rate constant predicted for a nonadiabatic electron transfer, while the line of circles shows the same plot for the adiabatic case. The currents were normalized to the same rate constant at the formal potential.

energy (1 eV). For both adiabatic and nonadiabatic cases, the electron transfer rate involving electronic states at the Fermi level decreases at potentials beyond the reorganization potential, λ_{Ox}/e. When the electron transfers occur only through states at the Fermi level, as assumed in the adiabatic case, the measured heterogeneous electron transfer is directly proportional to the density of electronic states of Ox within the reaction layer, which decreases in the inverted region. The nonadiabatic case gives an integrated response of all the filled electronic states and so gives approximately the integral of the adiabatic case.

Because the logarithms of the predicted rate constants within $\pm 0.5\lambda_{Ox}/e$ of the formal potential of the redox molecule have the same quadratic potential dependence in both the adiabatic and nonadiabatic cases [given by Eq. (19)], one can test the Marcus theory prediction and extract the reorganization energy and preexponential factor from a plot of $\log(k)$ vs. E. An experimental problem

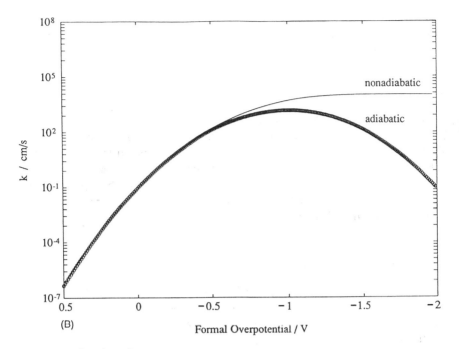

(B)

Formal Overpotential / V

FIGURE 6 Continued

with this approach has been the difficulty in obtaining rate constant data over a sufficiently wide range of potentials for the curvature in the $\log(k)$ vs. E plot predicted by the Marcus theory to be measurable [49]. The finite rate of mass transport of the redox molecule to the electrode surface and the influence of the electrode double layer on the electron transfer reaction can both hinder this analysis approach [32]. For the reduction of Ox shown in Figure 6, the experimentally accessible overpotential range may be about -0.3 to 0.1 V vs. the formal potential. At the negative limit the measured current becomes dominated by background currents arising from impurities in the solution and the oxidation of the reduction product. At 0.1 V, the heterogeneous electron transfer rate is controlled primarily by the rate of mass transfer.* Within this narrow potential range, the logarithm of the predicted heterogeneous electron transfer rate con-

---

*The positive voltage limit was calculated assuming the mass-transport-limited rate obtained at 1 ms in a potential step experiment $(D_{Ox}/\pi t)^{1/2}$. The diffusion coefficient of Ox was assumed to be 1×10^{-5} cm^2/s. For the reduction of Ox shown in Figure 6, the diffusion limited rate is one-tenth the heterogeneous electron transfer rate constant at 0.1 V vs. the formal potential of the redox couple.

stant vs. electrode potential plot displays minimal curvature, which can be masked by potential dependent changes in the diffuse-layer potential which alter the reaction-layer thickness.

III. Experimental Studies

In the remaining portion of this chapter we describe two active areas of experimental electrochemical research. In the first area we describe heterogeneous electron transfer measurements made at electrodes insulated by thin films that allow one to probe in some depth the reaction coordinate vs. instantaneous redox potential and interfacial and molecular structural effects on the electronic coupling between the electrode and redox molecule. In the second area of study we probe the influence of the dynamic properties of the solvent in determining the electron transfer rate.

A. Insulated Electrodes

The experimental difficulties associated with measuring electron transfer rates at bare electrodes, finite rates of mass transport, and double-layer effects can be greatly reduced by insulating the electrode surface [50]. By placing an ultrathin insulating layer at the electrode surface, one can limit the closest approach of the redox molecule to the metallic surface, reducing the electronic coupling between the metal and redox molecule. The electron transfer rate is reduced primarily via a decrease in the electronic transmission coefficient κ_{el}. By controlling the structure and thickness of the insulator, one can decrease the measured electron transfer rate by an arbitrary amount. This allows one to probe heterogeneous electron transfers at potentials at which the reaction would be too fast to measure at bare electrodes because of mass-transport limitation. Because a large fraction of the applied potential [measured relative to the potential of zero charge (PZC)] is usually dropped within the insulator, the diffuse-layer potentials can be reduced significantly, thereby reducing the double-layer effect.

1. Insulator Characteristics

To use such insulated electrodes for kinetic studies, the properties of the insulating film are of paramount importance. The electron transfer through the insulating film should not result in vibrational or electronic transitions of the film. In other words, the dominant mechanism for electron transfer through the film should be via an elastic mechanism. For such an elastic tunneling mechanism, the driving force for the electron transfer is identical to that which would occur at a bare electrode. The electron transfer can occur between an electrode and redox electronic state only when their energy levels are equal. The presence of significant inelastic transfer mechanisms loosens this isoenergetic condition, allowing the difference between the energy of the electrode and redox electronic state to be made up by energy transfer to the insulator. Such inelastic transfers can be used to study the vibrational and electronic properties of molecules and are the basis of inelastic tunneling spectroscopy [51]. However, for characterizing electron transfer reactions, such inelastic transfer serve only to complicate the observed electron transfer kinetics.

Ideally, one would like an insulating film that uniformly reduces the electronic coupling between the electrode and redox molecule at closest approach. The presence of significant numbers of "hot" spots at which the electronic coupling is much stronger than the bulk insulator could complicate the kinetic description in several ways. If the proportion of the measured current passing through these hot spots is significant, the preexponential factor describing the electron transfer rate will be larger than that characteristic of bulk insulating film, due to an increase in the transmission coefficient, ϵ_{el}. This increase in the preexponential factor will be muted by a decrease in the effective number of redox molecules in the reaction layer. In the limit that all the measured current proceeds through the hot spots, the number of redox molecules within the reaction layer would scale with the area occupied by the hot spots. If the electron transfer rate at these hot spots exceeds the mass transfer, diffusion limitations will be observed.

The insulator should be stable within a wide range of electrode potentials and not contribute to the background current via surface redox processes. The electronic coupling properties of the insulator must be known, particularly the potential dependence of the electronic coupling. Finally, it would be advantageous to be able to control this electronic coupling. Usually, such control would be achieved by controlling the thickness of the insulating film.

2. Self-Assembled Thiol Monolayer Insulators

Much of the early work using insulated electrodes has relied on oxide and semiconducting films [52–54]. These films have the advantage of being quite stable in a range of solvents and temperatures. The main limitation of these inorganic thin films has been in producing films with known electronic coupling properties, thickness, and uniformity. An alternative strategy has been to use spontaneously self-assembling organic monolayer films as electrode insulators. These films are produced through a simple adsorption of certain amphiphiles onto surfaces. For amphiphiles with headgroups that bind strongly to the surface and strong lateral associations between the amphiphiles, this adsorption can produce compact, highly ordered monolayer films [55–58].

While a range of self-assembled monolayers have been used to modify electrode surfaces, alkyl thiols absorbed onto Au surfaces due to their remarkable stability and ease of formation have been the most used for heterogeneous electron transfer rate studies [59–65]. Control of the heterogeneous electron transfer rate using these alkyl monolayers can be achieved simply by controlling the length of the alkyl thiol monolayer. Figure 7 shows the effect of increasing the length of ω-hydroxyalkylthiol monolayers for the reduction of $Fe(CN)_6^{3-}$ [66]. The addition of each methylene group to the ω-hydroxyalkylthiol results in a reproducible decrease in the heterogeneous electron transfer rate as indicated by the negative shift in the volummetric peaks. As the monolayer thickness increases beyond 14 methylene units in this series, the heterogeneous electron transfer rate constant decreases below the mass transfer rate so that no voltammetric peak is seen. For these thicker monolayers, the electron transfer rate remains largely kinetically controlled over the entire voltammetric range shown.

From the heterogeneous electron transfer rates extracted from the data shown in Figure 7, one can determine the electronic coupling properties for these ω-hydroxyalkylthiol monolayers. The strength of the electronic coupling

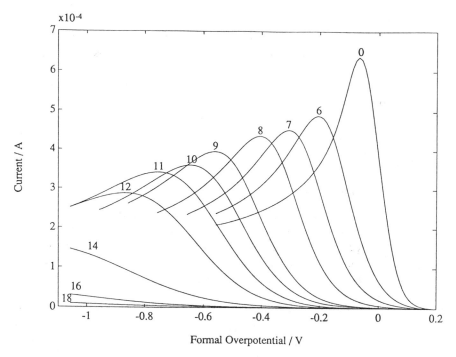

FIGURE 7 Cyclic voltammograms for the reduction of ferricyanide measured at electrodes insulated with monolayers of ω-hydroxyalkylthiols. The number of methylene units within the ω-hydroxyalkylthiol used to form the monolayer is shown above the voltammograms. The voltammogram measured at an underivatized electrode is marked with a zero. (Data from Ref. 66.)

is observed to decay exponentially with the number of methylene units within the ω-hydroxyalkylthiol used to form the monolayer so that

$$k = k_0 e^{-\beta n} \tag{21}$$

where k is the heterogeneous electron transfer rate constant measured at a particular electrode potential, k_0 the preexponential rate constant, β the tunneling coefficient, and n the number of methylene units within the ω-hydroxyalkylthiol used to form the insulating monolayer. One can obtain a value for the tunneling coefficient, β, by plotting the heterogeneous electron transfer rate extrapolated to the same potential as a function of the number of methylene units within the insulating monolayer film. Figure 8 shows the logarithm of the reduction-rate constant measured at a formal overpotential of -0.20 V plotted against the number of methylene groups obtained from the data in Figure 7. The plot is

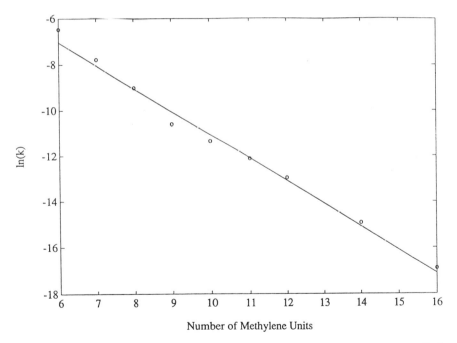

FIGURE 8 Plot of k vs. the number of methylene units within the ω-hydroxyalkyl-thiol monolayer. The heterogeneous electron transfer rate constants were extrapolated from the data shown in Figure 7 at a formal overpotential of -0.20 V. The line drawn through the experimental data represents the best-fit line. The data were corrected for diffusion limitations and double-layer effects as described in Ref. 66.

fitted with a single line whose slope gives a β value of 1.01 per methylene unit. One can avoid the extrapolation required to obtain Figure 8 by taking the ratio of the heterogeneous electron transfer rates measured for the same redox molecule at the same electrode potential at electrodes coated with monolayers differing in thickness by Δn methylene units. The tunneling coefficient becomes:

$$\beta = \ln\left(\frac{k_n}{k_{n+\Delta n}}\right)\frac{1}{\Delta n} \tag{22}$$

This pairwise comparison between two electrodes coated with monolayers that differ by 1 or 2 methylene units allows one to observe the potential dependence of the electronic coupling. Figure 9 is a plot of the potential dependence of this β for several redox probes [66]. These data were obtained using electrodes coated with the longer ω-hydroxyalkylthiol monolayers ($n = 12$ to 16) to allow the ratio of the heterogeneous electron transfer rate constants to be measured in as wide a voltage range as possible. From these data and others, Becka and Miller have determined the β coefficient to be 1.08 ± 0.20. This value is

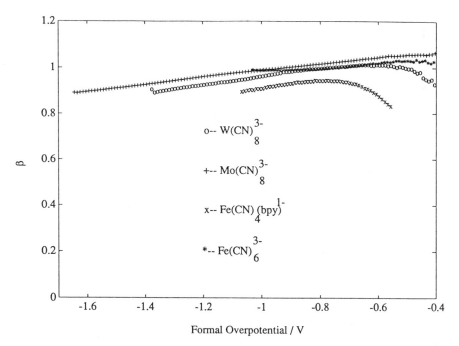

FIGURE 9 Plot of the tunneling coefficient, β, as a function of the electrode formal overpotential measured using the four redox probes indicated within the figure (see the text). (Adapted from Ref. 66.)

in close agreement with measurements made for other thiol monolayer systems [67,68] and from intramolecular electron transfer measurement of donor and acceptors separated by rigid hydrocarbon spacers [69–73]. The approximate independence of β to the electrode potential and the nature of the redox molecule is an ideal characteristic for an electrode insulator. The effect of the ω-hydroxy-alkylthiol monolayer is to reduce the electron transfer rate by a constant factor.

3. Comparisons with the Marcus Theory

The controllable attenuation of the electron transfer rate that is afforded by these monolayers can be used to test the Marcus theory prediction shown in Figure 6. Figure 10A shows a plot of the heterogeneous electron transfer rate measured at a 14-hydroxyl-1-tetradecylthiol monolayer coated Au electrode for the reduction of *tris*(4,4'-dimethyl-2,2'-bipyridyl)iron(III) complex (C. J. Miller, unpublished results).* The heterogeneous electron transfer rate constant displays the sigmoidal shape of the Marcus theory prediction for a nonadiabatic electron transfer. The reorganization energy is the potential at the inflection point on this curve. Beyond the reorganization energy, the heterogeneous electron transfer rate constant becomes less strongly dependent on the electrode potential as predicted by the Marcus theory for an electron transfer reaction within the inverted region. This inverted region behavior is more noticeable in a plot of the first derivative of the rate constant versus the electrode formal overpotential. When the density of electronic states within the metal is approximated by a step function at the Fermi level, the derivative of the rate constant is proportional to the density of electronic states distribution in the solution [66]. Figure 10B plots this derivative and compares it to a Gaussian best fit. The density of electronic states distribution is remarkably well fitted by the Gaussian supporting the harmonic free energy vs. reaction coordinate approximation of the Marcus theory.

Similar vindications of the Marcus theory predictions have been obtained by Chidsey's [67] and Finklea's [68,74] groups using monolayers having the redox molecule covalently attached to the terminus of the alkylthiol. The main advantage of these surface-attached redox molecules is that one can determine the number of reacting molecules (N_{Ox} or N_{Red}) in an independent voltammetric or coulometric experiment. All the redox active sites are fixed at the monolayer surface so that the concentration profile reduces to a delta function centered at the monolayer–electrolyte interface. This simplified concentration profile eliminates the need to introduce a reaction layer thickness κ.

The electron transfer kinetics for these electroactive monolayers is obtained most commonly in a potential pulse experiment. Upon application of a

---

*These data were corrected for diffusion limitations and the effects of the double layer as described in Ref. 66.

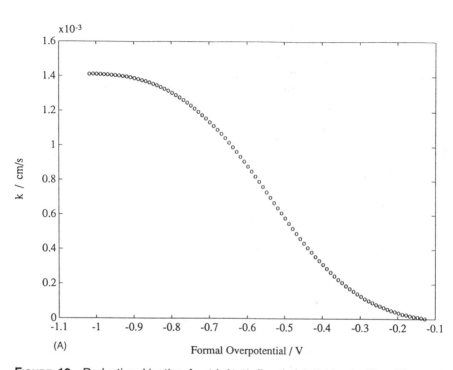

(A)   Formal Overpotential / V

FIGURE 10 Reduction kinetics for tris(4,4'-dimethyl-2,2'-bipyridyl)iron(III) complex measured at an Au electrode coated with a 14-hydroxy-1-tetradecylthiol monolayer in a 0.25 M CF$_3$COONa electrolyte at 0°C: (A) heterogeneous electron transfer rate constant (corrected for diffusion and double-layer effects) vs. the formal overpotential; (B) first derivative of (A), which is proportional to the density of electronic states distribution of the oxidized complex at the electrode surface. The solid line represents the best-fit Gaussian through the data.

given potential pulse, the observed current decays exponentially. The slope of ln(i) vs. time yields the unimolecular electron transfer rate. The observation of a single exponential decay indicates that all of the tethered redox molecules are reacting with the same electron transfer rate and strongly suggests that the redox molecules are being held at the same location and environment at the monolayer–electrolyte interface [67,68]. By measuring the electron transfer rate as a function of the pulse potential, the first-order rate constant for the oxidation and reduction of the redox species can be plotted and compared with the Marcus theory prediction. In Figure 11 are shown two examples of this analysis for a ferrocenyl [67] and a ruthenium pentamine pyridyl complex [74] bound to alkyl thiol monolayers on Au electrodes. The ln(k) vs. the electrode potential shows the same curvature predicted from the Marcus theory, allowing reorganization

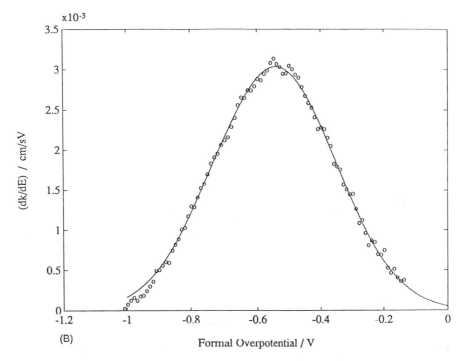

FIGURE 10 Continued

energy of 0.7 and 0.8 eV to be extracted for the ferrocenyl and ruthenium complexes, respectively.

The temperature dependence of the electron transfer reaction has also been measured for these monolayer bound redox molecules. Figure 12 shows $\ln(k)$ vs. electrode overpotential measured at two temperatures for the oxidation and reduction of the ferrocenyl [67] and ruthenium pentamine pyridyl [74] complexes. At low formal overpotentials, the electron transfer rates display the strongest temperature dependence. This dependence decreases at higher overpotentials. This is precisely in agreement with Marcus theory expectations. As the temperature is increased, the density of electronic states distributions in the metal and within the reaction layer broaden, particularly the distribution of the redox molecule. At electrode overpotentials much lower than the reorganization potentials, λ_{Ox}/e or λ_{Red}/e, this broadening of the density of electronic states increases the overlap between the metal and redox states, increasing the electron transfer rate observed. As the driving force for the electron transfer is increased toward the reorganization energies for the oxidized or reduced redox molecules,

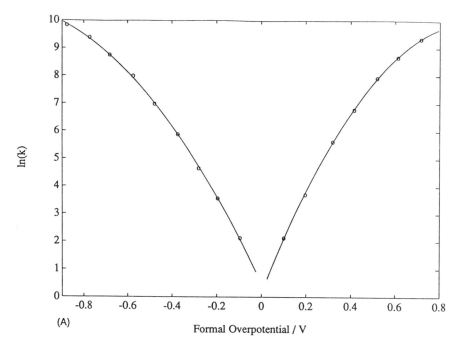

FIGURE 11   Heterogeneous electron transfer rates (in s^{-1}) for ferrocenyl (A) and ruthenium pentaammine pyridine (B) complexes attached covalently to thiol monolayers on Au electrodes. Each plot shows both the anodic and cathodic rates plotted vs. the formal overpotential of the complexes. The solid lines are the best parabolic fit through the data. (Data from Refs. 67 and 75.)

the temperature dependence decreases until at the reorganization energy the electron transfer rate become temperature independent. The independence of the electron transfer rate stems from the symmetric shape of the density of electronic states for the redox molecules. When the Fermi level of the electrode is held at the reorganization potential for Ox (or Red), the filled (or unfilled) metal electronic states overlap with half the redox molecules. Broadening the distributions by increasing the temperature does not affect this overlap, so that the electron transfer rate remains constant. In more traditional terms, the temperature dependence of the electron transfer rate constant is a direct consequence of the activated nature of the electron transfer. At low electrode overpotentials, the activation energy is highest, giving a stronger temperature dependence. For oxidation or reductions occurring at electrode potentials at the reorganization potentials for the reduced or oxidized form of the redox molecule, respectively,

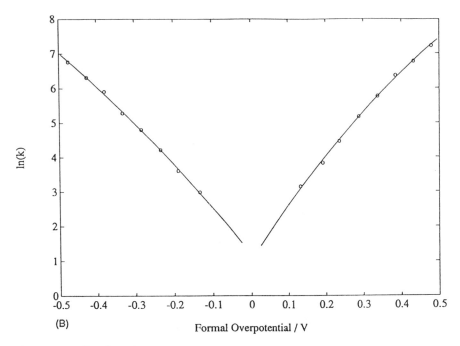

Formal Overpotential / V

FIGURE 11 Continued

the electron transfer proceeds without activation and is therefore temperature independent.

4. Characterization of Electroactive Molecules

In addition to quantitatively supporting the Marcus theory predictions, these monolayer insulated electrodes can be used to characterize redox reactivities and to determine the structure–reactivity relationships between various redox active molecules. The obvious initial use of the insulated electrode studies is to determine the reorganization energies (and hence the activation energies) for different redox molecules. Becka and Miller [66] determined that the reorganization energy of Ce^{4+} is approximately twice that observed for $Fe(CN)_6^{3-}$ (2.2 eV vs.1.1 eV, respectively). Such an observation is completely in line with the much slower formal rate constant of Ce^{4+} compared with $Fe(CN)_6^{3-}$ and could be predicted given the larger inner- and outer-shell reorganization components for the aquated cerric ion compared with the larger ferricyanide ion.

A more detailed structure–reactivity investigation is shown in Table 1 for a series of mixed cyano/bipyridyl (CN/bpy) iron(III) complexes using Au electrodes coated with monolayers of 14-hydroxy-1-tetradecanethiol [75]. As cyano

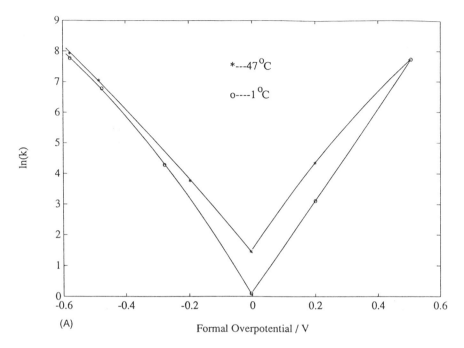

$$*\text{---}47\,^{\circ}\text{C}$$

$$o\text{----}1\,^{\circ}\text{C}$$

(A)   Formal Overpotential / V

FIGURE 12 Plots of the heterogeneous electron transfer rates vs. electrode formal overpotentials for the surface-attached ferrocenyl and ruthenium pentaammine pyridine complexes measured at different temperatures as indicated in the figures. (A) Rate constants for the ferrocenyl complex (from Ref. 67). The point corresponding to 0.0 V is the sum of the reduction and oxidation rate constants. (B) Rate constants for the ruthenium pentaammine pyridine complex (from Ref. 75). The solid lines through the points are the best parabolic fits.

ligands are replaced by bipyridyl ligands, the reorganization energies generally decrease. Qualitatively, this result would be expected because of the significant increase in the size of the tris-bipyridyl complex compared to the hexacyanoiron complex. The increasing size of the bipyridyl complex would reduce the outer-shell reorganization energy as seen from equation (14). Furthermore, the inner-shell reorganization energy should also decrease between the hexacyano and tris-bipyridyl complexes [34]. A more quantitative comparison between the data in Table 1 and the Marcus theory predictions is difficult to make due to a lack of accurate data for the structures and bond force constants for both the oxidized and reduced forms of these molecules.

In addition to the reorganization energy, the kinetics measured at the insulated electrodes allows one to probe the inherent adiabaticity of the redox

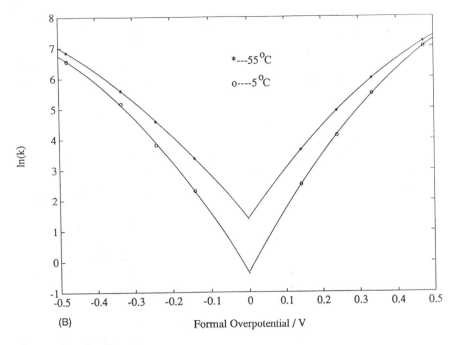

(B) Formal Overpotential / V

FIGURE 12 Continued

molecules listed in Table 1. This is achieved by determining the maximum heterogeneous rate constant, k_{max}, which occurs at electrode potentials far beyond the reorganization potential [66]. At these high overpotentials, the overlap between the oxidized redox states and the filled metallic states is complete, so that all the redox centers within the reaction layer have equal electron transfer rates independent of their particular extent of activation. This maximum heteroge-

TABLE 1 Kinetic Data for Fe(bipyridyl/cyano) Complexes

	$E^{\circ\prime}$ (V vs. SCE)	λ_{ox} (eV)	k_{max} (cm/s)
$Fe(CN)_6^{3-}$	0.223	1.12 ± 0.04	0.014 ± 0.002
$Fe(bpy)(CN)_4^-$	0.344	1.17 ± 0.13	0.18 ± 0.12
$Fe(bpy)_2(CN)_2^+$	0.547	0.76 ± 0.08	0.56 ± 0.02
$Fe(bpy)_3^{3+}$	0.827	0.56 ± 0.02	0.0014 ± 0.0002

Source: Ref. 76.

neous electron transfer rate depends on the preexponential factor [equations (9), (16), and (20)] and so depends on the reaction-layer thickness κ and the electronic transmission coefficient k_{el}. When κ can be assumed to be constant, a comparison between the k_{max} values for different redox molecules measured at the same insulated electrode can be used to probe which redox molecules couple more efficiently with the metallic electronic states.

From Table 1, the trend of the k_{max} values as the cyano ligands are replaced by bipyridyl ligands appears complex. The k_{max} values increase with the addition of one and two bipyridyl ligands followed by a precipitous decrease when the last two cyano groups are replaced by the third bipyridyl ligand. The initial rise in the k_{max} values is explained by a preconcentration of the hydrophobic bipyridyl complex at hydrophobic defects at the surface of the ω-hydroxyalkylthiol monolayer [75]. The rise in the k_{max} value then stems from an increase in the local concentration of the redox molecule at the monolayer surface, resulting in an increase in the reaction-layer thickness. This interpretation is supported by competitive binding studies in which the k_{max} value in the presence of 1-octanol shows a k_{max} value close in value to that observed for the $Fe(CN)_6^{3-}$ ion. Once the last cyano groups are replaced, the distance of closest approach of the iron complex with the monolayer surface increases due to the steric bulk of the three bipyridyl ligands. This steric bulk apparently eliminates the preconcentration of the iron complex at the monolayer surface so that one observes a sharp decrease in the k_{max} value. The bipyridyl ligands are less effective at coupling the redox molecule to the metal electronic states compared to the smaller cyano ligands. When the steric bulk of this tris-bipyridyl complex is increased by methylation at the 4 and 4′ positions, the k_{max} value is observed to decrease further by a factor of 2. It is interesting to note that although the $Fe(bpy)_3^{3+}$ complex has the lower reorganization energy, it is the $Fe(CN)_2(bpy)_2^{1+}$ complex that displays the most rapid heterogeneous electron transfer rate at the insulated electrode. One of the special abilities of these insulated electrode studies is that they can distinguish between the adiabaticity and activation energy components of a redox molecule's heterogeneous electron transfer reactivity.

B. Solvent Dynamic Effects

1. Theoretical Expectations

The dynamic properties of the solvent can play an important role in determining the rate of barrier crossing and hence the rate of electron transfer [37–40]. The thermal activation of the reactant along the reaction coordinate is a result of energy transfers between the reaction coordinate and the other vibrational and rotational modes of the redox molecule and solvent. The multitude of other modes of motion can be thought of as a thermal bath. Roughly speaking, this bath supplies the reaction coordinate with enough energy to allow passage of the reactant over the transition state. Once over the transition state, the bath then dissipates the energy held in the reaction coordinate, allowing the product to descend toward its most stable nuclear configuration [25]. However, because the bath injects and withdraws free energy from the reaction coordinate randomly, the trajectory of the reactant to product need not be direct. When the rate of passage through the transition-state region is slow relative to the rate of energy transfer between the reaction coordinate and the bath, the reactant motion through the transition state will become diffusional. In other words, the reactant will move through the transition-state region with an erratic motion as it suffers energy-transferring collisions from the thermal bath. The rate of barrier crossing will then depend strongly on this energy transfer rate.

One of the major mechanisms of energy exchange between a reaction coordinate and its surroundings is through charge–dipole interactions between the reactant and solvent. The characteristic time for these energy transfers is related to the longitudinal relaxation time of the solvent, τ_L [37,76]. For adiabatic electron transfer reactions that involve predominantly solvent reorganization rather than the faster inner-shell vibrations, the rate of passage of the reactant across the transition state can be determined by this solvent dynamic parameter. A simple form for the nuclear frequency factor in this case is [22,37,38].

$$\nu_n = \frac{1}{\tau_L}\left(\frac{\Delta G_{os}}{4\pi RT}\right)^{1/2} \tag{23}$$

where τ_L is the longitudinal relaxation time.

The longitudinal relaxation time, as well as the Pekar factor [$(1/\epsilon_{op})$ − $(1/\epsilon_s)$] of some common solvents used in electrochemical studies, are shown in Table 2 [77]. The τ_L values for a dynamically fast solvent, acetonitrile, are about 30 times smaller than for the slowest, hexamethylphosphoramide. Equation (23) therefore predicts that the heterogeneous electron transfer rate for certain redox reactions could vary dramatically when measured in these two solvents.

Solvent dynamics effects in the electron transfer rate of redox molecules whose activation free energy is dominated by the outer-shell component have been reported in a number of studies. A simplistic determination of the influence of the solvent dynamics involves plotting the logarithm of the heterogeneous electron transfer rate constant measured typically at the formal potential of the redox molecule, $\ln(k_{ex})$, vs. $\ln(\tau_L^{-1})$ for a series of solvents. From equation (23) one would predict a unity slope for this plot if solvent dynamics are important in determining the barrier-crossing frequency. For redox molecules with substantial inner-shell activation barriers, the frequency of passage of the reactant over the transition state is increased, muting the dependence of the frequency factor on the solvent dynamics. In this case the heterogeneous electron transfer rate is anticipated to depend on $(\tau_L^{-\alpha})$ where $0 < \alpha < 1$ [78].

TABLE 2 Dynamic and Static Solvent Parameters

Solvent	τ_L (ps)	$1/\epsilon_{op} - 1/\epsilon_s$
Debye solvents		
Acetone	0.3	0.495
Acetonitrile	0.2	0.529
Benzonitrile	5.8	0.390
Chloroform	2.4	0.276
Dichloroethane	1.6	0.384
Dichloromethane	0.9	0.382
Dimethoxyethane	0.8	0.371
Dimethylacetamide	1.5	0.459
Dimethylformamide	1.1	0.463
Dimethylsulfoxide	2.1	0.437
Hexamethylphosphoramide	8.8	0.438
Nitrobenzene	5.3	0.387
Nitromethane	0.2	0.498
N-Methylpyrolidone	2.5	0.435
Propionitrile	0.4	0.503
Pyridine	1.3	0.359
Tetrahydrofuran	1.7	0.388
Tetramethylurea	6	0.433
Non-Debye solvents		
Ethanol	4.1	0.500
Methanol	0.5	0.538
1-Propanol	16.9	0.474
1-Butanol	25.6	0.454
Propylene carbonate	3.0	0.480

Source: Ref. 78.

This simple approach to determining whether solvent dynamics are important for a particular electron transfer reaction suffers from several shortcomings [79]. The method does not account for the other ways the solvent can influence the observed electron transfer rate. A change in the solvent can affect both the transmission coefficient as well as the partition coefficient by changing the distance of closest approach of the solvated redox center to the electrode surface. The double-layer structure of the electrode and the extent of ion pairing within the solution can also be a function of the solvent. Each of these effects can induce changes in the measured heterogeneous electron transfer rates as the solvent is changed. These artifacts can be minimized by selecting redox molecules with sufficient electronic coupling to the electrode to ensure an adiabatic electron transfer in all the solvents studied. Ideally, the redox molecule should be initially neutral, to minimize the effect of changes in the electrode double layer [80,81]. Changes in the ion pairing of the redox site can be minimized by employing low concentrations of both the redox reagent and the electrolyte.

Another more serious deficiency of the $\ln(k_{ex})$ vs. $\ln(\tau_L^{-1})$ analysis is that it neglects the effect of the solvent on activation energy of the redox molecule [82]. To correct for solvent effects on the outer-shell activation free energy, one must correct the measured electron transfer rate constants for these changes in the activation energy. Because the electron transfer rate depends exponentially on the outer-shell activation energy, one can simply add ΔG_{os} to $\ln(k_{ex})$ to obtain the effective heterogeneous rate constant in the absence of the outer-shell activation. From the slope of a plot of $[\ln(k_{ex}) + \Delta G_{os}]$ vs. $\ln(\tau_L^{-1})$, one has an improved indicator of a solvent dynamic effect [2]. This improved method for determining the presence of a solvent dynamic effect relies on being able to determine ΔG_{os} accurately. Although one can use the dielectric continuum expression in equation (15), its uses involves some complications. The dielectric continuum model assumes a spherical redox center. Obtaining an effective radii for nonspherical redox molecules can involve some error [83]. The dielectric continuum model also requires one to input the distance between the redox center and the electrode surface to account for the screening of the reactant by its image charge in the metal electrode. This effect of the image charge on the electron transfer is difficult to assess quantitatively because of uncertainties in determining the distance R between the redox molecule and the electrode surface. Recent theoretical investigations of dynamic effects of the polarization within the metal suggest that for sufficiently close approaches of the redox center to the electrode surface, the image charge induced in the metal should result in an increase in the ΔG_{os} in sharp contrast to that predicted by equation (15) [42,84]. In response to these uncertainties, it is common practice to set R to infinity to eliminate the image charge correction from the ΔG_{os} expression.

Adding to the uncertainties in the input parameters for the ΔG_{os} expression, more sophisticated treatments of the outer-shell activation free energy suggest that the simple dielectric continuum expression overestimates the activation barrier in part because it neglects molecular interactions and assumes no dielectric saturation close to the redox center [85,86]. If the solvent dependence of the outer-shell activation barrier cannot be determined accurately, the observation of a solvent dynamic effect on the electron transfer rate becomes extremely difficult. Plotting the $\ln(k_{ex})$ corrected for the outer-shell activation energy vs. $\ln(\tau_L^{-1})$, one may obtain a linear relationship even in the absence of solvent dynamical effects. The spurious correlation between $\ln(k_{ex})$ and $\ln(\tau_L^{-1})$ is then a result of the overcorrection of the measured heterogeneous rate constants for the change in the inner-shell activation barriers. This systematic overestimation in the ΔG_{os} can give an apparent solvent dynamic effect when the optical properties of the solvent, $(1/\epsilon_{op} - 1/\epsilon_s)$, correlate with their dynamic properties τ_L^{-1} [87].

2. Experimental Results

In an impressive series of experiments, Opallo has obtained the heterogeneous electron transfer rate for 1,4-phenylenediamine at a Pt electrode in 12 solvents [88]. Figure 13 shows a plot of the logarithm of the formal rate constants as a function of $-\ln(\tau_L)$. There is a strong correlation between the measured heterogeneous electron transfer rate and the solvent longitudinal relaxation time indicative that solvent dynamics control the frequency factor for this reaction. For the solvents used in this experimental series, there is only a weak correlation between τ_L and $(1/\epsilon_{op} - 1/\epsilon_s)$, so that variations in the heterogeneous electron transfer rates due to changes in the outer-shell activation barrier should not affect this correlation. It was observed that the heterogeneous electron transfer rates for the protic solvents, methanol, *n*-methylformamide, 1-butanol, and 2-propanol

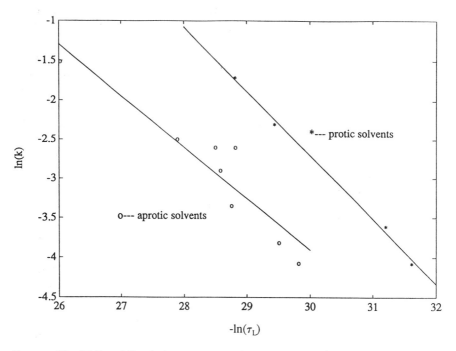

FIGURE 13 Plots of the heterogeneous electron transfer rate constant vs. the negative log of the solvent longitudinal relaxation time for 1,4-phenylenediamine measured at a Pt electrode. The two solid lines plot the best linear fits for the aprotic (circles) and protic (asterisks) solvents. (Data from Ref. 89.)

were significantly higher than for the other solvents. This increased electron transfer rate has been linked to non-Debye behavior for these solvents [2]. In addition to the longitudinal relaxation time used in the figure, these solvents are characterized by higher-frequency components in their dielectric loss spectra. The influence of these additional high-frequency modes in increasing the barrier crossing rate have been the subject of several studies [78,89–91].

As indicated above, the presence of a significant inner-shell activation barrier to an electron transfer is anticipated to mute this solvent dynamic effect. Only a few studied have focused on redox molecules with significant inner-shell activation barriers [80,81,92,93]. Nielson and Weaver [80] have measured an α value of 0.7 for a tris-dimethylglyoxime cobalt complex measured at Au and Hg electrodes. For this complex the inner-shell activation barrier, ΔG_{is}, was estimated to be about 15 to 18 kJ/mol, which is comparable to the ΔG_{os} calculated using equation (15). Fawcett and Opallo [91] found a smaller dependence of the heterogeneous electron transfer rate constants for tris(acetylacetonato) manganese(III) on the solvent logitudinal relaxation time than that for the corresponding tris(acetylacetonato)iron(III) complex. This was attributed to the manganese complex, which has larger ΔG_{is} than the iron complex. In a recent study by Mu and Schultz [92], a large solvent dynamic effect was observed for reduction of a chloro(tetraphenylporphinato)manganese(III) complex. The α exponent to the τ_L term was found to be approximately unity despite a substantial (16 kJ/mol) inner-shell activation barrier. For the six solvents used in this study, the optical properties of the solvent show some correlation with their dynamic properties, which could increase the α exponent that was determined.

An interesting use of these solvent dynamic effects has been a probe of both inner-shell activation barriers and reaction adiabaticity. A reaction that is insensitive to the dynamics properties of the solvent is expected to have either a large inner-shell barrier or a small electronic coupling with the electrode. In contrast, reactions that exhibit a strong dependence on the dynamic properties of the solvent can be said to be proceeding adiabatically with a low inner-shell activation barrier [9,30,94]. Use of this solvent dynamic probe of adiabaticity is particularly significant because of the extreme difficulty in determining the adiabaticity of a given redox reaction.

IV. Final Comments

The main focus of this chapter has been to develop a conceptual and mathematical description of heterogeneous electron transfer reactions which is particularly suited to follow recent experimental advances in the electron transfer reactions at metallic electrodes. Although not stressed in this chapter, the Marcus theory–transition state description presented here is completely parallel with what is now the standard homogeneous electron transfer model. The major differences between homogeneous and heterogeneous electron transfers stem from the differences between the electronic structures of the metal compared with solution species. Usually, the electronic properties of metals are discussed in terms of the band models of solid-state chemistry and physics, while a more molecular orbital approach is taken for electron transfers between molecules in solution. This difference in presentation formalism between solid-state and solution electron transfer makes the task of describing heterogeneous electron transfer a bit more difficult. The description of heterogeneous electron transfer kinetics requires some meeting between these different approaches. The incentive to present such a synthesis of the two approaches is the current research in heterogeneous electron transfer, which demonstrates the wealth of information that can be obtained via such electrochemical studies.

ACKNOWLEDGMENTS

Most of the ideas and research presented in this chapter are not of the author's creation but rather were borrowed from a great number of people who are listed in the references. The author has been particularly influenced by the works of R. A. Marcus, M. J. Weaver, W. R. Fawcett, K. J. Vetter, and H. Gerisher. The author would like to particularly acknowledge helpful discussions with Marshall D. Newton and W. Ronald Fawcett and preprints from Harry O. Finklea and Franklin A. Schultz.

REFERENCES

1. G. McClendon and R. Hake, Interprotein electron transfer, *Chem. Rev.*, *92*: 481 (1992).
2. M. J. Weaver, Dynamical solvent effects on activated electron-transfer reactions: principles, pitfalls, and progress, *Chem. Rev.*, *92*: 463 (1992).
3. T. M. McCleskey, J. R. Winkler, and H. B. Gray, Driving-force effects on the rates of biomolecular electron-transfer reactions, *J. Am. Chem. Soc.*, *114*: 6935 (1992).
4. R. M. Wightman, Microvoltametric electrodes, *Anal. Chem.*, *53*: 1125A (1981).
5. W. Obretenov, U. L. Schmidt, W. J. Lorenz, G. Staikov, E. Budevski, D. Carnal, U. Müller, H. Siegenthaler, and E. Schmidt, Underpotential deposition and electrocrystallization of metals: an atomic view by scanning tunneling microscopy, *J. Electrochem. Soc.*, *140*: 692 (1993).
6. A. J. Bard, M. V. Mirkin, P. R. Unwin, and D. O. Wipf, Scanning electrochemical microscopy. 12. Theory and experiment of the feedback mode with finite heterogeneous electron-transfer kinetics and arbitrary substrate size, *J. Phys. Chem.*, *96*: 1861 (1992).
7. A. J. Bard, F. F. Fan, D. T. Pierce, P. R. Unwin, D. O. Wipf, and F. Zhou, Chemical imaging of surfaces with the scanning electrochemical microscope, *Science*, *254*: 68 (1991).

8. S. Ching, J. T. McDevitt, S. R. Peck, and R. W. Murray, Liquid phase electrochemistry at ultralow temperatures, *J. Electrochem. Soc.*, *138*: 2308 (1991).
9. A. S. Baranski, K. Winkler, and W. R. Fawcett, New experimental evidence concerning the magnitude of activation parameters for fast heterogeneous electron transfer reactions, *J. Electroanal. Chem.*, *313*: 367 (1991).
10. L. K. Safford and M. J. Weaver, The evaluation of rate constants for rapid electrode reactions using microelectrode voltammetry: virtues of measurements at lower tempertures, *J. Electroanal. Chem.*, *331*: 857 (1992).
11. A. M. Kuznetsov, A theory of electron transfer at superconducting electrodes, *J. Electroanal. Chem.*, *278*: 1 (1990).
12. A. Ulman, *An Introduction to Ultrathin Organic Films: From Langmuir–Blodgett to Self-Assembly*, Academic Press, New York, 1991.
13. C. Miller, P. Cuendet, and M. Grätzel, Adsorbed ω-hydroxyl thiol monolayers on gold electrodes: evidence for electron tunneling to redox species in solution, *J. Phys. Chem.*, *95*: 877 (1991).
14. A. J. Bard and L. R. Faulkner, *Electrochemical Methods: Fundamentals and Applications*, Wiley, New York, 1980, p. 95.
15. R. A. Marcus, On the theory of electron-transfer reactions. VI. Unified treatment of homogeneous and electrode reactions, *J. Chem. Phys.*, *43*: 679 (1965).
16. H. Taube, *Electron Transfer Reactions of Complex Ions in Solution*, Academic Press, New York, 1970.
17. P. Delahay, Double layer and electrode kinetics, in *Advances in Electrochemistry and Electrochemical Engineering* (P. Delahay and C. W. Tobias, eds.), Wiley-Interscience, New York, 1965, Chap. 3.
18. N. Sutin, Nuclear, electronic, and frequency factors in electron-transfer reactions, *Acc. Chem. Res.*, *15*: 275 (1982).
19. M. J. Weaver, On the role of the bridging ligand in electrochemical inner-sphere electron transfer processes, *Inorg. Chem.*, *18*: 402 (1979).
20. J. O'M. Bockris and S. U. M. Khan, *Quantum Electrochemistry*, Plenum Press, New York, 1979, Chap. 8.
21. J. T. Hupp and M. J. Weaver, Experimental estimate of the electron-tunneling distance for some outersphere electrochemical reactions, *J. Phys. Chem.*, *88*: 1463 (1984).
22. L. D. Zusman, Outer-sphere electron transfer in polar solvents, *Chem. Phys.*, *49*: 295 (1980).
23. J. W. Schultze and K. J. Vetter, The influence of the tunnel probability on the anodic oxygen evolution and other redox reactions at oxide covered platinum electrodes, *Electrochim. Acta*, *18*: 889 (1973).
24. C. Kittel, *Introduction to Solid State Physics*, 5th ed., Wiley, New York, 1976.
25. D. Chandler, Roles of classical dynamics and quantum dynamics on activated processes occurring in liquids, *J. Stat. Phys.*, *42*: 49 (1986).
26. R. A. Marcus and N. Sutin, Electron transfers in chemistry and biology, *Biochim. Biophys. Acta*, *811*: 265 (1985).
27. R. A. Marcus, Theory and application of electron transfer at electrodes and in solutions, in *Special Topics in Electrochemistry* (P. A. Rock, ed.), Elsevier, Amsterdam, 1977, p. 161.

28. R. P. Feynman, R. B. Leighton, and M. Sands, *The Feynman Lectures on Physics*, Vol. 2, Addison-Wesley, Reading, Mass., 1977, Chap. 8.

29. J. T. Hupp and M. J. Weaver, The frequency factor for outer-sphere electrochemical reactions, *J. Electroanal. Chem.*, *152*: 1 (1983).

30. A. Gochev, G. E. McManis, and M. J. Weaver, Solvent dynamical effects in electron transfer: predicted influences of electronic coupling upon the rate-dielectric friction dependence, *J. Chem. Phys.*, *91*: 906 (1989).

31. M. J. Weaver and J. T. Hupp, Some comparisons between the energetics of electrochemical and homogeneous electron-transfer reactions, in *Mechanistic Aspects of Inorganic Reactions* (D. B. Rorabacher and J. F. Endicott, eds.) ACS Symposium Series 198, American Chemical Society, Washington, D.C., 1982, Chap. 8.

32. M. J. Weaver and F. C. Anson, Potential dependence of the electrochemical transfer coefficient: further studies of the reduction of chromium(III) at mercury electrodes, *J. Phys. Chem.*, *80*: 1861 (1976).

33. T. Saji, Y. Maruyama, and S. Aoyagui, Electrode kinetic parameters for the redox systems $Mo(CN)_8^{3-}/Mo(CN)_8^{4-}$, $IrCl_6^{2-}/IrCl_6^{3-}$, $Fe(phen)_3^{3+}/Fe(phen)_3^{2+}$ and $Fe(C_5H_5)_2^{+}/Fe(C_5H_5)_2$, *J. Electroanal. Chem.*, *86*: 219 (1978).

34. T. Saji, T. Yamada, and S. Aoyagui, Electron-transfer rate constants for redox systems of Fe(III)/Fe(II) complexes with 2,2'-bipyridine and/or cyanide ions as measured by the galvanostatic double pulse method, *J. Electroanal. Chem.*, *61*: 147 (1975).

35. R. A. Petersen and D. H. Evans, Heterogeneous electron transfer kinetics for a variety of organic electrode reactions at the mercury–acetonitrile interface using either tetraethylammonium perchlorate or tetraheptylammonium perchlorate electrolyte. *J. Electroanal. Chem.*, *222*: 129 (1987).

36. C. Miller and M. Grätzel, Electrochemistry at ω-hydroxy thiol coated electrodes. 2. Measurement of the density of electronic states distributions for several outer-sphere redox couples, *J. Phys. Chem.*, *95*: 5226 (1991).

37. D. F. Calef and P. G. Wolynes, Classical solvent dynamics and electron transfer. II. Molecular aspects, *J. Chem. Phys.*, *78*: 470 (1983).

38. D. F. Calef and P. G. Wolynes, Classical solvent dynamics and electron transfer, *J. Phys. Chem.*, *87*: 3387 (1983).

39. J. T. Hynes, The theory of reactions in solution, in *Theory of Chemical Reaction Dynamics*, Vol. 4 (M. Baer, ed.), CRC Press, Boca Raton, Fla., 1985, Chap. 4.

40. H. Sumi and R. A. Marcus, Dynamical effects in electron transfer reactions, *J. Chem. Phys.*, *84*: 4894 (1986).

41. R. A. Marcus and H. Sumi, Solvent dynamics and vibrational effects in electron transfer reactions, *J. Electroanal. Chem.*, *204*: 59 (1986).

42. P. G. Dzhavakhidze, A. A. Kornyshev, and L. I. Krishtalik, Activation energy of electrode reactions: the non-local effects, *J. Electroanal. Chem.*, *228*: 329 (1987).

43. M. J. Weaver, Redox reactions at metal–solution interfaces, in *Comprehensive Chemical Kinetics* (R. G. Compton, ed.), Elsevier, Amsterdam, 1987, p. 41.

44. M. J. Weaver, Interpretation of activation parameters for simple electrode reactions, *J. Phys. Chem.*, *80*: 2645 (1976).

45. M. J. Weaver, Activation parameters for simple electrode reactions: application to the elucidation of ion–solvent interactions in the transition state for heterogeneous electron transfer, *J. Phys. Chem.*, *83*: 1748 (1979).

46. H. Gerisher, *Z. Phys. Chem.* (*Munich*), *26*: 223 (1960).
47. D. A. Papaconstantopoulos, *Handbook of the Band Structure of Elemental Solids*, Plenum Press, New York, 1986.
48. R. A. Marcus, On the theory of oxidation–reduction reactions involving electron transfer. I, *J. Chem. Phys.*, *24*: 966 (1956).
49. Z. Samec and J. Weber, The effect of the double layer on the rate of the Fe^{3+}/Fe^{2+} reaction on a platinum electrode and the contemporary electron transfer theory. *J. Electroanal. Chem.*, *77*: 163 (1977).
50. A. J. Bennett, Proposed investigation of the density of electronic states in ionic solutions by tunneling at oxide coated electrodes, *J. Electroanal. Chem.*, *60*: 125 (1975).
51. P. K. Hansma, ed., *Tunneling Spectroscopy*, Plenum Press, New York, 1982.
52. R. Memming and R. Möller, *Ber. Bunsenges. Phys. Chem.*, *77*: 945 (1973).
53. K. Kobayashi, Y. Aikawa, and M. Sukigara, Tunnel electrode. I. Electron transfer process at highly doped SnO_2 electrode with Ce^{4+} in high overvoltage region, *Bull. Chem. Soc. Jpn.*, *55*: 2820 (1982).
54. H. Morisaki, H. Ono, and K. Yazawa, Electronic state densities of aquo-complex ions in water determined by electrochemical tunneling spectroscopy, *J. Electrochem. Soc.*, *136*: 1710 (1989).
55. J. Sagiv, Organized monolayers by adsorption. 1. Formation and structure of oleophobic mixed monolayers on solid surfaces, *J. Am. Chem. Soc.*, *102*: 92 (1980).
56. D. L. Allara and R. G. Nuzzo, Spontaneously organized molecular assemblies. 1. Formation, dynamics and physical properties of *n*-alkanoic acids adsorbed from solution on an oxidized aluminum surface, *Langmuir*, *1*: 45 (1985).
57. D. L. Allara and R. G. Nuzzo, Spontaneously organized molecular assemblies. 2. Quantitative infrared spectroscopic determination of equilibrium structures of solution-adsorbed *n*-alkanoic acids on an oxidized aluminum surface, *Langmuir*, *1*: 52 (1985).
58. R. Maoz and J. Sagiv, On the formation and structure of self-assembling monolayers. 1. A comparative ATR-wettability study of Langmuir–Blodgett and adsorbed films on flat substrates and glass microbeads, *J. Colloid Interface Sci.*, *100*: 465 (1984).
59. M. D. Porter, T. B. Bright, D. L. Allara, and C. E. D. Chidsey, Spontaneously organized molecular assemblies. 4. Structural characterization of *n*-alkyl thiol monolayers on gold by optical ellipsometry, infrared spectroscopy, and electrochemistry, *J. Am. Chem. Soc.*, *109*: 3559 (1987).
60. E. Sabatani, I. Rubinstein, R. Maoz, and J. Sagiv, Organized self-assembling monolayers on electrodes. I. Octadecyl derivatives on gold, *J. Electroanal. Chem.*, *219*: 365 (1987).
61. H. O. Finklea, S. Avery, M. Lynch, and T. Furtsch, Blocking oriented monolayers of alkyl mercaptans on gold electrodes, *Langmuir*, *3*: 409 (1987).
62. C. D. Bain, J. Evall, and G. M. Whitesides, Formation of monolayers by the coadsorption of thiols on gold: variation in the head group, tail group, and solvent, *J. Am. Chem. Soc.*, *111*: 7155 (1989).
63. A. Ulman, J. E. Eilers, and N. Tillman, Packing and molecular orientation of alkenethiol monolayers on gold surfaces, *Langmuir*, *5*: 1147 (1989).

64. R. G. Nuzzo, L. H. Dubois, and D. L. Allara, Fundamental studies of microscopic wetting on organic surfaces. 1. Formation and structural characterization of a self-consistent series of polyfunctional organic monolayers, *J. Am. Chem. Soc., 112*: 558 (1990).

65. M. M. Walczak, C. Chung, S. M. Stole, C. A. Widrig, and M. D. Porter, Structure and interfacial properties of spontaneously adsorbed *n*-alkanethiolate monolayers on evaporated silver surfaces, *J. Am. Chem. Soc., 113*: 2370 (1991).

66. A. M. Becka and C. J. Miller, Electrochemistry at ω-hydroxy thiol coated electrodes. 3. Voltage independence of the electron tunneling barrier and measurements of redox kinetics at large overpotentials. *J. Phys. Chem., 96*: 2657 (1992).

67. C. E. D. Chidsey, Free energy and temperature dependence of electron transfer at the metal–electrolyte interface, *Science, 251*: 919 (1991).

68. H. O. Finklea and D. D. Hanshew, Electron-transfer kinetics in organized thiol monolayers with attached pentaammine(pyridine)ruthenium redox centers, *J. Am. Chem. Soc., 114*: 3173 (1992).

69. K. D. Jordan and M. N. Paddon-Row, Analysis of the interactions responsible for long-range through-bond-mediated electronic coupling between remote chromophores attached to rigid polynorbornyl bridges, *Chem. Rev., 92*: 395 (1992).

70. S. S. Isied, M. Y. Ogawa, and J. F. Wishart, Peptide-mediated intramolecular electron transfer: long-range distance dependence, *Chem. Rev., 92*: 381 (1992).

71. G. L. Closs, L. T. Calcaterra, N. J. Green, K. W. Penfield, and J. R. Miller, Distance, stereoelectronic effects, and the Marcus inverted region in intramolecular electron transfer in organic radical anions, *J. Phys. Chem., 90*: 3673 (1986).

72. D. N. Beratan, Electron tunneling through rigid molecular bridges: bicyclo[2.2.2]octane, *J. Am. Soc. Chem. Soc., 108*, 4321 (1986).

73. M. D. Newton, Quantum chemical probes of electron-transfer kinetics: the nature of donor–acceptor interactions, *Chem. Rev., 91*: 767 (1991).

74. H. O. Finklea, M. S. Ravenscroft, and D. A. Snider, Electrolyte and temperature effects on long range electron transfer across self-assembled monolayers, *Langmuir, 9*: 223 (1993).

75. A. M. Becka, Insulated electrode voltammetry for the study of redox kinetics, Master's thesis, University of Maryland, College Park, Md., 1993.

76. D. Kivelson and H. Friedman, Longitudinal dielectric relaxation, *J. Phys. Chem., 93*: 7026 (1989).

77. W. R. Fawcett and C. A. Foss, Role of the solvent in the kinetics of heterogeneous electron and ion transfer reactions, *Electrochimica Acta, 36*: 1767 (1991).

78. W. Nadler and R. A. Marcus, Dynamical effects in electron transfer reactions. II. Numerical solution, *J. Chem. Phys., 86*: 3906 (1987).

79. W. R. Fawcett, Comparison of solvent effects in the kinetics of simple electron-transfer and amalgam formation reactions, *Langmuir, 5*: 661 (1989).

80. R. M. Nielson and M. J. Weaver, The role of solvent dynamics in electron transfer: electrochemical exchange kinetics of cobalt clathrochelates as a probe of reactant vibrational effects, *J. Electroanal. Chem., 260*: 15 (1989).

81. W. R. Fawcett and M. Opallo, Kinetic parameters for heterogeneous electron transfer to tris(acetylacetonato)manganese(III) and tris(acetylacetonato)iron(III) in aprotic solvents. *J. Electroanal. Chem., 331*: 815 (1992).

82. W. R. Fawcett and C. A. Foss, The analysis of solvent effects on the kinetics of simple heterogeneous electron transfer reactions, *J. Electroanal. Chem.*, *252*: 221 (1988).

83. W. R. Fawcett, M. Opallo, M. Fedurco, and J. W. Lee, Buckminsterfullerene as a model reactant for testing electron transfer theory, *J. Electroanal. Chem.*, *344*: 375 (1993).

84. D. K. Phelps, A. A. Kornyshev, and M. J. Weaver, Nonlocal electrostatic effects on electron-transfer activation energies: some consequences for and comparisons with electrochemical and homogeneous-phase kinetics. *J. Phys. Chem.*, *94*: 1454 (1990).

85. P. G. Wolynes, Linearized microscopic theories of nonequilibrium solvation, *J. Chem. Phys.*, *86*: 5133 (1987).

86. W. R. Fawcett and L. Blum, Estimation of the outer-sphere contribution to the activation parameters for homogeneous electron-transfer reactions using the mean spherical approximation, *Chem. Phys. Lett.*, *187*: 173 (1991).

87. W. R. Fawcett and C. A. Foss, Solvent effects on simple electron transfer reactions: a comparison of results for homogeneous and heterogeneous systems, *J. Electroanal. Chem.*, *270*: 103 (1989).

88. M. Opallo, The solvent effect on the electro-oxidation of 1,4-phenylenediamine: the influence of the solvent reorientation dynamics on the one-electron transfer rate, *J. Chem. Soc. Faraday Trans. 1*, *82*: 339 (1986).

89. G. E. McManis, M. N. Golovin, and M. J. Weaver, Role of solvent reorganization dynamics in electron-transfer processes: anomalous kinetic behavior in alcohol solvents, *J. Phys. Chem.*, *90*: 6563 (1986).

90. G. E. McManis and M. J. Weaver, Solvent dynamical effects in electron transfer: predicted consequences of non-Debye relaxation processes and some comparisons with experimental kinetics, *J. Chem. Phys.*, *90*: 912 (1989).

91. W. R. Fawcett and C. A. Foss, Solvent effects on simple electron transfer reactions: a comparison of results in Debye and non-Debye solvents, *J. Electroanal. Chem.*, *306*: 71 (1991).

92. X. H. Mu and F. A. Schultz, Influence of solvent dynamics on the heterogeneous kinetics of a reaction with a large inner-shell barrier: chloro(tetraphenylporphinato)manganese(III) reduction, *J. Electroanal. Chem.*, *353*: 349 (1993).

93. H. Fernandez and M. A. Zon, Solvent effect on the kinetics of heterogeneous electron transfer process: the TMPD/TMPD$^+$ redox couple, *J. Electroanal. Chem.*, *332*: 237 (1992).

94. G. E. McManis, R. M. Nielson, A. Gochev, and M. J. Weaver, Solvent dynamical effects in electron transfer: evaluation of electronic matrix coupling elements for metallocene self-exchange reactions, *J. Am. Chem. Soc.*, *111*: 5533 (1989).

3

Digital Simulations with the Fast Implicit Finite Difference Algorithm: The Development of a General Simulator for Electrochemical Processes

M. RUDOLPH
Friedrich-Schiller-Universität
Jena, Germany

I. INTRODUCTION

In electrochemistry the term *digital simulation* is almost synonymous with the name of Stephen Feldberg, who published the first paper on this topic in 1964 [1]. The *explicit finite difference* algorithm originally proposed by this author, as well as its *implicit* counterpart, have been described in detail, the classical paper of Feldberg [2] and the book of Britz [3] perhaps being the most widely used sources in this field. In recent years, another class of numerical methods for solving partial differential equations has attracted growing attention in electrochemistry. It is the *orthogonal collocation* [4] technique, based on polynomial curve fitting. To our knowledge, a general review of this method is being prepared by B. Speiser (personal communication). Consequently, it is not our objective in this chapter to give a general survey of digital simulation techniques and the literature published in this field. Rather, we concentrate on an efficient approach for solving the equations of the more advantageous, unconditionally stable *implicit* simulation techniques. In this way, the power of the *implicit finite difference* methods can be fully exploited, giving rise to a universal tool for the digital simulation of electrochemical kinetic-diffusion problems.

Because of this general applicability, the approach was used in developing a *general simulator*, a term used in this chapter to denote a simulation program that is not based on code libraries of known electrochemical mechanisms but is able to adapt the simulation algorithm automatically to almost any electrochemical mechanism given in the form of reaction equations. In this way, even new mechanisms can be simulated with a computational speed hitherto thought to be impossible without having to change the code of the program. It has been claimed for several simulation techniques that they can easily be adapted to a new system. In contradiction to that, many papers have been published showing the application of these methods to various electrochemical mechansims. When using the general simulator proposed in this chapter, the treatment of special electrochemical mechanisms becomes superfluous.

A. Mathematical Model

In general, the mathematical description of an electrochemical experiment leads to systems of coupled partial differential equations. Let us assume for the moment that we are interested in the behavior of only one species described by a single partial differential equation. To solve such an equation on a computer, a discretized mathematical model is required. For this purpose, the electrolyte solution near the electrode is divided in small, discrete volume elements, as shown in Figure 1. The concentration of the species in the individual space elements is represented by the average values $c_1, c_2, c_3, \ldots, c_n$ localized at points $\bar{x}_1, \bar{x}_2, \bar{x}_3, \ldots, \bar{x}_n$. The additional concentration point c_0 is introduced to express the concentration of the species at the electrode.* It accounts for the fact that in an electrochemical experiment the perturbation comes primarily from the charge-transfer process, which alters this concentration directly and results in transportation of particles through the boundaries of the space boxes. To describe this phenomenon in a quantitative way, fluxes are defined at the boundaries of each element (i.e., at $x_0, x_1, x_2, \ldots, x_n$), so that the number of moles moving through the kth boundary during the time interval Δt can be calculated immediately by multiplying the flux, f_k, with the area of the boundary, A_k. The mutual depen-

---

*The term *at the electrode* does not really mean the surface of the electrode itself but the point of closest approach where the electron transfer proceeds.

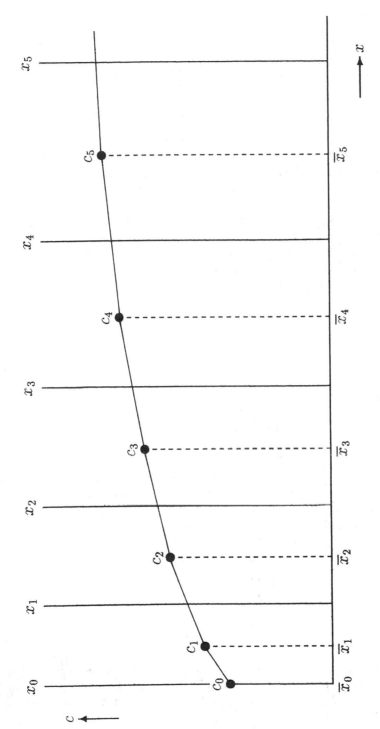

Figure 1 Discrete model of the spatial grid.

dence of fluxes and changes in concentration can now be expressed by the more-
or-less self-evident relation

$$\frac{\Delta c_k}{\Delta t} = - \frac{A_k f_k - A_{k-1} f_{k-1}}{V_k} \tag{1}$$

where V_k denotes the volume of the kth space box. This equation does not hold
true, however, if the concentration of the substance is changed not only by
diffusion but also by chemical reactions. In this case the corresponding kinetic
terms must be added to equation (1). For the moment, however, chemical kinetics will be disregarded.

Generally, the fluxes at the boundaries are not given explicitly in a simulation,
and it is convenient to perform all computations in terms of concentrations. That
can be accomplished easily if the mass transport is originated soley from diffusion
and the Nernst–Planck equation reduces to Fick's first law of diffusion:

$$f(x, t) = - D \frac{\partial c(x, t)}{\partial x} \tag{2}$$

or in discretized form, considering the right boundary of the kth box,

$$f_k = - D \frac{c_{k+1} - c_k}{\overline{x}_{k+1} - \overline{x}_k} \tag{3}$$

where D is the diffusion coefficient of the species.

Before substituting this relation into equation (1), the geometry of the
space grid must be specified. For most situations, an expanding space grid as
introduced by Joslin and Pletcher [5] and substantially improved by Feldberg
[6] can be recommended. In this approach the thickness of the volume elements
increases with the distance from the electrode so that the concentration profile
will be described in greater detail closer to the electrode surface and in less
detail at a greater distance from the surface where the perturbations are smaller
and their effect on the current becomes negligibly small. The exponentially
expanding space grid has been adapted to different electrode geometries [6,7].
For a planar electrode $A_{k-1} = A_k$ and $V_k = A_k (x_k - x_{k-1})$ with

$$k \geq 1: \qquad x_k = \Delta_x \frac{\exp(\beta k) - 1}{\exp(\beta) - 1} \tag{4}$$

$$k \geq 1: \qquad \overline{x}_k = \Delta_x \frac{\exp[\beta(k - \frac{1}{2})] - 1}{\exp(\beta) - 1} \tag{5}$$

The term Δx in these equations denotes the actual thickness of the first volume
element of the expanding space grid, and the degree of the exponential expansion is expressed by the parameter β. A uniform grid is the limiting case when
β tends to zero.

In the next step, (3), (4), and (5) are substituted into (1), and Δc_k is expressed as difference between "new" (unknown) concentrations c_k', assigned to the time level $t + \Delta t$, and "old" (known) concentrations c_k, assigned to the actual time t. This substitution yields

$$k \geq 1: \qquad c_k' = c_k + D_{2k}^* (c_{k+1} - c_k) - D_{1k}^* (c_k - c_{k-1}) \tag{6}$$

with

$$k \geq 1: \qquad D_{2k}^* = D^* \exp[2\beta \left(\tfrac{3}{4} - k\right)] \tag{7}$$

$$k \geq 2: \qquad D_{1k}^* = D^* \exp[2\beta \left(\tfrac{5}{4} - k\right)] \tag{8}$$

and

$$\beta > 0: \qquad D_{11}^* = D^* \frac{\exp(\beta) - 1}{\exp(\beta/2) - 1} \tag{9}$$

$$\beta = 0: \qquad D_{11}^* = 2D^* \tag{10}$$

The dimensionless *model diffusion coefficient* $D^* = D \, \Delta t/(\Delta x)^2$ summarizes the scaling factors for diffusion, time, and space, respectively. The choice of this parameter is determined by the real-world problem to which the simulation corresponds.

Before starting a simulation on a computer, it is necessary to estimate how many boxes are needed to describe the concentration profile in a particular experiment. A rule of thumb provides a safe answer: Any experiment that has proceeded for a time t will significantly alter the solution from its bulk character for a distance no larger than about $6 \, (Dt)^{1/2}$. If n denotes the number of boxes required in the jth time step, this distance is about

$$t = j \, \Delta t: \qquad x_n = 6 \sqrt{j \, \Delta t \, D} = 6 \, \Delta x \sqrt{jD^*} \tag{11}$$

and with the aid of (4)

$$\beta > 0: \qquad n = 1 + \frac{1}{\beta} \ln \{6[\exp(\beta) - 1] \sqrt{jD^*} + 1\} \tag{12}$$

$$\beta = 0: \qquad n = 1 + 6 \sqrt{jD^*} \tag{13}$$

The addition of 1 to the right-hand side of these equations ensures that even with truncation to the next integer value, n is always large enough.

To illustrate how a digital simulation works in the simplest case, we consider the Cottrell experiment [8]. The concentration of the species is assumed to be equal to the bulk concentration c^* anywhere in the cell at the time $t = 0$. The experiment is started by jumping the potential of the electrode to a value where a charge-transfer reaction proceeds with such high speed that the concentration of the species at the electrode surface is instantly forced to zero. Then the concentration c_0' is known explicitly, and the other concentrations c_k' ($k \geq$ 1) for $t = \Delta t$ can be calculated directly by means of (6) because all old concentrations are known from the *initial condition* $c_k = c^*$ at $t = 0$ for all k. The flux of the species to the electrode, which is proportional to the faradaic current, is calculated from the concentration profile using (3) for $k = 0$. The same procedure is then applied to the next time step, where the just calculated concentrations play the role of the old concentrations. In this way, the perturbation propagates over time and space.

Since any new concentration can be calculated explicitly from the old values, the method is called the explicit finite-difference (EFD) algorithm. This method was pioneered in electrochemistry by Feldberg, who also contributed much to its continuous development [1,2,9,10]. In the example above, the situation was especially easy because the value of c_0' was also given explicitly. In general, the concentrations of the species at the electrode surface are determined by including a corresponding number of additional equations which describe how these concentrations change as a function of the electrode potential, which itself may be a function of time, depending on the electrochemical method modeled.

The obvious advantage of the explicit method is its simplicity. However, this simplicity is achieved at the expense of performance resulting from the restriction that the model diffusion coefficient D^* is not allowed to be larger than 0.5 to ensure the numerical stability of the method. The restriction can be critical if, for instance, a fast chemical reaction effects a thin *reaction layer* [11] near the electrode. A fine space grid is then required to represent such a concentration profile, and the computation time may become unacceptably long since correspondingly small time steps must be used to fulfill the stability condition

$$D^* = \frac{D \, \Delta t}{(\Delta x)^2} \leq 0.5 \tag{14}$$

An unconditionally stable algorithm will be obtained by formulating Fick's first law (2) in terms of "new" fluxes:

$$f_k' = -D \frac{c_{k+1}' - c_k'}{\bar{x}_{k+1} - \bar{x}_k} \tag{15}$$

Substitution into the analog of (1) yields

$$1 \leq k \leq n: \qquad c_k' = c_k + D_{2k}^*(c_{k+1}' - c_k') - D_{1k}^* (c_k' - c_{k-1}') \tag{16}$$

or after sorting into new and old concentrations,

$$1 \leq k \leq n: \qquad -D_{1k}^* c_{k-1}' + D_{3k}^* c_k' - D_{2k}^* c_{k+1}' = c_k \tag{17}$$

with

$$D_{3k}^* = 1 + D_{1k}^* + D_{2k}^* \tag{18}$$

Unlike the explicit finite-difference algorithm described above, three unknown concentrations are now involved in each equation. Nevertheless, all information necessary for determining the unknowns is contained implicitly in (17). That is why the method is called the *fully implicit finite-difference* algorithm attributed to Laasonen [12] in the mathematical literature. It has rarely been applied in electrochemistry [13–15].

Again considering Cottrellian diffusion with $c_0' = 0$ and $c_{n+1}' = c^*$ for $t > 0$, formulas (17) build a system of n *tridiagonal* equations:

$$D_{31}^* c_1' - D_{21}^* c_2' = c_1 \tag{19}$$

$$-D_{12}^* c_1' + D_{32}^* c_2' - D_{22}^* c_3' = c_2 \tag{20}$$

$$-D_{13}^* c_2' + D_{33}^* c_3' - D_{23}^* c_4' = c_3 \tag{21}$$

$$-D_{14}^* c_3' + D_{34}^* c_4' - D_{24}^* c_5' = c_4 \tag{22}$$

$$\vdots$$

$$-D_{1n}^* c_{n-1}' + D_{3n}^* c_n' - D_{2n}^* c_{n+1}' = c_n \tag{23}$$

which must be solved in each time or potential step for determining the n unknown concentrations $c_1', c_2', c_3', \ldots, c_n'$. Tridiagonal equations are also obtained for another implicit simulation technique: Adding (6) to (17) gives the equations of the Crank–Nicolson algorithm [16]:

$$1 \le k \le n: \quad -D_{1k}^* c_{k-1}' + (2 + D_{1k}^* + D_{2k}^*)c_k' - D_{2k}^* c_{k+1}' \tag{24}$$

$$= D_{1k}^* c_{k-1} + (2 - D_{1k}^* - D_{2k}^*)c_k + D_{2k}^* c_{k+1}$$

Valuable contributions concerning the application of this method to electrochemical problems have been elaborated by Heinze and Britz [17–20].

For both implicit methods, the tridiagonal structure enables these equations to be solved efficiently by applying a special version of the Gauss elimination algorithm with formulas adjusted to mention only the nonzero entries of the coefficient matrix [3]. The procedure requires about two to three times the computation time of the EFD algorithm, but the implicit methods win when large homogeneous rate constants require correspondingly large values of D^* to optimize computational speed and accuracy.

Unfortunately, the tridiagonal structure of these equations is lost when the reaction scheme is complicated by kinetic couplings between different species. Previously, such systems were thought not to be amenable to the more advantageous implicit simulation techniques because solving the equations in the usual way by applying the general form of the Gauss elimination algorithm to the entire coefficient matrix may become a very time (and memory)-consuming procedure. Explicit simulation techniques [9,10] can then be even faster and easier to handle on a PC-level computer. Presumably for this reason, the implicit methods have not been widely used in electrochemistry.

There is, however, a simple way to overcome this problem [21]. It will be shown that regardless of the complexity of the equations obtained for the individual species, almost any electrochemical problem can be described mathematically by a set of tridiagonal equations when writing the implicit formulas in a matrix notation. Consequently, the real power of the implicit methods can be exploited by extending the efficient approach of solving tridiagonal equations to the case of matrix equations. This is the underlying idea of the *fast implicit finite-difference* (FIFD) algorithm [21]. The FIFD approach can be applied to both the Laasonen and Crank–Nicolson formulas. Although the latter provide the higher accuracy, we focus attention on the fully implicit method, which yields the more straightforward equations. Moreover, it seems that the higher accuracy of the Crank–Nicolson method is achieved at the expense of stability when dealing with second-order phenomena. On the other hand, only a slight modification of the Laasonen algorithm improves its accuracy substantially, virtually matching the accuracy of the Crank–Nicolson algorithm. For simplicity, we continue using the unmodified Laasonen formulas in the following chapters, bearing in mind that the accuracy of the method can be improved following the rules outlined in Section III.A.1.

II. BASIC EQUATIONS OF THE FAST IMPLICIT FINITE-DIFFERENCE ALGORITHM

A. Electrochemical Mechanisms with First- and Second-Order Chemical Reactions

The explicit finite-difference algorithm described in Section I provides a useful tool for the digital simulation of electrochemical processes involving charge-transfer steps and slow first- or second-order chemical reactions. Hence there is no need to treat those mechanisms with more sophisticated simulation techniques such as implicit finite-difference methods. The situation changes, however, when using the *fast implicit finite-difference* (FIFD) algorithm [21]. It leads to matrix equations that can be solved almost as efficiently as the equations obtained for explicit difference algorithms without being restricted to small time (or potential) steps if mechanisms with fast chemical reactions are to be simulated.

The FIFD approach is first demonstrated for the simple quasi-reversible charge-transfer mechanism

$$B + e \underset{k_{hb}}{\overset{k_{hf}}{\rightleftharpoons}} R \qquad \text{(mechanism 1)} \tag{25}$$

and will then be extended step by step to more general cases. Using the notation $b = [B]$ and $r = [R]$, the change in the concentration of these species is to be calculated for the following initial and boundary conditions:

$$t = 0, x \geq 0: \qquad b = b^* \qquad r = r^* \tag{26}$$

$$t > 0, x \to \infty: \qquad b = b^* \qquad r = r^* \tag{27}$$

$$t > 0, x = 0: \qquad D_B \frac{\partial b(x, t)}{\partial x} = -D_R \frac{\partial r(x, t)}{\partial x} = \frac{I}{zFA} \tag{28}$$

$$I = zFA(k_{hf}\, b - k_{hb}r) \tag{29}$$

where D_B and D_R are the diffusion coefficients and b^* and r^* the bulk concentrations of the corresponding species. The quantities k_{hf} and k_{hb} represent the actual forward and backward heterogeneous rate constants of the charge-transfer reaction at a given potential. They are usually related to the electrode potential E and the standard $E°$ by the Butler–Volmer equations

$$k_{hf} = k_s° \exp\left[\frac{-\alpha F}{RT} (E - E°)\right] \tag{30}$$

$$k_{hb} = k_s° \exp\left[\frac{(1 - \alpha)F}{RT} (E - E°)\right] \tag{31}$$

where the intrinsic kinetics of the charge-transfer process is characterized by the standard heterogeneous rate constant $k_s°$ and the transfer coefficient α. Provided for simplicity, the diffusion coefficients of both species are identical ($D_B = D_R = D$). The fully implicit formulas (17) derived in Section I lead to the following systems of tridiagonal equations*:

$$1 \leq k \leq n: \qquad -D_{1k}^* b'_{k-1} + D_{3k}^* b'_k - D_{2k}^* b'_{k+1} = b_k \tag{32}$$

$$1 \leq k \leq n: \qquad -D_{1k}^* r'_{k-1} + D_{3k}^* r'_k - D_{2k}^* r'_{k+1} = r_k \tag{33}$$

Additional relations are obtained by expressing the boundary conditions (27), (28), and (29) in a discretized form

$$b'_{n+1} = b^* \qquad r'_{n+1} = r^* \tag{34}$$

$$\frac{D}{\Delta \bar{x}_1} (b'_1 - b'_0) = -\frac{D}{\Delta \bar{x}_1} (r'_1 - r'_0) = \frac{I'}{zFA} \tag{35}$$

$$I' = zFA\, (k_{hf}b'_0 - k_{hb}\, r'_0) \tag{36}$$

---

*The inclusion of unequal diffusion coefficients makes no difficulties but complicates the notation of the formulas since additional indices are necessary for characterizing the diffusional terms associated with each species.

According to (12), n is large enough to ensure that concentrations in the $(n + 1)$ the space box are essentially the bulk concentrations. The quantity $\Delta \bar{x}_1 = \bar{x}_1 - \bar{x}_0$ marks the distance of the first concentration point from the electrode, and the prime (′) indicates that the respective quantities involve *unknown* concentrations assigned to the time level $t + \Delta t$. Although not marked explicitly by a prime, the heterogeneous rate constants k_{hf} and k_{hb} are also related to the time $t + \Delta t$, but unlike the primed concentrations, the rate constants do not play the role of unknowns in the simulation. They can be calculated explicitly beforehand for any time from (30) and (31) and the potential program modeled. Combining these relations yield the two equations

$$\left(\frac{D}{\Delta \bar{x}_1} + k_{hf}\right) b_0' - \frac{D}{\Delta \bar{x}_1} b_1' - k_{hb} r_0' = 0 \tag{37}$$

$$\frac{D}{\Delta \bar{x}_1} (b_1' - b_0') + \frac{D}{\Delta \bar{x}_1} (r_1' - r_0') = 0 \tag{38}$$

still necessary for determining the $2n + 2$ unknown concentrations $b_0', b_1', b_2', \ldots, b_n'$ and $r_0', r_1', r_2', \ldots, r_n'$ involved in (32) and (33).

In the present case, *both* systems of linear equations (32) and (33) can be solved very quickly, making good use of the tridiagonal structure of these equations. To avoid distraction we do not elaborate on it here, since the tridiagonal structure of the equation is lost if the mechanism becomes a bit more complicated. For example, adding only one reaction more to the reaction scheme above leads to the first-order *catalytic* mechanism

$$B + e \underset{k_{hb}}{\overset{k_{hf}}{\rightleftharpoons}} R \qquad \text{(mechanism 2)} \tag{39}$$

$$R \underset{k_b}{\overset{k_f}{\rightleftharpoons}} B \tag{40}$$

The initial and boundary conditions are assumed to be identical with those of mechanism 1. However, because of the chemical reaction (40), the concentrations will now be changed not only by diffusion but also by chemical kinetics:

$$\left(\frac{\partial c}{\partial t}\right)_{total} = \left(\frac{\partial c}{\partial t}\right)_{diff} + \left(\frac{\partial c}{\partial t}\right)_{kin} \qquad c = b, r \tag{41}$$

Within the framework of the fully implicit method, the part resulting from the chemical reaction in the kth space box can be discretized as

$$\frac{\Delta b_k}{\Delta t} = k_f r_k' - k_b b_k' \tag{42}$$

$$\frac{\Delta r_k}{\Delta t} = -k_f r_k' + k_b b_k' \tag{43}$$

and the fully implicit formulas (17) now are

$$1 \le k \le n: \quad -D_{1k}^* b_{k-1}' + (D_{3k}^* + k_b^*) b_k' - D_{2k}^* b_{k+1}' - k_f^* r_k' = b_k \tag{44}$$

$$1 \le k \le n: \quad -D_{1k}^* r_{k-1}' + (D_{3k}^* + k_f^*) r_k' - D_{2k}^* r_{k+1}' - k_b^* b_k' = r_k \tag{45}$$

where the term k_{index}^* is an abbreviation for $k_{\text{index}} \Delta t$ throughout.

Seemingly, the tridiagonal structure of the equations is destroyed by the additional kinetic terms $k_f^* r_k'$ and $k_b^* b_k'$, respectively. Nevertheless, the equation couple to each value of k can be written in tridiagonal form when using a matrix notation

$$
\begin{aligned}
&- D_{1k}^* \begin{bmatrix} b_{k-1}' \\ r_{k-1}' \end{bmatrix} + \left(\begin{bmatrix} D_{3k}^* & 0 \\ 0 & D_{3k}^* \end{bmatrix} - \begin{bmatrix} -k_b^* & k_f^* \\ k_b^* & -k_f^* \end{bmatrix} \right) \begin{bmatrix} b_k' \\ r_k' \end{bmatrix} \\
&- D_{2k}^* \begin{bmatrix} b_{k+1}' \\ r_{k+1}' \end{bmatrix} = \begin{bmatrix} a_k \\ r_k \end{bmatrix}
\end{aligned}
\tag{46}
$$

Except for the factor Δt, the matrix composed of the rate constants k_f^* and k_b^* is identical with that representing the kinetic couplings (42) and (43) in matrix form:

$$\frac{\Delta}{\Delta t} [\mathbf{C}_k] = \mathbf{K} \mathbf{C}_k' \tag{47}$$

with

$$
\mathbf{C}_k = \begin{bmatrix} b_k \\ r_k \end{bmatrix} \qquad \mathbf{K} = \begin{bmatrix} -k_b & k_f \\ k_b & -k_f \end{bmatrix} \qquad \mathbf{C}_k' = \begin{bmatrix} b_k' \\ r_k' \end{bmatrix}
\tag{48}
$$

Consequently, the matrix equations (46) can easily be built up if the first-order kinetics involved in the mechanism has been expressed in form of the coupling matrices $\mathbf{K}$ or $\mathbf{K}^*$, respectively, where $\mathbf{K}^*$ is obtained by multiplying each element of $\mathbf{K}$ with Δt.

Here and in the following, noncursive boldface letters indicate the matrix or vector character of the corresponding quantities. The notation $\frac{\Delta}{\Delta t} [\mathbf{C}_k]$ was used to express the fact that the operator $\frac{\Delta}{\Delta t}$ has to be applied to each element of the vector $\mathbf{C}_k$, thus producing a new vector object.

With the aid of the definitions

$$
\mathbf{B}_0 = \begin{bmatrix} (D/\Delta \bar{x}_1) + k_{hf} & -k_{hb} \\ D/\Delta \bar{x}_1 & D/\Delta \bar{x}_1 \end{bmatrix} \qquad \mathbf{B}_1 = \begin{bmatrix} D/\Delta \bar{x}_1 & 0 \\ D/\Delta \bar{x}_1 & D/\Delta \bar{x}_1 \end{bmatrix}
\tag{49}
$$

the boundary conditions can also be embodied in matrix form, resulting in a mathematical description of the entire mechanism in terms of the tridiagonal matrix equations

$$\mathbf{B}_0\mathbf{C}_0' - \mathbf{B}_1\mathbf{C}_1' = 0 \tag{50}$$

$$1 \le k \le n: \quad -D_{1k}^*\mathbf{C}_{k-1}' + \mathbf{D}_k\mathbf{C}_k' - D_{2k}^*\mathbf{C}_{k+1}' = \mathbf{C}_k \tag{51}$$

According to (46), the matrices $\mathbf{D}_k$ are defined by

$$1 \le k \le n: \quad \mathbf{D}_k = D_{3k}^*\mathbf{I} - \mathbf{K}^* \tag{52}$$

where $\mathbf{I}$ denotes the *identity* matrix. The analogy to the single-species equations (19) to (23) is obvious and provides the foundation for an effective solution of these matrix equations.

The approach presented here for a particular first-order mechanism can be extended to any other mechanism, comprising not only first- but also second-order chemical reactions. In the presence of second-order reactions, the matrices describing the second-order kinetics involved in the mechanism will no longer be composed solely of rate constants but of products of rate constants and concentration terms. Since concentrations vary from space box to space box, the second-order coupling matrix of the kth space box will become a function of the concentration vector $\mathbf{C}_k'$. We write simply $\mathbf{K}_k'$ (or $\mathbf{K}_k^{*\prime}$) to indicate that the unknown concentrations (marked by the prime) of the kth space box will now be involved in this matrix. Consequently, an electrochemical mechanism with second-order chemical reactions can also be described mathematically by the matrix equations (50) and (51), but the coefficient matrix $\mathbf{D}_k$ will become a function of the unknown concentrations, as indicated by the prime in the following definition:

$$1 \le k \le n: \quad \mathbf{D}_k' = D_{3k}^*\mathbf{I} - \mathbf{K}^* - \mathbf{K}_k^{*\prime} \tag{53}$$

In other words, the matrix equations of a second-order mechanism are tridiagonal again, but they are nonlinear and cannot be solved directly in this form.

A standard procedure for solving nonlinear equations is the Gauss–Newton procedure. It is known from the mathematical literature that an n-dimensional system of nonlinear equations

$$\mathbf{F}(\mathbf{x}) = 0 \qquad \mathbf{F}(\mathbf{x}) = \begin{pmatrix} f_1(\mathbf{x}) \\ f_2(\mathbf{x}) \\ \vdots \\ f_n(\mathbf{x}) \end{pmatrix} \qquad \mathbf{x} = \begin{pmatrix} x_1 \\ x_2 \\ \vdots \\ x_n \end{pmatrix} \tag{54}$$

can be solved as follows: Starting with a first guess $\mathbf{x}^{(0)}$ of the exact solution, one obtains successively better approximations $\mathbf{x}^{(1)}$, $\mathbf{x}^{(2)}$, ... by solving the following system of linear equations;

$$\mathbf{J}(\mathbf{x}^{(i)})\,\mathbf{x}^{(i+1)} = \mathbf{J}(\mathbf{x}^{(i)})\mathbf{x}^{(i)} - \mathbf{F}(\mathbf{x}^{(i)}) \tag{55}$$

recursively. The Jacobian matrix, $\mathbf{J}$, contains the partial derivatives

$$
J = \begin{pmatrix}
\dfrac{\partial f_1}{\partial x_1} & \dfrac{\partial f_1}{\partial x_2} & \cdots & \dfrac{\partial f_1}{\partial x_n} \\[2ex]
\dfrac{\partial f_2}{\partial x_1} & \dfrac{\partial f_2}{\partial x_2} & \cdots & \dfrac{\partial f_2}{\partial x_n} \\[2ex]
\vdots & \vdots & \ddots & \vdots \\[2ex]
\dfrac{\partial f_n}{\partial x_1} & \dfrac{\partial f_n}{\partial x_2} & \cdots & \dfrac{\partial f_n}{\partial x_n}
\end{pmatrix}
\tag{56}
$$

The equations (55) are linear because the coefficient matrix $\mathbf{J}$ for calculating the first improvement $\mathbf{x}^{(1)}$ contains the known, more-or-less arbitrary starting values $\mathbf{x}^{(0)}$, and the computation of any other improved solution vector $\mathbf{x}^{(i+1)}$ requires only the vector $\mathbf{x}^{(i)}$, obtained as the result of the previous iteration.

With regard to the second-order analog of (50) and (51), meaning the equations obtained by replacing $\mathbf{D}_k$ through $\mathbf{D}_k'$, we search for the concentration vectors that make the following functions equal to zero:

$$
-\mathbf{F}(\mathbf{C}_0', \mathbf{C}_1') = \mathbf{B}_1 \mathbf{C}_1' - \mathbf{B}_0 \mathbf{C}_0' = 0
\tag{57}
$$

$$
1 \le k \le n: \quad -\mathbf{F}(\mathbf{C}_{k-1}', \mathbf{C}_k', \mathbf{C}_{k+1}')
$$
$$
= \mathbf{C}_k + D_{1k}^* \mathbf{C}_{k-1}' - \mathbf{D}_k' \mathbf{C}_k' + D_{2k}^* \mathbf{C}_{k+1}' = 0
\tag{58}
$$

Since the kth equation in (58) involves only concentrations of the kth box and its neighboring boxes, most of the partial derivatives in the Jacobian matrix (56) become zero. According to (53) the concentration-dependent term in $\mathbf{D}_k'$ is the second-order coupling matrix $\mathbf{K}_k^{*\prime}$ and the application of (55) to (57) and (58) therefore yields

$$
\mathbf{B}_0 \mathbf{C}_0'^{(i+1)} - \mathbf{B}_1 \mathbf{C}_1'^{(i+1)} = 0
\tag{59}
$$

$$
1 \le k \le n: \quad -D_{1k}^* \mathbf{C}_{k-1}'^{(i+1)} + \mathbf{J}_k'^{(i)} \mathbf{C}_k'^{(i+1)} - D_{2k}^* \mathbf{C}_{k+1}'^{(i+1)} = \mathbf{C}_k - \mathbf{S}_k'^{(i)} \mathbf{C}_k'^{(i)}
\tag{60}
$$

with (in symbolic notation)

$$
\mathbf{J}_k' = \frac{\partial [\mathbf{D}_k' \mathbf{C}_k']}{\partial \mathbf{C}_k'} = \mathbf{D}_k' - \frac{\partial \mathbf{K}_k^{*\prime}}{\partial \mathbf{C}_k'} \mathbf{C}_k' = \mathbf{D}_k' - \mathbf{S}_k'
\tag{61}
$$

The matrices $\mathbf{S}_k'$ summarize the partial derivatives of the second-order coupling matrices with respect to the individual concentrations. Since each element of $\mathbf{K}_k^{*\prime}$ may depend on any concentration involved in the kth space box, the element in the lth row and jth column of the kth matrix is given by

$$
(\mathbf{S}_k')_{lj} = \frac{\partial (K_k^{*\prime})_{l1}}{\partial c_{j,k}'} c_{1,k}' + \frac{\partial (K_k^{*\prime})_{l2}}{\partial c_{j,k}'} c_{2,k}' + \cdots + \frac{\partial (K_k^{*\prime})_{lm}}{\partial c_{j,k}'} c_{m,k}'
\tag{62}
$$

where $c'_{j,k}$ denotes the concentration of species j in the kth space box and m is the number of species involved in the mechanism.

In this way a general mathematical description of any electrochemical mechanism comprising charge-transfer processes as well as first- and second-order chemical reactions has been achieved in terms of linear tridiagonal matrix equations which can be uniformly solved on a computer. The only mechanism-specific part in the computer program is the generation of the matrices $\mathbf{B}_0$, $\mathbf{B}_1$, $\mathbf{J}'_k$, and $\mathbf{S}'_k$. So we have the favorable situation that changing a mechanism requires only modifications in that part of the computer program that generates these matrices. Provided that we have an efficient algorithm for solving these matrix equations, further treatment would be the same for any mechanism. That opens the door to the development of a general simulator, a program that is able to translate the reaction equations of a given electrochemical mechanism into the matrices needed for carrying out the simulation. Unlike other approaches based on code libraries of known electrochemical mechanisms [22], this program enables the user to design a new mechanism simply by typing in the reaction scheme and the kinetic parameters. Because of this general applicability, the matrix equations will be referred to as the *basic equations* of the fast implicit finite-difference (FIFD) algorithm.

B.   Basic Principles for Developing a General Simulator

As pointed out in Section II.A, the FIFD approach offers some favorable features for developing a general, mechanism-independent simulation program. One point is the extraordinary numerical stability of the fully implicit algorithm, which enables the method to cope with rate constants of almost any order of magnitude without having to change the program. Another feature is the uniform mathematical description of common electrochemical experiments by the basic equations (59) and (60).

In the presence of second-order kinetics, the matrices involved in the basic equations will contain concentration terms that differ from space box to space box, and from time step to time step, so that the matrices must be regenerated continually in the course of the simulation. Hence the performance of a general simulation program will depend mainly on how efficiently the reaction scheme can be translated into these matrices and how efficiently the resulting equations can be solved.

For efficient generation of the matrix equations, we take advantage of the following fact: In the same way that a mechanism embodies the sum of all involved chemical (or electrochemical) reactions written in the form of reaction equations, the respective matrices can be composed additively of submatrices, each representing the contribution of an individual reaction equation to the corresponding summary matrix. For instance, if a mechanism involves a total of m

species, the part reflecting only the jth first-order chemical reaction

$$A_j \overset{k_{f_j}}{\underset{k_{b_j}}{\rightleftharpoons}} B_j \tag{63}$$

can be written analogously to (47) and (48):

$$\frac{\Delta}{\Delta t}
\begin{bmatrix} c_{1,k} \\ \vdots \\ c_{a_j,k} \\ \vdots \\ c_{b_j,k} \\ \vdots \\ c_{m,k} \end{bmatrix}
=
\begin{bmatrix}
0 & \cdots & 0 & \cdots & 0 & \cdots & 0 \\
\vdots & \ddots & \vdots & \ddots & \vdots & \ddots & \vdots \\
0 & \cdots & -k_{f_j} & \cdots & k_{b_j} & \cdots & 0 \\
\vdots & \ddots & \vdots & \ddots & \vdots & \ddots & \vdots \\
0 & \cdots & k_{f_j} & \cdots & -k_{b_j} & \cdots & 0 \\
\vdots & \ddots & \vdots & \ddots & \vdots & \ddots & \vdots \\
0 & \cdots & 0 & \cdots & 0 & \cdots & 0
\end{bmatrix}
\begin{bmatrix} c'_{1,k} \\ \vdots \\ c'_{a_j,k} \\ \vdots \\ c'_{b_j,k} \\ \vdots \\ c'_{m,k} \end{bmatrix} \tag{64}$$

In the simulation program the species are no longer characterized by names but by indices, and it is assumed here that a_j denotes the index assigned internally to the species A_j, and b_j that of species B_j, so that the concentrations of these species in the kth space box are given by $c_{a_j,k}$ and $c_{b_j,k}$, respectively. After multiplication with Δt, the matrix on the right-hand side of (64), contains the elements of the first-order coupling matrix resulting from the jth first-order reaction. Since most of the elements in this matrix are zero, we mention in the following only the significant elements, writing the jth submatrix $\mathbf{K}_j^*$ in the form

$$\mathbf{K}_j^* =
\begin{bmatrix}
& a_j & b_j \\
a_j & \left(-k_{f_j}^* \right. & k_{b_j}^* \\
b_j & k_{f_j}^* & \left. -k_{b_j}^* \right)
\end{bmatrix} \tag{65}$$

and the (first-order) coupling matrix for a mechanism with N_F first-order chemical reactions is obtained by adding up all submatrices:

$$\mathbf{K}^* = \sum_{j=1}^{N_F} \mathbf{K}_j^* \tag{66}$$

This should be interpreted as follows: Starting with a zero matrix of order $(m \times m)$, the (positive or negative) values of the first-order rate constants are added to the elements indicated by the index couples a_j and b_j. Since the rate constants have to be specified by the user before starting the simulation, the submatrices, and thereby the summary coupling matrix $\mathbf{K}^*$, can be generated easily if the indices a_j and b_j connected with the jth first-order reaction have been stored in a suitable way.

The treatment of second-order reactions leads to a similar situation. Assuming that the mechanism involves N_s chemical second-order reactions of the general type

$$D_j + E_j \underset{k_{sb_j}}{\overset{k_{s_j}}{\rightleftharpoons}} F_j + G_j \tag{67}$$

the submatrices $\mathbf{K}_{k,j}^{*'}$ characterizing the effect of the jth second-order reactions in the kth space box are given by*

$$\mathbf{K}_{k,j}^{*'} = \begin{bmatrix} & d_j & e_j & f_j & g_j \\ d_j & \begin{pmatrix} -k_{sf_j}^* c_{e_j,k}' & 0 & k_{sb_j}^* c_{g_j,k}' & 0 \\ e_j & 0 & -k_{sf_j}^* c_{d_j,k}' & k_{sb_j}^* c_{g_j,k}' & 0 \\ f_j & k_{sf_j}^* c_{e_j,k}' & 0 & -k_{sb_j}^* c_{g_j,k}' & 0 \\ g_j & k_{sf_j}^* c_{e_j,k}' & 0 & 0 & -k_{sb_j}^* c_{f_j,k}' \end{pmatrix} \end{bmatrix} \tag{68}$$

and the application of (62) to eqn. (68) produces the submatrices

$$\mathbf{S}_{k,j}' = \begin{bmatrix} & d_j & e_j & f_j & g_j \\ d_j & \begin{pmatrix} 0 & -k_{sf_j}^* c_{d_j,k}' & 0 & k_{sb_j}^* c_{f_j,k}' \\ e_j & -k_{sf_j}^* c_{e_j,k}' & 0 & 0 & k_{sb_j}^* c_{f_j,k}' \\ f_j & 0 & k_{sf_j}^* c_{d_j,k}' & 0 & -k_{sb_j}^* c_{f_j,k}' \\ g_j & 0 & k_{sf_j}^* c_{d_j,k}' & -k_{sb_j}^* c_{g_j,k}' & 0 \end{pmatrix} \end{bmatrix} \tag{69}$$

Again, these submatrices and the summary matrices

$$\mathbf{K}_k^{*'} = \sum_{j=1}^{N_s} \mathbf{K}_{k,j}^{*'} \tag{70}$$

$$\mathbf{S}_k' = \sum_{j=1}^{N_s} \mathbf{S}_{k,j}' \tag{71}$$

can be built up immediately if the indices d_j, e_j, f_j, and g_j are known for each second-order reaction.

The generalized reaction equation (67) includes second-order dimerization reactions. Then $d_j = e_j$ or $f_j = g_j$ and the corresponding kinetic term is simply added twice at the same position. However, a special trick is used to enable the

---

*Because of the second-order terms, there is more than one possible notation for the matrices (68) and (69).

formalism above to be applied to second-order reactions of the general type

$$D_j \underset{k\,sb_j}{\overset{k\,sf_j}{\rightleftharpoons}} F_j + G_j$$

or

$$D_j + E_j \underset{k\,sb_j}{\overset{k\,sf_j}{\rightleftharpoons}} F_j$$

If a second-order reaction involves only three species, a virtual species assigned to the reserved index 0 is introduced at the empty position ($e_j = 0$ in the first equation or $g_j = 0$ in the second one), but the routines for solving the basic equations start from the species index 1, so that the concentration of species 0, which is se to $c_{0,k} = 1$ in all space boxes, does not change in the course of the simulation.

Before dealing with the submatrices of the charge-transfer reactions, the following fact should be taken into consideration: If a mechanism does not contain chemical reactions, the coupling matrices $\mathbf{K}^*$ and $\mathbf{K}_k^{*\prime}$ are zero, and according to (53), the system is correctly described by the diffusion terms. In the same way, if a mechanism did not involve charge-transfer processes, the boundary condition (59) should reduce to that of an electrochemically inactive system:

$$\frac{D}{\Delta\bar{x}_1}(\mathbf{C}_1' - \mathbf{C}_0') = 0 \tag{72}$$

Of course, an electrochemical mechanism without a charge-transfer reaction makes no sense, but defining $\mathbf{B}_0$ and $\mathbf{B}_1$ in the form

$$\mathbf{B}_0 = \frac{D}{\Delta\bar{x}_1}\mathbf{I} + \sum_{j=1}^{N_C}\mathbf{B}_{0,j} \tag{73}$$

$$\mathbf{B}_1 = \frac{D}{\Delta\bar{x}_1}\mathbf{I} + \sum_{j=1}^{N_C}\mathbf{B}_{1,j} \tag{74}$$

where N_C denotes the number of charge-transfer reactions, ensures that each species not subjected to a charge-transfer step be treated correctly as electrochemically inactive. The submatrices reflecting the jth charge-transfer reaction of the general type

$$P_j + e \underset{kh\,b_j}{\overset{kh\,f_j}{\rightleftharpoons}} R_j \tag{75}$$

are then defined by

$$\mathbf{B}_{0,j} = \begin{bmatrix} & p_j & r_j \\ p_j & \begin{pmatrix} k_{hf_j} & -k_{hb_j} \\ r_j & \begin{pmatrix} D/\Delta\bar{x}_1 & 0 \end{pmatrix} \end{pmatrix} \end{bmatrix} \qquad \mathbf{B}_{1,j} = \begin{bmatrix} & p_j \\ r_j & (D/\Delta\bar{x}_1) \end{bmatrix} \qquad (76)$$

These definitions no longer hold true if the species produced by the charge-transfer reaction undergoes a further charge-transfer step. The treatment of multistep charge-transfer reactions requires more sophisticated considerations. To avoid distraction, we disregard such mechanisms in the following. It should be emphasized, however, that the prototype of the general simulator is able to simulate such mechanisms without restrictions.

With the aid of the submatrices, the basic equations (59) and (60) for any mechanism expressible by the foregoing types of reaction equations can easily be established provided that the following information has been extracted from the reaction equations:

1. The type of each reaction equation (first-order chemical reaction, second-order chemical reaction, or charge-transfer step) is determined and the counters N_F, N_S, and N_C are set correspondingly.
2. The number of species involved in the mechanism is stored in the variable m and the name of each species is associated with a species index.
3. The indices of species that are coupled by a chemical or electrochemical reaction are stored in index tables $\mathbf{I}_F$, $\mathbf{I}_S$, and $\mathbf{I}_C$, respectively. The entries in these tables correspond to labels indicating the position of the elements in the submatrices.

For illustration we consider the following *ladder scheme*:

$$M + e \underset{k_{hb_1}}{\overset{k_{hf_1}}{\rightleftharpoons}} R$$

$$k_{f_1} \updownarrow k_{b_1} \qquad k_{f_2} \updownarrow k_{b_2}$$

$$MX + e \underset{k_{hb_2}}{\overset{k_{hf_2}}{\rightleftharpoons}} RX \quad \text{(mechanism 3)} \qquad (77)$$

$$k_{f_3} \updownarrow k_{b_3} \qquad k_{f_4} \updownarrow k_{b_4}$$

$$MXX + e \underset{k_{hb_3}}{\overset{k_{hf_3}}{\rightleftharpoons}} RXX$$

with the inclusion of the second-order cross reactions

$$RX + M \underset{ksb_1}{\overset{ksf_1}{\rightleftharpoons}} MX + R \tag{78}$$

$$RXX + M \underset{ksb_2}{\overset{ksf_2}{\rightleftharpoons}} MXX + R \tag{79}$$

$$RXX + MX \underset{ksb_3}{\overset{ksf_3}{\rightleftharpoons}} MXX + RX \tag{80}$$

The ladder scheme plays an important role, especially in the electrochemistry of coordinatively unsaturated complex compounds [23–26] which are able to undergo ligand addition/elimination reactions in different oxidation states.

Assuming that the concentrations of the species were assigned as follows,

$$c_1 = [M], \quad c_2 = [MX], \quad c_3 = [MXX],$$
$$c_4 = [R], \quad c_5 = [RX], \quad c_6 = [RXX]$$

the examination of the mechanism (simply typed in in form of the 10 reaction equations without kinetic parameters, i.e., M = MX; MX = MXX; M + e = R; ...) ends up with $N_C = 3$, $N_F = 4$, $N_S = 3$, $m = 6$, and

$$I_C = \begin{bmatrix} 1 & 4 \\ 2 & 5 \\ 3 & 6 \end{bmatrix} \quad I_S = \begin{bmatrix} 5 & 1 & 2 & 4 \\ 6 & 1 & 3 & 4 \\ 6 & 2 & 3 & 5 \end{bmatrix} \quad I_F = \begin{bmatrix} 1 & 2 \\ 2 & 3 \\ 4 & 5 \\ 5 & 6 \end{bmatrix} \tag{81}$$

The meaning of these index tables is easily understood. For instance, the *first* row of I_C expresses that in the *first* charge-transfer reaction ($j = 1$), species 1 is reduced to species 4, and the *third* row of I_S indicates that in the *third* second-order reaction ($j = 3$) species 6 and 2 react to species 3 and 5. In other words, the indices for generating the submatrices of the first charge-transfer reaction according to (76) are $p_1 = 1$, $r_1 = 4$, and those of the third second-order reaction according to (68) and (69) are $d_3 = 6$, $e_3 = 2$, $f_3 = 3$, and $g_3 = 5$.

If the examination of the reaction scheme has been completed, the program asks the user to type in the general simulation parameters and the kinetic constants associated with each reaction equation. The program now has all necessary information to build up the matrices involved in the basic equations in each time/potential step. With the aid of the index tables above, the underlying subroutines can be written in terms of a self-explanatory (pseudo) programming language. For instance, the second-order coupling matrices for any mechanism with $N_S > 0$ are obtained by executing the following subroutine:

REM $K^{*\prime}(i,j,k)$ denotes the element in the ith row and jth column of the coupling matrix connected with the kth space box;
REM all elements $K^{*\prime}(i,j,k)$ were set to zero before starting the following loops:
FOR $j = 1$ **TO** N_S
   $d = I_S(j,1)$: $e = I_S(j,2)$: $f = I_S(j,3)$: $g = I_S(j,4)$
   FOR $k = 1$ **TO** n

$$K^{*\prime}(d, d, k) = K^{*\prime}(d, d, k) - k_{sf}^{*}(j) \times c^{\prime}(e, k)$$
$$K^{*\prime}(d, f, k) = K^{*\prime}(d, f, k) + k_{sb}^{*}(j) \times c^{\prime}(g, k)$$
$$K^{*\prime}(e, e, k) = K^{*\prime}(e, e, k) - k_{sf}^{*}(j) \times c^{\prime}(d, k)$$
$$K^{*\prime}(e, f, k) = K^{*\prime}(e, f, k) + k_{sb}^{*}(j) \times c^{\prime}(g, k)$$
$$K^{*\prime}(f, d, k) = K^{*\prime}(f, d, k) + k_{sf}^{*}(j) \times c^{\prime}(e, k)$$
$$K^{*\prime}(f, f, k) = K^{*\prime}(f, f, k) - k_{sb}^{*}(j) \times c^{\prime}(g, k)$$
$$K^{*\prime}(g, d, k) = K^{*\prime}(g, d, k) + k_{sf}^{*}(j) \times c^{\prime}(e, k)$$
$$K^{*\prime}(g, g, k) = K^{*\prime}(g, g, k) - k_{sb}^{*}(j) \times c^{\prime}(f, k)$$

   NEXT k
NEXT j

The first-order coupling matrix and the S-matrices are constructed analogously. In reality, the coupling matrices themselves are not generated, but the corresponding elements are directly added to the matrices $J_k^{\prime}$ which have been filled beforehand with the diffusional terms and the elements of $S_k^{\prime}$ according to (53) and (61).

As mentioned above, only single-step charge-transfer reactions are considered in the present chapter. Then the following simple subroutine can be applied for generating the **B**-matrices.

FOR $j = 1$ **TO** N_C
   $p = I_C(j, 1)$: $r = I_C(j, 2)$
   $B_0(p, p) = B_0(p, p) + k_{hj}(j)$: $B_0(p, r) = B_0(p, r) - k_{hb}(j)$
   $B_0(r, p) = B_0(r, p) + D/\Delta\bar{x}_1$: $B_1(r, p) = B_1(r, p) + D/\Delta\bar{x}_1$
NEXT j

According to (73), the diagonal elements of B_0 and B_1 must be set to $D/\Delta\bar{x}_1$, before executing this subroutine. Nondiagonal elements were set to zero. The basic equations are solved as described in Section II.C.1 and the current is calculated with the aid of the following subroutine resulting from a generalization of boundary condition (29).

> **REM** $I(i)$ denotes the current in the ith time/potential step
> $sum = 0$
> **FOR** $j = 1$ **TO** N_C
> $\quad p = I_C(j, 1): r = I_C(j, 2)$
> $\quad sum = sum + k_{hf}(j) \times c'(p, 0) - k_{hb}(j) \times c'(r, 0)$
> **NEXT** j
> $I(i) = zFA \times sum$

Although not all details of the realization of the general simulator have been treated in this section, we have shown how the goal can be reached in principle. Before dealing with the solution of the basic equations, one of the trickiest problems should be mentioned.

It was assumed that the kinetic parameters will be specified by the user if the program had finished examination of the reaction scheme. Since the program then knows how many reactions of each type are involved in the mechanism, it seems to be a straightforward task to write a general subroutine that asks the user to type in the forward and backward rate constants for each reaction (or the equilibrium constant/standard potential and a kinetic parameter that is more convenient with regard to the charge-transfer processes). A serious problem arises, however, in the case of overdetermined equilibria. The *ladder scheme* above is such an overdetermined system because 10 reactions can be formulated with only six species. Since an additional mass balance must also be fulfilled, it becomes clear that only five independent equilibrium constants/standard potentials are needed for a correct thermodynamic description of the mechanism. The remaining equilibrium constants can also be expressed by these five constants, but not every combination of five equilibrium constants gives a correct thermodynamic description. To allow the user to enter any valid combination, the general simulation program must prevent the input of equilibrium constants leading to a thermodynamically incorrect situation. For instance, after entering the standard potentials of both charge-transfer reactions

$$M + e \rightleftharpoons R \quad (E_1^\circ)$$

$$MX + e \rightleftharpoons RX \quad (E_2^\circ)$$

the input of the equilibrium constant of the second-order cross reaction

$$RX + M \rightleftharpoons MX + R \quad (K_{s_1})$$

must not be allowed, since K_{s_1} is already determined by

$$K_{s_1} = \exp\left[\frac{F}{RT}(E_1^\circ - E_2^\circ)\right]$$

and has to be calculated from the context. The description of the general logic that the computer follows to cope with such situations for any imaginable mechanism is beyond the scope of this chapter and will be presented elsewhere [27].

C. Computational Aspects

1. Solution of the Basic Equations of the Fast Implicit Finite-Difference Algorithm

The solution of the basic equations of the FIFD algorithm was described in detail in Ref. 21, and with respect to nonlinear equations, in Ref. 28. To avoid the reader having to resort to these papers frequently, we give here a summary of the relevant parts of both in a slightly modified form.

For solving the basic equations of the FIFD algorithm, it is of primary importance that the Gauss–Newton approach is based on a linearization strategy that produces *linear* equations in every case. Consequently, all concentrations involved in the matrices and on the right-hand side of (60) are known explicitly either from the previous iteration* or, in the first iteration, from properly guessed starting values. Hence primes will be used in the following to indicate only those quantities that are really unknown in the respective iteration so that equations of the general type

$$\mathbf{B}_0 \mathbf{C}_0' - \mathbf{B}_1 \mathbf{C}_1' = \mathbf{F}_0 \tag{82}$$

$$1 \le k \le n: \qquad -D_{1k}^* \mathbf{C}_{k-1}' + \mathbf{J}_k \mathbf{C}_k' - D_{2k}^* \mathbf{C}_{k+1}' = \mathbf{F}_k \tag{83}$$

must be solved in each time or potential step. The right-hand sides are written simply here as $\mathbf{F}_k$, thus taking into consideration that the solution algorithm should also be applicable to the Crank–Nicolson algorithm, which would produce expressions different from the Laasonen formulas used above. For the nonlinear phenomena treated in this chapter, the right-hand term $\mathbf{F}_0$ in (82) was simply zero, but it will become nonzero if the Gauss–Newton approach is applied to a nonlinear boundary condition. This is demonstrated in Section III.B.

Bearing in mind that the concentrations in the $(n + 1)$th space box are not affected by the electrochemical experiment if n has been determined by means of (12), the vector $\mathbf{C}_{n+1}'$ can be set equal to

$$\mathbf{C}_{n+1}' = \mathbf{C}^* \tag{84}$$

where $\mathbf{C}^*$ contains the bulk concentrations of all species. With the aid of

$$\mathbf{d}_n = \mathbf{J}_n \tag{85}$$

$$\mathbf{G}_n = \mathbf{F}_n + D_{2n}^* \mathbf{C}^* \tag{86}$$

---

*To avoid confusion we use the term *iteration* only in connection with the iterative solution of nonlinear equations but not in the sense of time or potential step.

the nth matrix equation in (83) can be written as

$$1 \le k \le n: \qquad -D^*_{1n}C'_{n-1} + d_nC'_n = G_n \tag{87}$$

or

$$C'_n = [d_n]^{-1}G_n + D^*_{1n}[d_n]^{-1}C'_{n-1} \tag{88}$$

The notation $[d_n]^{-1}$ indicates the inverse matrix of d_n. Substituting (88) into (83) for $k = n - 1$ yields the relation

$$-D^*_{1n-1}C'_{n-2} + (J_{n-1} - D^*_{2n-1}D^*_{1n}[d_n]^{-1})C'_{n-1}$$
$$= F_{n-1} + D^*_{2n-1}[d_n]^{-1}G_n \tag{89}$$

in which the unknown concentration vector C'_n could be removed. Successive repetition of this procedure for decreasing values of k leads to reduced matrix equations containing only two instead of three unknown concentration vectors in each equation:

$$1 \le k \le n: \qquad -D^*_{1k}C'_{k-1} + d_kC'_k = G_k \tag{90}$$

In these equations the matrix d_n and the vector G_n are given by (85) and (86) respectively, and the meaning of the other d_k and G_k can easily be deduced by comparing the corresponding terms in (89) and (90):

$$(n - 1) \ge k \ge 1: \qquad d_k = J_k - D^*_{2k}D^*_{1k+1}[d_{k+1}]^{-1} \tag{91}$$

$$(n - 1) \ge k \ge 1: \qquad G_k = F_k + D^*_{2k}[d_{k+1}]^{-1}G_{k+1} \tag{92}$$

In the next step, the reduced equation for $k = 1$ can be substituted into the boundary condition (82). The resulting equation,

$$d_0C'_0 = G_0 \tag{93}$$

with

$$d_0 = B_0 - D^*_{11}B_1[d_1]^{-1} \tag{94}$$

$$G_0 = F_0 + B_1[d_1]^{-1}G_1 \tag{95}$$

is solved by multiplying (93) by the inverse matrix of d_0. Once C'_0 has been determined, the concentrations in the other space boxes can be calculated by applying the forward recursive relation

$$1 \le k \le n: \qquad C'_k = [d_k]^{-1}(G_k + D^*_{1k}C'_{k-1}) \tag{96}$$

which requires only the previously calculated inverse matrices.

The simulation becomes especially easy if the J_k are diagonal matrices. The inverse matrices are then obtained simply by computing the reciprocal values of the diagonal elements, and only the diagonal elements are relevant to the matrix multiplications. The matrices J_k will be diagonal for mechanisms in which the species are not mutually coupled by homogeneous chemical reactions. More complicated mechanisms with kinetic couplings between different species will produce nonzero elements outside the diagonal of each matrix $\mathbf{J}_k$. Then a standard procedure such as that of Gauss–Jordan should be used for calculating the inverse matrices, but the matrices that must be inverted are only of order m, and the computation time therefore increases proportionally to $(n + 1) \times m^3$. Previously, such mechanisms have been treated by solving the whole-system coefficient matrix. In this case the computation time increases proportionally to $[(n + 1)m]^3$ and Britz [3] concluded to give up using implicit simulation techniques at this point. Now such mechanisms become tractable even on PC-level computers, since assessing that the number of space boxes varies mostly between 10 and a few hundred in a simulation, the implicit simulation methods are speeded up thoroughly by a factor of 10^2 or even 10^5 when computing the unknown concentrations as presented here.

In comparison with the FQEFD algorithm [10], that is, the most efficient *explicit* simulation technique for complex mechanisms which also has to solve $n + 1$ linear equations of order m in each iteration, the relative merits of both methods can be estimated considering the potential steps applicable in a simulation. While the FIFD method tolerates potential steps of 1 mV in any case, the FQEFD algorithm requires 0.1 mV for dimensionless rate constants of about 10^4 and 0.001 mV for values of 10^{10}. Thus both methods are comparable for mechanisms with slow or moderately fast chemical reactions, but the FIFD approach is superior if the rate constants become larger. This is a rough estimation, of course, since the computation time of the FIFD simulation is not only affected by the matrix inversions but also by the matrix multiplications; on the other hand, that is compensated in many cases where potential steps greater than 1 mV can be used in the FIFD method.

As far as second-order mechanisms are concerned, the solution of (59) and (60) requires starting values for the unknowns and a criterion to decide how many iterations have to be performed before moving the potential to the next value. Generally, a good choice is to use the concentrations that have been calculated during the previous potential step as a first approximation for the unknown concentrations. In other words, the first iteration starts with $\mathbf{C}_k'^{(0)} = \mathbf{C}_k$, and these concentrations are used in the matrices $\mathbf{S}_k'^{(0)}$ and $\mathbf{J}_k'^{(0)}$. The solution procedure results in concentration vectors $\mathbf{C}_k'^{(1)}$ which taken altogether are a better solution of the second-order problem than the starting vectors. Yet more accurate concentration vectors $\mathbf{C}_k'^{(2)}$ can be achieved by repeating the entire approach with $\mathbf{C}_k'^{(1)}$ instead of $\mathbf{C}_k'^{(0)}$, and so on. The question as to whether a further iteration should or should not be performed cannot be answered so easily. On the one hand, only a few additional iterations can smooth the concentration profile to a much greater extent than performing only one iteration with reduced time or potential steps [28]. On the other hand, as far as the current is concerned, the improvements are usually small. However, the usefulness of improving iteratively the solution vector becomes evident for the simulation of the nonlinear boundary value problem presented in Section III.B.

The iterative refinement of the concentration profile also provides a useful tool to cope with negative concentrations which are produced frequently in the first iteration, if the rate constants of second-order reactions approach the limit of diffusion control. In this situation, all negative concentrations are set to zero and an additional iteration is carried out to revert to positive values and to maintain the mass balance. While the fully implicit algorithm regains positive concentrations in almost any situation, the Crank–Nicolson scheme behaves somewhat critically, since, despite its higher accuracy, this method is more likely to produce negative concentrations and the iterative solution often does not revert to positive values. On the other hand, as shown in Section III.A.1, the accuracy of the fully implicit algorithm can be improved remarkably, so that both methods become comparable in this respect. This is why the *modified* fully implicit algorithm was chosen for developing the general simulation program.

2. Testing the General Simulator

Before concluding this section, the results of a few test simulations will be presented. The simulations were carried out with the prototype of the general simulator. This program is based on the *modified* implicit formulas that will be derived in Section III.A.1, but it generates the basic equations (59) and (60) in the same way as pointed out in Section II.B. Solely, for increasing the versatility of the program, the treatment of the charge-transfer processes was generalized so that not only single-step but also multistep charge-transfer processes can be simulated without any restrictions. The resulting matrix equations are then solved with the backward–forward substitution procedure described in Section III.C.1.

Figure 2 shows cyclic voltammograms of the ladder scheme (77) simulated with the following dimensionless quantities:

$$\psi_1 = \frac{k_{s1}^\circ}{\sqrt{\pi v F D / RT}} = 100 \tag{97}$$

$$\psi_2 = \frac{k_{s2}^\circ}{\sqrt{\pi v F D / RT}} = 100 \tag{98}$$

$$\psi_3 = \frac{k_{s3}^\circ}{\sqrt{\pi v F D / RT}} = 100 \tag{99}$$

$$K_1 = \frac{k_{f_1}}{k_{b_1}} = 10 \tag{100}$$

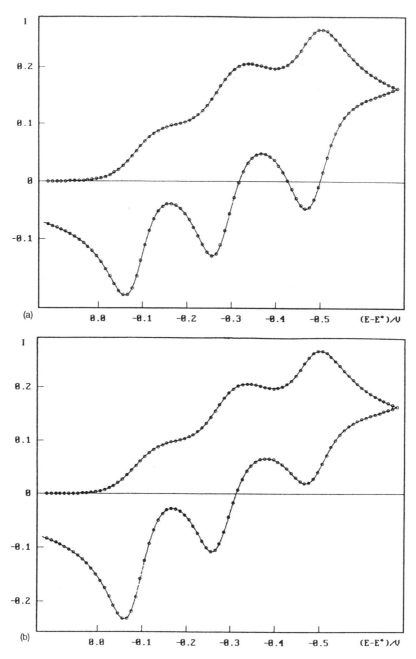

(a)

(b)

FIGURE 2 Digital simulation of the ladder scheme using the modified fully implicit algorithm with potential steps of 0.1 mV (——) and 5 mV (○ ○ ○ ○): (a) $k_{sf_1}^{\#} = k_{sf_2}^{\#} = k_{sf_3}^{\#} = 0$; (b) $k_{sf_1}^{\#} = k_{sf_2}^{\#} = k_{sf_3}^{\#} = 0.01$; (c) $k_{sf_1}^{\#} = k_{sf_2}^{\#} = k_{sf_3}^{\#} = 1$; (d) $k_{sf_1}^{\#} = k_{sf_2}^{\#} = k_{sf_3}^{\#} = 10^6$. Other simulation parameters are given by (107) to (110).

106

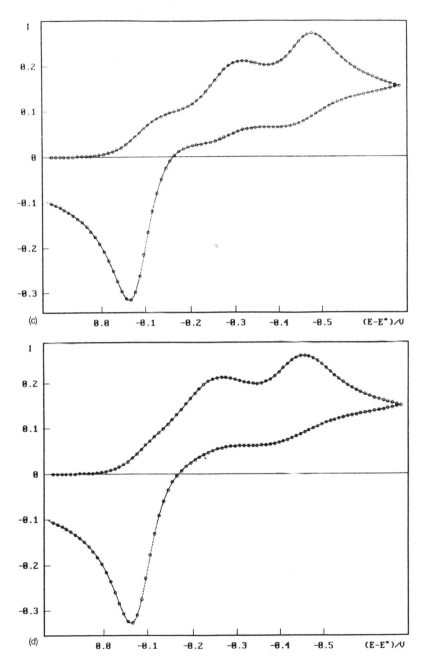

FIGURE 2 Continued

$$k_{f_1}^{\#} = \frac{k_{f_1}}{vF/RT} = 1000 \tag{101}$$

$$K_2 = \frac{k_{f_2}}{k_{b_2}} = 0.002 \tag{102}$$

$$k_{f_2}^{\#} = \frac{k_{f_2}}{vF/RT} = 10^{-6} \tag{103}$$

$$K_3 = \frac{k_{f_3}}{k_{b_3}} = 20 \tag{104}$$

$$k_{f_3}^{\#} = \frac{k_{f_3}}{vF/RT} = 100 \tag{105}$$

$$K_4 = \frac{k_{f_4}}{k_{b_4}} = 2 \times 10^{-6} \tag{106}$$

$$k_{f_4}^{\#} = \frac{k_{f_4}}{vF/RT} = 10^{-8} \tag{107}$$

$$k_{sf_1}^{\#} = \frac{k_{sf_1}c^*}{vF/RT} = 0 \cdots 10^6 \tag{108}$$

$$k_{sf_2}^{\#} = \frac{k_{sf_2}c^*}{vF/RT} = 0 \cdots 10^6 \tag{109}$$

$$k_{sf_3}^{\#} = \frac{k_{sf_3}}{vF/RT} = 0 \cdots 10^6 \tag{110}$$

The diffusion coefficients of the species were assumed to be identical and a space grid with $\beta = 0.25$ was chosen. The dimensionless model diffusion coefficient was automatically set by the program as proposed in Refs. 10 and 21.

The cyclic voltammograms in Figure 2 show the increasing effect of the homogeneous electron transfer processes on the current curve if the rate constants of the second-order cross-reactions become larger. Interestingly, even in the case of identical diffusion coefficients, the cross-reactions can play a dominant role for this mechanism. Unlike the ladder scheme, the electron coupling of diffusional pathways has no effect on the cyclic voltammetric response of multistep charge-transfer processes if the diffusion coefficients of the invovled species are equal. Figure 3 shows voltammograms simulated for the *three-step* charge-transfer mechanism

$$A + e \quad \underset{k_{hb_1}}{\overset{k_{hf_1}}{\rightleftharpoons}} \quad B \tag{111}$$

$$B + e \quad \underset{k_{hb_2}}{\overset{k_{hf_2}}{\rightleftharpoons}} \quad C \tag{112}$$

$$C + e \quad \underset{k_{hb_3}}{\overset{k_{hf_3}}{\rightleftharpoons}} \quad D \tag{113}$$

where the following cross-reactions may occur:

$$C + A \quad \underset{k_{sb_1}}{\overset{k_{sf_1}}{\rightleftharpoons}} \quad B + B \tag{114}$$

$$D + B \quad \underset{k_{sb_2}}{\overset{k_{sf_2}}{\rightleftharpoons}} \quad C + C \tag{115}$$

$$D + A \quad \underset{k_{sb_3}}{\overset{k_{sf_3}}{\rightleftharpoons}} \quad B + C \tag{116}$$

The voltammograms depicted in Figure 3 reveal that as predicted for similar mechanisms [29–31], the effect of the homogeneous electron transfer couplings was larger the more the ratio of the diffusion coefficients differed from unity. No effect of the second-order cross-reactions on the current function could be observed if diffusion coefficients of the species were identical. Finally, it should be mentioned that the multistep charge-transfer mechanisms belong to the few systems where an improperly chosen value of β not only affects the accuracy of the simulations, but may produce serious artifacts in the current curve. Since the *general simulator* is not based on code libraries of known electrochemical mechanisms, it has no information about the best choice of the simulation parameters for these mechanisms and may fail in a critical situation, for instance, if a strongly expanding space grid with $\beta = 0.5$ (the default value) was used in combination with large second-order rate constants and unequal diffusion coefficients. As shown in Figure 3b, such critical mechanisms could also be simulated with the general simulation program without needing to change the code of the program. Only a few test simulations had to be performed for finding the optimum value for the expansion factor of the space grid. The value $\beta = 0.1$ was used in the simulations of the three-step charge-transfer mechanism described above.

III. REFINEMENT OF THE GENERAL SIMULATOR

A. Improving the Accuracy of the Fully Implicit Algorithm

In combination with nonlinear data-fitting routines, the FIFD method offers some interesting features for extracting thermodynamic and kinetic data from

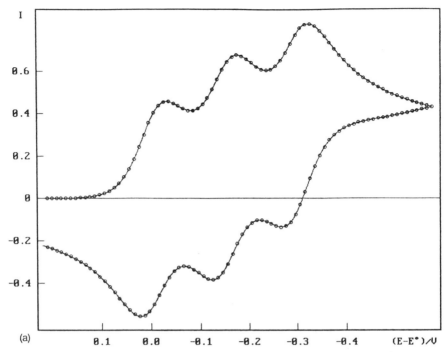

FIGURE 3 Digital simulation of the three-step charge-transfer mechanism using the modified fully implicit algorithm with potential steps of 0.1 mV (——) and 5 mV (∘ ∘ ∘ ∘): (a) $D_A = D_B = D_C = D_D = D$; $k^\#_{sf_1} = k^\#_{sf_2} = k^\#_{sf_3} = 0$ (——); $k^\#_{sf_1} = k^\#_{sf_2} = k^\#_{sf_3} = 10^6$ (∘ ∘ ∘ ∘); (b) $D_A = D$; $D_B/D = D_C/D = D_D/D = 3.0$; $k^\#_{sf_1} = k^\#_{sf_2} = k^\#_{sf_1} = 0$ (——); $k^\#_{sf_1} = k^\#_{sf_2} = k^\#_{sf_3} = 10^6$ (open circles and the corresponding solid line). The other simulation parameters are $E^\circ_1 = E^\circ_0 - 0.150$ V; $E^\circ_2 = E^\circ_0 - 0.300$ V; $\psi_1 = \psi_2 = \psi_3 = 100$.

electrochemical experiments. To our knowledge Arena and Rusling [32] pioneered the use of digital simulation in nonlinear regression analysis for fitting entire experimental cyclic voltammetric curves to a model mechanism. Unlike other approaches based on working curves [33,34], with this method, not only are a few selected points such as peak currents and/or peak potentials analyzed, but so is the entire experimental curve. This yields more reliable results, especially when complex multiparameter mechanisms are investigated. For increasing the versatility of the general simulator presented in Section II.B, the program was combined with a nonlinear data-fitting program so that the user can easily

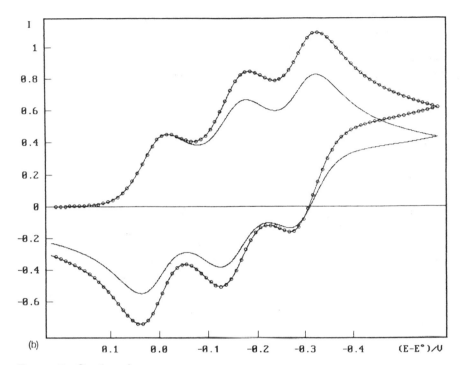

(b)

FIGURE 3 Continued

vary the mechanism until the best agreement between experimental and simulated curves has been attained.

The nonlinear regression routine was based on the Gauss–Newton procedure described in Section II.A for solving the (nonlinear) basic equations. The kinetic parameters play the role of the unknowns, but their effect on the current function cannot be expressed by a known mathematical function so that the derivatives in the Jacobian matrix must be calculated numerically [35]. This requires a large number of simulations, and the speed of the algorithm becomes an important factor, as do stability and flexibility. In a sense, a simulation algorithm with higher accuracy may speed up the data fitting, since greater potential steps can then be used without affecting the results significantly. With the aid of the FIFD approach, even complicated electrochemical mechanisms can be simulated in seconds on a PC-level computers, but the accuracy of the fully implicit formulas does not reach that of the Crank–Nicolson algorithm. It will be shown below how the accuracy of the fully implicit method can be improved without affecting its extraordinary stability.

1. Use of the General n-Point Finite-Difference Equations for
    the Approximation of Time Derivatives

One reason for the reduced accuracy of the fully implicit algorithm is obvious
from the discretization of the partial differential equation

$$\frac{\partial b(x,t)}{\partial t} = D \, \frac{\partial b(x,t)}{\partial x^2} \tag{117}$$

by means of the finite-difference expressions

$$1 \le k \le n: \qquad \frac{\Delta b_k}{\Delta t} = \frac{D}{\Delta x^2} \, (b'_{k-1} - 2b'_k + b'_{k+1}) \tag{118}$$

formulated here for an equally spaced grid ($\beta = 0$). While the right-hand side
of (118) approximates the second-order derivative at $t + \Delta t$ in the best possible
manner for a tripoint equation, the left-hand side of (118) provides a better
approximation for the time derivative at the middle point of the time interval
(i.e., for $t + 0.5\Delta t$). Hence the question arises as to which potential the current
value has to be assigned to in the ith time step.

Figure 4 shows two cyclic voltammograms simulated with potential steps
of 0.1 mV and 10 mV, respectively, where the current was assigned to the
potential connected with the endpoint of the time interval [i.e., $I(i) \rightarrow E(i\Delta t)$].
Resulting from the asymmetry mentioned above, the curves simulated with
different potential steps do not exactly coincide. Since the other assignment, $I(i) \rightarrow$
$E[(i - 0.5)\Delta t]$, does not produce better results, the best we can do is assign the
current to the average potential (i.e., $I(i) \rightarrow E[(i - 0.25)\Delta t]$). Indeed, the curves
coincide then, but it is not a satisfactory solution of this problem.

The lack of symmetry in the equations does not occur with the discretization of the Crank–Nicoloson algorithm

$$\frac{\Delta b_k}{\Delta t} = \frac{1}{2} \frac{D}{\Delta x^2} \, [(b_{k-1} - 2b_k + b_{k+1}) + (b'_{k-1} - 2b'_k + b'_{k+1})] \tag{119}$$

The effect of averaging old and new concentrations on the right-hand side of
(119) effectively produces a time shift to $t + 0.5\Delta t$ so that all derivatives now
refer to the middle point of the time interval. Due to the inherent symmetry of
the discretization, the method yields very accurate results. Why, then, did we
not use the Crank–Nicolson formulas in the general simulation program? The
answer is: The algorithm behaves critically in case of mechanisms with fast
second-order reactions, and the results are hardly predictable.

For instance, simulations of the square scheme with the Crank–Nicolson
algorithm and the parameter set published in Ref. 28 produced a large number
of negative concentrations if the rate constant of the second-order cross-reaction
approached the limit of diffusion control. Unlike the fully implicit algorithm,

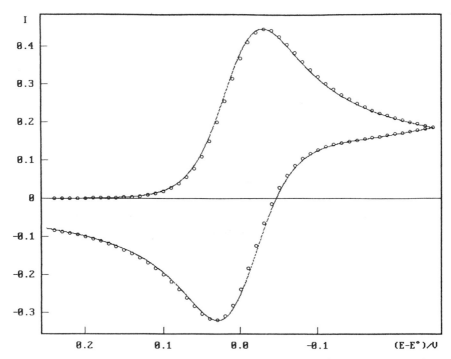

FIGURE 4 Digital simulation of the reversible charge-transfer mechanism using the unmodified fully implicit algorithm with potential steps of 0.1 mV (——) and 10 mV (○ ○ ○ ○), respectively.

the negative concentrations frequently could not be removed by applying the Gauss–Newton procedure iteratively. Thus to avoid an endless loop in such a situation, the procedure was terminated after several iterations and all negative concentrations were set to zero. Seemingly, this works well for the square scheme since the simulated voltammograms were almost identical with those described in Ref. 28. However, quite another situation was observed for the following catalytic mechanism:

$$
B + e \underset{h_{hb}}{\overset{k_{hf}}{\rightleftharpoons}} R \quad \text{(reversible charge transfer)} \tag{120}
$$

$$
R + A \underset{k_{sb}}{\overset{k_{sf}}{\rightleftharpoons}} B + C \tag{121}
$$

where the electrochemically active species is regenerated in a second-order reaction.

Figure 5 shows cyclic voltammograms simulated with the following set of dimensionless parameters:

$$K_s = \frac{k_{sf}}{k_{sb}} = 10^5 \tag{122}$$

$$k_{sf}^{\#} = \frac{k_{sf} a^* RT}{vF} = 1410 \tag{123}$$

$$\frac{a^*}{b^*} = 1 \tag{124}$$

using potential steps of 0.1 mV (solid curve) and 5 mV (open circles), respectively. Although the Crank–Nicolson algorithm produced about 630 negative concentrations when using the larger potential steps, the agreement with the 0.1-m/V curve was very good. However, increasing the rate constant of the

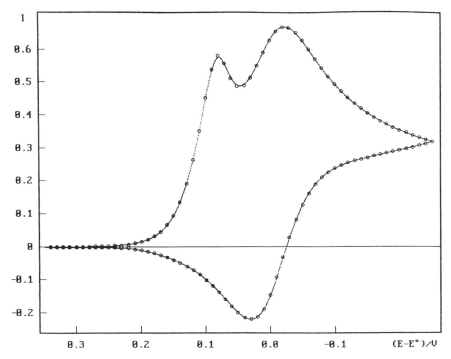

FIGURE 5 Digital simulation of the second-order catalytic mechanism using the Crank–Nicolson algorithm with potential steps of 0.1 mV (——) and 5 mV (o o o o). The simulation parameters are given by (122) to (124).

second-order reaction only by 1% caused a growing oscillation in the current function and the simulation ended with a numerical overflow.

The fully implicit method does not suffer from such instabilities, Figure 6 compares cyclic voltammetric curves obtained with potential steps of 0.1 mV and 5 mV, again. While the curve emphasized by the full circles corresponds to the set of parameters above, the open circles indicate a voltammogram simulated with a dimensionless rate constant $k_{sf}^{\#} = 10^6$ and potential steps of 5 mV. Except for the few points around the top of the sharp prepeak, both voltammograms coincide very well. Actually, the algorithm worked stably with dimensionless rate constants up to $k_{sf}^{\#} = 10^{18}$, and even then only about 15 negative concentrations were produced, which could be fixed every time by an additional iteration of the Gauss–Newton procedure. In fact, we have never observed stability prob-

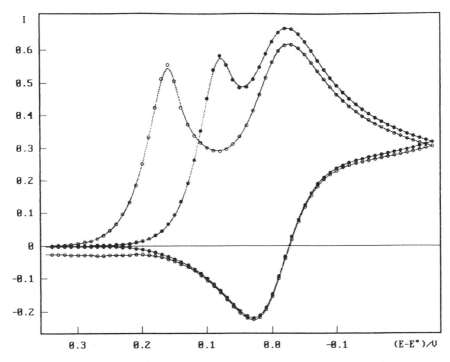

FIGURE 6 Digital simulation of the second-order catalytic mechanism using the modified fully implicit algorithm with potential steps of 0.1 mV (——) and 5 mV (○ ○ ○ ○). The current points indicated by the filled circles and the corresponding solid curve were obtained with $k_{sf}^{\#} = 1410$. The open circles and the corresponding solid curve refer to $k_{sf}^{\#} = 10^6$. Other simulation parameters are given by (122) to (124).

lems for any mechanism with rate constants of a physically sensible order of magnitude. The only problems that occurred sometimes resulted from the exponentially expanding space grid when β was chosen too large for accurately representing the concentration profiles of some second-order reactions. This problem can possibly be circumvented in future by implementing a space grid that is readjusted automatically in the course of the simulation [36].

Alternatively to the Crank–Nicolson discretization, the asymmetry in the fully implicit formulas can be overcome by using a finite difference equation that approximates the time derivative at the endpoint of the time interval (i.e., at $t + \Delta t$) in the best possible manner. In the case of the simple quasi-reversible charge-transfer mechanism, the three-point approximation [37]*

$$\left[\frac{\partial b(x,t)}{\partial t}\right]_{t+\Delta t} \approx \frac{3b(x,t + \Delta t) - 4b(x,t) + b(x,t - \Delta t)}{2\Delta t} \tag{125}$$

gave satisfactory results, but for more complicated mechanisms, four- or five-point equations should be used. The simulations depicted in Figure 6 were obtained with the five-point approximation

$$\left[\frac{\partial b(x,t)}{\partial t}\right]_{t+\Delta t} \approx$$

$$\frac{25b(x,t + \Delta t) - 48b(x,t) + 36b(x,t - \Delta t) - 16b(x,t - 2\Delta t) + 3b(x,t - 3\Delta t)}{12\Delta t} \tag{126}$$

In no case was the extraordinary stability of the fully implicit algorithm compromised, and the computation time is almost unaffected (typically less than 1%).

Only slight modifications are necessary for implementing one of the foregoing more-point formulas in an existing computer program written on the basis of a two-point equation. Consequently, all relations derived in Chapter 3 remain true if the following quantities are replaced:

1. The definition of D_{3k}^* must be changed:
   Two-point equation: $D_{3k}^* = 1 + D_{1k}^* + D_{2k}^*$
   Three-point equation: $D_{3k}^* = 1.5 + D_{1k}^* + D_{2k}^*$
   Five-point equation: $D_{3k}^* = \frac{25}{12} + D_{1k}^* + D_{2k}^*$
2. When using more-point approximations, the vector $\mathbf{C}_k$ in the basic equations (60) becomes a linear combination of old concentration vectors related to different time levels:

---

*I was informed by S. W. Feldberg (Brookhaven National Laboratories, Upton, N.Y.) that, independently, Jan Mocek (Slovakian Technical University, Bratislava) suggested the use of the Richtmeyer modification for improving the accuracy of the FIFD approach.

Two-point equation: $\mathbf{C}_k = \mathbf{C}_{k,t}$

Three-point equation: $\mathbf{C}_k = 2\mathbf{C}_{k,t} - \frac{1}{2}\mathbf{C}_{k,t-\Delta t}$

Five-point equation: $\mathbf{C}_k = 4\mathbf{C}_{k,t} - 3\mathbf{C}_{k,t-\Delta t} + \frac{4}{3}\mathbf{C}_{k,t-2\Delta t} - \frac{1}{4}\mathbf{C}_{k,t-3\Delta t}$

In the remaining part of this section we compare the accuracy of simulation techniques based on different implicit formulas. In the literature, the accuracy of a simulation technique is usually discussed on the basis of peak currents and/or peak potentials. Of course, reduction of the entire experimental curve to a few isolated points leads to an enormous loss of information. If the digital simulation is used in a data-fitting routine, the question becomes important how an inaccuracy in the current function is passed to the parameters that have to be determined. For this reason, the following tests have been carried out. Cyclic voltammetric reference curves were simulated with the Crank–Nicolson algorithm using relatively small potential steps of 0.1 mV. Since the iterative solution of the nonlinear equations did not work successfully for the Crank–Nicolson algorithm, the solution was always performed in a single iteration and only moderately fast rate constants were chosen to prevent any instability that could compromise the accuracy of this method. Then it was tested whether the parameter set of the reference curves could be reproduced by means of a nonlinear regression method using significantly larger potential steps in the fitting routine.

In the case of the quasi-reversible charge-transfer mechanism, the reference voltammogram was simulated with the following parameters:

$$D = 1.0 \times 10^{-6} \text{ cm}^2/\text{s} \tag{127}$$

$$E° = -250 \text{ mV} \tag{128}$$

$$\psi = \frac{k_s°}{\sqrt{\pi v F D / R T}} = 1.0 \tag{129}$$

using a space grid with $\beta = 0.5$.

Table 1 contains the parameters regained by subjecting the reference curve to nonlinear data-fitting routines based on different implicit formulas. The only method that yielded unsatisfactory results for this mechanism was the unmodified fully implicit algorithm (method 4). In this case the inaccuracy results mainly from the ambiguity in localizing the current points on the potential scale and could be overcome by assigning the current points, $I(i)$, to the average potential, $E[(i - 0.25)\Delta t]$ (method 3). The accuracy of the current function itself was satisfactory for this mechanism.

The use of a nonlinear fitting routine may not be very convincing for the simple quasi-reversible charge-transfer mechanism since the standard potential, the diffusion coefficient, and the heterogeneous rate constant can be more easily determined from the height and position of the current peak. Considering, how-

TABLE 1 Digital Simulation of the Quasi–Reversible Charge–Transfer Mechanism[a]

	Method 1	Method 2	Method 3	Method 4
Algorithm	Crank–Nicolson	Laasonen	Laasonen	Laasonen
Approximation	Two-point	Five-point	Two-point	Two-point
Assignment of $I(i)$	$E\ (i\ \Delta t)$	$E\ (i\ \Delta t)$	$E\ [(i - 0.25)\ \Delta t]$	$E\ (i\ \Delta t)$
$D \times 10^6$ (cm^{2}s^{-1})	0.999	1.000	1.000	0.991
$E°$ (mV)	-250.0	-250.0	-250.0	-250.25
ψ	1.004	1.006	0.989	1.307

[a]Parameters are regained by subjecting the reference curve to nonlinear data-fitting routines based on different implicit formulas. Potential steps of 10 mV were used in the simulation procedures of the fitting routines. Reference parameters are given by (127) to (129).

ever, a more complicated mechanism such as the ladder scheme (77), it becomes clear that because of a large number of possible combinations of parameters, the peak currents and peak potentials alone do not provide an unambiguous criterion for enlightening quantitatively the kinetics of such a mechanism. On the other hand, there is a real chance to determine these parameters by using a nonlinear data-fitting strategy. This will first be demonstrated for the square scheme, a limiting case of the ladder scheme (77), for which all rate constants connected with the species MXX and RXX are zero.

The reference voltammogram for this mechanism was simulated with the following set of (dimensionless) thermodynamic and kinetic parameters [28]:

$$\psi_1 = \frac{k_{s1}°}{\sqrt{\pi v F D/RT}} = 100 \tag{130}$$

$$\psi_2 = \frac{k_{s2}°}{\sqrt{\pi v F D/RT}} = 100 \tag{131}$$

$$k_{sf_1}^{\#} = \frac{k_{sf_1} c^*}{v F/RT} = 10.0 \tag{132}$$

$$K_1 = \frac{k_{f_1}}{k_{b_1}} = 20 \tag{133}$$

$$K_2 = \frac{k_{f_2}}{k_{b_2}} = 2.0 \times 10^{-4} \tag{134}$$

$$k_{f_1}^{\#} = \frac{k_{f_1}}{vF/RT} = 2.0 \tag{135}$$

$$k_{f_2}^{\#} = \frac{k_{f_2}}{vF/RT} = 2.0 \times 10^{-6} \tag{136}$$

using a space grid with $\beta = 0.25$.

Of course, a kinetic parameter cannot be determined if its effect on the current function becomes negligibly small. In the example above, the charge-transfer reactions are almost reversible. Hence these rate constants were assumed to be known and the exact values were used in the fitting routine. The fit of the remaining parameters is shown in Table 2. Only a fully implicit algorithm with a five-point approximation (method 2) was able to compete with the Crank–Nicolson algorithm (method 1). Method 3 yielded unexpectedly large errors for some of the model paramters [(132) to (136)], although the standard deviation, taken over all current points with respect to the reference curve, was only 0.125% of the cathodic peak current. Consequently, the use of highly precise simulation techniques makes no sense if the experimental data are fitted by trial and error—still a standard practice in the electrochemical literature. In contrast, even when using large potential steps of 5 to 10 mV in the simulation procedure of the fitting routine, it seems hardly possible to have experimental data of such a high quality that the modified fully implicit algorithm limits the accuracy of the results.

Besides the potential steps, the choice of the space grid and insofar as second-order reactions are involved, the computational effort in solving itera-

TABLE 2 Digital Simulation of the Square Scheme[a]

	Method 1	Method 2	Method 3
Algorithm	Crank–Nicolson	Laasonen	Laasonen
Approximation	Two-point	Five-point	Two-point
Assignment of $I(i)$	$E (i \, \Delta t)$	$E (i \, \Delta t)$	$E [(i - 0.25) \Delta t]$
K_1	20.00	20.00	19.64
$k_{f_1}^{\#}$	2.005	2.002	1.911
$K_2 \times 10^4$	2.000	2.000	1.963
$k_{f_2}^{\#}$	2.019	1.999	1.578
$k_{sf_1}^{\#}$	9.937	9.983	10.90

[a]Parameters are regained by subjecting the reference curve to nonlinear data-fitting routines based on different implicit formulas. Potential steps of 5 mV were used in the simulation procedures of the fitting routines. Reference parameters are given by (132) to (136).

tively the nonlinear equations may also affect the accuracy of a simulation. These points are discussed below for the ladder scheme (77). With the inclusion of second-order cross-reactions (78) to (80), a total of 11 parameters (100) to (110) has to be fitted for this mechanism. Because of the intricate couplings of the species in this reaction scheme, two or more parameters can produce similar effects on the current curve. For this reason, no attempt was made to fit all 11 parameters to a single voltammogram, but rather, to three different reference curves calculated with the scan rates $v = 0.2 \times v_{ref}$, $v = v_{ref}$, and $v = 5 \times v_{ref}$, respectively. In this way, the common practice of varying the scan rate in a real experiment was modeled. The simulation parameters were chosen such as to produce the dimensionless quantities (97) to (107) and $k_{sf_1}^{\#} = k_{sf_2}^{\#} = k_{sf_3}^{\#} = 0.01$ with the reference scan rate of v_{ref}. Because of the critical behavior of the Crank–Nicolson method observed for some second-order mechanisms, the reference voltammograms were computed with the modified fully implicit algorithm using a less expanding space grid with $\beta = 0.05$ and potential steps of 0.1 mV. Despite the small potential steps, five iterations were performed for solving the nonlinear equations. Table 3 contains the parameters regained by the data-fitting routines where the degree of the exponential growing of the space grid and the number of iterations was varied systematically in the simulation procedure of the fitting routine.

TABLE 3 Digital Simulation of the Ladder Scheme[a]

β: Iterations:	0.1 1	0.25 1	0.5 1	0.5 2
K_1	10.00	10.04	10.09	10.06
$k_{f_1}^{\#}$	1.000	1.000	1.000	0.999
$K_2 \times 10^3$	2.000	2.001	2.002	2.001
$k_{f_2}^{\#} \times 10^6$	1.005	1.038	1.102	1.089
K_3	19.99	19.90	19.82	19.88
$k_{f_3}^{\#} \times 10$	0.999	0.990	0.982	0.988
$K_4 \times 10^6$	2.000	1.999	1.998	1.999
$k_{f_4}^{\#} \times 10^8$	0.990	1.005	1.115	1.122
$k_{f_1}^{\#} \times 10^2$	1.003	0.997	0.962	0.962
$k_{sf_1}^{\#} \times 10^2$	1.000	0.998	1.000	1.003
$k_{sf_1}^{\#} \times 10^2$	1.000	1.000	1.001	1.002

[a]Parameters are regained by subjecting three reference curves simulated with $v = 0.2 \times v_{ref}$, $v = v_{ref}$ and $v = 5 \times v_{ref}$ to nonlinear data-fitting routines. The simulation procedure in the fitting routine operated with potential steps of 5 mV. The expansion of the space grid and the number of iterations for solving the nonlinear equations were varied. Reference parameters for $v = v_{ref}$ are given by (100) to (110).

It is obvious from these data that an inaccuracy due to an improperly chosen space grid cannot be compensated by refining the solution of the non-linear equations. The results achieved for this mechanism with $\beta = 0.5$ demonstrate that regardless of the number of iterations, almost identical parameters were regained. The deviations in the parameters become smaller the more the value of β was diminished. In principle, the reference parameters could be reproduced with $\beta = 0.1$. A smaller value, as used for calculating the reference curves, is not really necessary, but even the value of $\beta = 0.5$ produces parameter errors that can hardly be attained in a real experiment.

2. Use of General *n*-Point Finite Difference Equations for Approximating the Flux at the Electrode

Because of the satisfactory results that could be attained with the modified fully implicit algorithm, we did not expect much effort to get still more precise formulas, for instance, by using better (general *n*-point) current approximations as proposed by Britz [19]. In a recent paper [38], the same author described the implementation of higher *n*-point current approximations in a Crank–Nicolson algorithm based on the matrix technique of the FIFD approach. Britz vigorously defended the usefulness of the more-point equations—my own experience, however, was rather disappointing. Nevertheless, use of three-point flux equations in the boundary conditions is outlined briefly in the following. In accordance with the idea of the general simulator, the corresponding submatrices will be derived so that the three-point boundary conditions for any system can be immediately written in the form of a matrix equation. The extension to other *n*-point approximations makes no difficulties.

Using the general space grid with unequal space elements depicted in Figure 1, the concentration gradient at the electrode surface can be expressed by the following three-point equation:

$$\left[\frac{\partial c(x,t')}{\partial x} \right]_{x=x_0} \approx -\gamma_0 c_0' + \gamma_1 c_1' - \gamma_2 c_2' \tag{137}$$

with

$$\gamma_0 = \frac{(\bar{x}_1 - \bar{x}_0) + (\bar{x}_2 - \bar{x}_0)}{(\bar{x}_1 - \bar{x}_0)(\bar{x}_2 - \bar{x}_0)} \tag{138}$$

$$\gamma_1 = \frac{(\bar{x}_2 - \bar{x}_0)}{(\bar{x}_1 - \bar{x}_0)(\bar{x}_2 - \bar{x}_1)} \tag{139}$$

$$\gamma_2 = \frac{(\bar{x}_1 - \bar{x}_0)}{(\bar{x}_2 - \bar{x}_0)(\bar{x}_2 - \bar{x}_1)} \tag{140}$$

The formulation of each boundary condition analogous to (35) by means of the three-point equation above instead of a two-point equation results in a new matrix representation of the general boundary condition (59):

$$\mathbf{B}_0\mathbf{C}_0' + \mathbf{B}_1\mathbf{C}_1' + \mathbf{B}_2\mathbf{C}_2' = 0 \tag{141}$$

Following the rules described in Section II.B, the involved **B** matrices can again be composed of submatrices

$$\mathbf{B}_0 = D\gamma_0\mathbf{I} + \sum_{j=1}^{N_C} \mathbf{B}_{0,j} \tag{142}$$

$$\mathbf{B}_1 = D\gamma_1\mathbf{I} + \sum_{j=1}^{N_C} \mathbf{B}_{1,j} \tag{143}$$

$$\mathbf{B}_2 = D\gamma_2\mathbf{I} + \sum_{j=1}^{N_C} \mathbf{B}_{2,j} \tag{144}$$

Again disregarding mechanisms with multistep charge-transfer reactions, the definitions of these submatrices are easily deduced from (76)

$$\mathbf{B}_{0,j} = \begin{bmatrix} p_j & r_j \\ p_j\left(k_{hf_j} & -k_{hb_j}\right) \\ r_j\left(D\gamma_0 & 0\right) \end{bmatrix} \tag{145}$$

$$\mathbf{B}_{1,j} = \begin{bmatrix} p_j \\ r_j\,(D\gamma_1) \end{bmatrix} \qquad \mathbf{B}_{2,j} = \begin{bmatrix} p_j \\ r_j\,(D\gamma_2) \end{bmatrix}$$

The solution of the basic equations (60) with (141) instead of (59) is performed as described in Section II.C.1. In principle, only the quantities involved in (93) must be redefined:

$$\mathbf{d}_0 = \mathbf{B}_0 - D_{11}^* \left(\mathbf{B}_1 - D_{12}^*\mathbf{B}_2[\mathbf{d}_2]^{-1}\right)[\mathbf{d}_1]^{-1} \tag{146}$$

$$\mathbf{G}_0 = \left(\mathbf{B}_1 - D_{12}^*\mathbf{B}_2[\mathbf{d}_2]\right)^{-1}\mathbf{G}_1 - \mathbf{B}_2[\mathbf{d}_2]^{-1}\mathbf{G}_2 \tag{147}$$

In our experience, the results obtained with the three-point flux equations were worse rather than better. While simulations of the simple charge-transfer mechanism using the two-point formula produced accurate results for any combination of potential steps (smaller than 10 mV) and model diffusion coefficients (greater than about 1.0), the three-point equation gave comparable results only by enlarging the model diffusion coefficient or by reducing the potential steps significantly. Both lead to a finer space grid. Presumably the general n-point equations fail if due to the exponentially expanding space grid, not all n concentrations points lie close enough to the electrode surface. A finer space grid, however, extends the computation time, and this erased our motivation to experiment with five- or higher n-point equations.

B. Inclusion of IR Drop and Double-Layer Charging

In the nonlinear phenomena so far treated in Section II.A., the nonlinearity was part of the equations describing the system in the electrolyte phase. We now deal with a problem where the nonlinearity is part of the boundary condition. Nonlinear boundary conditions are obtained when adsorption processes are described by nonlinear adsorption isotherms or when the effect of uncompensated ohmic resistance has to be included in a digital simulation. From the mathematical point of view, the treatment of both phenomena is quite similar. Since a detailed treatment of adsorption processes is beyond the scope of this chapter, the digital simulation of cyclic voltammetric experiments with inclusion of IR drop and double-layer charging was chosen to demonstrate the approach.

In principle, the effect of uncompensated ohmic resistance can be eliminated, or at least minimized, by the use of three-electrode potentiostats with positive feedback circuits [39,40], but in many cases it is difficult to stabilize the potentiostat without producing serious artifacts such as phase shifts, additional noise, ringing, and overshoot. Total compensation is usually impractical. A uncompensated ohmic resistance between the reference and the working electrode causes a voltage drop proportional to the current flow. Hence the effect of uncompensated ohmic resistance in a cyclic voltammetric experiment is twofold. First, the voltammogram is displaced on the potential scale. Second, the IR drop causes the potential scan to differ from the ideal triangular form.

The mathematical treatment of the problem leads to the following situation. On the one hand, the potential at the point $\bar{x}_0$ (where the charge-transfer process takes place) must be known for calculating the heterogeneous rate constants by means of the Butler–Volmer equations (30) and (31). On the other hand, the current (and consequently, the heterogeneous rate constants) are required for calculating the real potential at this point.

The problem was solved by Nicolson [41] for a reversible charge-transfer mechanism through Laplace transformations and numerical solution of the nonlinear integral equations. According to our concept of the general simulator described in Section II.B, we present here a more general numerical solution of the problem by deriving the submatrices needed for generating the boundary condition for any mechanism with inclusion of IR drop and double-layer charging. As in the previous sections, mechanisms with multistep charge-transfer processes and species with different diffusion coefficients will be disregarded for simplicity. The prototype of the general simulator, however, is able to simulate such mechanisms taking the effects of IR drop and double-layer charging into account. Since the program is used in combination with a nonlinear data-fitting program, we did not expend effort to describe the model in terms of dimensionless model parameters.

In principle, the boundary condition for an electrochemical mechanism with inclusion of IR drop and double-layer charging can be written again in the form of the matrix equation (59), but to take the voltage drop caused by the uncompensated ohmic resistance into account, the heterogeneous rate constants k_{hf_j} and k_{hb_j} involved in the matrix $\mathbf{B}_0$ must be replaced by

$$k'_{hf_j} = k_{hf_j} \exp(I'R_{f_j}) \tag{148}$$

$$k'_{hb_j} = k_{hb_j} \exp(I'R_{b_j}) \tag{149}$$

with

$$R_{f_j} = \frac{-\alpha_j z F R_u}{RT} \tag{150}$$

$$R_{b_j} = \frac{(1 - \alpha_j) z F R_u}{RT} \tag{151}$$

where R_u represents the uncompensated ohmic resistance. Unlike k_{hf_j} and k_{hb_j}, which can be calculated beforehand by means of the Butler–Volmer equations (30) and (31), the corrected heterogeneous rate constants k'_{hf_j} and k'_{hb_j} are really unknowns since the unknown current I' is involved in these quantities. Nevertheless, the problem can be formulated in such a way that no other unknowns except concentrations have to be determined. If both effects, IR drop and double-layer charging, are to be included in a digital simulation, the total current I' can be expressed as

$$I' = I'_f - C_d v - C_d R_u \frac{dI'}{dt} \tag{152}$$

where C_d represents the differential capacitance of the double layer and v is the scan rate used in the experiment. The last term on the right-hand side of (152) takes into account that the IR drop affects the linearity of the triangular potential scan so that the charging current is no longer given alone by $C_d v$. The term I'_f is an abbreviation for the faradaic current. With regard to (35) and (36), there are two possibilities to express the faradaic current in terms of new concentrations. Distinctly more straightforward equations will be obtained by using (35) in a generalized form. Provided that the mechanism involves N_C charge-transfer reactions, and the same notation as in Section II.B is chosen, the faradaic current I'_f can be written as

$$I'_f = \frac{zFAD}{\Delta \bar{x}_1} \sum_{j=1}^{N_C} (c'_{p_j,1} - c'_{p_j,0}) \tag{153}$$

For the following considerations it is adequate to approximate the time derivative on the right-hand side of (152) by a two-point equation,

$$\frac{dI'}{dt} \approx \frac{I' - I}{\Delta t} \tag{154}$$

so that the discretized representation of the total current

$$I' = \left[\frac{zFAD}{\Delta \bar{x}_1} \sum_{j=1}^{N_C} (c'_{P_j,1} - c'_{P_j,0}) - C_d v + \frac{C_d R_u}{\Delta t} I \right] \Big/ \left(1 + \frac{C_d R_u}{\Delta t} \right) \qquad (155)$$

does not contain any other unknowns except the primed concentrations. Consequently, only the primed concentrations must be determined by applying the Gauss–Newton procedure (55) to each couple of equations reflecting the contribution of the *j*th charge-transfer reaction (75):

$$f_{0,p_j} = \left(\frac{D}{\Delta \bar{x}_1} + k'_{hf_j} \right) c'_{P_j,0} - \frac{D}{\Delta \bar{x}_1} c'_{P_j,1} - k'_{hb_j} c'_{r_j,0} = 0 \qquad (156)$$

$$f_{0,r_j} = \frac{D}{\Delta \bar{x}_1} \left(c'_{P_j,0} - c'_{P_j,1} \right) + \frac{D}{\Delta \bar{x}_1} (c'_{r_j,0} - c'_{r_j,1}) = 0 \qquad (157)$$

Next, we concentrate on the nonlinear equation f_{0,p_j}. According to the definition of the current function (155), the heterogeneous rate constants k'_{hf_j} and k'_{hb_j} may depend on concentrations not involved explicitly in (156). Thus not only must the derivatives with respect to $c'_{P_j,0}$ $c'_{P_j,1}$, and $c'_{r_j,0}$ be taken into account for generating the Jacobian matrix of the problem

$$\frac{\partial f_{0,p_j}}{\partial c'_{P_j,0}} = \left(\frac{D}{\Delta \bar{x}_1} + k'_{hf_j} \right) + \frac{\partial k'_{hf_j}}{\partial c'_{P_j,0}} c'_{P_j,0} - \frac{\partial k'_{hb_j}}{\partial c'_{P_j,0}} c'_{r_j,0} \qquad (158)$$

$$\frac{\partial f_{0,p_j}}{\partial c'_{P_j,1}} = - \frac{D}{\Delta \bar{x}_1} + \frac{\partial k'_{hf_j}}{\partial c'_{P_j,1}} c'_{P_j,0} - \frac{\partial k'_{hb_j}}{\partial c'_{P_j,1}} c'_{r_j,0} \qquad (159)$$

$$\frac{\partial f_{0,p_j}}{\partial c'_{r_j,0}} = -k'_{hb_j} \qquad (160)$$

but also those with respect to concentrations of other species possibly involved in the charge-transfer processes of the mechanism:

$$1 \le i \le Nc; \, p_i \ne p_j: \qquad \frac{\partial f_{0,p_j}}{\partial c'_{P_i,0}} = \frac{\partial k'_{hf_j}}{\partial c'_{P_i,0}} c'_{P_j,0} - \frac{\partial k'_{hb_j}}{\partial c'_{P_i,0}} c'_{r_j,0} \qquad (161)$$

$$1 \le i \le Nc; \, p_i \ne p_j: \qquad \frac{\partial f_{0,p_j}}{\partial c'_{P_i,1}} = \frac{\partial k'_{hf_j}}{\partial c'_{P_i,1}} c'_{P_j,0} - \frac{\partial k'_{hb_j}}{\partial c'_{P_i,1}} c'_{r_j,0} \qquad (162)$$

Because of the symmetry of the current function (155) with respect to $c'_{P_j,1}$ and $c'_{P_j,0}$ for all species P_j undergoing a charge-transfer reaction, the corresponding derivatives differ only in the sign

$1 \leq j \leq N_C; \ 1 \leq i \leq N_C:$

$$\frac{\partial k'_{hf_j}}{\partial c'_{p_i,1}} = \frac{\partial k'_{hf_j}}{\partial I'} \frac{\partial I'}{\partial c'_{p_i,1}} = \frac{zFADR_{f_j} k'_{hf_j}}{(1 + C_d R_u/\Delta t)\, \Delta \bar{x}_1} = -\frac{\partial k'_{hf_j}}{\partial c'_{p_i,0}} \tag{163}$$

$1 \leq j \leq N_C; \ 1 \leq i \leq N_C:$

$$\frac{\partial k'_{hb_j}}{\partial c'_{p_i,1}} = \frac{\partial k'_{hb_j}}{\partial I'} \frac{\partial I'}{\partial c'_{p_i,1}} = \frac{zFADR_{b_j} k'_{hb_j}}{(1 + C_d R_u/\Delta t)\, \Delta \bar{x}_1} = -\frac{\partial k'_{hb_j}}{\partial c'_{p_i,0}} \tag{164}$$

Consequently, after defining the submatrices

$$\mathbf{B}'_{0,j} = \begin{bmatrix} & p_j & r_j \\ p_j & \left(k'_{hf_j} & -k'_{hb_j} \right. \\ r_j & \left. D/\Delta \bar{x}_1 & 0 \right) \end{bmatrix} \qquad \mathbf{B}_{1,j} = \begin{bmatrix} p_j \\ r_j (D/\Delta \bar{x}_1) \end{bmatrix} \tag{165}$$

and

$$\mathbf{S}'_{0,j} = \frac{zFAD(R_{b_j}\, k'_{hb_j}\, c'_{r_j,0} - R_{f_j}\, k'_{hf_j}\, c'_{pj,0})}{(1 + C_d R_u/\Delta t)\, \Delta \bar{x}_1}$$

$$\begin{bmatrix} & p_1 & p_2 & p_3 & \cdots & p_{N_C} \\ p_j & (1 & 1 & 1 & \cdots & 1 \;) \end{bmatrix} \tag{166}$$

the result of applying the Gauss–Newton procedure to the (nonlinear) whole-system boundary condition can be conveniently written in form of the (linear) matrix equation

$$\mathbf{B}'^{(i)}_0\, \mathbf{C}'^{(i+1)}_0 - \mathbf{B}'^{(i)}_1\, \mathbf{C}'^{(i+1)}_1 = \mathbf{S}'^{(i)}_0\, (\mathbf{C}'^{(i)}_0 - \mathbf{C}'^{(i)}_1) \tag{167}$$

with

$$\mathbf{S}'_0 = \sum_{j=1}^{j=N_C} \mathbf{S}'_{0,j} \tag{168}$$

$$\mathbf{B}'_0 = \sum_{j=1}^{j=N_C} \mathbf{B}'_{0,j} + \mathbf{S}'_{0,j} \tag{169}$$

$$\mathbf{B}'_1 = \sum_{j=1}^{j=N_C} \mathbf{B}_{1,j} + \mathbf{S}'_{0,j} \tag{170}$$

Instead of (59), equation (167) must now be solved iteratively together with (60). Because of its principal conformity with (82), the backward–forward substituion algorithm presented in Section II.C.1 can be applied unchanged, but a higher computational effort is necessary to cope with the nonlinearity in the boundary condition.

The effect of the increasing refinement in solving the nonlinear equations on the current function will be demonstrated for the square scheme using the same criteria as in Section III.A.1. The reference voltammogram (solid curve in Fig. 7) was simulated with the parameter set (1.30) to (136) and the following dimensionless values for the uncompensated ohmic resistance and the double-layer capacity:

$$R_u^* = R_u F A c^* \left(\frac{F}{RT}\right)^{3/2} \sqrt{\pi v D} = 0.075 \tag{171}$$

$$C_d^* = \frac{C_d}{F A c^*} \sqrt{\frac{RTv}{\pi FD}} = 0.075 \tag{172}$$

Instead of (154), a five-point approximation was used in the computer program. The simulation was performed with potential steps of 0.1 mV and the nonlinear

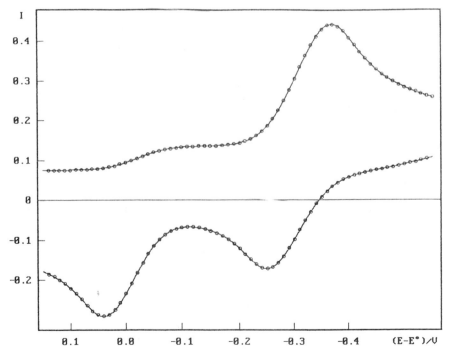

FIGURE 7 Digital simulation of the square scheme with inclusion of IR drop and double-layer charging using potential steps of 0.1 mV (——) and 10 mV (○ ○ ○ ○). $R_u^* = C_d^* = 0.075$; other simulation parameters are given by (132) to (136).

TABLE 4   Digital Simulation of the Square Scheme with Inclusion of IR Drop and Double-Layer Charging[a]

Potential steps:	5 mV	2.5 mV	1 mV	0.2 mV	10 mV	5 mV	10 mV
Iterations:	1	1	1	1	2	2	3
K_1	19.46	19.69	19.86	19.97	19.99	20.00	19.99
$k_{f_1}^{\#}$	1.869	1.924	1.967	1.993	2.000	2.002	2.000
$K_2 \times 10^4$	1.923	1.956	1.981	1.996	2.000	2.000	2.000
$k_{f_2}^{\#} \times 10^6$	2.478	2.264	2.112	2.023	1.994	1.996	1.993
$k_{sf_1}^{\#}$	9.229	9.589	9.827	9.965	10.01	10.00	10.01

[a]Parameters are regained by subjecting the reference curve to nonlinear data-fitting routines. The potential steps and the number of iterations for solving the nonlinear equations were varied in the simulation procedures of the fitting routine. $R_u^{\star} = C_d^{\star} = 0.075$; other reference parameters are given by (132) to (136).

boundary condition was solved with five iterations every time. The modified fully implicit algorithm was used in all simulations.

The reference curve was then subjected to a data-fitting routine. While the exact values of $R_u^{\star}$, $C_d^{\star}$, ψ_1, and ψ_2 were used in the simulation procedure of the fitting routine, the potential steps and the number of iterations were varied systematically. The regained parameter sets are summarized in Table 4. Since doubling the number of iterations for solving the nonlinear boundary condition has only a small effect on the computation time, it is obvious from these data that an additional iteration in the Gauss–Newton procedure yields much better results than reducing the potential steps. While the parameter set of the reference curve could be reproduced even with potential steps of 10 mV when the nonlinear equations were solved in two iterations, potential steps of about 0.2 mV were necessary to achieve a comparable accuracy with a single-iteration solution.

REFERENCES

1. S. W. Feldberg and C. Auerbach, *Anal. Chem.*, *36*: 505 (1964).
2. S. W. Feldberg, Digital simulation: a general method for solving electrochemical diffusion-kinetic problems, in *Electroanalytical Chemistry*, Vol. 3 (A. J. Bard, ed.), Marcel Dekker, New York, 1969, p. 199.
3. D. Britz, *Digital Simulation in Electrochemistry*, Springer-Verlag, Berlin, 1988.
4. L. F. Whiting and P. W. Carr, *J. Electroanal. Chem.*, *81*: 1 (1977).
5. T. Joslin and D. Pletcher, *J. Electroanal. Chem.*, *49*: 171 (1974).
6. S. W. Feldberg, *J. Electroanal. Chem.*, *127*: 1 (1981).
7. J. V. Arena and J. F. Rusling, *Anal. Chem.*, *58*: 1481 (1986).
8. F. G. Cottrell, *Z. Phys. Chem.*, *42*: 385 (1902).

9. I. Ruzić and S. W. Feldberg, *J. Electroanal. Chem.*, *50*: 153 (1974).
10. S. W. Feldberg, *J. Electroanal. Chem.*, *290*: 49 (1990).
11. K. Wiesner, *Chem. Listy*, *41*: 6 (1947).
12. P. Laasonen, *Acta Math.*, *81*: 309 (1949).
13. N. Winograd, *J. Electroanal. Chem.*, *43*: 1 (1973).
14. J. L. Anderson and S. Moldoveanu, *J. Electroanal. Chem.*, *179*: 107 (1984).
15. T. R. Brumleve and R. P. Buck, *J. Electroanal. Chem.*, *90*: 1 (1978).
16. J. Crank and P. Nicolson, *Proc. Cambridge Phil. Soc.*, *43*: 50 (1947).
17. J. Heinze, M. Störzbach, and J. Mortensen, *J. Electroanal. Chem.*, *165*: 61 (1984).
18. D. Britz, J. Heinze, J. Mortensen, and M. Störzbach, *J. Electroanal. Chem.*, *240*: 27 (1988).
19. D. Britz, *Anal. Chim. Acta*, *193*: 277 (1987).
20. D. Britz and K. N. Thomsen, *Anal. Chim. Acta*, *194*: 317 (1987).
21. M. Rudolph, *J. Electroanal. Chem.*, *314*: 13 (1991).
22. B. Speiser, *Comput. Chem.*, *14*: 127 (1990).
23. D. H. Evans, *Chem. Rev.*, *90*: 739 (1990).
24. K. M. Kadish, *Prog. Inorg. Chem.*, *34*: 435 (1981).
25. D. Lexa and J. M. Saveant, *J. Am. Chem. Soc.*, *98*: 2652 (1976).
26. E. Eichlorn, A. Rieker, and B. Speiser, *Angew. Chem.*, *104*: 1246 (1992).
27. W. Luo, S. W. Feldberg and M. Rudolph, *J. Electroanal. Chem.*, *368*: 109 (1994).
28. M. Rudolph, *J. Electroanal. Chem.*, *338*: 85 (1992).
29. C. P. Andrieux, P. Hapiot, and J. M. Saveant, *J. Electroanal. Chem.*, *172*: 49 (1984).
30. C. P. Andrieux, P. Hapiot, and J. M. Saveant, *J. Electroanal. Chem.*, *186*: 237 (1985).
31. K. Hinkelmann and J. Heinze, *Ber. Bunsenges. Phys. Chem.*, *91*: 243 (1987).
32. J. V. Arena and J. F. Rusling, *Anal. Chem.*, *58*: 1481 (1986).
33. B. Speiser, *Software-Development in Chemistry*, Vol. 4 (J. Gasteiger, ed.), Springer-Verlag, Berlin, 1990.
34. B. Speiser, *Anal. Chem.*, *57*: 1390 (1985).
35. K. D. Herdt, *Intelligent Instrum. Comput.*, *5*: 78 (1987).
36. L. K. Bieniasz, *J. Electroanal. Chem.*, *360*: 119 (1993).
37. R. D. Richtmeyer, *Difference Methods for Initial Value Problems*, Wiley-Interscience, New York, 1957.
38. D. Britz, *J. Electroanal. Chem.*, *352*: 17 (1993).
39. D. E. Smith, *Crit. Rev. Anal. Chem.*, *2*: 247 (1971).
40. D. Britz, *J. Electroanal. Chem.*, *88*: 309 (1978).
41. R. S. Nicolson, *Anal Chem.*, *37*: 1351 (1965).

4

Electrochemistry at Ultramicroelectrodes

CHRISTIAN AMATORE
Ecole Normale Supérieure, URA CNRS 1679
Paris, France

I. INTRODUCTION

What are *ultramicroelectrodes*, and do they actually differ from electrodes and *microelectrodes*? Etymology [*ulter-mikros-electron-hodos*, from: *ulter* (Latin): situated beyond; *mikros* (Greek): small; *elektron* (Greek): amber (i.e., the charged particle obtained by rubbing amber); *hodos* (Greek): way, path, channel] defines them as "electron channels" (i.e., electrodes) that are smaller than small electrodes (i.e., *microelectrodes*) and therefore much smaller than regular electrodes. Most electrochemists will agree on this definition. However, the notion of an usual electrode differs according to the particular origin of electrochemists. For electroanalytical chemists, a regular electrode is of millimetric dimensions; thus the term *microelectrodes* is appropriate to define electrodes of micrometric dimensions (or below). Molecular electrochemists, used to dealing with electrolysis, think that a normal electrode is more like those used in laboratory electrolysis cells; millimetric electrodes were thus termed *microelectrodes* decades ago, which therefore leads to *ultramicroelectrodes* for electrodes of even smaller size; hereafter, we use this definition.

Such definitions, based on historical–anthropomorphic criteria, may appear useless since they better define the origin of their users than the object itself, which is always an electrode of micrometric dimensions (or below). However, electrochemists may disagree and argue on the proper word to use, but they all agree on the dimensional limits that classify electrodes into various classes according to their particular functions and properties.

Electrodes used in laboratories for electrolysis have centimetric or decimetric dimensions because they are designed to convert a maximum of material in a minimum amount of time. However, their time constants (RC) and associated ohmic drop (iR) are too large to make them practicable for voltammetric or electroanalytical purposes [1,2]. In an electrolysis cell the current crosses over a solution volume of roughly cylindrical geometry that is delimited by two electrodes of identical surface area A, facing each other and separated by a distance l. The cell resistance (R) is then proportional l/A, whereas the cell capacitance (C) and current (i) are proportional to A, RC and iR are then proportional to l and are independent of the electrode dimensions. Consider now that the dimension of one of the two electrodes (the working electrode) is decreased so as to become a disk of radius r_0 with $\pi r_0^2 \ll A$. The current is forced to flow through the conical volume delimited by the two electrodes. The cell resistance is then proportional to the reciprocal of r_0 and independent of l, but the current and cell capacitance remain proportional to the working electrode surface area (i.e., to r_0^2). Thus RC and iR are now proportional to r_0 and decrease when r_0 is decreased (Fig. 1). This is true provided that the current remains proportional to the working electrode surface area. However, when r_0 becomes small enough, edge effects and nonplanar diffusion (see below) control the transport of molecules to the electrode surface, with the result that the current becomes proportional to the electrode dimension (i.e., to r_0) rather than to its surface area (see below) [3]. Then iR is again independent of the electrode dimension (Fig. 1) [4,5]. The capacitance of the electrode that originates in the double layer (i.e., the interfacial region that extends only over a few tens of angstroms from the electrode surface) [2] is not affected by such edge effects provided that the electrode size remains much greater than the thickness of the double layer. Thus the capacitance of the electrode still varies like the electrode surface area and RC remains proportional to r_0 (Fig. 1).

Three domains (i.e., three classes of electrodes) may then be defined based on the relationship between the electrode radius and ohmic drop (but only two classes based on the relationship between the electrode radius and its time constant). For classical electrochemical conditions [1] the limits of these domains are located around a few millimeters (electrodes–microelectrodes) and around 10 μm (microelectrodes–ultramicroelectrodes).

For analytical and voltammetric applications, millimetric electrodes (microelectrodes) were generally preferred up to the beginning of the 1980s, that

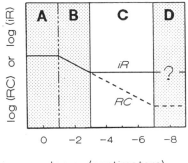

FIGURE 1 Schematic representation of ohmic drop (*iR*; solid line) or time constant (*RC*; dashed line) of an electrochemical half-cell as a function of the radius of the working electrode. See the text for definitions of the various zones indicated (A, electrodes; B, microelectrodes; C, ultramicroelectrodes; D, [?]-electrodes). Note that the ordinate is represented using an arbitrary scale.

is, before conducting materials of a much smaller dimension became easily available. Indeed, they consisted in the optimal compromise between reduced ohmic drop and a small time constant on the one hand, and difficulty of fabrication on the other hand. However, ultramicroelectrodes have been manufactured and used since the 1940s, for example to measure oxygen inside biological tissues [6], or were used as part of commercial devices such as Clark electrodes. The increasing availability of carbon or metal fibers of micrometric (and under) dimensions and of microlithographic techniques has allowed ultramicroelectrodes to become more widely used. In this respect it is also worthwhile to mention that a few commercial sources of ultramicroelectrodes already exist today.

Owing to these increasing facilities for manufacturing smaller and smaller electrodes, several groups have already begun to explore the properties of submicrometric (or even less) electrodes [7–18]. Will these electrodes require a new class to be defined? If we retain our definition of a class based on identity of physicochemical behavior (rather than on degree of miniaturization), ultramicroelectrodes will remain *ultramicroelectrodes* up to the point where one of their dimensions will reach a molecular level (i.e., around ca. 100 to 10 Å). Below this point several physical features should modify their physicochemical characteristics. Because the thicknesses of double layers [2] will be comparable or larger than the electrode dimension, the capacitance of these electrodes will vary as their radius rather than as their surface area. Furthermore, at these electrodes the dimensions of diffusion layers will be close to those of double layers and may even become smaller. New theories [19,20] are then required to describe coupling between diffusion and migration within double layers. Moreover, the structural reorganization of double layers (i.e., a few picoseconds [21–25]) may then not be considerably faster than diffusional times. Also, the size of these electrodes will be comparable to those of large electroactive molecules [26], thus breaking the postulate of infinite accessibility of point-sized molecules to the electrode surface that is used implicitly in classical electrochemical theories. It is noteworthy that a term (*nanodes* [27]) has already been coined [etymologically: *nanos hados* (Greek); dwarf path or dwarf channel] for these electrodes, although most of the singularities above have not been explicitly recognized by its author.

From what precedes, the class defined by *ultramicroelectrodes* should encompass electrodes that have at least one dimension, ranging from a few tens of micrometer to 100 Å or so (Fig. 1). Within this class the physicochemical properties of electrodes should be identical. However, the upper limit is given here for the usual electrochemical conditions and may vary under different conditions. For example, an electrode of 1 μm radius placed in a usual solvent will behave as an ultramicroelectrode, whereas it may behave as a microelectrode when placed in a viscous medium (e.g., a polymer or a gel [28]), where diffusion coefficients are much smaller than 10^{-6} to 10^{-5} cm^2 s^{-1}, as will be made clear in the following.

Several excellent reviews [3,29] and a recent and well-documented collective book [30], including contributions from most of the leading experts in the area, exist on the topics of ultramicroelectrodes. It is therefore not our purpose to duplicate or summarize here these works to which the reader is highly encouraged to refer directly. We simply want to take the opportunity here to present and discuss some of the conceptual and basic aspects that make electrochemistry at ultramicroelectrodes a particular field in electrochemistry.

II. DIFFUSION AT ULTRAMICROELECTRODES

Electrochemistry is by nature an interfacial technique, since materials dissolved in a three-dimensional volume (the cell) are reacting at a two-dimensional interfacial surface (the electrode). To proceed, electrolysis requires that new material be brought to the electrode surface continuously and products carried away. In a liquid, the three elementary processes by which molecules are transported are diffusion, migration, and convection [2]. Diffusion is a consequence of the trend to restore uniform concentration. Migration results from the electrical force that is applied to a charged particle placed in any electrical field. Convection originates from movements of fluid packets of micrometric size that carry the molecules contained within [31]. In electroanalytical or voltammetric techniques, one generally suppresses migrational transport by using a large excess of dissociated inert salt [1], the supporting electrolyte (see below).

Under most circumstances the electrode is embedded in an insulating wall that affects the fluid velocity in its near vicinity, with the result that convective transport necessarily vanishes close to the electrode surface [32]. Therefore, diffusion always governs the final approach of an electroactive molecule toward the electrode surface, and a diffusional flux is created. This flux decays when the distance from the electrode surface increases; it eventually reaches negligible values when the concentration of the considered species approaches its bulk concentration. The converse happens for laminar convective fluxes [32] that increase quadratically with the distance from the wall where the electrode is embedded. Based on this simple description it should easily be understood that convection always prevails at large distances from the electrode surface. Within an intermediate range of distances from the electrode surface, the convective and diffusional fluxes have comparable magnitudes, as schematized in Figure 2 for chronoamperometry at a planar electrode.

It is seen from this figure that the diffusional flux increases and applies over a shorter distance from the electrode surface when the duration of the experiment is decreased. If the experiment duration is short enough, convection becomes negligible in the region where concentrations differ significantly from their bulk values (Fig. 2); thus transport occurs mainly by diffusion. Conversely, for sufficient duration times, convection is the major transport mode, except very near the electrode surface. The very same effect may also be achieved upon raising the convective flux by imposing and controlling a large fluid velocity. This is the basis of hydrodynamic electrochemical methods that involve forced

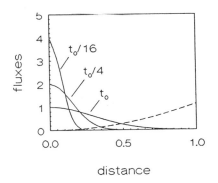

Figure 2 Schematic representation of diffusional fluxes (solid lines) and convective fluxes at an electrode. Scales are arbitrary, but diffusional fluxes at t_0, $t_0/4$, and $t_0/16$ are represented on an identical scale.

convection [1,2]. Transport of molecules to and from the electrode surface is then independent of time, and steady-state voltammograms are obtained.

When convection arises only from thermal motions of the fluid within the cell and from the usual mechanical vibrations, convective fluxes are small enough for diffusion to be the major mode of transport up to durations of a few seconds, that is, up to distances from the electrode surface that range from a few tens of micrometers to less than a few hundred micrometers. This *convection-free domain* may be even larger when the fluid viscosity is much larger than that of the usual electrochemical solvents. We have restricted this presentation to conditions where diffusion prevails over convection.

At electrodes of millimetric dimensions, the thickness δ_{conv} of the convection-free domain is necessarily negligible vis-à-vis the dimensions of the electrode. Therefore, only planar diffusion can be observed at these electrodes, and electrochemical data are independent of the electrode shape, currents being only a function of the electrode surface area. Conversely, if at least one dimension of the electrode is made smaller than δ_{conv}, diffusion may extend without significant interference of convection over distances that exceed this particular dimension of the electrode. Then nonplanar diffusion may be observed at these electrodes provided that the experiment duration is sufficient (see below). Under these conditions, electrochemical data become affected by the electrode shape because this shape affects diffusional fluxes (see below). This is the basic origin of the limit between domains B and C in Figure 1, that is, with our definition the limit between microelectrodes, whose dimensions are all larger than δ_{conv}, and ultramicroelectrodes, which have at least one dimension smaller than δ_{conv}. Note that since in structured solutions such as gels or polymers δ_{conv} is virtually infinite, any microelectrode may behave as an ultramicroelectrode provided that the duration of the experiment is sufficient. Also, in media of the usual viscosity, but at times short enough for diffusion to occur only over distances that are much less than the electrode dimension, any ultramicroelectrode behaves as a microelectrode. Both examples stress once more the futility of trying to propose an absolute definition of ultramicroelectrodes based on the objects themselves. The same electrode may be a microelectrode or an ultramicroelectrode, depending on the hydrodynamics or on the viscosity of the medium in which it is used, or upon changing the time scale of an experiment performed within the same medium.

A. Diffusion at Electrodes of Spherical and Cylindrical Geometry

1. Spherical Electrodes

Diffusion at the spherical electrode is the paragon of diffusion at ultramicroelectrodes since it exemplifies most of the characteristics that are observed at electrodes of other geometries. Moreover, because of the seminal role of the dropping mercury electrode in the development of modern molecular or analytical electrochemistry, spherical diffusion at electrodes has been solved in the context of polarography since the end of the 1930s [33].

Under pure diffusional conditions, the concentration $C(r,t)$ of a species whose diffusion coefficient is D, present at a concentration C^0 in the bulk and whose concentration at $t \geq 0$ is made virtually zero at the electrode surface ($r = r_0$) by application of a potential step of sufficient magnitude, varies as a function of time t and distance r from the electrode surface according to [2].

$$\frac{C(r,t)}{C^0} = 1 - \frac{r_0}{r} \, \text{erfc}\left[\frac{r - r_0}{(4Dt)^{1/2}}\right] \tag{1}$$

which derives from integration of Fick's second law [2],

$$\frac{\partial C(r,t)}{\partial t} = D \, \nabla^2 C(r,t) = D \left[\frac{\partial^2 C(r,t)}{\partial r^2} + \frac{2}{r}\frac{\partial C(r,t)}{\partial r}\right] \tag{2}$$

taking into account the following boundary conditions:

$$t < 0, \, r \geq r_0: \qquad C(r,t) = C^0 \tag{3}$$

$$t \geq 0, \, r = r_0: \qquad C(r_0,t) = 0 \tag{4}$$

$$t \geq 0, \, r \to \infty: \qquad C(r,t) \to C^0 \tag{5}$$

Figure 3a gives the variations of $C(r,t)$ as a function of the distance from the electrode surface at different times. It is seen from these variations that when the time increases, the concentration profile tends asymptotically toward the limiting concentration profile indicated by a dashed line. This limit is given by

$$\frac{C(r,\infty)}{C^0} = 1 - \frac{r_0}{r} \tag{6}$$

as is deduced from equation (1) since $\text{erfc}[(r - r_0)/(4Dt)^{1/2}] \to 1$ when $(r - r_0)/(Dt)^{1/2} \to 0$. The current intensity is proportional to the gradient of $C(r,t)$ at the electrode surface (n is the number of electrons exchanged and A is the electrode surface area); then [2]

$$i = nFAD \left[\frac{\partial C(r,t)}{\partial r}\right]_{r=r_0} = \frac{nFADC^0}{r_0}\left[1 + \frac{r_0}{(\pi Dt)^{1/2}}\right] \tag{7}$$

It tends toward a constant value when $r_0/(Dt)^{1/2} \to 0$ (Fig. 4):

$$i \to i_{\text{st.state}} = \frac{nFADC^0}{r_0} \tag{8}$$

It is noteworthy that the steady-state concentration profile in equation (6) extends over distances that are several times larger than the electrode radius.

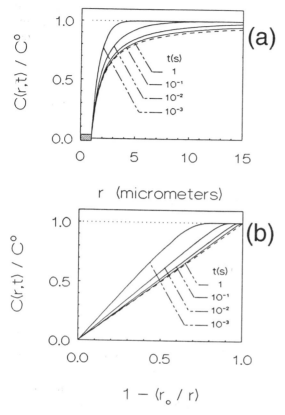

FIGURE 3 Diffusion at an electrode of spherical geometry in the true space (a) or in a transformed space (b), for different times indicated ($r_0 = 1$ μm, $D = 10^{-5}$ cm^2 s^{-1}). The dashed curve in (a) or (b) is the limit at infinite time.

For example, at $r = 20r_0$ the concentration is only 95% of its bulk value. From equation (1) it is seen that attainment of a steady-state concentration profile at these large distances ($r \gg r_0$) requires durations of experiments which are large enough for $(r - r_0)/(Dt)^{1/2} \to 0$. Conversely, much smaller duration times are sufficient for achieving a steady-state current, since one requires only that $r_0/(Dt)^{1/2} \to 0$ [equation (7)]. Completion of a steady-state concentration profile over the entire range where $C(r,t)$ differs significantly from its bulk value thus requires much larger times than achievement of a steady-state current. This is illustrated in Figure 3a, where it is seen that the concentration profile is close enough to its limit at infinite times near the electrode surface, whereas it differs significantly from this limit at larger distances.

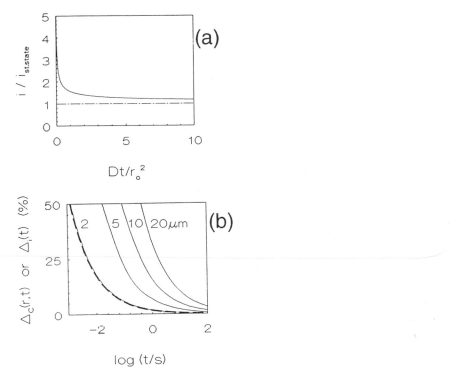

FIGURE 4 Diffusion at an electrode of spherical geometry. Deviations of the current (a) or of concentrations (b) from their steady-state values as a function of time. See the text for the definitions of $\Delta_c(r,t)$ (solid curves) or of $\Delta_i(t)$ (heavy dashed curve) in (b). Curves in (b) are given for $r_0 = 1$ μm, and $D = 10^{-5}$ cm^2 s^{-1}.

This effect is made even more visible when the concentration profile is plotted as a function of $(1 - r_0/r)$, as shown in Figure 3b [34], or when $\Delta_c(r,t) = [C(r,t) - C(r,\infty)]/[C^0 - C(r,\infty)]$ (i.e., the relative deviation from the limiting concentration profile at distance r) is compared to $\Delta_i(t) = [i(t) - i_{st.state}]/i_{st.state}$ (i.e., the relative deviation of the current from its steady-state limit) [35]. Both relative deviations are plotted as a function of time in Figure 4b, showing that, for example, for $r_0 = 1$ μm and $D = 10^{-5}$ cm^2 s^{-1}, durations in excess of 100 s are required to attain a steady-state concentration profile over the entire range where $C(r,t)$ is less than 95% of its bulk value. For $r_0 = 5$ μm, a size more often used, times of about 1 h and distances exceeding 100 μm are required. In common solvents under normal electrochemical conditions, development of a diffusion layer over such distances is obviously impossible because convective transport arising from thermal motions of the fluid, vibrations and so on, dominates transport at such distances (compare Fig. 2). However, these long-distance perturbations are almost nonreflected in current measurements, because the current is essentially governed by diffusion in the close vicinity of the electrode. This is akin to what is shown in Figure 3a and 4b, where it is seen that the concentration gradient at the electrode surface is almost independent of the concentration variations that still occur far away from the electrode surface.*

It is thus concluded that although steady-state currents may be achieved at spherical ultramicroelectrodes of usual sizes, steady-state concentration profiles cannot be attained without at least a partial involvement of convection. However, since the current reflects primarily concentration profiles over distances from the electrode surface that are comparable to the electrode radius, this effect is scarcely observable in classical electrochemical experiments (see below for ECL experiments).

2.   Cylindrical Electrodes

A cylindrical electrode is the paragon of electrodes in which one dimension (the radius r_0) is considerably smaller than the other (the length l which is of millimetric size. Under these conditions, edge effects due to both ends of the cylinder are negligible, and diffusion is identical in any cross-section plane perpendicular to the cylinder axis. In such a plane, diffusion depends only on the distance r from the cylinder axis and on the duration t of the experiment. Concentration profiles are given by integration of Fick's second law:

$$\frac{\partial C(r,t)}{\partial t} = D \; \nabla^2 C(r,t) = D \left[\frac{\partial^2 C(r,t)}{\partial r^2} + \frac{1}{r} \frac{\partial C(r,t)}{\partial r} \right] \tag{9}$$

taking into account the pertinent boundary conditions, which are identical to those in equations (3) to (5) for a chronoamperometric experiment. Under these conditions the problem is identical to that solved for the first time in relation with heat conduction [36], and the current is then given by

$$i = nFAD \left[\frac{\partial C(r,t)}{\partial r} \right]_{r=r_0} = \frac{nFADC^0}{r_0} \; \mathscr{C} \left(\frac{Dt}{r_0^2} \right) \tag{10}$$

---

*After this chapter was written, Royce Engstrom and his group reported (J. Vitt, C. Dick, and R. C. Engstrom, *184th Electrochemical Society Meeting, Microelectrodes and Microenvironments Symposium*, New Orleans, October 10–13, 1993) an investigation of diffusion at disk electrodes using an electrochemical reaction leading to a fluorescent product whose concentration may then be monitored as a function of space and time by photoimaging techniques. In such an ingenious experiment the time expansion of the diffusion layer with time is then observed directly. Thus all the effects described above (non-steady-state diffusion layer, convection) are detected, at least qualitatively.

where A is the electrode surface area and $\mathscr{C}(\xi)$ a function that depends only on $\xi = Dt/r_0^2$:

$$\mathscr{C}(\xi) = \frac{4}{\pi^2} \int_0^\infty \frac{\exp(-\xi\eta^2)}{\eta[J_0^2(\eta) + Y_0^2(\eta)]}\, d\eta \qquad (11)$$

where J_0 and Y_0 are Bessel functions of zero order of the first and second kind, respectively. $\mathscr{C}(Dt/r_0^2)$ may be determined numerically from equation (11). Compact empirical expressions approaching the variations of $\mathscr{C}(Dt/r_0^2)$ better than a specified accuracy have also been proposed and are extremely satisfactory for most analytical purposes [37–41] (see, e.g., Table 1).

Series expansion of $\mathscr{C}(Dt/r_0^2)$ for small values of $\mathscr{C}(Dt/r_0^2)$ affords the short time limit,

$$i = \frac{nFADC^0}{r_0}\left[0.5 + \frac{r_0}{(\pi Dt)^{1/2}}\right] \qquad (12)$$

which is very reminiscent of the current at the spherical electrode given in equation (7). Conversely, series expansion of $\mathscr{C}(Dt/r_0^2)$ for large values of

TABLE 1 Chronoamperometric Currents Determined at Ultramicroelectrodes of Usual Geometries: $i = (nFADC^0)\,[1/\Delta(t)]^a$

Electrode	$1/\Delta(t)$	Empirical approximations[b]
Sphere/hemisphere (radius r_0)	$(1/r_0) + (\pi Dt)^{-1/2}$	—
Disk (radius r_0)	$(4/\pi r_0) \times \mathscr{D}(Dt/r_0^2)$	$\mathscr{D}(\zeta) \approx 0.7854 + 0.25 \times (\pi/\zeta)^{1/2} + 0.2146 \times \exp(-0.3912 \times \zeta^{-1/2})$ (better than 0.6%)
Cylinder/hemicylinder (radius r_0)	$(1/r_0) \times \mathscr{C}(Dt/r_0^2)$	$\mathscr{C}(\zeta) \approx (\pi\zeta)^{-1/2} \times \exp[-0.1\,(\pi\zeta)^{1/2}] + 1/\ln(5.2945 + 1.4986 \times \zeta^{1/2})$ (better than 1.3%)
Band (width w)	$(1/w \times \mathscr{B}(Dt/w^2)$	For $\zeta \le 0.4$: $\mathscr{B}(\zeta) \approx 1 + (\pi\zeta)^{-1/2}$ (better than 1.3%) For $\zeta \ge 0.4$: $\mathscr{B}(\zeta) \approx 0.25 \times (\pi/\zeta)^{1/2} \times \exp[-0.4\,(\pi\zeta)^{1/2}] + \pi/\ln(5.2945 + 5.9944 \times \zeta^{1/2})$ (better than 1.3%)

[a] A, Surface area; $\Delta(t)$, time-dependent apparent diffusion layer (δ, the steady-state or quasi-steady-state values achieved at infinite times are given in Table 2).
[b] $\zeta = Dt/r_0^2$ for functions $\mathscr{C}$ and $\mathscr{D}$, or $\zeta = Dt/w^2$ for function $\mathscr{B}$. The approximations given for these functions are empirical and adapted from Refs. 41 ($\mathscr{B}$ and $\mathscr{C}$) and 53 ($\mathscr{D}$). Definitions of these functions and rigorous asymptotic formulations that are valid for $\zeta \to 0$ or $\zeta \to \infty$ are also given in the text or in the references quoted therein.

(Dt/r_0^2) gives the long time limit,

$$i = \frac{nFADC^0}{r_0} \frac{2}{\ln(4Dt/r_0^2)}$$ (13)

It is noteworthy that this long time limit is time dependent, whereas that for the spherical electrode is independent of time. This behavior is reflected by the fact that the concentration profiles extend over much larger distances from the electrode surface than for the spherical electrode, as is easily shown by comparing Figures 5a (cylinder) and 3a (sphere). On this basis it must be concluded that steady-state diffusion does not exist at cylindrical electrodes, although it should be pointed out that the time dependence of a current varying

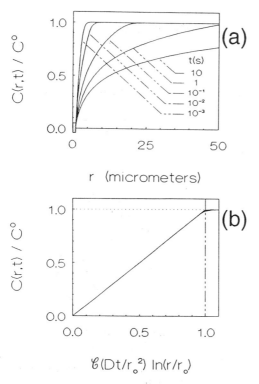

FIGURE 5 Diffusion at an electrode of cylindrical geometry in the true space (a) or in a transformed space (b), for various times indicated (r_0 = 1 μm, D = 10^{-5} cm^2 s^{-1}). Note that in (b) the curves for t = 0.01, 0.1, 1, and 10 s are all superimposed.

as the reciprocal of the logarithm of time may be difficult to characterize experimentally (see below).

The structure of equation (13) is very reminiscent of that giving the current at an electrode of surface area A within the Nernst approximation used in hydrodynamic electrochemical methods [42]:

$$i = \frac{nFADC^0}{\delta} \tag{14}$$

with a diffusion-layer thickness $\delta = (r_0/2) \ln(4Dt/r_0^2)$. This analogy is not factual but reflects fundamental causes. This is apparent upon effecting the changes of dimensionless variables [43] $\rho = \ln(r/r_0)$ and $\tau = Dt/r_0^2$, which make it possible to transform equation (9) into

$$\frac{\partial C(\rho,\tau)}{\partial \tau} = \exp(-2\rho) \frac{\partial^2 C(\rho,\tau)}{\partial \rho^2} \tag{15}$$

The concentration profiles are then given by integration of equation (15), taking into account the following boundary conditions:

$$\tau < 0, \rho \geq 0: \qquad C(\rho,\tau) = C^0 \tag{16}$$

$$\tau \geq 0, \rho = 0: \qquad C(0,\tau) = 0 \tag{17}$$

$$\tau \geq 0, \rho \to \infty: \qquad C(\rho,\tau) \to C^0 \tag{18}$$

Because the exponential term on the right-hand side of equation (15) rapidly decays toward zero when ρ becomes large (i.e., when long-duration experiments are considered), $\partial C(\rho,\tau)/\partial \tau$ tends toward zero under these conditions, except at any particular location where $\partial^2 C(\rho,\tau)/\partial \rho^2$ may become infinite. Therefore, for experiments with long duration times ($\tau \gg 1$), the solution of equation (15) tends toward [43]

$$C(\rho,\tau) = \rho \times \left[\frac{\partial C(\rho,\tau)}{\partial \rho}\right]_{\rho=0} = \rho \times \frac{C^0}{\delta_\rho} \tag{19}$$

where δ_ρ is defined as $\delta_\rho = [\partial C(\rho,\tau)/\partial \rho]_{\rho=0}/C^0$. However, such a result is necessarily invalid for $\rho > \delta_\rho$ because owing to conservation of matter, $C(\rho,t)$ is necessarily smaller than (or equal to) C^0. Thus

$$\rho > \delta_\rho: \qquad C(\rho,\tau) = C^0 \tag{20}$$

Equation (19) is thus valid only for $0 \leq \rho < \delta_\rho$. At $\rho = \delta_\rho$ is a point of singularity for the concentration profile: $\partial^2 C(\rho,\tau)/\partial \rho^2$ becomes infinite at this location, owing to the discontinuity of the concentration profile slope. Therefore, $\partial C(\rho,\tau)/\partial \tau$ cannot be neglected at this particular point of space. This necessitates that δ_ρ be time dependent [i.e., $\delta_\rho = \delta_\rho(\tau)$]. A time dependence is then introduced through this bias for the concentration profile. Indeed, integration of equation

(15) over the domain around $\rho = \delta_\rho$ [43] shows that [see equation (11) for the definition of $\mathscr{C}(Dt/r_0^2)$]

$$\delta_\rho(\tau) = \frac{1}{\lim\limits_{Dt/r_0^2 \to \infty} \mathscr{C}(Dt/r_0^2)} \approx \frac{\ln(4Dt/r_0^2)}{2} \tag{21}$$

This leads to a formulation identical to that in equation (13) for the current. Indeed, from equation (19), the current is

$$i = nFAD\left[\frac{\partial C(r,t)}{\partial r}\right]_{r=r_0} = \frac{nFAD}{r_0}\left[\frac{\partial C(\rho,\tau)}{\partial \rho}\right]_{\rho=0} = \frac{nFADC^0}{r_0\delta_\rho} \tag{22}$$

The formulation in equations (19) to (22) demonstrates that when $Dt/r_0^2 \longrightarrow \infty$, the concentration profiles that develop at cylindrical electrodes possess all the intrinsic characteristics of steady-state concentration profiles, except that they are contained within a diffusion layer that slowly expands toward infinity while time increases. This result is particularly obvious when the concentration profiles represented in Figure 5a are plotted as a function of $\mathscr{C}(Dt/r_0^2)\ln(r/r^0)$, as is done in Figure 5b, and compared to those obtained at the spherical electrode in Figure 3b. Because of this behavior, cylindrical electrodes have been said to give *quasi-steady-state diffusion* [43] at long times.

Putting terminology aside, this quasi-steady-state notion is extremely useful for determining concentration profiles at the cylindrical electrode under conditions that are not amenable to simple analytical formulations. Indeed, for any conditions that do not affect the time dependance of $\delta_\rho(\tau)$, concentration profiles at the cylindrical electrode can be determined by considering steady-state diffusion [i.e., $\partial C(\rho,\tau)/\partial \tau = 0$], as is done for the spherical electrode. The time dependance is easily reintroduced into the final results via that of $\delta_\rho(\tau)$, that is, along a procedure much reminiscent of that used in Ilkovic's model of polarography at the expanding mercury drop [2].

As pointed out for steady-state diffusion at the spherical electrode, achievement of this quasi-steady-state regime requires that the concentration profiles extend over distances from the electrode surface that considerably exceed the radial dimension of the electrode. It is therefore highly doubtful that such concentration profiles may be immune to convection and be described by considering pure diffusional transport only. However, as for the spherical electrode, interference of natural convection results in only slight modifications of the solutions above in the region near the electrode surface and will be difficultly observable in current measurements. It is noteworthy that this has been confirmed experimentally by spatially resolved spectroelectrochemistry using a cylindrical electrode with 6 μm radius, placed under vibration-free conditions [44]. However, the concentration profiles above are severely distorted so as to affect the time variations of the current when moderate forced convection is involved, as has been established experimentally by mounting a cylindrical electrode on a vibrator tip [45].

B. Diffusion at Disk and Band Electrodes

Although hemispherical and cylindrical ultramicroelectrodes can be manufactured, their fabrication and use are rather difficult vis-à-vis those of disk and band ultramicroelectrodes, which can be constructed by a cross section of conducting wires or sheets sealed in an insulator. Moreover, the surface of the latter electrodes may be cleaned or renewed easily simply by polishing, as is done with electrodes of millimetric dimensions. Such important facilities explain the considerable success of these electrodes for experimental applications in analytical or molecular electrochemistry. However, these advantages are earned at the expense of increased theoretical difficulties in solving diffusion at these electrodes. Indeed, because all points of the electrode surface are not equivalent, nonuniform diffusion and nonuniform current densities have to be considered at these electrodes [46–48]. Obviously, these effects are not apparent at extremely short time scales since the thicknesses of the diffusion layers that develop at these electrodes, being on the order of $(Dt)^{1/2}$ [2], are considerably smaller than the electrodes smallest dimension. The electrode behaves then as an electrode of infinite dimensions, and planar diffusion operates. Therefore, complications in theory of diffusion at these electrodes arise when diffusion layers become comparable or exceed the smallest dimension of the electrode [i.e., when steady-state (or quasi-steady-state) diffusion is approached]. Under such conditions the concentration profiles that develop at disk electrodes resemble those obtained at the hemispherical electrode except within distances from the electrode surface that are comparable to (or less than) the electrode radius. In the electrode vicinity nonuniform diffusion must be considered. Similarly, concentration profiles that develop at band electrodes resemble those observed at hemicylinders. These analogies are particularly evident when isoconcentration lines determined at these four electrodes are compared, as is done in Figure 6a (hemisphere/disk) and b (hemicylinder/band).

Based on this observation, experimental data obtained under steady-state or quasi-steady-state diffusion at disk or band electrodes have often been treated using *equivalent* hemispheres or hemicylinders, respectively (see, e.g., Refs. 47, 49, and 50 for disks or Ref. 51 for bands). However, such procedures may be highly unsafe because they are based only on the identity of diffusional fields that exist far away from the electrode surface and do not consider the possible effect of nonuniform diffusion in the region close to the electrode surface. Such heterogeneities may, however, significantly affect homogeneous chemical reactions and heterogeneous electrochemical kinetics. In return, these alter the local current density observed at these electrodes. It is thus easily guessed that no

(a)

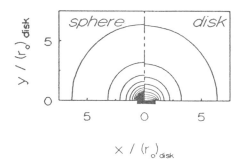

(b)

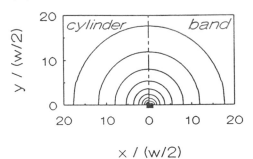

(c)

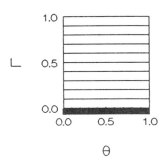

FIGURE 6 Comparison of steady-state or quasi-steady-state diffusion at spherical (a, left), disk (a, right), cylindrical (b, left), and band (b, right) electrodes. In (a), $(r_0)_{sphere} = (2/\pi)\, (r_0)_{disk}$; in (b), $(r_0)_{cyl} = w/4$. The curves indicated are isoconcentration lines for $C/C^0 = 0.9$ (outermost), 0.8, . . . , 0.2, 0.1, and 0 (electrode surface). (c) Equivalent representation of the four sets of isoconcentration lines in (a) and (b) in each (Γ, θ) pertinent transformed space (see the text). In (c), $C/C^0 = \Gamma$.

simple and general equivalences could exist, but that different equivalence relationships (if any) may be found between a given disk (or band) electrode and hemispherical (or hemicylindrical) electrode as a function of the particular electrochemical mechanism occurring at this electrode.

1. Disk Electrodes

Steady-state diffusion at the disk electrode was first solved [52] based on the analogous problem of heat conduction [36] and the steady-state current was shown to be identical to that observed at an hemispherical electrode of radius $(r)_{\text{hemisphere}} = (2/\pi)(r_0)_{\text{disk}}$ [47]:

$$i_{\text{st.state}} = 4nFDC^0(r_0)_{\text{disk}} \tag{23}$$

This equivalence is also apparent when isoconcentration lines at a disk and at its equivalent hemisphere electrode are compared as in Figure 6a under conditions where steady-state diffusion is achieved. It is seen that whereas these lines necessarily differ near each electrode surface because of obvious differences in electrode shapes, they tend to be identical at distances larger than a few electrode radii. This confirms that the equivalence relationship based on the identity of the steady-state currents [equations (8) and (23)] [47] is not ad hoc but truly reflects the identity of the steady-state diffusional fields created by the two electrodes at distances that exceed a few times their radius.

A rigorous solution [53] describing the transition between planar diffusion (observed at short times) and steady-state diffusion (observed at long times) was proposed more recently for chronoamperometry [$r_0 = (r_0)_{\text{disk}}$]:

$$i = 4nFDC^0 r_0 \mathcal{D}(Dt/r_0^2) \tag{24}$$

where $\mathcal{D}(Dt/r_0^2)$ is a function determined from mathematical series expansions [53] or from empirical expressions [54–57] (see, e.g., Table 1).

The situation is then analogous to that encountered above for the cylindrical electrode. Except for particular situations where an accurate analytical solution for diffusion is more simple to use as, for example, for calibration of electrodes, it is our feeling that finite-difference integration [2] of Fick's laws is preferable. Moreover, such numerical [58–60] approaches easily allow the incorporation of any mechanistic complication, such as heterogeneous kinetics or homogeneous chemical reactions.

There are, however, several difficulties in devising numerical procedures for electrodes at which current densities are nonuniform and may become infinite at the edges of the electrode (Fig. 7a). Indeed, regular or classical finite-element grids [2] based on the real physical space are inefficient to allow Taylor's developments of Fick's laws with an identical accuracy at each point of the space grid. This key difficulty may easily be bypassed by performing the numerical integration in a transformed space where the lines of isoconcentration approach

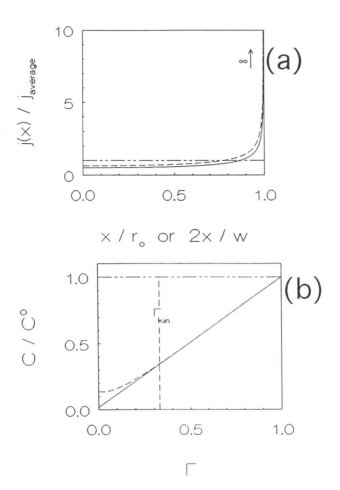

FIGURE 7 (a) Comparison of the current densities $j(x)$ at a disk (solid curve) and at a band electrode (dashed curve) as a function of the distance x from the center of the electrode, for fast electron transfer under steady-state (disk) or quasi-steady-state (band) conditions; the horizontal dashed line corresponds to an infinitely slow kinetics of electron transfer at both electrodes under identical conditions. (b) Concentration profiles at a disk electrode for a slow charge-transfer kinetics, in the transformed (Γ, θ) space defined by equations (25) and (26) at $\theta = 0$ (dashed curve) or $\theta = 1$ (solid curve). See the text for the definition of Γ_{kin}. In (b), $(Dt)^{1/2}/r_0 = 100$ and $kr_0/D = 1$. (Adapted from Ref. 60.)

equidistant parallel lines, as schematized in Figure 6c. If such a transformed space may be found, a coarse simulation grid is sufficient to perform a numerical integration of Fick's laws with high accuracy. This approach is conceptually analogous to those developed in electrostatics or in electromagnetism that are based on conformal mapping techniques [61].

Conformal mapping [62] transformations have been proposed to allow such simplifications in analytical or numerical integration of the Laplace equation (i.e., $\nabla^2 C = 0$ here) for a wide variety of boundary conditions. It must be emphasized that even if these methods are most efficient when $\nabla^2 C = 0$ (i.e., when steady-state diffusion prevails), their use is obviously not restricted to these conditions (see below). Besides the considerable simplifications they introduce in numerical simulations, these techniques are particularly adequate for solving electrochemical problems because they respect the orthogonality of iso-concentration lines and flux lines in the transformed space (see below).

However, these transformations are devised for two-dimensional spaces and apply to three-dimensional space only when the problem obeys cylindrical geometry that is when diffusion is identical in any cross-sectional plane perpendicular to the third-dimensional axis of the structure considered. The disk electrode is obviously not amenable to such an approach. However, the following transform [60], adapted from Newman's seminal work [46], relates the coordinates of the physical half-space $(x,y; x \geq 0, y > 0)$ shown in Figure 6a to those (Γ, θ) of the transformed one, in Figure 6c, which possesses similar properties for simulation purposes:

$$\frac{x}{(r_0)_{\text{disk}}} = \frac{(1 - \theta^2)^{1/2}}{\cos(\Gamma\pi/2\beta)} \tag{25}$$

$$\frac{y}{(r_0)_{\text{disk}}} = \theta \tan\frac{\Gamma\pi}{2\beta} \tag{26}$$

Note that β is a constant parameter whose value may be adjusted as a function of t_{max}, the maximal duration of the simulated experiment, as explained in the following. The perfect adequacy of the transform above is evidenced by a comparison of Figure 6a (right panel) and c. The highly curved steady-state isoconcentration lines at the disk electrode shown in Figure 6a in (x,y) half-space become straight parallel equidistant lines in (Γ, θ) half-space built from equations (25) and (26) (with $\beta = 1$ [60]).

Use of $\beta = 1$ as in Figure 6c is adequate under steady-state conditions because then the (Γ, θ) transformed half-space consists of a closed box ($0 \leq \Gamma \leq 1$; $0 \leq \theta \leq 1$), with $\Gamma \longrightarrow 1$ corresponding to x or $y \longrightarrow \infty$. However, for smaller values of time the resulting concentration profiles tend to be confined within a diffusion layer whose thickness shrinks as $(Dt)^{1/2}/(r_0)_{\text{disk}}$. This can easily be accounted for by decreasing the size of the finite elements along the Γ axis and reducing the size of the grid: that is, by assuming that $\Gamma \leq \Gamma_{max} = n(Dt_{max})^{1/2}/(r_0)_{disk}$, where n is a function of the desired accuracy and t_{max} is the maximal duration of the experiment [58–60]. This can be performed more conveniently by adjusting the parameter β as a function of $(Dt_{max})^{1/2}/(r_0)_{disk}$. For example, the expression

$$\beta = 1 + \frac{(r_0)_{disk}}{n(Dt_{max})^{1/2}} \tag{27}$$

makes it possible to obtain a diffusion layer with a thickness that remains almost independent of t_{max} in (Γ,θ) space. Values of $n = 10$ to 20 lead to accuracies that are sufficient for most purposes [60].

The original expression of Fick's second law valid in (x,y) space [equation (28)] is converted into equation (29) [60], which depicts diffusion in (Γ,θ) space:

$$\frac{\partial C}{\partial t} = D\left(\frac{\partial^2 C}{\partial x^2} + \frac{\partial^2 C}{\partial y^2} + \frac{1}{x}\frac{\partial C}{\partial x}\right) \tag{28}$$

$$\frac{\partial C}{\partial t} = D*\left[\left(\frac{2\beta}{\pi}\cos\frac{\Gamma\pi}{2\beta}\right)^2\frac{\partial^2 C}{\partial\Gamma^2} + (1 - \theta^2)\frac{\partial^2 C}{\partial\theta^2} - 2\theta\frac{\partial C}{\partial\theta}\right] \tag{29}$$

$D*$ (in s^{-1}) represents the diffusion coefficient in (Γ,θ) dimensionless space:

$$D* = \frac{D/(r_0)^2_{disk}}{\theta^2 + [\tan(\Gamma\pi/2\beta)]^2} \tag{30}$$

It varies with Γ and θ to account for the local compression or dilatation of space. When (Γ/β) tends toward unity, $D*$ tends toward zero, with the result that $\partial C/\partial t$ also tends to zero. This establishes that steady-state diffusion is approached when $\Gamma/\beta \rightarrow 1$:

$$0 = \left(\frac{2\beta}{\pi}\cos\frac{\Gamma\pi}{2\beta}\right)^2\frac{\partial^2 C}{\partial\Gamma^2} + (1 - \theta^2)\frac{\partial^2 C}{\partial\theta^2} - 2\theta\frac{\partial C}{\partial\theta} \tag{31}$$

The boundary conditions at $\theta = 0$ or 1 are $\partial C/\partial\theta = 0$ since they correspond either to the symmetry axis of the system ($\theta = 1$) or to the insulator ($\theta = 0$). An obvious solution of equation (31) that also fulfills the other boundary conditions [i.e., $C = 0$ at the electrode surface (at $\Gamma = 0$) and $C = C^0$ at infinity (at $\Gamma = 1$)] is independent of θ:

$$C(\Gamma,\theta) = \Gamma C^0 \tag{32}$$

The expression of the current is readily obtained from equation (32). Indeed, one has (for $\beta = 1$)

$$i = 2\pi nFD\int_0^{(r_0)_{disk}}\left(\frac{\partial C}{\partial y}\right)_{y=0} x\,dx = [4nFD(r_0)_{disk}]\int_0^1\left(\frac{\partial C}{\partial\Gamma}\right)_{\Gamma=0} d\theta \tag{33}$$

from which $i = 4nFDC^0(r_0)_{\text{disk}}$ [equation (23)] is readily obtained for the steady-state current at a disk electrode, without requiring sophisticated analytical procedures.

When heterogeneous kinetics also have to be considered, concentrations are no longer constant at the electrode surface, and a dependence of $C(\Gamma,\theta)$ on θ is thus introduced [60]. The same is true when fast homogeneous kinetics are involved [63,64]. Such complex situations may easily be solved numerically through finite-element procedures based on transformed spaces [58−60]. It must be emphasized that since the transformed space defined in equations (25) and (26) is closely mapping the concentration variations, only a limited number of grid points are required for high accuracies, so that simulations are fast and easily performed on desk computers.

Note also that since the transformation concerns only space variables, formulation of homogeneous kinetics remains independent of the space considered. This is particularly interesting because adapting a simulation program to incorporate any desired homogeneous kinetics does not require modification of the diffusion simulator. Therefore, a wide variety of kinetic situations can easily be treated using an identical diffusion simulator algorithm based on equation (29) and tabulated values of D^* [equation (30)] at each point of the finite-element grid.

2. Band Electrodes

Exact analytical solutions have been proposed for chronoamperometric currents at band electrodes of width w and length l [65,66]:

$$i = nFDC^0 l \mathcal{B}(Dt/w^2) \tag{34}$$

where $\mathcal{B}(Dt/w^2)$ is a function of the experiment duration and is available through series expansions or empirical fitting expressions (Table 1) [41]. The long time solution

$$i = nFDC^0 l \frac{2\pi}{\ln(64Dt/w^2)} \tag{35}$$

is identical to that obtained for the hemicylindrical electrode [equation (13)] with $(r_0)_{\text{cyl}} = w/4$ [41,51,66]. As for the disk−hemisphere equivalence, this result is not ad hoc but reflects that when Dt/w^2 is sufficiently large (i.e., when the quasi-steady-state diffusional regime is attained), the diffusional fields created by a band electrode of width w or by an hemicylinder of radius $w/4$ are identical at distances from the electrode surface that exceed its width or radius [51]. This analogy is particularly apparent when the isoconcentration lines developing at the two "equivalent" electrodes are compared, as is done in Figure 6b.

As with the disk electrode, more complicated kinetics are difficult to handle by these analytical procedures or by classical finite-difference simulations because of nonuniform current densities (Fig. 7a). However, the symmetry of the band electrode is perfectly adapted for application of conformal mapping techniques, since the diffusion problem is identical in any cross-section plane perpendicular to the band axis. Introducing the transformations [67]

$$x = \frac{w}{2} \cosh \frac{\Gamma}{\beta} \cos \frac{\theta\pi}{2} \tag{36}$$

$$y = \frac{w}{2} \sinh \frac{\Gamma}{\beta} \sin \frac{\theta\pi}{2} \tag{37}$$

allows us to transform the half-space ($x \geq 0$, $y \geq 0$) into the closed box ($0 \leq \Gamma \leq 1$, $0 \leq \theta \leq 1$) represented in Figure 6c. As in he case of the disk electrode, β is a constant parameter that is a function of t_{max}, the maximal duration of the experiment, and is adjusted so that the diffusion-layer thickness remains almost independent of t_{max} in (Γ,θ) space. The following empirical expression, with $n = 10$ to 20 according to the desired accuracy, was found adequate for most purposes [67]:

$$\beta = \frac{w}{4n(Dt_{max})^{1/2}} + \frac{1}{\ln[2 + 8(Dt_{max})^{1/2}/w]} \tag{38}$$

The performance of the transformation in equations (36) to (38) is illustrated by the parallel and equidistant isoconcentration lines that are obtained in Figure 6c for the band electrode under conditions of quasi-steady-state diffusion. The inherent simplification introduced by this change of space coordinates is particularly patent when this figure is compared to its counterpart in real (x,y) space shown in Figure 6b for the same band electrode under the same experimental conditions. It is also noteworthy that Figure 6c is identical for disk and band electrodes provided that Γ and θ are defined by equations (25) to (27) in the first case or by equations (36) to (38) in the second (see below).

Fick's second law, which is expressed by

$$\frac{C}{\partial t} = D\left(\frac{\partial^2 C}{\partial x^2} + \frac{\partial^2 C}{\partial y^2}\right) \tag{39}$$

in (x,y) space, conserves an almost identical formulation in (Γ,θ) space:

$$\frac{\partial C}{\partial t} = D^*\left[\frac{\partial^2 C}{\partial \Gamma^2} + \left(\frac{2}{\beta\pi}\right)^2 \frac{\partial^2 C}{\partial \theta^2}\right] \tag{40}$$

yet as noted in the disk case, because of the compression or dilatation of space due to the transformation in equations (36) to (38), D^*, the *pseudo* diffusion coefficient (in s^{-1}) in the (Γ, θ) space, is space dependent [67]:

$$D^* = \frac{D\,(2\beta/w)^2}{[\sinh(\Gamma/\beta)]^2 + [\sin(\theta\pi/2)]^2} \tag{41}$$

When time increases, β decreases [equation (38)] and equation (40) simplifies to $\partial^2 C/\partial \Gamma^2 = 0$. This shows that in (Γ, θ) space, concentrations become time and θ independent. This corresponds to the conditions under which quasi-steady-state diffusion is achieved. The concentration is then given by $C = \Gamma C^0$, from which the current is readily obtained. Because the conformal mapping transformation conserves fluxes through surfaces, one has at any value of time:

$$i = 2nFDl \int_0^{w/2} \left(\frac{\partial C}{\partial y}\right)_{y=0} dx = (2nFDl)\,\frac{\pi\beta}{2} \int_0^1 \left(\frac{\partial C}{\partial \Gamma}\right)_{\Gamma=0} d\theta \tag{42}$$

Introduction of $(\partial C/\partial \Gamma)_{\Gamma=0} = C^0$ and $\beta = 1/\ln[8(DT)^{1/2}/w]$, as obtained from equation (38) for $DT/w^2 \gg 1$, yields the expression of the current already given in equation (35) without the need of sophisticated analytical derivation.

When the duration of the experiment is not sufficient for quasi-steady-state diffusion to be achieved, concentration profiles can be determined in (Γ, θ) space by finite-difference numerical integration of equation (40) [67]. The corresponding currents are then obtained using equation (42). As already emphasized for the disk electrode, since (Γ, θ) space closely fits the concentration profile when DT_{max}/w^2 is changed, rather coarse finite-element grids are sufficient to produce extremely good accuracies in current determinations. Also, since the space transformation affects only space variables, formulation of homogeneous kinetics remains identical to that used classically in (x, y) space. The same is true for heterogeneous kinetics [51] because the conformal mapping conserves fluxes [62]. This allows treatment of any complicated mechanism based on a single common subroutine (i.e., a diffusion simulator) to account for diffusion.

3. Equivalency of Diffusion at Spherical, Disk, Cylindrical, and Band Electrodes Under Steady-State or Quasi-Steady-State Conditions

As already emphasized, isoconcentration lines for disk and band electrodes are identical in (Γ, θ) space, provided that the transformed space is defined by the pertinent set of equations (25) to (27) or (36) to (38). Moreover, if one also defines θ so that $\theta\pi/2$ is the apical angular position [i.e., measured from the vertical y axis of the (x, y) plane] and Γ such as [34]

$$\Gamma = 1 - \frac{r}{(r_0)_{sphere}} \tag{43}$$

for the hemispherical electrode or as [43]

$$\Gamma = \frac{2 \ln[r/(r_0)_{cyl}]}{\ln[4Dt/(r_0)^2_{cyl}]} \tag{44}$$

for the hemicylindrical one, the four different sets of isoconcentration lines shown in Figure 6a and b become rigorously identical in (Γ, θ) space and are represented by the set of isoconcentration lines shown in Figure 6c. This is always true whenever steady-state (disk or hemisphere) or quasi-steady-state (band or hemicylinder) diffusion is achieved (i.e., provided that $Dt/a^2 \gg 1$, where a is the radius or the half-width of the electrode, respectively). Irrespective of the electrode shape, the concentration profiles then depend only on Γ and are given by

$$C = C_{el} + (C^0 - C_{el})\Gamma \tag{45}$$

where C_{el} is the concentration imposed at the electrode. The current is obtained by integrating the constant flux at $\Gamma = 0$ over the electrode surface. In (Γ, θ) space this integration involves a factor, $\sigma(\theta)$, dependent on θ, which formulation is a function of the electrode shape and dimension:

$$i = nFAD \int_0^1 \left(\frac{\partial C}{\partial \Gamma}\right)_{\Gamma=0} \sigma(\theta) \, d\theta = nFAD(C^0 - C_{el}) \int_0^1 \sigma(\theta) \, d\theta \tag{46}$$

which can be reformulated as

$$i = \frac{nFAD(C^0 - C_{el})}{\delta} \tag{47}$$

with

$$\delta = \frac{1}{\displaystyle\int_0^1 \sigma(\theta) \, d\theta} \tag{48}$$

The formulation in equation (47) shows that δ represents the equivalent diffusion-layer thickness associated with a given electrode [42]. Table 2 summarizes the values of δ for the electrodes considered here.

From what precedes it is seen that under steady-state or quasi-steady-state conditions, the diffusional fields created by all the electrodes considered in Table 1 are equivalent in (Γ, θ) space. This property can be transposed to any other electrode, provided that an adequate space transformation is designed (see below). The main interest of such equivalences is that any property that can be established under steady-state or quasi-steady-state conditions at one of these electrodes may easily be transposed to the other electrodes' shape through correspondence of their respective (Γ, θ) spaces (see below, for example, for

TABLE 2 Equivalent Diffusion Layers (δ) for Common Ultramicroelectrodes Under Steady-State or Quasi-Steady-State Regimes Within a Nernst Layer [42] Formulation: $i_{\text{st.state}} = nFADC^0/\delta$

Electrode	Characteristic dimension(s)	Area A	Diffusion layer, δ
Sphere/hemisphere	r_0 (radius)	$4\pi r_0^2$ or $2\pi r_0^2$	r_0
disk	r_0 (radius)	πr_0^2	$\pi r_0/4$
Cylinder/hemicylinder	r_0 (radius), l (length)[a]	$2\pi r_0 l$ or $\pi r_0 l$	$r_0 \ln[2(Dt)^{1/2}/r_0]$
band	w (width), l (length)[a]	wl	$(w/\pi) \ln[8(Dt)^{1/2}/w]$

[a]The length, being millimetric, is much greater than that of common diffusion layers.

diffusion–migration). However, these equivalences have been established for electrochemical phenomena that are controlled by diffusion only. Introduction of other kinetic controls may break these equivalences whenever they introduce a dependence of the concentration profiles over the variable θ, and therefore on the electrode shape.

This is particularly obvious when heterogeneous kinetics are considered. Indeed, heterogeneous rate laws couple surface concentrations (i.e., at $\Gamma = 0$) and local current densities [2]. For any electrode at which the current density is not uniform (Fig. 7a), the boundary condition at $\Gamma = 0$ then depends on θ, introducing a dependence of the concentration profile on this variable [51,60,67] (compare Fig. 7b). Note that this is true for any electrode shape, except for cylindrical or spherical electrodes because they have uniform current densities. It is noteworthy, however, that because of the structure of the diffusional operators in (Γ,θ) space, these perturbations are limited to a small fraction of the space [51,60,67] (i.e., for $\Gamma \leq \Gamma_{\text{kin}}$) when steady-state or quasi-steady-state diffusion prevails. The linearity of the concentration profiles with Γ and their independence of θ are indeed restored for $\Gamma > \Gamma_{\text{kin}}$, as shown, for example, in Figure 7b. Under such conditions, equivalence between two electrodes requires that they give the same values of concentration and flux at $\Gamma = \Gamma_{\text{kin}}$ [51,60], in addition to equivalent diffusional fields at large distances from the electrode surfaces.

This means that equivalent rate constants must be considered in addition to equivalent radii or widths. For example, the hemicylinder equivalent to a band electrode of width w was shown above to have radius $r_0 = w/4$. This relationship is sufficient when pure diffusional control operates. If an heterogeneous kinetics with a rate constant k_s^{band} is also operative at the band electrode, the equivalent hemicylinder is still with a radius $r_0 = w/4$ (so as to maintain the same diffusional field far away from the electrode surface), but in addition an equivalent heterogeneous rate constant $k_s^{cyl} = (4/\pi)k_s^{band}$ [51] must be used at the cylindrical electrode, so as to result in identical values of concentration and flux at $\Gamma = \Gamma_{kin}$.

Similar relationships can be developed for very fast homogeneous kinetics that correspond to extremely thin kinetic layers vis-à-vis the electrode dimension. Homogeneous kinetics then operate in a region of space where deviation from planar diffusion is negligible and impose a coupling between surface concentrations and local current densities. Thus a situation somewhat analogous to that described above for heterogeneous kinetics applies [63,64]. Note, however, that since the reaction orders may be quite different from those observed for true heterogeneous kinetics, the relationship between equivalent rate constants may differ from that given above for heterogeneous rate constants.

Conversely, when the homogeneous reactions that are considered are slow (so that they operate within most of the diffusion layer rather than in a thin kinetic layer as considered above) they compete with diffusion in regions where a spherical or cylindrical diffusion field prevails irrespective of the disk or band shape of the electrode. Then the equivalency between kinetics corresponds to the same relationship as that used for equivalency of diffusion. This means, obviously, that no general equivalence relationship exists (except ad hoc) in the transition between slow and very fast homogeneous kinetics.

It is seen that these notions of equivalence may be extremely delicate to handle and should therefore be used with caution whenever extraction of accurate rates and kinetics from experimental data are desired. Moreover, their intrinsic value for such applications has been depreciated considerably by the availability of fast algorithms based on conformal maps that allow simulation of diffusion at ultramicroelectrodes with nonuniform current densities on PCs. It is noteworthy that such simulations will often require less time than is required by a search for the proper set of equivalent relationships to be used.

C. Arrays of Ultramicroelectrodes

Arrays of microelectrodes [68, 69, and references therein] consist of ensembles of ultramicroelectrodes, which may be regular (i.e., periodical arrangement of identical electrodes with identical dimensions) or random (i.e., disordered arrangements of identical electrodes, or ensembles of electrodes with identical shape but uneven dimensions, or statistical arrangements of irregularly shaped electrodes), as schematized in Figure 8.

Regular arrays with well-defined geometries are generally constructed by microlithographic techniques, yet other methods have also been reported [34,68–84]. They consist generally of parallel band electrodes or of disk electrodes that are arranged in a hexagonal array [82–84]. These two geometries

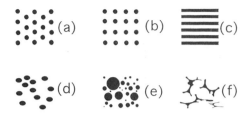

FIGURE 8 Schematic examples of arrays of ultramicroelectrodes. A: Periodical arrays of electrodes of identical shapes and sizes: hexagonal (a) or squared (b) arrays of disks; (c) array of parallel band electrodes. B: Random arrays of (d) electrodes of identical shape and identical size (slanted cross section of insulated cylindrical fibers), of (e) electrodes of identical shape but random sizes, or (f) totally disordered array.

are indeed preferred because their symmetries allow a more facile modeling of diffusion at these ensembles. Their period (i.e., the interelectrode distance), as well as the dimensions of electrodes, are generally a few micrometers [69]. However, ingenious stacking procedures [78] may be used to produce arrays of parallel bands with submicrometric interelectrode gaps, yet still with a micrometric period.

Statistical arrays may achieve smaller dimensions, yet their geometries are poorly defined [68,85–97]. For example, arrays of disks can be constructed from a cross section of ensembles of insulated conducting fibers embedded in an insulating matrix (epoxy, glass, etc.) [89–92]. Random arrays of closely calibrated disks of extremely small dimensions have also been constructed by electrodepositing platinum into the calibrated cylindrical pores of an insulating membrane film deposited over a conducting substrate or by related procedures [93–95]. Statistical arrays of microspheres with random radii have also been constructed by depositing mercury on a substrate that is inactive vis-à-vis the electrochemical reaction to be investigated at these arrays [96,97] or by controlled fractional covering of a gold electrode surface by a monolayer of an organic insulating molecule [98,99]. In this respect it should be noted that the most common disordered array consists of a dirty electrode on which an insulating material that has been deposited during the course of an experiment or during electrode preparation partially blocks the surface of the electrode [100]. To the best of our knowledge, the earliest theories of diffusion at arrays of electrodes have indeed been developed to rationalize the distortions of electrochemical signals due to such partial blocking of large electrode surfaces [82–84,100–102].

Rigorous solutions of diffusion at statistical arrays are obviously impossible. The general solutions for periodical arrays (see below) [82–84, 98,99,101–108] are generally used and may be adapted by the introduction of statistical corrections [68]. Accordingly, in the following we restrict our presentation to regular arrays.

1. Arrays Where All Electrodes Operate at an Identical Potential

In the most common mode of operation of arrays, all the electrodes operate at an identical potential. Under these conditions, the total response of the array depends on the relative sizes of the array characteristic dimensions vis-à-vis the thickness of the diffusion layer that develops at each element of the array.

This is easily understood upon considering the four limiting situations featured in Figure 9. In Figure 9a, the duration of the experiment is sufficiently small for the diffusion layer at one electrode element of the array to be extremely small with respect to the size (a_{el}) of this element. Planar diffusion is then observed at each element and the current is proportional to the overall electroactive surface area, A_{el} (i.e., to the sum of the surface area of the active elements of the array). For example, in chronoamperometry,

$$(Dt)^{1/2} \ll a_{el}: \qquad i = \frac{(nFDC^0)A_{el}}{(\pi Dt)^{1/2}} \tag{49}$$

When the duration of the experiment increases, so that the diffusion layer is large with respect to a_{el} but small vis-à-vis the distance p between two electrode elements, steady-state or quasi-steady-state diffusion occurs at each array element, depending on its shape (Fig. 9b). The current remains proportional to the active surface area, but the proportionality factor depends on the shape of each active element. This occurs because the current function effective at each array element now depends on its shape (compare Tables 1 and 2):

$$a_{el} \ll (Dt^{1/2} \ll p: \qquad i = \frac{(nFDC^0)A_{el}}{\delta(a_{el})} \tag{50}$$

Validity of equation (50) obviously requires that $a_{el} \ll p$. If this is not valid, nonplanar diffusion layers developing at adjacent electrode elements will interpenetrate before they reach a sufficient size for steady-state or quasi-steady-state diffusion to be observable. A similar situation also occurs when $a_{el} \ll p$, since the diffusion layers developing at each individual electrodes also merge together at long times. In both situations, the overlap of individual diffusion layers results in the creation of an apparent planar diffusion layer that extends all over the array (Fig. 9c). Under these conditions, the array behaves like a large electrode whose surface area is the total surface area of the array [i.e., including the active

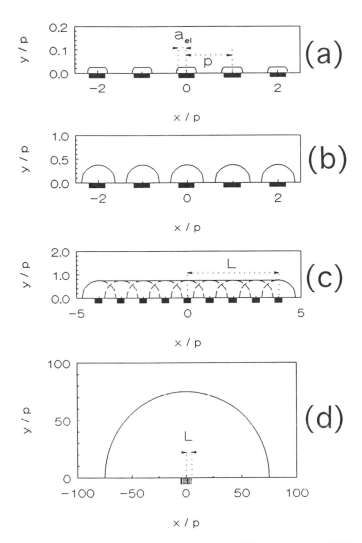

FIGURE 9 Schematic representation of diffusion layers ($C/C^0 = 0.95$) at a periodical array of disk (Fig. 8a) or band (Fig. 8c) electrodes. a_{el}, Radius or width of each electrode element; p, period of the array. (a) Planar or (b) spherical (or, respectively, cylindrical) diffusion at each electrode element; note the vertical scale expansion ($\times 10$) in (a). (c, d) Reconstruction of a (c) planar or (d) spherical (or, respectively, cylindrical) diffusion layer for the entire array by overlap of individual diffusion layers due to each element.

(A_{el}) and nonactive (A_{insul}) surface area]. This remains true provided that the diffusion layer is smaller than the overall dimension, L, of the array:

$$p \ll (Dt)^{1/2} \ll L: \qquad i = \frac{(nFDC^0)(A_{el} + A_{insul})}{(\pi Dt)^{1/2}} \qquad (51)$$

Eventually, at much larger times the diffusion layer size exceeds L (Fig. 9d), and steady-state or quasi-steady-state diffusion is again observed. The current is then given by an expression akin to that in equation (50). The surface area to be considered is the overall surface area of the array, and the equivalent diffusion layer depends on L and on the shape of the entire array (compare Tables 1 and 2):

$$L \ll (Dt)^{1/2}: \qquad i = \frac{(nFDC^0)(A_{el} + A_{insul})}{\delta(L)} \qquad (52)$$

All the foregoing limiting behaviors and their transitions have been observed experimentally at arrays of electrodes (see, e.g., Refs. 83, 84, 94, 99, and 107 to 110). However, it must be emphasized that the expressions above (and the corresponding equivalencies) are valid only when diffusion-controlled currents are measured at these arrays. When other kinetic control are involved, other phenomena are also apparent. For example, under the situation where equation (51) applies, the considerable torsion of the lines of flux near each electrode element imposes a local rate of diffusion that is considerably larger than that which applies far from the electrode element, where planar diffusion prevails [100,102]. Therefore, although the array may be considered as a single large electrode at which planar diffusion occurs, it behaves as each of its individual elements for what concerns heterogeneous kinetics [102]. If this effect is not taken into account, the measurement of rate of electron exchange performed at such arrays will be significantly affected, with the result that a rate constant much smaller than the true rate will be determined [98–100,102]. The same is true for fast homogeneous kinetics, that is, when the corresponding kinetic layers are smaller than the interelectrode distance. In this respect it is worthwhile to recall the well-known result that rate constants determined at poorly polished or dirty (i.e., partially blocked) electrodes are always smaller than those determined at clean electrodes, although the overall current amplitudes observed at these electrodes are often not drastically affected by cleaning or polishing.

The analysis above shows that when the diffusion layers developing at each electrode are smaller than the interelectrode distance, an array behaves as the sum of its individual electrodes [107–110]. The current is then exactly equal to that observed at one electrode element times the number of elements. Therefore, under these conditions the array plays only the role of a current amplifier with a gain factor that is equal to the number of electrodes of the array. For

experiments of larger duration times, the diffusion layers at each electrode overlap and the current amplification is less than above. This occurs because diffusion at each electrode element is shielded by adjacent electrodes [80,107] that
act as a kind of *Faraday-diffusional cage*. The current observed at the array is
then considerably smaller than that which would be observed for an isolated
electrode element times the number of elements and is identical to that which
would be observed at a single large electrode whose size is that of the array.
This shielding effect is illustrated in Figure 10a (bottom panel), where it is seen
that the current flowing at an array consisting of two parallel band electrodes is
less than the double of the current flowing at one of these electrodes operated
alone under identical conditions (top panel) but is almost identical to that at a

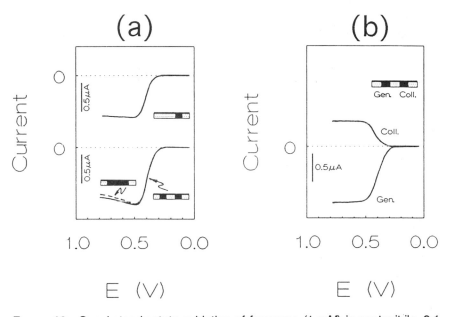

FIGURE 10 Quasi-steady-state oxidation of ferrocene (1 m*M*) in acetonitrile, 0.1
M NBu$_4$PF$_6$, at a double-band electrode (*w* = 4.6 μm, *g* = 4 μm); only forward
scans are shown for clarity. (a) Operation of one band alone (top panel) or of
both bands (bottom panel) set at an identical potential; dashed curve in bottom
panel, simulated response for a single-band electrode with *w'* = 2*w* + *g* = 13.2
μm, under identical conditions. (b) Operation of the double-band assembly in a
collector–generator mode; the potential of the collector (cathode) is set at 0.0 V
while the potential of the generator (anode) is scanned; the collector and generator currents (same current scale) are represented as a function of the generator
potential. Potentials in (a) or (b) are given in volts vs. SCE. (Adapted from Ref.
80.)

single band electrode whose width is equal to the overall width of the array (bottom panel, dashed line) [80].

Based on what precedes, one may really wonder about any possible advantage of electrochemistry at arrays where all electrodes are operated at the same potential. Indeed, in the short time limit, amplification of the current at one element necessarily occurs with a finite and fixed gain because the number of elements of the array is finite and constant. The array's performance are then of a poor quality vis-à-vis those achieved by classical electronic means. In the long time limit the current is equal to that observed at a single electrode of a dimension equal to that of the array, a situation that seems to present absolutely no interest at all vis-à-vis direct use of a large electrode. However, and surprisingly, under the latter circumstances, arrays present their major interest. Indeed, noise pickup in electrochemical experiments is proportional to the surface area of conductors. It is then necessarily less at an array than at a regular electrode of identical overall dimension. Theoretically, the signal-to-noise ratio is improved at arrays by a factor close to $(1 + A_{insul}/A_{el})$, where A_{el} is the total concrete surface area of electrodes and A_{insul} that of insulators. The same is true for capacitive currents since these are also proportional to the true electrode surface area, showing that at arrays the ratio between faradaic and capacitive currents is increased by the same factor $(1 + A_{insul}/A_{el})$. Both advantages have been used to allow detection of low currents and for the design of electrochemical detectors adapted to small quantities of material (e.g., in chromatography) [111–116].

To conclude this section, we want to discuss one aspect that is often advanced as an advantage of arrays in relation to preparative-scale electrolysis in resistive media [68,117]. It is based on the observation that ohmic drop (iR) at an ultramicroelectrode is much less than that at electrodes of millimetric or centimetric size (compare Fig. 1). This property has allowed analytical and kinetic applications of ultramicroelectrodes under very resistive conditions where other electrodes could not have given any practicable electrochemical response (see below). However, it should be emphasized that the resistance (R) at an ultramicroelectrode is necessarily much larger than that at a larger electrode when both are used under identical conditions. The reduced ohmic drop stems from the fact that the current (i.e., the quantity of material electrochemically converted, is almost negligible). This obviously prohibits the use of ultramicroelectrodes for preparative-scale applications where a stoichiometric quantity of electrons is required.

A few authors have advanced that arrays may cumulate both advantages [68,117]: each element of the array would perform under very resistive conditions where a large electrode would not be able to perform, yet the low conversion at each electrode element would be compensated by the large number of electrode elements present in the array. Naïvely, if one element converts only 0.1% of the material, an array with 1000 elements should give 100% conversion.

Unfortunately, this is not true. Indeed, the electrical resistance at an array is always larger than that at a single electrode with an identical overall dimension, because of the important narrowing of the electrical field lines in close proximity of each electrode element. Thus whenever one considers identical overall currents (i.e., identical quantities of material converted *per* unit of time) ohmic drop at an array is necessarily larger than that effective at a single electrode of identical overall dimension. In other words, from the strict point of view of ohmic drop, using an array for preparative-scale electrolysis is worse than using a plain electrode of identical overall dimension.

However, from other points of view, as, for example, the effect of increased local diffusion rates on electrochemical selectivity and the cost of an electrocatalytic conducting electrode material, arrays may present advantages over classical electrodes. Similarly, arrays of interdigitated electrodes where adjacent electrodes are, alternatively, a cathode or an anode may present specific advantages for electrosynthesis; for example, tandem electrolysis of two reactants may lead to unstable intermediates that react together over the interelectrode gaps (see section II.C.2.b) to afford the desired product. Such an interdigitated electrode could be used in very resistive media because the current need only be transported over a very limited range (compare corrosion) and would present the important advantage that very unstable intermediates could be used since lifetimes of a few milliseconds would be sufficient for high efficiency of the cell.

2. Arrays with Electrodes Operated in a Collector–Generator Mode

Consider an electrochemical reaction taking place at an ultramicroelectrode under steady-state or quasi-steady-state conditions. Whenever the product of the electrochemical reaction is chemically stable, its concentration profile is similar to one of those represented in Figure 6a and b with $C_{product} \approx C^0 - C_{reactant}$ (the rigorous equality is achieved only when the reactant and product diffusion coefficients are identical). From these concentration profiles it is seen that if another ultramicroelectrode is implanted near the first one and its potential set at a value such that the product is electroactive, the second ultramicroelectrode acts as a detector for the product electrogenerated by the first (see, e.g., Refs 34, 69, 72, 73, 78 to 80, 107, and 118 to 122). This is illustrated in Figure 10b [80], where the detector potential is set on the plateau of the reduction wave of ferricinium, while the generator potential is scanned over the ferrocene oxidation wave. Such a generator–collector pair has functions and overall properties that are reminiscent of those of the classical rotated ring-disk electrode (RRDE) or of thin-layer cells [1,2,123]. However, besides miniaturization, the following important differences should be noted. On the one hand, at paired ultramicroelectrodes transport of molecules depends almost exclusively on diffusion (as in the thin-layer cell) and not on forced convection as with the RRDE (see below). On the other hand, the ultramicroelectrode pair conserves its immunity to ohmic drop that is common to single ultramicroelectrodes because the flux of material that spills over from such an assembly to the bulk of solution is necessarily smaller than that which flows at the generator electrode operated alone (see below).

Intuitively, the ideal geometry for high collection efficiency seems to be that of a disk generator electrode surrounded by a ring collector electrode analogous to the RRDE. However, there are obvious technological difficulties in fabricating such a precisely centered arrangement of electrodes at a micrometric (or less) scale [124–129]. Good surrogates to this ideal assembly consist of a band generator flanked by one or two parallel band identical collector(s) [34,69,72,73,78–80,107,118–120]. Indeed, such structures can be made rather easily by sandwich-type techniques [80] analogous to those used for the manufacture of single-band electrodes [3]. Use of microlithographic techniques allows the fabrication of interdigitated band arrays (IDAs) similar to that in Figure 8c, where the parallel electrodes are alternatively generator and collector [69], with interelectrode gaps as small as 0.5 μm [78]. As already mentioned above, from our point of view an important disadvantage of microlithographic structures relates to the difficulty of renewing electrode surfaces in between runs, whereas this is easily done by polishing sandwich-type structures. Yet if the collector–generator structure is to be used for standardized assay analysis rather than for kinetic laboratory studies, the cheap manufacturing costs and high reproducibility of microlithographic techniques allow the use of a disposable IDA for each measurement. It is therefore understood that the notion of ''best assembly'' is not innate but is highly dependent on the application to which the structure is devoted.

The intrinsic difficulties of modeling diffusion at band electrodes (see above) are obviously increased at these generator–structures based on parallel bands, because at least two nonuniform diffusional fields operate together and necessarily crosstalk. On the other hand, one of the main interests of these structures is their capability to address kinetic problems, since collection efficiencies appear to be related to the geometry of the structure itself [68] (see below for steady-state regimes) but obviously also depend on how much material produced at the generator chemically survives the diffusional flight [118] over the gap between the generator and the collector. Therefore, the development of reliable and practical techniques for modeling diffusion at these assemblies is essential. In addition, these techniques should be adjustable easily and rapidly to any particular kinetic situation under investigation. It is our feeling that this can be performed by simulation only.

Simulations performed in the real space are extremely difficult to handle [79,107] because of the nonuniform current densities at the band electrode

[65–67] (Figure 7a). As for the single-band electrode, these important difficulties may easily be bypassed if numerical integration of Fick's laws is performed in a space where current densities are nearly uniform and concentrations vary almost linearly with only a few smooth curvatures. The geometry of the assemblies is perfectly suited for the use of conformal mapping techniques. As illustrated in Figure 11, Schwarz–Christoffel transformation [62] allows virtually

(a)

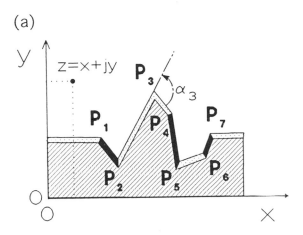

(b)

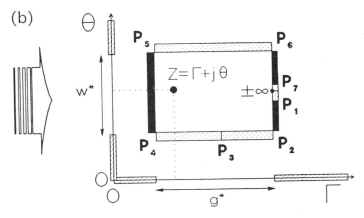

FIGURE 11 Schematic representation of the effect of Schwarz–Christoffel transformation. The true array whose cross section is shown in (a) is transformed into the thin-layer cell equivalent in (b).

any device based on parallel band electrodes to be transformed into an equivalent thin-layer cell where isoconcentration lines (and their orthogonal flux lines) are parallel and equidistant. They may then be determined easily and rapidly on a desk computer with high accuracy by classical finite-difference numerical algorithms.

Schwarz–Christoffel transformation allows the transformation of a space (x,y) into a desired space (Γ,θ), with the important advantage of conserving the angle between two transformed curves [62]. Thus two orthogonal curves in (x,y) space are transformed into two orthogonal ones in (Γ,θ) space. When applied to electrochemical problems this property is of extreme importance because it conserves the orthogonality between isoconcentration lines and flux lines. An immediate consequence is that the structure of the Laplacian operator is conserved and currents evaluated at the transformed electrodes in (Γ,θ) space exactly equal to those at the real electrodes. The transformation formulates [62] by expressing the mathematical relationship between the complex number $Z = \Gamma + j\theta$ (with $j^2 = -1$) representing a transformed point of (Γ,θ) space and the complex number $z = x + jy$ representing the corresponding original point of (x,y) space:

$$Z = T + K \int_0^z \left[\prod_{k=1}^n (\zeta - p_k)^{\alpha_k/\pi} \right] d\zeta \tag{53}$$

where T and K are any suitable constant complex numbers that allow, respectively, an adequate translation and a scaling/rotation of the transformed space; $p_k = x_k + jy_k$ ($k = 1, \ldots, n$) are the n poles of inversion, that is, n points selected in (x,y) space, where a counterclockwise rotation of an angle α_k is effected in the transformed space (see Fig. 11). [Note that in equation (53), ζ is the integration variable used in (x,y) space.]

Application of the relation in equation (53) allows transformation of the complex array in Figure 11a, where, for example, electrodes located in between poles P_1 and P_2 or in between P_6 and P_7 are generators and that in between P_4 and P_5 is a collector, into its thin-layer cell equivalent shown in Figure 11b. At steady state, that is, when the flux exiting the box in Figure 11b through the point indicated as $\pm\infty$ tends toward zero, isoconcentration lines in Figure 11b approach parallel equidistant lines, except for a slight distortion in the region located in between the two generators. Indeed, because of the properties of the transformation in equation (53) [62], Fick's second law, expressed as

$$\frac{\partial C}{\partial t} = D \left[\left(\frac{\partial^2 C}{\partial x^2} \right) + \left(\frac{\partial^2 C}{\partial y^2} \right) \right] \tag{54}$$

in the physical space (x,y), becomes

$$\frac{\partial C}{\partial t} = D^* \left[\left(\frac{\partial^2 C}{\partial \Gamma^2} \right) + \left(\frac{\partial^2 C}{\partial \theta^2} \right) \right] \tag{55}$$

in (Γ,θ) transformed space, D^* being a space-dependent diffusion coefficient accounting for local dilatation or compression of space [compare, e.g., equation (41)]. Under steady-state conditions, $\partial C/\partial t = 0$ by definition of steady state. Because of equation (55) this indicates that the fluxes are conservative in the box represented in Figure 11b. A first conclusion is that the current that flows away from the two generators on the right-hand side of the box necessarily flows out of the box through the collector on the left-hand side. In other words, at steady state, the collection efficiency is necessarily 100%. Also, because of the symmetry of the system in Figure 11b, it is deduced that except in a small domain around the intergenerator gap (P_1–P_7), concentrations approach the asymptotic solution $C = (\partial C/\partial \Gamma)_{\Gamma_{col}} \Gamma + \text{constant}$, where Γ_{col} is the abscissa of the collector electrode (P_2–P_5) in (Γ,θ) space. A good approximation for $(\partial C/\partial \Gamma)_{\Gamma_{col}}$ is then readily deduced:

$$\left(\frac{\partial C}{\partial \Gamma} \right)_{\Gamma_{col}} = \frac{C_{gen} - C_{col}}{g^*} \tag{56}$$

where $g^* = \Gamma_{gen} - \Gamma_{col}$ is the gap of the thin-layer cell in (Γ,θ) space; C_{gen} and C_{col} are the concentrations on the generator and collector electrodes, respectively. Because of the property of flux conservation of the transform in equation (53), the current flowing through the system under steady-state conditions is obtained by direct integration of $(\partial C/\partial \Gamma)_{l\,col}$ over, for example, the collector surface area in (Γ,θ) space; that is,

$$i = nFDl(C_{gen} - C_{col}) \frac{w^*}{g^*} \tag{57}$$

where w^* is the thin-layer width in (Γ,θ) space and l the length of the band electrodes in (x,y) physical space. Therefore, the problem of evaluating the steady-state limiting current amounts only to determining the value of w^*/g^*, which is readily performed from equation (53) (see below).

This simple example illustrates most of the advantages of conformal mapping techniques [61,62]. For simplicity we have considered here only an asymptotic solution that is valid only when the intergenerator gap is small vis-à-vis w^* (see, e.g., Ref. 120). However, a straightforward numerical integration of equation (55) (with $\partial C/\partial t = 0$) performed in the (Γ,θ) space of Figure 11b would easily provide the exact steady-state solution (compare [120] for solution of a similar problem).

As noted above for single electrodes, although this change of space is best suited for steady-state conditions (those for the best operation of these generator–collector assemblies), integration of non-steady-state Fick's law [equation (55)] is also feasible [34,120]. It is also worthwhile to emphasize that these time-dependent integrations are performed with fewer technical difficulties than in the natural (x,y) space [34,120,130], except perhaps when true planar diffusion is approached. Under these non-steady-state conditions, numerical integration of diffusion equations requires knowledge of D^* values at each point of the (Γ,θ) simulation grid. Although these values may be determined analytically or numerically from the transform in equation (53), their precise evaluation at each point of a simulation grid in (Γ,θ) space may be cumbersome or at least tedious and is always time (human or computer) consuming. Therefore, it is often advisable to compromise between the simplicity of the transformed space and the ease of evaluation of diffusion coefficients in the transformed space. An example of such a compromise is given next for a double-band generator–collector device [34,120].

A cross section of a double-band generator–collector assembly is shown in Figure 12a in the real space (x,y). For calculations under steady-state conditions, the best transformed space, shown in Figure 12b, is obtained through application of the following transformation, where $Z^* = \Gamma^* + j\theta^*$ is the transformed point of $z = x + jy$ [62]:

$$Z^* = \int_0^z \left\{ \left[\zeta^2 - \left(\frac{g}{2}\right)^2 \right] \left[\zeta^2 - \left(w + \frac{g}{2}\right)^2 \right] \right\}^{-1/2} d\zeta \tag{58}$$

and ζ is the integration variable in (x,y) space. Note that this transformation corresponds to a rotation of $\pm\pi/2$, respectively, at each edge of each electrode where the four poles of inversion are located. The transformed device is then a thin-layer cell with an infinitely small diffusional leak at $\pm\infty$ through which the thin-layer cell communicates with the bulk solution infinitely far away. At steady state no diffusional leakage occurs through this point and the collection efficiency is then necessarily 100% because fluxes are conservative (see above). Moreover, since this thin-layer cell is totally symmetrical, the solution given in equation (57) is now exact [34]. The term w^*/g^* is readily determined from the transform in equation (58) and is given by the ratio of elliptic integrals that can be evaluated for any value of w/g, the true width-to-gap ratio, from classical numerical integration or from tables [34]:

$$\frac{w^*}{g^*} = \frac{\int_{g/2}^{w+g/2} \{[\zeta^2 - (g/2)^2][(w + g/2)^2 - \zeta^2]\}^{-1/2} d\zeta}{\int_{-g/2}^{g/2} \{[\zeta^2 - (g/2)^2][\zeta^2 - (w + g/2)^2]\}^{-1/2} d\zeta} \tag{59}$$

Based on this transformation, the extremely difficult derivation of the steady-state solution in the (x,y) real space is replaced by an easy evaluation of classic

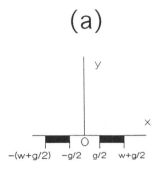

(a)

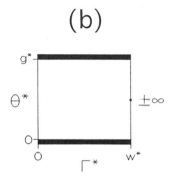

(b)

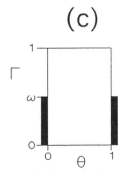

(c)

FIGURE 12 Transformation of a double-band electrode (a) into its thin-layer closed cell in (b) or thin-layer open cell in (c). See the text for the pertinent transformations used.

elliptic integrals in equation (59). When one wants to investigate how this steady-state solution is approached, Fick's second law has to be integrated without considering that $\partial C/\partial t = 0$. This requires the difficult evaluation of D^* values at each point of (Γ^*, θ^*) space. We found [34,120] that the best compromise between suppression of technical difficulties in numerical procedures (i.e., suppression of the infinite current densities and infinite curvature of isoconcentration lines at the edges of each electrode) on the one hand, and simple formulation of D^* on the other hand, consisted of using a transformation based on that in equations (36) to (38), developed previously for the band electrode [67]. Thus use of equations (36) to (38) (with g instead of w) makes it possible to transform the physical (x,y) space shown in Figure 12a into another (Γ, θ) conformal space represented in Figure 12c. The time-dependent concentrations and currents are then given by integration of Fick's second law [equation (40), where D^* is given in equation (41) (with g instead of w). Although this transformed space is not the most suitable when steady state is approached, it affords a set of isoconcentration lines (Figure 13a) that is considerably more regular than that obtained in the real (x,y) space shown in Figure 13b [34]. A rather coarse simulation grid is then sufficient to allow a highly accurate finite-difference integration of Fick's second law. This is to be contrasted with the complexity of the simulation grid to be used for a similar accuracy if the solution was to be determined numerically in real (x,y) space (compare, e.g., [79]). This is particularly patent when one compares Figure 13a and b. Note, for example, that the highly curved isoconcentration lines in Figure 13b (gap region) are replaced in Figure 13a by a set of almost parallel and equidistant lines that are very reminiscent of those obtained in a thin-layer cell.

Another interest of Schwarz–Christoffel transformation is that it allows easy comparison of the properties of collector–generator devices based on different configurations of electrodes. The three most used configurations are the double band examined above; the triple band, where a generator is flanked by two parallel symmetrical collectors; and interdigitated arrays (IDAs), where a large series of identical and equidistant parallel electrodes are operated alternatively as generators or collectors [34,68,69,72,73,79,80,107,118–120]. Based on symmetry considerations and on the use of Schwarz–Christoffel transformations it can be shown that these three devices are equivalent to a double-band configuration, irrespective of their differences in the physical space. This is illustrated in Figure 14. Because of the symmetry around the vertical axis passing through the center of the generator (Fig. 14a), the triple band is equivalent to twice the half-cell unit in Figure 14b. Rotation of $\pi/2$ at the outer edge of the half-collector, performed through Schwarz–Christoffel transformation, makes it possible to obtain the equivalent double band represented in Figure 14c. Similarly, when the number N of paired electrodes is sufficiently large for edge effects at the array's ends to be negligible, the IDA in Figure 14d is

(a)

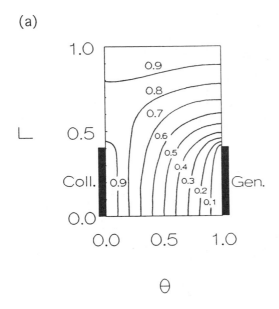

(b)

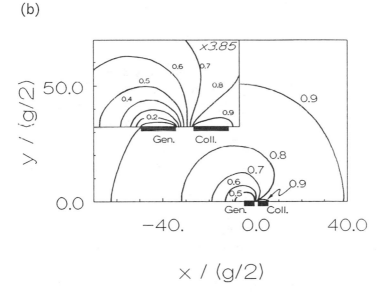

FIGURE 13 Isoconcentration lines at a double-band generator–collector device in the (Γ,θ) transformed space or in the (x,y) real space. Numbers on the curves are the values of C/C^0. Representation is given under non-steady-state conditions, for $w/g = 2$ and $(Dt)^{1/2}/g = 10$. (Adapted from Ref. 34.)

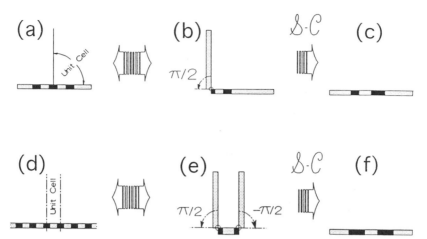

FIGURE 14 Equivalencies between a triple band (A) or an interdigitated array (B) and a double band. The true devices (a, d) are represented together with their elementary unit cell (b, e) in true (x,y) space. The unit cells (b, e) are then transformed into their double-band equivalents (c, f) using the Schwarz–Christoffel transformation.

equivalent to N times the unit cell represented in Figure 14e. Use of the Schwarz–Christoffel transformation to effect a rotation of $\pm\pi/2$, respectively, at each half-electrode external edge allows Figure 14e to be transformed in the double-band equivalent shown in Figure 14f. From these simple mathematical transforms it is shown that all these configurations are indeed equivalent to double-band ones. The same is also true for stacked parallel bands such as that in Ref. 78. The nature of the original configuration is reflected only by the relative sizes of the equivalent gap and of the widths of the two equivalent electrodes. Cast in other words, this means that there is always a double-band configuration that is equivalent to a triple band or to an interdigitated array. This is illustrated, for example, in Figure 15a, which compares the theoretical steady-state currents per generator element at a double band (DB), a triple band (TB), and an IDA constructed in real space with identical electrodes of width w separated by identical gaps with $g = w$ [120]. Note that the collection efficiencies are not shown because they are all equal to 100% under steady-state conditions (see above). From this figure and the abacus in Figure 15b [34], it is seen that a triple band with $w/g = 1$ is equivalent to a double band with $w/g = 4.54$, whereas an IDA with $w/g = 1$ is equivalent to a double band with $w/g = 2.36$.

   This property has interesting consequences with experimental and theoretical values. Indeed, from an experimental point of view it indicates that the search for complicated configurations is not really rewarding whenever only

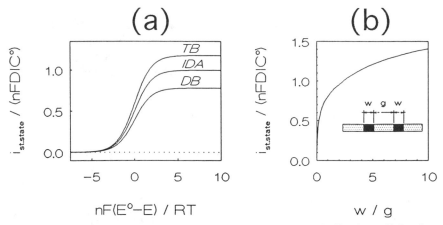

FIGURE 15 (a) Comparison of the steady-state currents (collection efficiencies are 100% in each case) at a double band (DB), a triple band (TB), or an interdigitated array (IDA) with $w/g = 1$ in each case. For IDA the current shown corresponds to a single generator element. (b) Relationship between the steady-state current at a double-band electrode and the width-to-gap ratio w/g. (Adapted from Ref. 120.)

steady-state applications are of interest. Indeed, the nature of the exact configuration is reflected only through the rapidity with which this steady-state behavior is attained: for example, considering identical widths and gaps, a triple band achieves steady state faster than a double band, yet the limit corresponds to 100% collection efficiency in both cases [120]. From a theoretical point of view, the equivalence above establishes that any theoretical result based on, for example, double bands under steady state are directly transposable to TB, IDA, or other equivalent configurations, thus avoiding the necessity of an almost infinite number of theories that should be developed to account for an almost infinite number of assemblies.

To conclude this section devoted to diffusion at a generator–collector device, we want to address two other important points. A first one deals with the steady-state nature of the current observed at these assemblies. It is seen by comparison of the voltammograms in Figure 10a and b (bottom trace) that the character of steady state is larger for the current at the double band operated in the generator–collector mode (Fig. 10b) than for a single band of identical width w (Fig. 10a, top trace). This reflects the local feedback of material imposed by the collector into the diffusion layer of the generator [79]. This is compelling when one observes the isoconcentration lines represented in the (Γ, θ) space of Figure 13a. Indeed, as noted above, in the interelectrode gap, isoconcentration lines are similar to those observed in a thin-layer cell, being almost parallel and equidistant. Yet since the figure does not correspond to a sufficiently large experiment duration, a diffusional leak still exists toward the solution bulk (i.e., toward $\Gamma = 1$), as evidenced by the trend of the isoconcentration lines to become parallel to the θ axis when Γ approaches unity. The corresponding flux of material toward the solution is obviously smaller than that describing the direct exchange between the two electrodes, as evidenced by the larger spacing between isoconcentration lines. In other words, the steady-state crosstalk between the two electrodes overcomes and prevents the time-dependent diffusion toward the bulk solution. The current is then necessarily larger at a generator–collector pair than at a single band, and its time variations are then smaller than those at a single band (Table 1) because they stem only from those of the reduced flux of material that leaks out of the thin-layer cell [34,120].

This leads to the second point that we wished to address. The description above supposes implicitly that a strong correlation exists between feedback, collection efficiency, and steady-state character of the current at these devices [34,72,78,120]. Figure 16a demonstrates that for a given assembly, such a correlation actually exists between these different characteristics: when time increases, the experimental points proceed along the solid theoretical curve from zero feedback ($i_{band}/i_{gen} = 1$) and zero collection efficiency ($i_{coll}/i_{gen} = 0$) to infinite feedback ($i_{band}/i_{gen} = 0$) and 100% collection efficiency ($i_{coll}/i_{gen} = 1$). However, the latter limit is attained only at infinite time because it corresponds to complete annihilation of the diffusional flux toward the bulk solution. Because cylindrical geometry predominates at large distances from the electrodes, far-outer-range diffusion is cylindrical. This is, for example, manifested by the isoconcentration line shown for $C/C^0 = 0.9$ in Figure 13b, which is extremely close to an half-circle. It is then inferred that the diffusional flux toward the bulk solution decays as the current at a single-band electrode [i.e., as the reciprocal of the logarithm of time (Table 2)] [67]. In this respect it is noteworthy that as soon as Dt exceeds g^2 a few times, the slow variations of i_{band}/i_{gen} when time increases (vertical axis of Fig. 16a) reflect primarily the progressive decay of i_{band} because i_{gen} is almost independent of time under these conditions, being almost at its steady-state limit. Similarly and under the same conditions, the variations in the collection efficiency (i_{coll}/i_{gen}; horizontal axis of Fig. 16a) feature primarily those of i_{coll}.

This occurs because i_{gen} represents primarily the status of diffusion in close proximity of the gap (i.e., in the region of space where most of the flux lines flow) (Fig. 13). Therefore, provided that the duration of the experiment is sufficient for the diffusion layer at each electrode to extend over the gap and overlap, the current that flows at the generator is almost at its steady-state limit for infinite times. Conversely, the collector current still increases with time because it keeps on collecting those molecules that flew (and still fly) from the outer edge of the generator and eventually attain the outer edge of the collector after

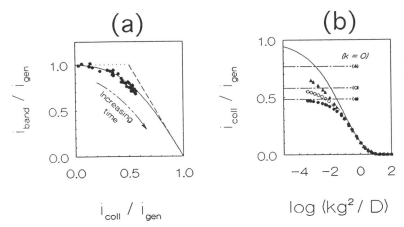

FIGURE 16 Double-band electrode operated in a generator–collector mode. (a) Relationship between the feedback (i_{gen}/i_{band}) and the collection efficiency (i_{coll}/i_{gen}) as a function of the experiment duration and in the absence of a chemical reaction. Points correspond to experimental systems (oxidations of ferrocene or 9,10-diphenylanthracene in acetonitrile). Solid line to simulated currents. Dotted and dashed lines, asymptotic theoretical behavior at zero time (dotted line) or infinite time (dashed line). (b) Effect of a chemical reaction (first-order rate constant k) on the collection efficiency as a function of kg^2/D, for different values of time and width-to-gap ratio. Triangles, $(Dt)^{1/2}/g = 100$, $w/g = 2$; open circles, $(Dt)^{1/2}/g = 5$, $w/g = 1$; filled circles, $(Dt)^{1/2}/g = 2.5$, $w/g = 0.5$. Solid line is the collection efficiency as a function of kg^2/D only, according to the empirical equation (61). Horizontal dot-dashed lines are the values of collection efficiencies at $k = 0$ for the systems represented by identical symbols. (Adapted from Ref. 34.)

long diffusional trajectories (compare Fig. 13). Because diffusion is cylindrical these external trajectories become larger and larger when time increases (Fig. 5a) and collection efficiency increases with a smaller and smaller rate. Indeed, the instant amount of material that flies over such trajectories (either from the external edge of the generator or to the external edge of the collector) decays as the current at a band does. This is the intrinsic origin of the asymptotic limit at large times shown by the slanted dashed line in Figure 16a and whose equation is [34,120]

$$\frac{i_{coll}}{i_{gen}} = 1 - \eta \times \frac{i_{band}}{i_{gen}} \tag{60}$$

with $\eta = 0.5$ for a double-band configuration [34]. For a triple band (TB) η is close to 0.25 [120] and is even smaller for an IDA [78], reflecting that less and less material may "escape" over these long trajectories when the generator is flanked by two collectors (TB) or when the space available for diffusion is restricted (IDA) so that diffusion toward the bulk solution becomes planar (compare, e.g., Fig. 9c and 14d and e). Because the time dependence is introduced through the variations of i_{band}, the lower η is, the faster steady-state diffusion is approached [120].

The description above shows that at these arrays a special diffusional regime is achieved as soon as the diffusion layer developing at the generator exceeds the interelectrode gap(s) a few fold. Under these conditions, the level of crosstalk between the generator and collector(s) is almost identical to that which will be achieved at infinite times. This regime is evidenced experimentally by the fact that voltammograms performed at these assemblies under these conditions present the primary characteristic of steady-state voltammograms (i.e., a flat plateau region) [1,2] (compare, e.g., Fig. 10b). However, this regime is not true steady state: the collection efficiency is not 100% (compare, e.g., Fig. 10b), because in regions that are much farther from the generator, diffusion is still a function of time. These regimes thus feature a "quasi-steady-state" limit that is akin to that at a single-band electrode. Note that because the true steady-state limit is achieved only at an infinite time, most arrays are generally used under this quasi-steady-state regime.

The two following experiments illustrate unambiguously that the rationalization above is not mere theoretical speculation. To achieve this goal, one must find a way experimentally to "disconnect" short-range diffusion [i.e., that operating over distances that are close to the gap(s) width(s)] from long-range diffusion [i.e., that active over distances that exceed to a considerable extent the gap(s) width(s)]. In other words, one wants to confirm experimentally the dichotomized nature of diffusion that we have identified above on theoretical grounds. This can be performed thanks to homogeneous chemical kinetics.

a. Determination of Homogeneous Kinetics at Arrays

Let us consider that the product formed at the generator surface has a finite lifetime, which corresponds to a first-order homogeneous rate constant k. If the lifetime is comparable to (or less than) the time corresponding to diffusion over the gap width (g), all the product molecules that undergo diffusion over the long-range trajectories discussed above are chemically annihilated before they have any chance to reach the collector surface. Under such conditions, collection efficiency necessarily depends only on diffusion–reaction effects over the interelectrode gap. They should therefore be only a function of kg^2/D that compares the relative values of the lifetime (i.e., approximately $1/k$) vis-à-vis the time of flight over the gap (i.e., approximately g^2/D) [34,118–120]. This should be true whatever the nature of the device or the width(s) of the electrodes or whatever is the duration of the experiment, provided that the latter is sufficient for the diffusion layer developing at the generator to extend over the collector(s).

Conversely, if k is made smaller and smaller, so that more and more molecules undergoing long trajectories may survive diffusion to the collector, the collection efficiency should also reflect the status of diffusion far from the electrode surfaces. Therefore, (i_{coll}/i_{gen}) should still be a function of k but should also become dependent on the nature of the device, on the duration of the experiment, and on w/g, the gap-to-width ratio of the device, since these parameters control long-range diffusion trajectories [34,120] (compare Fig. 15b for the effect of w/g at a double band).

Figure 16b shows that these predictions are effectively true. At a double band, for $kg^2/D > 0.1$, collection efficiencies are independent of the duration t_{max} of the experiment or of w/g, but depend only on kg^2/D [34]. However, when k is decreased so that $kg^2/D \leq 0.1$, collection efficiencies still depend on kg^2/D but also become a function of w/g and of Dt_{max}/g^2. Under the latter conditions, experimental determinations of rate constants using a double-band generator–collector assembly require that a pertinent working curve is available for the precise values of w/g and Dt_{max}/g^2 used in the experiment. Conversely, when $kg^2/D > 0.1$ a single working curve is sufficient, irrespective of w/g and Dt_{max}/g^2. Empirical fitting of the theoretical results in Figure 16b, and others not displayed, shows that this working curve (shown as the solid curve in Fig. 16b) is approximated with very good accuracy by

$$\frac{i_{coll}}{i_{gen}} = \exp\left[-\left(\frac{4\pi kg^2}{D}\right)^{1/\pi} \right] \tag{61}$$

for a double-band collector–generator assembly (note that a different expression is reported in Ref. 34, yet both empirical expressions are almost identical for the range of kg^2/D values where they apply). It is noteworthy that equation (61) reflects only diffusion over the gap region, that is, when crosstalk and current exchange involve only small domains located at the inner edges of the generator and collector electrodes. Indeed, under these conditions, every product molecule escaping from the outer edge of the generator is chemically annihilated before reaching the collector. Therefore the supply of material at the outer edge of the generator becomes rapidly negligible, as is that at the outer edge of the collector. The empirical equation (61) then reflects only the geometry of the assembly in the gap vicinity. Although to the best of our knowledge this has not yet been established, equation (61) should then be almost independent of the nature of the assembly and should be valid for double-band, triple-band, IDA, and other configurations and be independent of the electrode widths.

b. Electrochemical Luminescence at Double-Band Electrodes

The basic principles of electrochemical luminescence (ECL) are well documented [2] and the reader is encouraged to refer to excellent reviews on this subject [131,132]. For our purpose here, it suffices to consider that when a strong electron donor (D⁻) exchanges homogeneously an electron with a strong electron acceptor (A⁺), a fraction of the electron transfer reactions proceeds via Marcus's inverted region (for a review, see, e.q., Ref. 133) and affords the reduced form of the electron acceptor into an electronically excited state $A*$ which may then deactivate thermally into vibrations but also through emission of a visible photon:

$$D^- + A^+ \xrightarrow{\Delta G^0 \ll 0} D + (1 - \epsilon)A + \epsilon(A* \xrightarrow{h\nu} A) \tag{62}$$

The rate constants of these processes, being close to the diffusion limit [134], are then sufficiently large to be considered as infinite within the time scales considered here. In other words, A⁺ and D⁻ cannot coexist within the same volume of space because they instantly annihilate each other via the process in equation (62).

The same principle obviously applies when A⁺ and D⁻ are, respectively, the oxidized and reduced forms of an identical starting material, M. Upon using a double-band device, M can be reduced to M⁻ at the band cathode and oxidized to M⁺ at the parallel band anode [135]. M⁺ and M⁻ then recombine within the median plane of the double band according to

$$M^- + M^+ \xrightarrow{\Delta G^0 \ll 0} (2 - \epsilon)M + \epsilon(M* \xrightarrow{h\nu} M) \tag{63}$$

to afford a photon (with an overall quantum yield ϕ) and two M molecules that diffuse again toward the anode and cathode bands, where they are recycled to M⁺ and M⁻, respectively. In other words, a double band operated in ECL mode is equivalent to the juxtaposition of two double-band systems, each composed of a real generator (the cathode or anode, respectively) and a virtual collector constituted by the median plane of the double-band device as outlined in Figure 17a and b. In each half-system the generator current is the real current flowing at the anode or cathode, respectively, whereas the virtual current at the virtual collector is measured by the photon flux corrected for the quantum yield of the overall process in equation (63). This analogy is even more apparent when the system is represented in the $(\Gamma*, \theta*)$ conformal space defined by the transformation given in equation (58) (compare Fig. 12b), as shown in Figure 17c. In this space the median plane of the device is also the median plane between the electrodes and is represented by the horizontal dashed line in Figure 17b. This plane separates the original thin-layer cell into two identical thin-layer cells whose gaps are exactly half that, $g*$, of the original cell. Yet the width, $w*$, of the electrodes remains unchanged. From this figure and equation (57), it is easily inferred that the current flowing under steady-state conditions at a double band operated in an ECL mode is exactly twice that observed under the same conditions at the same double band when it is operated in a classical generator–collector mode [35]. Also, because the equivalent gap is reduced by half, the steady-state regime is achieved much faster in an ECL mode than in a

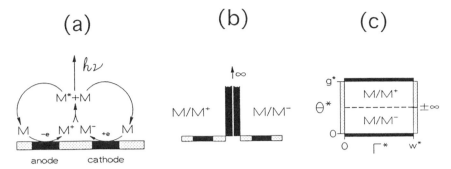

FIGURE 17 Double-band electrode operated in ECL mode (see the text): (a) schematic representation of the device operation; (b) equivalent representation in the form of two half-cells operating in a collector–generator mode; (c) Schwarz–Christoffel transformation of the cell in (a) into its thin-layer closed-cell equivalent (see the text).

collector–generator mode. Both predictions [35] are confirmed experimentally [35,135], as evidenced by the plots in Figure 18a.

Figure 18b compares the time variations of the current to those of the photon flux. It shows that whereas the current has almost reached its steady-state limit, the photon flux still lags behind. The fact that such a discrepancy is not due to some experimental problems but is an intrinsic characteristic of the system is confirmed by the fact that theory [35] predicts the same behavior (Fig. 18c). In fact, as will be made clear in the following, in such experiments the current represents primarily the status of diffusion in the immediate vicinity of the electrodes, whereas the photon flux represents both diffusion in this region and diffusion far away from the electrode surfaces. The first component (i.e., short-range diffusion over the gap region) corresponds to the initial fast rise of the photon flux that is comparable to the current decay toward its steady-state limit; both are completed after a duration on the order of a few g^2/D. The second component is shown by the slow rise toward the steady-state limit, which is approached as the reciprocal of the logarithm of time. It represents the effect of outer range diffusion. Indeed, since the virtual collectors at the median plane extend to infinity, they can collect eventually all those molecules that are missing initially in the balance between the current and the photon flux (Fig. 18c) because they underwent and still proceed along very long diffusional trajectories. Transport of these "missing molecules" over these long trajectories proceeds along cylindrical diffusion because of the intrinsic geometry of the device at long distances from the electrodes (compare Fig. 13b). This requires much larger times than diffusion over the gap region [i.e., about g^2/D; see above and

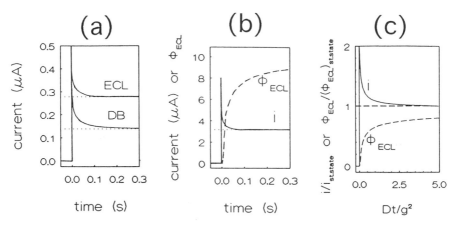

FIGURE 18 Theoretical and experimental ECL at a double-band electrode (w = 5 μm, g = 5.3 μm, l = 3 mm). (a) Comparison of the experimental currents at the anode when the cell is operated in an ECL mode (ECL) or in generator–collector mode (DB). Ru(bpy)$_3$(PF$_6$)$_2$, 0.6 mM in acetonitrile, 0.1 M NBu$_4$PF$_6$. Anode potential, 1.57 V vs. Ag wire; cathode potential, −1.33 V (ECL) or 0.0 V (DB) vs. Ag wire. (b) Comparison between time responses of current (i, μA) or of photon flux (ϕ_{ECL}, arbitrary units) during ECL of 9,10-diphenylanthracene, 10 mM, in 50:50 v:v acetonitrile: toluene with 0.1 M NBu$_4$PF$_6$. (c) Theoretical responses of current and photon flux during ECL for w/g = 1; note that both responses are normalized toward their asymptotic values at infinite time, shown as a common dashed horizontal line. (Adapted from Ref. 35.)

[34,118–120]), thus constraining the photon flux to lag much behind the current: its distance from its steady-state limit decreases along a time variation akin to that of the quasi-steady-state current at a single band [i.e., as $1/\ln(t)$] (Table 2).

D. Conclusions

ECL experiments and homogeneous chemistry afford a visual representation of the dichotomized nature of diffusion at ultramicroelectrodes: on the one hand, short-range diffusion over distances comparable to the electrode width or radius a that achieves steady-state behavior within times on the order of a few a^2/D and is primarily responsible for the current exchanged at the electrode surface; and on the other hand, a long-range diffusion that never achieves steady state because it corresponds to zero fluxes at infinite time [34,120] but is almost not reflected in the current exchanged at the electrode surface (compare, e.g., Fig. 3a and 4).

Moreover, as already noted at several instances in this chapter, under most experimental circumstances, this long-range diffusion is often purely speculative since it involves diffusional fluxes that fade out when time or distance from the electrode increase (Fig. 5a). Such vanishingly small fluxes may then easily be overwhelmed by convection (that increases when the distance from the electrode increases; Fig. 2) and by any slow chemical reaction that otherwise would be negligible and undetected. The latter point is perfectly evidenced by the fact that whereas the experimental variations of the current and photon flux reported in Figure 18b for 9,10-diphenylanthracene follow the theoretical predictions given in Figure 18c over a time range of several g^2/D units, the steady-state limit of the photon flux is slightly less than predicted and is achieved faster (and it is noteworthy, with a large standard deviation from run to run) than predicted [35]. This illustrates that in its time-course expansion, the diffusion layer necessarily reaches a region where difficultly reproducible convective transport becomes faster and dominates transfer of molecules. This involvement of convection is also patent by the fact that although theory predicts that the current at an electrode of cylindrical geometry [equation (13) or (35)] or at a planar electrode [2] should eventually tend to zero when the time goes to infinity, this limit is never attained: experimental currents always tend toward a constant limit because convection takes the relay of diffusion when diffusional fluxes would become vanishingly small. The same effect also occurs at the spherical electrode, although it is masked because the diffusional current also tends toward a constant limit that is much larger than the convective contribution.

A second point we wish to emphasize here deals with "ohmic drop at an electrode." As pointed out in Section I, resistance at a microelectrode or at an ultramicroelectrode depends on the electrode dimension and not on its surface area. This occurs because the electrostatic resistance expression follows from integration of the Laplace equation, $\nabla^2 V = \rho/\epsilon$, where V is the electrostatic potential, ρ the spatial charge density, and ϵ the dielectric constant of the medium. In an electrochemical cell, the solution is electroneutral except within a few tens of angstroms from the electrode surface (the double layer [2], *vide infra*). Thus one has $\nabla^2 V = 0$ over the entire domain of interest. It is noteworthy that this equation is identical to $\nabla^2 C = 0$, which gives the concentration at an electrode when steady-state diffusion prevails (see above). Because of this identity, the ohmic drop, iR, at an electrode under diffusional steady state is constant and independent of the electrode shape and dimension [4]. This corresponds to the domain labeled C in Figure 1, which was used in Section I to define the *ultramicroelectrode* class.

However, this section on diffusion at ultramicroelectrodes has evidenced that the property above is not an intrinsic property of an electrode of a given shape and given dimension(s). Achievement of a steady-state or quasi-steady-state current depends not only on the size (a) of the electrode, but also on the duration of the experiment (t) and on the diffusion coefficient (D) of the electroactive material in the particular medium considered. Indeed, establishment of a steady-state or quasi-steady-state current requires that $Dt/a^2 \gg 1$ (compare Table 1, with $a = r_0$ or w). When the converse inequality is true (i.e., $Dt/a^2 \ll 1$), planar diffusion is observed instead and the current is proportional to the surface area (A) of the electrode and to the reciprocal of $(Dt)^{1/2}$ (compare the limits given in Table 1 for $\zeta \rightarrow 0$). Being independent of diffusion transport, resistance conserves its value, deduced from integration of $\nabla^2 V = 0$. Ohmic drop is then proportional to the reciprocal of $t^{1/2}$ and increases when the duration of the experiment is made shorter and shorter. It also increases when the electrode dimension is larger and larger. This second behavior corresponds formally to the domain labeled B in Figure 1, which was used for defining the *microelectrode* class.

This shows that the limit between domains B and C in Figure 1 is rather fuzzy because it depends not only on the electrode dimension and shape but also on the electroactive material and on the medium investigated (D) or on the duration of the experiment. For example, as already pointed out, a disk electrode with a 0.1-μm radius, placed in a polymer solution where $D = 10^{-10}$ cm^2 s^{-1} [28], behaves as a microelectrode (class B in Fig. 1) for times less than a second but becomes an ultramicroelectrode (class C in Fig. 1) for times that are larger. Conversely, a much larger electrode, of radius 5 μm, placed in an acetonitrile solution where $D = 10^{-5}$ cm^2 s^{-1}, behaves as an ultramicroelectrode (class C in Fig. 1) as soon as the duration of the experiment exceeds 25 ms.

III. APPLICATIONS OF ULTRAMICROELECTRODES

In addition to the applications that exploit their very small size for detection or production of chemicals in restricted volumes, most other specific applications of ultramicroelectrodes are based on one or several of the three intrinsic characteristics that have been outlined above: (1) reduced ohmic drop, (2) reduced time constants, and (3) steady-state or quasi-steady-state diffusional currents without necessity of using forced convection.

Almost an infinite variety of applications based on the reduced size of ultramicroelectrodes can be conceived (see, e.g., Sections 5 and 7 in Ref. 30). Indeed, under any circumstances where detection or measurement of concentrations of an electroactive species is required with a spatial resolution of a few micrometers, ultramicroelectrodes constitute the best tool available. In this chapter we have given examples of this kind: in collector–generator systems or in ECL experiments one electrode is used to probe and examine the chemical content of the diffusion layer of another. Several applications of similar principle have been developed in which a mobile ultramicroelectrode is used to detect chemical emission or consumption by another object (i.e., another electrode [121,122], but also a living tissue or a single cell [136–142] or localized chemical events such as corrosion [143]) with micrometric resolution. Micrometric cartographies of chemically inert objects placed in a solution containing an electroactive molecule can also be achieved by scanning an ultramicroelectrode over their surfaces because the objects's presence and local shape obstructs diffusion to the ultramicroelectrode and thus affects the current response [18,144]. Also, an ultramicroelectrode can be used to monitor local hydrodynamics of a solution near a surface or in a small channel with micrometric spatial resolution [145]. Similarly, their small size and large signal-to-noise ratio make them ideally suited for use as electrochemical detectors in liquid capillary chromatography [146–149].

Applications based on the possibility of recording steady-state currents at ultramicroelectrodes obviously resemble those developed with hydrodynamic electrochemical methods at electrodes of millimetric dimensions (i.e., polarography, rotating disk electrode, etc.). Although this certainly corresponds to the most frequent use of ultramicroelectrodes, we encourage the reader to refer to classical textbooks describing hydrodynamic methods [1,2,42,123]. Indeed, in these applications the only particularity of ultramicroelectrodes is that the steady-state currents are imposed by diffusional means rather than by forced convection, a topic developed and discussed extensively in Section II. We just want to emphasize one specific application of ultramicroelectrodes that is based on this property and is related to the determination of *absolute* electron stoichiometries in microelectrolysis [150].

Although that may appear surprising at first glance, electron stoichiometries are among the most difficult data to determine experimentally without resorting to exhaustive electrolysis of a solution. This occurs because one may determine easily the charge passed in a solution, yet without knowing how many molecules have led to this charge consumption. Indeed, the volume electrolyzed depends on the size of the diffusion layer and therefore on the value of the diffusion coefficient. This is apparent, for example, in Table 2, where it is seen that nD only (but not n and D) can be determined from the steady-state current at a disk electrode. If n (i.e., the chemistry involved in the experiment) is unknown, D is generally also unknown; moreover, n often depends on the time scale T_c of the experiment because several mechanisms may compete: $n = n(T_c)$. The time scale corresponding to a microelectrolysis experiment corresponds to the time required for a molecule to crossover the diffusion layer: $T_c = \Delta^2/D$, where Δ is given in Table 1. For steady-state diffusion at a disk electrode, $T_c = \delta^2/D \approx r_0^2/D$, as deduced from Table 2, and is independent of the true duration (t) of the experiment. At a millimetric electrode, where planar diffusion prevails, T_c is related to the duration t of the experiment: $T_c = \pi t$, as is deduced from Table 1 from the limit of Δ when $Dt/a^2 \to 0$ ($a = r_0$ or w). Thus by choosing an identical time scale T_c [i.e., so that $n(T_c)$ is maintained at identical values in both experiments], $n(D)^{1/2}$ can be determined from chronoamperometric measurements at a millimetric electrode (compare Table 1) and nD from the steady-state current at an ultramicroelectrode. From these independent measurements of nD and $n(D)^{1/2}$, the individual values of $n(T_c)$ and D are easily determined. Obviously, any other couple of methods allowing determination of nD^{α} and nD^{β} (with $\alpha \neq \beta$) could be used instead of chronoamperometry and steady-state voltammetry at a disk ultramicroelectrode. However, this couple is that which gives better accuracy in the determinations of n and D since it involves the largest differences of exponents (i.e., $\alpha - \beta = \frac{1}{2}$) for D [150].

In the following sections we present and discuss some more specific applications of ultramicroelectrodes that are based on reduced ohmic drop and small time constants.

A. Electrochemistry in Highly Resistive Media

These applications mainly take advantage of reduced ohmic drop, since they are generally developed under steady-state conditions. In any electrochemical methods one may tolerate a slight distortion due to a slight ohmic-drop contribution. As for any other physicochemical measurements, the limit is a function of the desired quality of the data to be extracted from the experiment and often results from experimental compromises. Because ohmic drop is intrinsically much lower at an ultramicroelectrode than at a millimetric electrode, one may explore media with higher resistivities while keeping the ohmic drop within the same range as that obtained at millimetric electrodes for classical electrochemical conditions. This increased resistivity may result from the fact that the solvent has a larger viscosity (gel [151], polymer [28,69,152], epoxy resin, frozen matrix, etc., but also living tissues [136–141], food or nutrients [153], etc.) or a smaller dielectric constant (e.g., hexane [154], toluene [155,156], and other low-dielectric-constant solvents [157] as well as oils and lubricants [158], etc.) than the usual electrochemical solvents. Both aspects have led to important applications in chemistry itself as well as in trace analysis in unconventional media. However, besides their experimental uniqueness and the fact that they have opened new and vast frontiers to electrochemistry, these experiments obey the classical laws of electrochemistry, and in particular are diffusion controlled. As such, it is not necessary that they be particularized in this chapter.

In the following part of this section, we wish to focus on electrochemistry under conditions of low ionic strength. In electrochemistry, current transport through most of the solution is performed by migration, that is, by displacement of ions within the potential difference created by the two electrodes interfaces [2]. Use of a large excess of an inert (i.e., not electroactive) electrolyte makes it possible to reduce the overall electrical resistance of a solution, but also prevents diffusion of the electroactive molecules to and from the electrodes being interfered with significantly by migration (see below). Within the diffusion layer of each electrode, the migrational flux of inert ions adjusts as a function of distance from the electrode surface to compensate the decay of diffusional fluxes, so that the overall flux of ions (i.e., the current) remains constant. When a large excess of inert ions is used, this concerted mixed transport of current (migration of inert ions and diffusion of electroactive ones) within diffusion layers is limited by diffusion of the electroactive molecules (see below). This is why migration was not considered earlier.

When the concentration of the inert electrolyte is decreased (or when the concentration of the electroactive material is increased) so as to become comparable to—or even less than—that of the electroactive material, intimate coupling between diffusion and migration has to be considered since the two modes of transport then interact with each other. Thus each ion present in the diffusion layer participates in the current transport via migration and diffusion. Under steady-state or quasi-steady-state conditions, its concentration is given as a function of time and space by (vectors are noted in bold fonts) [2]

$$\frac{\partial C}{\partial t} = -\text{div } \mathbf{J} = \text{div}\left[D(\textbf{grad } C) + \frac{zDF}{RT} \times C\mathbf{E} \right] \tag{64}$$

with $\partial C / \partial t = 0$; $\mathbf{J}$ is the local flux of species C, $\mathbf{E}$ the electrical field effective at this point of space at the time considered, and z the electrical charge (in electron absolute charge units, i.e., $z = q/|e|$, where q is the true charge in coulombs) of the ion or molecule considered.

We established above that the changes of variables in equations (25) and (26), (36) and (37), (43), or (44) make it possible to transform the true physical spaces surrounding, respectively, a disk, a band, a hemisphere, or an hemicylinder electrode into the same (Γ,θ) space as that shown in Figure 6c (i.e., akin to a thin-layer cell). It follows that diffusion–migration at all these electrodes is equivalent, as was diffusion alone, and may be evaluated in this more convenient space. Moreover, equation (64) expresses the fact that under steady-state or quasi-steady-state conditions the vector $\mathbf{J}$ has a conservative flux since div $\mathbf{J} = 0$ because $\partial C/\partial t = 0$. In the transformed space (Γ,θ) of Figure 6c, this establishes that the flux of $\mathbf{J}$ through any surface such as $\Gamma = $ constant $(0 \leq \theta \leq 1)$ is constant and therefore that $\mathbf{J}$ is constant. It is thus inferred that for any electrochemical method where the boundary condition at the electrode surface is that of a constant concentration, all concentrations and the electrical field are only a function of Γ and independent of θ (see above for diffusion alone). Thus the transport equation of any species (i.e., electroactive or inert) is given by

$$\frac{\partial C}{\partial \Gamma} + \frac{F}{RT} \times (zC)E = \left(\frac{\partial C}{\partial \Gamma}\right)_0 + \left(\frac{F}{RT}\right) \times (zC_0)E_0 \tag{65}$$

where C is the local concentration of this species; the subscript 0 denotes a value at $\Gamma = 0 - E$ is readily evaluated by taking into account the electroneutrality law* [2,159] $\Sigma(zC) = 0$ (Σ represents a summation extended to all species present in the solution). Indeed, multiplication of both members of equation (65) by the pertinent charge z and summation over all species gives readily

$$E = E_0 \frac{\Sigma(z^2 C_0)}{\Sigma(z^2 C)} \tag{66}$$

Introducing this relation into equation (65) yields

$$\frac{\partial C}{\partial \Gamma} + \frac{zC}{\Sigma(z^2 C)} \times \left[\frac{FE_0}{RT} \Sigma(z^2 C_0) \right] = \left(\frac{\partial C}{\partial \Gamma} \right)_0 + zC_0 \times \frac{FE_0}{RT} \tag{67}$$

On the other hand, the current i exchanged at the electrode surface is obtained by integrating the sum of the fluxes of all species times their charge over the electrode surface. In (Γ, θ) space this leads to

$$i = FAD \int_0^1 \left\{ \Sigma z \left[\left(\frac{\partial C}{\partial \Gamma} \right)_0 + zC_0 \times \frac{FE_0}{RT} \right] \right\} \sigma(\theta) \, d\theta \tag{68}$$

where $\sigma(\theta)$ is a function depending on the space transformation used, that is on the electrode shape and dimension(s) in the true space [compare equations (46) to (48)]. The function $\sigma(\theta)$ is identical to that used for diffusion only since it is characteristic of the space transformation and not of the mode of transport of molecules and ions. Therefore, from the definition of δ in equation (48) and noting that $\Sigma(zC) = 0$ because of the electroneutrality law, equation (68) is rewritten (D is assumed to be identical for all species) [160,161]:

$$i = \frac{FAD}{\delta} \times \left[\frac{FE_0}{RT} \times \Sigma(z^2 C_0) \right] \tag{69}$$

where δ is given in Table 2 for different electrode geometries and A is the electrode surface area in the (x,y) physical space. Introduction of this expression into equation (67) affords the following general diffusion–migration equation,

---

*Using the electroneutrality law is convenient here, yet this is only an approximation. A full and rigorous resolution requires the use of Poisson's law [i.e., div $\mathbf{E} = (\Sigma zFC)/\epsilon$] instead of the electroneutrality law. Using the electroneutrality law is tantamount to imposing a conservative electrical field condition (i.e., div $\mathbf{E} = 0$), which is mathematically inconsistent with the predictions of (66). However, provided that the electrode size, w, and the supporting electrolyte excess γ (see below) are not too small and that ϵ is large enough, the error between the exact electrical field (i.e., that determined using Poisson's law) and that obtained in (66) (i.e., that determined using the electroneutrality law) is negligible, except under special circumstances where the solution approximated would predict infinite electrical fields (compare, e.g., Ref. 161). After this chapter was written, this point was thoroughly clarified and the limits (i.e., those on w, γ, and ϵ) where the "electroneutrality approximation" breaks down have been evaluated by C. P. Smith and H. White (*Anal. Chem.* in press).

which is valid irrespective of the shape and dimension(s) of the electrode [161]:

$$\frac{\partial C}{\partial \Gamma} = \left(\frac{\partial C}{\partial \Gamma}\right)_0 + \left[\frac{zC_0}{\Sigma(z^2C_0)} - \frac{zC}{\Sigma(z^2C)}\right] \times \frac{i\delta}{FAD} \tag{70}$$

This shows that if a species concentration is such that $zC \ll \Sigma(z^2C)$, its concentration profile is linear as a function of Γ in (Γ,θ) space because the term in brackets on the right-hand side of equation (70) becomes negligible. Thus transport of this species is controlled by diffusion only. This demonstrates that in most experiments where a large excess of an inert supporting electrolyte is used, electroactive species are not affected by migration. This is also true under any condition for any neutral species ($z = 0$). For other situations, one needs to evaluate the concentration profiles of each species by analytical or numerical integration of equation (70). This requires an independent evaluation of $i\delta/FAD$. This is obtained as follows [160,161].

The current is the sum of each individual current due to each electroactive couple, (i.e., $i = \Sigma_{AB} i_{AB}$), where i_{AB} is the contribution from the electroactive couple A^{z_A}/B^{z_B}, which exchanges $n_{AB} = z_A - z_B$ electrons with the electrode. These individual contributions are obtained from the flux of A (or that of B) at the electrode surface [2]: for example,

$$i_{AB} = n_{AB}FAD \int_0^1 \left[\left(\frac{\partial C^A}{\partial \Gamma}\right)_0 + z_A C_0^A \times \frac{FE_0}{RT}\right] \sigma(\theta)\, d\theta \tag{71}$$

From equations (48) and (69), one deduces that

$$\frac{i\delta}{FAD} = \sum_{AB} \frac{n_{AB}\,(\partial C^A/\partial \Gamma)_0}{1 - n_{AB}z_A C_0^A/\Sigma(z^2C_0)} \tag{72}$$

Analytical or numerical (i.e., via iteration procedures) combination of this equation with equation (70) affords the concentration profile of each species as well as $(\partial C/\partial \Gamma)_0$ and C_0 for each electroactive species and therefore yields the current through use of equation (72).

Figure 19 presents the results of the foregoing analysis for experiments in which only one electroactive species of charge z is initially present in solution at concentration C^0, and exchanges $n = \pm 1$ electron at the electrode ($n = 1$: reduction; $n = -1$: oxidation) in the presence of an inert 1:1 electrolyte at concentration γC^0 (note that the counterion of the electroactive species is supposed to be a monoanion or a monocation, respectively, its bulk concentration thus being zC^0). The results in Figure 19a are given in the form $i_\gamma/i_{\gamma=\infty}$, that is, the ratio between the current observed for an excess γ of inert electrolyte and the diffusional current (i.e., observed for $\gamma \to \infty$; compare Table 2) when the potential is set on the plateau of the electrochemical wave. It shows that migration results in an increase in the current exchanged for reduction ($n = 1$) of a

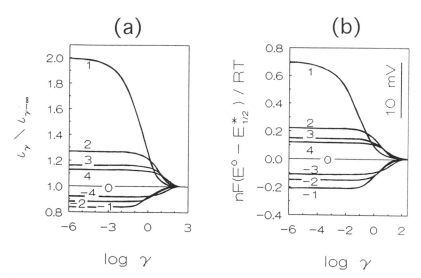

FIGURE 19 Effect of reduced ionic strength (1:1 inert electrolyte) on an electrochemical wave for $n = \pm 1$ under steady-state conditions. Variations of the limiting current (a) or of the half-wave potential (b). Note that in (b) half-wave potentials are corrected for ohmic-drop contribution. γ, Concentration ratio between the inert 1:1 electrolyte and the electroactive material. Numbers on the curves are values of zn, where z is the charge ($q/|e|$) of the electroactive material. (Adapted from Ref. 161.)

cation or oxidation ($n = -1$) of an anion, but in a decrease of the current for reduction of an anion or oxidation of a cation. However, the effect is rather modest except for reduction of a monocation or oxidation of a monoanion, in which cases the current in the absence of an inert electrolyte is twice that observed in the presence of a large excess of electrolyte.

We want also to emphasize the following point. Figure 19a shows that significant variations of $i_\gamma/i_{\gamma=\infty}$ occur only within a restricted range of γ values, which extends approximately from $\gamma = 100$ to $\gamma = 0.01$. We have already explained the reason for the upper limit: it corresponds to the fact that the factor of i in equation (70) becomes negligible; transport is therefore achieved almost exclusively by diffusion. The existence of a lower limit may appear more puzzling. It arises primarily because of the electroneutrality law: each ion consumed or created in the diffusion layer by the electrochemical reaction(s) must drag an equivalent concentration of charge in or out of the diffusion layer so that the diffusion layer remains electrically neutral at each point [160–163]. Thus as soon as the inert electrolyte concentration in the bulk solution becomes negligible vis-à-vis that of the electroactive molecule, the boundary condition at the external end of the diffusion layer is almost unaffected by a further decrease in the electrolyte concentration. It follows that diffusion–migration within the diffusion layer does not sense that the ionic content of the bulk solution is decreased as soon as the value of γ is lower than about 10^{-2} [161]. Although reversed in terms of boundary conditions, the basic reason for the limit at low γ values is thus analogous to that which gives limiting current plateaus in steady-state electrochemical methods [2].

An interesting conclusion may also be drawn from this observation. Indeed, whenever the ionic strength is decreased to study its effect on an electrochemical reaction, it is worthless to decrease the ionic concentration much below that of the electroactive material. Indeed, whenever an electrochemical reaction occurs, the electroactive materials (and their products) drag an equivalent concentration of inert ions in the diffusion layer. In other words, all the chemical and electrochemical events that occur in the diffusion layer take place within an environment where the ionic concentration is close to that of the electroactive material, even if the inert electrolyte concentration is much lower in the bulk solution. In other words, it is chemical and electrochemical nonsense to decrease purposely the concentration of the supporting electrolyte much below about 100 times that of the electroactive material for the sole purpose of studying electrochemistry ''under nearly zero ionic strength.'' Indeed, these under-nearly-zero-ionic-strength conditions *never* exist in the very region where the electrochemical events are taking place unless the electroneutrality law is broken (see below). However, they contribute to severe distortion of the voltammograms because of the increased resistance of the solution bulk. This conclusion obviously does not apply when extremely low supporting electrolyte ratios are imposed for other chemical reasons (see, e.g., Ref. 164) or are an experimental necessity [30].

Figure 19b presents the effect of migration on the half-wave potential of an electrochemical wave [161]. It is noteworthy that the corresponding potential shifts are extremely modest, being a few millivolts except for the case where $nz = 1$. They are also limited to a rather small domain of γ values, for the same reason as discussed above. Note, however, that these variations consider only the effects on half-wave potentials that result from distortion of the concentration profiles due to involvement of migration in the diffusion layer and do not consider ohmic-drop contributions. However, decreasing γ at a constant C^0 is necessarily associated with an increasing ohmic drop. The difficulty in evaluating this contribution is related primarily to the fact that it is specific to each experiment since in addition to γ, it depends on other parameters, such as the electrode geometry, its dimension(s) and the concentration of the electroactive species. Furthermore, because of the electroneutrality law, each ion that is forced in or out of the diffusion layer by the electrochemical reaction draws with itself counterion(s) that balance its charge [159], as explained above (however, see below when considering smaller distances that are comparable to ionic and molecular dimensions). Therefore, the electrochemical reaction that takes place at each electrode enriches or depletes the ionic content of its diffusion layer. This results in a variation in the resistance of the solution due to each diffusion layer vis-à-vis the resistance determined in the absence of electrochemical reaction. (Note that this is a well-known phenomenon in corrosion, with obvious important consequences on rates of corrosion.)

A model has been proposed for the spherical electrode, to account for these variations in the overall resistance of a cell as a function of the electrochemical current [162,163], but it does not accurately predict the experimental observations [165]. We believe that this is due in part to the fact that the model is based on diffusion layers that may extend up to infinity [162]. We have shown at several instances in this chapter that this cannot be true because convection is necessarily involved at large distances from the electrode. Presumably, then, the model overestimates the role of modification of the ionic content in diffusion layers because it considers that these modifications exist in regions where, in truth, homogeneous bulk conditions are restored because of convection.

The theory that has been summarized here (equations (64) to (72) [161]) has been tested at several instances using sanctioned reversible electrochemical systems (see, e.g., Refs. 160, 161, and 165) and has been found to describe experiments with reasonable accuracy. It may also be easily adapted to more complex situations in which homogeneous chemical kinetics are involved [166]. To conclude this section, let us emphasize again that this model is not related to a particular shape or dimension of electrodes provided that a steady-state or quasi-steady-state regime is achieved. This has been established here for the four most common geometries of ultramicroelectrodes. However, this is evidently true for any other geometry that is amenable to the thin-layer cell equivalent shown in Figure 6c via any adequate transformation of space. Similarly, the results in Figure 19 obviously apply to diffusion–migration at generator–collector assemblies (double-band, triple-band, or interdigitated arrays). Indeed, when they operate under steady-state or quasi-steady-state conditions, all these devices are equivalent to a double band (compare Fig. 14), which, in turn, is equivalent to a single electrode [compare (Γ^*, θ^*) space in Fig. 12b and (Γ, θ) space in Fig. 6c]. Indeed, for a single electrode the extremity of the diffusion layer (i.e., $\Gamma = 1$ in Fig. 6c) may be considered as a virtual collector. The same is also obviously true when these generator–collector assemblies are operated in ECL mode [135], because of the equivalence shown in Figure 17c.

It must, however, be pointed out that the theory described above is based on four assumptions [160,161]. Two are minor and not dooming. One deals with the equality of diffusion coefficients that is necessary for the factorization in equation (69). The second neglects variations of diffusion coefficients with ionic strength, although these necessarily vary within the diffusion layer because its ionic strength varies [160,167,168]; however, the corresponding perturbations are small for the dilute solutions generally used in such experiments [160,161]. Both assumptions are not drastic for most experimental systems and are used only to delineate general trends that can be further refined and particularized so as to be adapted to specific circumstances. A third assumption, which is surely restrictive for the model, is related to the validity of the electroneutrality law [i.e. $\Sigma(zC) = 0$] that is used for evaluation of the electrical field intensity E in equation (66) and deriving its relationship to the current in equation (69) (see above). Finally, the fourth assumption is that Fick's laws still apply, which supposes that no other phenomena (e.g., nucleation, precipitation, segregation, etc. [170], or even long-distance electron transfers [171] occur upon reducing the ionic strength. Obviously, this type of specific phenomenon cannot be considered within the framework of a general theory but requires a specific model for each experimental situation.

B. Electrochemistry in the Submicrosecond Time Scale

In what precedes we have examined applications of electrochemistry to experimental conditions in which the *natural* reduction of ohmic drop at ultramicroelectrodes is used to examine media of high resistivity. That may also be applied to situations where the concentrations of electroactive materials are much larger than under the usual electrochemical conditions, so as to approach those found in electrosynthetic cells [117,172], or consist of the solvent itself [173]. In other words, the benefit of an intrinsic decrease in the iR term may be invested by increasing either R (increased resistivity) or i (increased concentration). Another way to invest this benefit also involves an increase in current but is now related to investigation of short time scales rather than large concentrations. Indeed, under non-steady state conditions when the current is controlled by planar diffusion, Cottrell's law is followed [1,2]. The current is then proportional to the reciprocal square root of time, $i \propto (t)^{-1/2}$ (Table 1) and increases when the duration of the experiment is decreased.

Exploration of short time scales also requires that RC, the time constant of the cell, is smaller than (or at least comparable to) the duration of the experiment so as to avoid significant filtering of the electrochemical perturbation and signal. Reduced time constants (Fig. 1) and relative immunity to ohmic drop are thus two crucial properties of ultramicroelectrodes that have opened submicrosecond frontiers to transient electrochemistry. Obviously, instrumental distortions must be minor [174] to be easily corrected [175]. Instrumental limitations will not be considered below, however, because in our view they are not intrinsic to the electrochemical phenomena and may be improved by the design of better potentiostats [174–178].

1.   Ultrafast vs. Ultrasmall

Before pursuing this topic we wish to address the following question: Why submicrosecond transient electrochemistry instead of steady-state electrochemistry at very small ultramicroelectrodes? Indeed, in electrochemistry the time scale is not necessarily the duration of the experiment, as in transient experiments, but is always imposed by the time T_c required for diffusion of a species from the bulk solution to the electrode surface (or vice versa) to occur. The principle of any electrochemical kinetic determination is to oppose (or adjust) this time of diffusion to the half-life of the chemical (homogeneous kinetics) or electrochemical (heterogeneous kinetics) event to be measured [63]. In transient methods T_c is related directly to the duration of the electrochemical perturbation [2]. Under steady-state conditions at a disk or hemispherical electrode of radius r_0, T_c is not related to the true duration of the experiment because the diffusional field depends only on the electrode dimension. One then has $T_c \approx r_0^2/D$, since the thickness δ of the equivalent diffusion layer is on the order of r_0 (Table 2). For example, with $D = 10^{-5}$ cm^2 s^{-1}, exploration of submicrosecond timescales under steady-state conditions requires that $r_0 \leq 30$ nm.

Today, the technical difficulties required for manufacturing such small electrodes or for developing ultrafast potentiostats are certainly comparable in terms of their complexities and *savoir faire*. Evidence for that is given by the fact that only a few research groups are active worldwide in each area. The choice is then not based on technical difficulty. It is no longer based on theoretical aspects because what may occur in transient or steady-state methods is necessarily similar because the size of the diffusion layers are necessarily comparable whenever comparable time scales are investigated. Extraction of data from experimental current–potential curves involves similar difficulties for both methods and follows well-established procedures [2].

It is thus understood that the preference for *ultrafast* transient electrochemistry or for steady state at *ultrasmall* electrodes could not be dictated by any of the foregoing considerations. Could it then simply be a matter of choice? We do not believe so. Indeed, in our experience of a molecular electrochemist, used to deal with investigation of *unknown* mechanisms, we found that transient electrochemistry, particularly cyclic voltammetry [2], brings unique information. This occurs because the forward potential scan allows the creation in the diffusion layer of intermediates that may be detected upon scan reversal. Therefore, transient electrochemistry does not really compare to steady state at a single electrode, but rather to generator–collector assemblies (see above) in which the generator and collector functions of the electrode would be temporally disconnected and assumed by the same electrode.

In terms of mechanistic investigations, the true steady-state equivalent for submicrosecond time-resolved cyclic voltammetry is therefore not an ultrasmall electrode with a 30-nm radius (or less), but a collector–generator device with a gap of less than 30 to 100 nm (see above and compare Fig. 16b). Obviously, this important property of cyclic voltammetry or of reverse pulse methods [179,180], and therefore the latter equivalence, is relevant only when electro- chemical characterization of intermediate(s) or product(s) of an electrochemical reaction is desired, for example, to demonstrate its chemical existence or to establish a reaction mechanism. Conversely, when submicrosecond transient electrochemistry is used only for measurement of the rate constant of a well- established kinetic process, the collector–generator equivalence is accessory and submicrosecond transient methods may then be compared to steady state at a single electrode with a dimension less than about 30 nm.

2. Practical and Experimental Limits

In steady-state methods at ultrasmall electrodes, the two main difficulties are related to the fabrication of well-defined electrodes with nanometric dimensions and to the measurement of ultrasmall currents. Both aspects have already been mastered by a few research groups. For example, well-characterized band elec- trodes [16,17] with widths up to 2 nm have been fabricated by sealing a de- posited metal film between insulators; 30-nm conical well-defined electrodes (which resemble a spherical electrode in terms of diffusion) have also been manufactured [18]. Although they still remain open to question, disk-shaped electrodes of a few tens of angstroms have been reported [27]. Under electro- chemical conditions, the detection of ultrasmall currents in the picoampere range is certainly difficult, yet since only steady-stae currents have to be determined, solutions have been found [16–18,27].

In the submicrosecond domain the situation is exactly reversed. Thus cur- rents in the microampere range have to be detected at electrodes with radii of a few micrometer. However, the electrochemical perturbation must be applied to the faradaic impedance (i.e., to the electrochemical reaction), and the ensuing current detected with minimal distortion, within a submicrosecond time scale [181–183]. Although such requirements seem rather easy to fulfill under the usual conditions with modern electronic components, the specificities of an elec- trochemical cell make them a real challenge.

The simplest electrical circuit equivalent to an electrochemical cell is rep- resented in Figure 20a for a three-electrode configuration (for a two-electrode configuration, the reference branch is not connected). Formally, such a circuit depends on three faradaic impedances, three capacitances, and three resistances. However, use of a potentiostat [1,2] and of an adequate reference electrode [1,2] makes it possible to apply a desired electrochemical perturbation to the working electrode circuit element alone (Fig. 20b; note the change in notation: Z_F now stands for Z_{WE}, C_d for C_{WE}, and R_u for R_{WE}, to indicate, respectively, the faradaic impedance, the double-layer capacitance, and the uncompensated resistance).

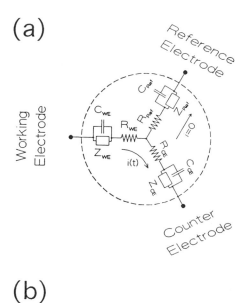

(a)

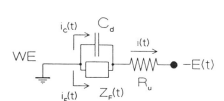

(b)

FIGURE 20 Electrical equivalent of an electrochemical cell: (a) global representation of a three-electrode cell; (b) equivalency to be used when the three-electrode cell is controlled by a potentiostat (see the text).

This is not fully satisfactory since one should really be able to apply the electrochemical perturbation to the time-dependent faradaic impedance and be also able to measure the faradaic current i_F flowing through this impedance only. In cyclic voltammetry or in double-step methods the electrochemical perturbation consists of an imposed potential variation, $E(t)$, that is applied to the circuit in Figure 20b [1,2]. Thus one has

$$E(t) = Z_F i_F + R_u i(t) \tag{73}$$

where $i = i_F + i_C$ is the measured current and

$$i_C(t) = C_d \frac{d(Z_F i_F)}{dt} = C_d \frac{d[E(t) - R_u i(t)]}{dt} \tag{74}$$

Extraction of the true faradaic current function $i_F(t) = f[E(t) - R_u i(t)]$ (i.e., of the *voltammogram*) is possible by application of equations (73) and (74), provided that R_u and C_d are known but also (*and only*) if $Z_F(t)$ is known independently. The extraction may then be performed either by simulation of the distorted voltammogram [184] or by restoration of an undistorted voltammogram by convolution integral procedures [185]. It should be pointed out, however, that both approaches are based on a circular-loop procedure, because knowing $Z_F(t)$ amounts to knowing the mechanism and therefore the result. Therefore, these methods are merely a posteriori verification procedures: a given mechanism [i.e., a given $Z_F(t)$] is postulated that allows us to solve equations (73) and (74) and verify if the $i_F(t)$ experimental function agrees with the initial mechanistic hypothesis. There is nothing criticizable per se in such hypothesis-verification test procedures, but in our view they present an important disadvantage for the study of unknown kinetics. Moreover, and still within this perspective, the fact that they require computation delays makes them hardly compatible with the necessity of instantaneous adjustment of experimental strategies required in the investigation of unknown complex mechanisms.

Because of our specific interest in (unknown) chemical kinetics we have tried to develop another strategy to solve the problem borne by equations (73) and (74). This is based on positive feedback [2] and is as follows. Knowing R_u and measuring $i(t)$ sufficiently fast, one is able to apply $E(t) + (1 - \epsilon) \times R_u i(t)$ to the cell instead of $E(t)$ as before. Note that ϵ may be small enough for $|E(t)| \gg \epsilon \times |R_u i(t)|$ but cannot be made equal to zero to avoid severe electronic instabilities [176,178]. Equations (73) and (74) then become

$$E(t) = Z_F i_F + \epsilon R_u i(t) \approx Z_F i_F \tag{75}$$

$$i_F(t) = i(t) - C_d \frac{d[E(t) - \epsilon R_u i(t)]}{dt} \approx i(t) - C_d \frac{dE(t)}{dt} \approx i(t) - i_c(t) \tag{76}$$

allowing one to extract the experimental values of $i_F(t)$ and $Z_F(t)$ without supposing a priori any specific mechanism. Furthermore, since $i_c(t)$ in equation (76) is now independent of the faradaic impedance, it may be recorded independently and thus be subtracted from the experimental current $i(t)$ to afford an almost *on-line* display of the corrected voltammogram. An example of this procedure is given in Figure 21 for reduction of a pyrylium cation (C. Amatore, S. Arbault, and C. Lefrou, unpublished results, October 1992). Thus the on-line corrected voltammogram in Figure 21b clearly shows the reoxidation wave of a fraction

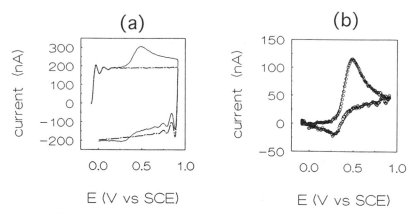

FIGURE 21 Ultrafast cyclic voltammetry (v = 207,000 V s^{-1}) of the reduction of a pyrylium cation [equation (77), Ar = C_6H_5, 5mM] at a platinum disk electrode (r_0 = 5 μm) in acetonitrile 0.3 M NBu$_4$BF$_4$, at 20°C. (a) Voltammogram in the absence (dot-dashed line) or in the presence (solid line) of the pyrylium cation. (b) Symbols: subtracted voltammogram obtained by subtraction of curves in (a); note the almost complete disappearance of oscillations present in (a), although the current scale in (b) is expanded about three times. Solid curve, theoretical voltammogram (α = 0.4, k^0 = 2 cm s^{-1}, E^0 = -0.38 V vs. SCE, k_{dim} = 10^9 M^{-1} s^{-1}) simulated without considering ohmic-drop or time-constant distortions.

of the pyrylium radicals (Ar = C_6H_5) that have been generated during the forward scan and have survived the rapid dimerization in equation (77) [186–188]:

The validity of the ohmic-drop compensation procedure is evidenced by the "rectangular" aspect of the capacitive current in Figure 21a. Note that the oscillations in Figure 21a disappear in the subtracted voltammogram of Figure 21b. They are indeed due to instabilities resulting from the compensation procedure and affect the faradaic potential only slightly [178]; note that they appear to be important in Figure 21a because the double-layer capacitance acts as a derivator circuit (compare equation (76) [178]). The adequacy of the method is also shown by the very good agreement between the experimental voltammogram in Figure 21b and the theoretical voltammogram that is simulated without considering any iR, RC, or instrumental distortions [189]. It is also noteworthy that the rate constant $k_{dim} = (0.9 \pm 0.3) \times 10^9\ M^{-1}\ s^{-1}$ (C. Amatore, S. Arbault, and C. Lefrou, unpublished results, October 1992) of the dimerization in equation (77), is almost at the diffusion limit; it could have been estimated previously only by flash-photolysis techniques [188].

The example above clearly illustrates the value of the approach developed in our group on the basis of positive feedback [2] for on-line electronic compensation of ohmic drop at ultramicroelectrodes. However, it should be emphasized that these advantages are earned at the expense of designing more delicate electronic circuitry. Presently, this has been achieved with minimal instrumental distortions up to scan rates of 550,000 V s^{-1} [178], which correspond formally to half-lives in the range of a few tens of a nanosecond. It is noteworthy that this present experimental limit also corresponds to an ultimate limit for transient electrochemistry, as explained below.

3. Theoretical Limits

Without ohmic-drop compensation, transient electrochemical methods require that a minimal ohmic drop is effective; otherwise, the "raw" voltammogram is so distorted that further correction is almost impossible experimentally. Although larger ohmic-drop terms may be involved, the same remains true when on-line ohmic-drop correction is used, because the potential correction injected via the feedback loop may not exceed a few hundreds of millivolts in normal practice because ϵ in equation (75) cannot be made totally negligible [178]. It is thus concluded that the intrinsic ohmic drop must be less than a given value imposed by the desired accuracy. For our purpose here the ohmic drop may be considered as the sum of two independent components: a "faradaic" one, $R_u i_F$, and a "capacitive" one, $R_u i_C$, the interactions [compare equations (73) and (74)] between these two quantities being neglected. For a disk electrode, R_u is proportional to the reciprocal of r_0 [162]. When $R_u C_d$ is negligible, i_F is proportional to the electrode surface (i.e., to r_0^2) and to the square root of the reciprocal of the duration of the experiment, that is, to the square root of the scan rate ($v = dE/dt$) [2]. From equation (74) and in the absence of a faradaic signal, i_C is equal to $C_d(dE/dt) = C_d v$ and is thus proportional to $r_0^2 v$ [2]. Neglecting the interaction between faradaic and capacitive currents, the ohmic drop may be approximated by

$$R_u i = (\alpha v^{1/2} + \beta v)r_0 \tag{78}$$

where α and β are two constants that depend on the particular experiment considered but are independent of r_0 and v. This establishes that for ohmic drop to be less than a given value, $(iR)_{max}$, r_0 must simultaneously be less than both $(iR)_{max}/\alpha v^{1/2}$ and $(iR)_{max}/\beta v$.

Two other conditions must also be fulfilled. First, the time constant, $R_u C_d$, which is proportional to r_0, must be less than a specified fraction of the duration of the experiment (i.e., proportional to $1/v$). This means that r_0 must also be less than γ/v, where γ is a constant that is independent of v or r_0 but depends on the experiment. Note that this limit is proportional to that corresponding to $(iR)_{max}/\beta v$; thus only the more demanding of the two is to be retained. Second, to obtain minimal distortion by nonlinear diffusion, r_0 must be much larger than the thickness of the diffusion layer that is proportional to the square root of the reciprocal of time (i.e., to $v^{1/2}$); one must therefore also have $r_0 \geq \lambda v^{1/2}$, where λ is a constant that depends on the experiment.

This set of four simultaneous conditions defines a domain within which the radius of the electrode must lie in order that the experiment may be performed with an accuracy better than a specified limit. This domain is represented most conveniently in a $\log(r_0)$ vs. $\log(v)$ diagram as shown in Figure 22 [190]. It is then deduced that a given electrode is best fitted to a given range of scan rates; for example, under the conditions of Figure 22a, an electrode with $r_0 = 1$ μm, allows us to record voltammograms with an ohmic drop of less than 15 mV and less than 10% distortion on peak currents due to nonplanar diffusion for scan rates that range from 3000 to 100,000 V s^{-1}; note that within this range no distortion arises from the cell time constant. The upper limit may be much

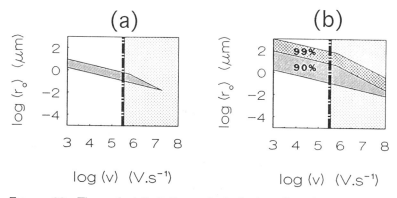

FIGURE 22 Theoretical limitations of ultrafast cyclic voltammetry. The shaded area (slanted lines) represent the area where the radius r_0 of a disk electrode must be located for a given scan rate so that ohmic drop is less than 15 mV and distortions due to nonplanar diffusion are less than 10% on peak currents. (a) Without ohmic drop compensation; (b) with 90% or 99% ohmic drop compensation by positive feedback. The dotted area in (a) and (b) represents the region where transport within the double layer affects voltammograms (see Section III.C). Limits are indicative and correspond approximately to anthracene 5 mM in acetonitrile, 0.3 M NBu$_4$BF$_4$, at a gold disk electrode. (Adapted from Refs. 178 and 190.)

larger upon compensating 90 or 99% of the ohmic drop (Fig. 22b). Another interest of this diagram is that it shows that there is a theoretical limit to the benefit of using electrodes of smaller and smaller dimension. For example, under the conditions of Figure 22a, in the absence of ohmic-drop compensation, the endpoint corresponds to a radius of about 10 nm and a scan rate of about 20,000,000 V s^{-1}. Both values are not so far from what may be achieved experimentally today. However, to reach this limit one would trespass a fundamental intrinsic limit of electrochemistry that is common to *ultrafast* and *ultrasmall* methods.

C. Ultimate Electrochemistry

As explained above, upon decreasing the time scale T_c of an electrochemical experiment, either by shortening the duration t of the electrochemical perturbation in transient electrochemistry ($T_c = t$) or by minaturizing the size a of the electrode in steady-state electrochemistry ($T_c = a^2/D$), one acts in fact on the size of the diffusion layer (compare Table 1). This size, being on the order of $\Delta_{diff} = (DT_c)^{1/2}$, will eventually become comparable and less than that of the double layer [19,20]. The double layer is an interfacial region adjacent to the electrode surface that corresponds to the solution part, where a large electrical field applies because of the electrode–solution difference of potential [2]. Under the usual electrochemical conditions this electrical field is screened over a distance Δ_{dl} of a few tens of angstroms because of an excess of positive (or negative, accordingly) charges in this interfacial region [2]. Note that for solutions with low electrolyte concentration, this interfacial region (i.e., the double layer) may be wider [2].

In classical theories of electrochemistry, it is assumed that $\Delta_{diff} \gg \Delta_{dl}$, so that the effect of diffusion is almost negligible within the double layer and the effect of migration due to the electrical field of the double layer is almost negligible within most parts of the diffusion layer. Within the double layer it is considered that the concentrations of charged particles obey a Boltzmann distribution imposed by the electrical potential of the double layer. Outside the double layer, molecules and ions diffuse. Based on this dichotomous approximation [21–25], known as the *Frumkin correction* [2], one may relate true electrode concentrations to those existing at the end of the double layer. The latter concentrations are then used as the electrode boundary conditions for solving diffusion, although they truly represent the solution composition at the end of the double layer and not at the electrode surface. To the best of our knowledge, all electrochemical theories (except two recent ones [19,20]; see below) are based on this approximation.

The Frumkin approximation necessarily breaks down when the size of the diffusion layer and that of the double layer become comparable, because the latter can no longer be a boundary of the former. Considering an average diffusion coefficient of 10^{-5} cm^2 s^{-1} and an average thickness of 10 Å [2] for the double layer, the latter would represent 10% of the diffusion layer for $T_c = 0.1$ μs. This corresponds to scan rates of about 300,000 V s^{-1} in cyclic voltammetry (*ultrafast* electrochemistry) or to dimensions of about 10 nm in steady-state methods (*ultrasmall* electrochemistry). It is worthwhile to emphasize that these limits are not unrealistic and are achieved routinely by a couple of groups in each area.

Beyond these limits, the Frumkin approximation [2] is no more valid, and migration within the double layer and diffusion will interfere. Under these conditions the system obeys equation (64) (with $\partial C/\partial t \neq 0$ for transient methods and $\partial C/\partial t = 0$ for steady-state methods) where the electrical field is that prevailing in the double layer [2]. Two models have been developed almost simultaneously to examine these situations, one for steady-state voltammetry at an hemispherical electrode [20], the other for cyclic voltammetry at a planar electrode [19]. In agreement with the identical nature of the phenomena involved, they both predict very similar behavior. The general trend is that electrochemical waves may be controlled by the rate of diffusion–migration within the diffusion layer, similar to the way in which homogeneous electron transfers may be controlled by the diffusion limit [63,191,192]. In fact, both phenomena (i.e., the homogeneous and the electrochemical) are then very similar conceptually and experimentally. Indeed, the size of double layers or that of diffusion layers within these time scales are not very different from those that exist around one molecule [2,192]. The only difference between an electrode and a molecular oxidant or reductant is then that a molecular electron transfer reagent is *exhausted* after one electron (or possibly a few electrons) have been transferred, whereas an electrode may virtually transfer an infinity of electrons to an infinity of molecules. In other words, statistics performed on single events taking place at an Avogadro number of centers (each homogeneous molecule) [191,192] is replaced by statistics dealing with an Avogadro number of events taking place at a single center (the electrode) [19,20].

In complete agreement with this conceptual analogy, it is found that coupling between diffusion and migration in the electrode double-layer field results in the appearance of kinetic control by an apparent heterogeneous electron transfer rate constant [19]. However, and contrary to true electrochemical heterogeneous rate constants that increase to infinity when the potential difference is increased [2], this pseudo rate constant eventually reaches a maximum value at which it remains poised. This limit features the maximum rate at which the reactant may cross the double layer (note that this is very reminiscent of the control of a homogeneous reaction by the diffusion limit rate constant [191] and is therefore independent of the potential except for any modification of the double layer with this parameter [2]).

Because of this peculiar behavior, current plateaus in steady-state methods or peak shapes in cyclic voltammetry do not obety the usual electrochemical laws and may even be drastically affected [19,20]. It is thus understood that classical electrochemistry [1,2] ceases as soon as these phenomena take place. This is the origin of the limit between domains C (ultramicroelectrodes) and D (nanodes?) in Figure 1 for ultrasmall electrochemistry and of the vertical limit shown in Figure 22 for ultrafast electrochemistry.

IV. CONCLUSIONS

In this chapter we have attempted to persuade the reader that ultramicroelectrodes offer considerable advantages over electrodes of classical dimensions. They improve greatly the quality of electrochemical data, allowing experiments that were almost inconceivable several years ago to be performed today on a routine basis. These improved characteristics have opened new frontiers to electrochemistry in many other fields of science but simultaneously, have opened new fields in electrochcmistry. These applications stem in part from the fact that by miniaturizing an electrode, one may investigate in true time and in their true environment fundamental processes of chemistry or of life that occur at a micrometric scale. Yet the same process of miniaturization makes the electrode nearly immune to ohmic drop and decreases its time constant, allowing investigations in highly resistive media, concentrated solutions, or at a nanosccond time scale. Since these aspects are very well described in several other excellent reviews on the subject, here we have preferred to focus on the basic concepts that sustain all the forcgoing applications.

Among these is the specificity of the transport of molecules at ultramicroelectrodes and thus of the currents or that of any related electrochemical data obtained at these electrodes. At electrodes of classical sizes, transport of molecules occurs by planar diffusion and is independent of the exact shape of the electrode. At ultramicroelectrodes several regimes may be observed: transport of molecules depends on the shape and dimension of the electrode as well as on the electroactive material, the viscosity of the medium, or the time scale of the experiment.

This may be considered as an important disadvantage, since that supposes that a specific theory is available for nearly each specific situation considered experimentally. We have tried to show that under most conditions use of adequate space transforms allows theories developed for one electrode shape or one device based on ultramicroelectrodes to be transposed easily to systems that are equivalent in their transformed spaces (although they may be quite different in our physical space). This means that models may be developed only for those electrodes or devices that have more adequate geometries for theory but are extremely difficult to manufacture. They may then be transposed to electrodes that may be more easily manufactured. Coupling these techniques to dimensionless analysis should lead to the development of fast simulation programs usable on desktop computers and able to encompass virtually any common experimental situation.

ACKNOWLEDGMENTS

It is our pleasure to thank all our co-workers whose names appear as coauthors in the reference list. Among them. Mark Deakin (Indiana University) played a key role in initiating the research based on conformal maps, at a moment where we had not yet realized the full power (and generality) of this tool for solving diffusion at ultramicroelectrodes. This subject was then elaborated further by several other co-workers, among whom Bruno Fosset (ENS) played a major role. Christine Lefrou (ENS) and Gérard Simonneau (CNRS-ENS) played similar decisive roles in the development of ultrafast cyclic voltammetry in our group. It is also our pleasure to thank Mark Wightman (University of North Carolina at Chapel Hill) for our intense collaborative research effort in this area over the past 10 years and the mutual enrichment that has resulted.

REFERENCES

1. P. T., Kissinger and W. R., Heineman, eds., *Laboratory Techniques in Electroanalytical Chemistry*, Marcel Dekker, New York, 1984.
2. A. J. Bard and L. R. Faulkner, *Electrochemical Methods*, Wiley, New York, 1980.
3. R. M. Wightman and D. O. Wipf, in *Electroanalytical Chemistry* (A. J. Bard, ed.), Marcel Dekker, New York, 1989, Vol. 15, Chap. 3, pp. 267–353.
4. S. Bruckenstein, *Anal. Chem.*, *59*: 2098 (1987).
5. K. B. Oldham, *J. Electroanal. Chem.*, *237*: 303 (1987).
6. P. W. Davies and F. Brink, *Rev. Sci. Instrum.*, *13*: 524 (1942).
7. M. Fleischmann, F. Lasserre, J. Robinson, and D. Swan, *J. Electroanal. Chem.*, *177*: 97 (1984).
8. R. M. Penner, M. J. Heben, and N. S. Lewis, *Anal. Chem.*, *61*: 1630 (1989).
9. B. D. Pendley and H. D. Abruña, *Anal. Chem.*, *62*: 782 (1990).
10. C. Lee, C. J. Miller, and A. J. Bard, *Anal. Chem.*, *63*: 78 (1991).
11. M. Armstrong-James, K. Fox, and J. Millar, *J. Neurosci. Methods*, *2*: 431 (1980).
12. A. Meulemans, B. Poulain, G. Baux, L. Tauc, and D. Henzel, *Anal. Chem.*, *58*: 2088 (1986).
13. R. A. Saraceno and A. G. Ewing, *J. Electroanal. Chem.*, *257*: 83 (1988).
14. K. T. Kawagoe, J. A. Jankowski, and R. M. Wightman, *Anal. Chem.*, *63*: 1589 (1991).
15. T. G. Strein and A. G. Ewing, *Anal. Chem.*, *64*: 1368 (1992).
16. K. R. Wehmeyer, M. R. Deakin, and R. M. Wightman, *Anal. Chem.*, *57*: 1913 (1985).
17. R. B. Morris, D. J. Franta, and H. S. White, *J. Phys. Chem.*, *91*: 3559 (1987).

18. M. V. Mirkin, F.-R. F. Fan, and A. J. Bard, *Science, 257*: 364 (1992).
19. C. Amatore and C. Lefrou, *J. Electroanal. Chem., 296*: 335 (1990).
20. J. D. Norton, H. S. White and S. W. Feldberg, *J. Phys. Chem., 94*: 6772 (1990).
21. A. N. Frumkin, *Z. Phys. Chem. A, 164*: 121 (1933).
22. R. Parsons, *Trans. Faraday Soc., 47*: 1332 (1951).
23. J. E. B. Randles, *Trans. Faraday Soc., 48*: 828 (1952).
24. D. M. Mohilner and P. Delahay, *J. Phys. Chem., 67*: 588 (1963).
25. P. Delahay, *Double Layer and Electrode Kinetics*, Insterscience, New York, 1965, pp. 153–167.
26. J. D. Seibold, E. R. Scott, and H. S. White, *J. Electroanal. Chem., 264*: 281 (1989).
27. R. M. Penner, M. J. Heben, T. L. Longin, and N. S. Lewis, *Science, 250*: 1118 (1990).
28. R. W. Murray, ed., *Molecular Design of Electrode Surfaces, Techniques of Chemistry*, Vol. 22, Wiley-Interscience, New York, 1992.
29. A. C. Michael and R. M. Wightman, Microelectrodes, in *Laboratory Techniques in Electroanalytical Chemistry*, 2nd ed. (P. T. Kissinger, and W. R. Heineman, eds.), Marcel Dekker, New York in press.
30. M. I. Montenegro, M. A. Queiros, and J. L. Daschbach, eds., *Microelectrodes: Theory and Applications*, NATO ASI Series. Kluwer Academic Publishers, Dordrecht, The Netherlands, 1991.
31. M. Moreau, and P. Turq, eds., *Chemical Reactivity in Liquids: Fundamental Aspects*, Plenum Press, New York, 1988, Chap. 5, pp. 561–606.
32. V. G. Levich, *Physicochemical Hydrodynamics*, Prentice Hall, Englewood Cliffs, N. J., 1962.
33. D. MacGillavry and E. K. Rideal, *Recl. Trav. Chim. Pays-Bas, 56*: 1013 (1937).
34. B. Fosset, C. A. Amatore, J. E. Bartelt, A. C. Michael, and R. M. Wightman, *Anal. Chem., 63*: 306 (1991).
35. C. Amatore, B. Fosset, K. M. Maness, and R. M. Wightman, *Anal. Chem., 65*: 2311 (1993).
36. H. S. Carslaw and J. C. Jaeger, *Conduction of Heat in Solids*, 2nd ed., Clarendon Press, Oxford, 1959.
37. K. Aoki, K. Honda, K. Tokuda, and H. Matsuda, *J. Electroanal. Chem., 186*: 79 (1985).
38. K. Aoki, K. Honda, K. Tokuda, and H. Matsuda, *J. Electroanal. Chem., 195*: 51 (1985).
39. S. Sujaritvanichpong, K. Aoki, K. Tokuda, and H. Matsuda, *J. Electroanal. Chem., 199*: 271 (1986).
40. K. Aoki, K. Tokuda, and H. Matsuda, *J. Electroanal. Chem., 206*: 47 (1986).
41. A. Szabo, D. K. Cope, D. E. Tallman, P. M. Kovach, and R. M. Wightman, *J. Electroanal. Chem., 217*: 417 (1987).
42. Z. Galus, *Fundamentals of Electrochemical Analysis*, Ellis Horwood, Chichester, West Sussex, England, 1976.
43. C. A. Amatore, M. R. Deakin, and R. M. Wightman, *J. Electroanal. Chem., 206*: 23 (1986).
44. H. P. Wu and R. L. McCreery, *Anal. Chem., 61*: 2347 (1989).

45. S. A. Schuette and R. L. McCreery, *Anal. Chem.*, *58*: 1778 (1986).
46. J. Newman, *J. Electrochem. Soc.*, *113*: 501 (1966).
47. K. B. Oldham, *J. Electroanal. Chem.*, *122*: 1 (1981).
48. J. Albery and S. Bruckenstein, *J. Electroanal. Chem.*, *144*: 105 (1983).
49. Z. Galus, J. O. Schenk, and R. N. Adams, *J. Electroanal. Chem.*, *135*: 1 (1982).
50. J. O. Howell and R. M. Wightman, *Anal. Chem.*, *56*: 524 (1984).
51. C. Amatore, M. R. Deakin, R. M. Wightman, and B. Fosset, *J. Electroanal. Chem.*, *225*: 33 (1987).
52. Y. Saito, *Rev. Polarogr. Japan.*, *15*: 177 (1968).
53. K. Aoki and J. Osteryoung, *J. Electroanal. Chem.*, *122*: 19 (1981).
54. D. Shoup and A. Szabo, *J. Electroanal. Chem.*, *140*: 237 (1982).
55. K. Aoki and J. Osteryoung, *J. Electroanal. Chem.*, *160*: 335 (1984).
56. M. Kakihana, H. Ikeuchi, G. P. Sato, and K. Tokuda, *J. Electroanal. Chem.*, *108*: 381 (1980).
57. M. Kakihana, H. Ikeuchi, G. P. Sato, and K. Tokuda, *J. Electroanal. Chem.*, *117*: 201 (1981).
58. A. C. Michael, R. M. Wightman, and C. A. Amatore, *J. Electroanal. Chem.*, *267*: 33 (1989).
59. I. Lavagnini, P. Pastore, F. Magno, and C. A. Amatore, *J. Electroanal. Chem.*, *316*: 37 (1991).
60. C. A. Amatore and B. Fosset, *J. Electroanal. Chem.*, *328*: 21 (1992).
61. K. J. Binns and P. J. Lawrenson, *Analysis and Computations of Electric and Magnetic Field Problems*, Pergamon Press, Oxford, 1963.
62. M. R. Spiegel, *Theory and Problems of Complex Variables*, Schaum's Outline Series, McGraw-Hill, New York.
63. C. Amatore, in *Organic Electrochemistry* (H. Lund and M. M. Baizer, eds.), Marcel Dekker, New York, 1991, Chap. 2, pp. 11–119.
64. C. P. Andrieux and J.-M. Savéant, in *Investigation of Rates and Mechanisms of Reactions* (C. F. Bernasconi, ed.), Wiley, New York, 1986, Vol. 6/41E, Part 2, Chap. 7.
65. S. Coen, D. K. Cope, and D. E. Tallman, *J. Electroanal. Chem.*, *215*: 29 (1986).
66. K. Aoki, K. Tokuda, and H. Matsuda, *J. Electroanal. Chem.*, *225*: 19 (1987).
67. M. R. Deakin, R. M. Wightman, and C. A. Amatore, *J. Electroanal. Chem.*, *215*: 49 (1986).
68. B. J. Scharifker, Ensembles of microelectrodes, in Ref. 30, Section 4, pp. 227–239, and references therein.
69. R. I. McCarley, M. G. Sullivan, S. Ching, Y. Zhang, and R. W. Murray, Lithographic and related microelectrode fabrication techniques, in Ref. 30, Section 4, pp. 205–226, and references therein.
70. T. Gueshi, K. Tokuda, and H. Matsuda, *J. Electroanal. Chem.*, *89*: 247 (1978).
71. B. J. Seddon, H. H. Girault, and M. J. Eddowes, *J. Electroanal. Chem.*, *266*: 227 (1989).
72. A. J. Bard, J. A. Crayston, G. P. Kittlesen, T. Varco Shea, and M. S. Wrighton, *Anal. Chem.*, *58*: 2321 (1986).
73. C. E. Chidsey, B. J. Feldman, C. Lungren, and R. W. Murray, *Anal. Chem.*, *58*: 601 (1986).

74. L. E. Fosdick and J. L. Anderson, *Anal. Chem.*, *58*: 2431 (1986).
75. E. W. Paul, A. J. Ricco, and M. S. Wrighton, *J. Phys. Chem.*, *89*: 1441 (1985).
76. W. Thormann, P. van den Bosch, and A. M. Bond, *Anal. Chem.*, *57*: 2764 (1985).
77. K. Aoki, M. Morita, O. Niwa, and H. Tabei, *J. Electroanal. Chem.*, *256*: 269 (1988).
78. O. Niwa, M. Morita, and H. Tabei, *J. Electroanal. Chem.*, *267*: 291 (1989).
79. T. Varco Shea and A. J. Bard, *Anal. Chem.*, *59*: 2101 (1987).
80. J. E. Bartelt, M. R. Deakin, C. Amatore, and M. R. Wightman, *Anal. Chem.*, *60*: 2167 (1988).
81. L. J. Magee and J. Osteryoung, *Anal. Chem.*, *61*: 2124 (1989).
82. T. Gueshi, K. Tokuda, and H. Matsuda, *J. Electroanal. Chem.*, *89*: 247 (1978).
83. T. Gueshi, K. Tokuda, and H. Matsuda, *J. Electroanal. Chem.*, *101*: 29 (1979).
84. K. Tokuda, T. Gueshi, and H. Matsuda, *J. Electroanal. Chem.*, *102*: 41 (1979).
85. D. E. Weisshaar, D. E. Tallman, and J. L. Anderson, *Anal. Chem.*, *5*: 1809 (1981).
86. L. Falat and H. Y. Cheng, *Anal. Chem.*, *54*: 2109 (1982).
87. J. Wang and B. A. Freiha, *J. Chromatogr.*, *298*: 79 (1984).
88. N. Sleszynski, J. Osteryoung, and M. Carter, *Anal. Chem.*, *56*: 130 (1984).
89. W. L. Caudill, J. O. Howell, and R. M. Wightman, *Anal. Chem.*, *54*: 2531 (1982).
90. R. L. Deutscher and S. Fletcher, *J. Electroanal. Chem.*, *239*: 17 (1988).
91. R. C. Paciej, G. L. Cahen, G. E. Stoner, and E. Gileadi, *J. Electroanal. Chem.*, *132*: 1307 (1985).
92. M. Ciszkiwska and Z. Stojek, *J. Electroanal. Chem.*, *191*: 101 (1985).
93. R. M. Penner and C. R. Martin, *Anal. Chem.*, *59*: 2625 (1987).
94. I. F. Cheng, L. D. Whiteley, and C. R. Martin, *Anal. Chem.*, *61*: 762 (1989).
95. I. F. Cheng, and C. R. Martin, *Anal. Chem.*, *60*: 2163 (1988).
96. P. Bindra, A. P. Brown, M. Fleischmann, and D. Pletcher, *J. Electroanal. Chem.*, *58*: 31 (1975).
97. P. Bindra and J. Ulstrup, *J. Electroanal. Chem.*, *140*: 131 (1982).
98. E. Sabatani, I. Rubinstein, R. Maoz, and J. Sagiv, *J. Electroanal. Chem.*, *219*: 365 (1987).
99. E. Sabatani and I. Rubinstein, *J. Phys. Chem.*, *91*: 6663 (1987).
100. C. Amatore, J.-M. Savéant, and D. Tessier, *J. Electroanal. Chem.*, *146*: 37 (1983).
101. O. Contamin and E. Levart. *J. Electroanal. Chem.*, *136*: 259 (1982).
102. C. Amatore, J.-M. Savéant, and D. Tessier, *J. Electroanal. Chem.*, *147*: 39 (1983).
103. J. Cassidy, J. Ghoroghchian, F. Sarfarazi, J. J. Smith, and S. Pons, *Electrochim. Acta*, *31*: 629 (1986).
104. D. Shoup and A. Szabo, *J. Electroanal. Chem.*, *160*: 19 (1984).
105. H. Reller, E. Kirowa-Eisner, and E. Gileadi, *J. Electroanal. Chem.*, *138*: 65 (1982).
106. H. Reller, E. Kirowa-Eisner, and E. Gileadi, *J. Electroanal. Chem.*, *161*: 247 (1984).
107. A. J. Bard, J. A. Crayston, G. P. Kittlesten, T. Varco Shea and M. S. Wrighton, *Anal. Chem.*, *58*: 2321 (1986).
108. W. Thormann, P. van den Bosch, and A. M. Bond, *Anal. Chem.*, *57*: 2764 (1985).
109. K. Aoki and J. Osteryoung, *J. Electroanal. Chem.*, *125*: 315 (1981).
110. W. L. Caudill, J. O. Howell, and R. M. Wightman, *Anal. Chem.*, *54*: 2532 (1982).

111. V. Y. Filinovsky, *Electrochim. Acta, 25*: 309 (1980).
112. S. Moldoveanu and J. L. Anderson, *J. Electroanal. Chem., 185*: 239 (1985).
113. J. L. Anderson, T. Y. Ou, and S. Moldoveanu, *J. Electroanal. Chem., 196*: 213 (1985).
114. D. K. Cope and D. E. Tallman, *J. Electroanal. Chem., 205*: 101 (1986).
115. L. E. Fosdick and J. L. Anderson, *Anal. Chem., 58*: 2481 (1986).
116. L. E. Fosdick, J. L. Anderson, T. A. Baginski, and R. C. Jaeger, *Anal. Chem., 58*: 2750 (1986).
117. M. I. Montenegro, Application of microelectrodes in electrosynthesis, in Ref. 30, Section 7, pp. 429–443.
118. V. Cammarata, D. R. Talham, R. M. Crooks, and M. S. Wrighton, *J. Phys. Chem., 94*: 2680 (1990).
119. B. J. Feldman, S. W. Feldberg, and R. W. Murray, *J. Phys. Chem., 91*: 6558 (1987).
120. B. Fosset, C. Amatore, J. Bartelt, and R. M. Wightman, *Anal. Chem., 63*: 1403 (1991).
121. R. C. Engstrom, M. Weber, D. J. Wunder, R. Burgess, and S. Winquist, *Anal. Chem., 58*: 844 (1986).
122. H.-Y. Liu, F.-R. F. Fan, C. W. Lin, and A. J. Bard, *J. Am. Chem. Soc., 108*: 3838 (1986).
123. W. J. Albery and M. L. Hitchman, *Ring Disc Electrodes*, Clarendon Press, Oxford, 1971.
124. M. Fleischmann, S. Bandyopadhyay, and S. Pons, *J. Phys. Chem., 89*: 5537 (1985).
125. M. Fleischmann and S. Pons, *J. Electroanal. Chem., 222*: 107 (1987).
126. A. Szabo, *J. Phys. Chem., 91*: 3108 (1987).
127. D. R. MacFarlane and D. K. Y Wong, *J. Electroanal. Chem., 185*: 197 (1985).
128. Y.-T. Kim, D. M. Scarnulis, and A. G. Ewing, *Anal. Chem., 58*: 1782 (1986).
129. A. Meulemans, B. Poulain, G. Baux, L. Tauc, and D. Henzel, *Anal. Chem., 58*: 2091 (1986).
130. P. Pastore, F. Magno, I. Lavagnini, and C. Amatore, *J. Electroanal. Chem., 301*: 1 (1991).
131. L. R. Faulkner and A. J. Bard, in *Electroanalytical Chemistry* (A. J. Bard, ed.), Marcel Dekker, New York, 1977, Vol. 10, p. 1.
132. H. Tachikawa and L. R. Faulkner, Electrochemiluminescence, in Ref. 1, Chap. 23.5, pp. 660–674.
133. N. Sutin, in *Progress in Inorganic Chemistry* (S. J. Lippard, ed.), Wiley-Interscience, New York, 1983, Vol. 30, pp. 441–498.
134. R. P. Van Duyne and S. F. Fischer, *Chem. Phys., 5*: 183 (1974).
135. J. E. Bartelt, S. M. Drew, and R. M. Wightman, *J. Electrochem. Soc., 139*: 70 (1992).
136. R. N. Adams, *Anal. Chem., 48*: 1126A (1976).
137. J.-L. Ponchon, R. Cespuglio, F. Gonon, M. Jouvet, and J.-F. Pujol, *Anal. Chem., 51*: 1483 (1979).
138. C. Amatore, R. S. Kelly, E. W. Kristensen, W. G. Kuhr, and R. M. Wightman, *J. Electroanal. Chem., 213*: 31 (1986).

139. R. N. Adams, *Prog. Neurobiol.*, *35*: 297 (1990).
140. J. B. Chien, R. A. Wallingford, and A. G. Ewing, *J. Neurochem.*, *54*: 633 (1990).
141. R. M. Wightman, R. T. Kennedy, D. J. Wiedemann, K. T. Kawagoe, J. B. Zimmerman, and D. J. Leszczyszyn, Microelectrodes in biological systems, in Ref. 30, Section 7, pp. 453–462, and references therein.
142. T. J. Schroeder, J. A. Jankowski, K. T. Kawagoe, R. M. Wightman, C. Lefrou, and C. Amatore, *Anal. Chem.*, *64*: 3077 (1992) [corrected: *Anal. Chem.*, *65*: 2711 (1993)].
143. D. E. Williams, Microelectrodes in the study of localized corrosion, in Ref. 30, Section 7, pp. 445–451.
144. J. Kwak and A. J. Bard, *Anal. Chem.*, *61*: 1221 (1989).
145. E. W. Kristensen, R. L. Wilson, and R. M. Wightman, *Anal. Chem.*, *58*: 986 (1986).
146. L. A. Knecht, E. J. Guthrie, and J. W. Jorgenson, *Anal. Chem.*, *56*: 479 (1984).
147. J. G. White, R. L. St. Claire, and J. W. Jorgenson, *Anal. Chem.*, *58*: 293 (1986).
148. J. G. White, and J. W. Jorgenson, *Anal. Chem.*, *58*: 2992 (1986).
149. M. Goto and K. Shimada, *Chromatographia*, *21*: 631 (1986).
150. C. Amatore, M. Azzabi, P. Calas, A. Jutand, C. Lefrou, and Y. Rollin, *J. Electroanal. Chem.*, *288*: 45 (1990).
151. A. M. Bond, M. Fleischmann, and J. Robinson, *J. Electroanal. Chem.*, *180*: 257 (1984).
152. J. F. Parcher, C. J. Barbour, and R. W. Murray, *Anal. Chem.*, *61*: 584 (1989).
153. D. E. Williams, Microclectrodes in analysis, in Ref. 30, Section 7, pp. 415–427.
154. L. Geng, A. G. Ewing, J. C. Jernigan, and R. W. Murray, *Anal. Chem.*, *58*: 852 (1986).
155. L. Geng and R. W. Murray, *Inorg. Chem.*, *25*: 3115 (1986).
156. C. Amatore and F. Pflüger, *Organometallics*, *9*: 2276 (1990).
157. J. O. Howell and R. M. Wightman, *J. Phys. Chem.*, *88*: 3915 (1984).
158. S. S. Wang, *J. Electrochem. Soc.*, *136*: 713 (1989).
159. R. P. Buck, *J. Membrane Sci.*, *17*: 62 (1984).
160. M. R. Deakin, R. M. Wightman, and C. A. Amatore, *J. Electroanal. Chem.*, *225*: 49 (1987).
161. C. Amatore, B. Fosset, J. Bartelt, M. R. Deakin, and R. M. Wightman, *J. Electroanal. Chem.*, *256*: 255 (1988).
162. K. B. Oldham, *J. Electroanal. Chem.*, *250*: 1 (1988).
163. J. B. Cooper and A. M. Bond, *J. Electroanal. Chem.*, *315*: 143 (1991).
164. C. Amatore, J.-N Verpeaux, and P. J. Krusic, *Organometallics*, *7*: 2426 (1988).
165. S. M. Drew, R. M. Wightman, and C. A. Amatore, *J. Electroanal. Chem.*, *317*: 117 (1991).
166. J. D. Norton, W. E. Benson, H. S. White, B. D. Pendley, and H. D. Abruña, *Anal. Chem.*, *63*: 1909 (1991).
167. L. Onsager and R. Fuoss, *J. Phys. Chem.*, *36*: 2689 (1932).
168. J. O'M. Bockris and A. K. N. Reddy, *Modern Electrochemistry*, Plenum Press, New York, 1970, p. 384.
169. A. M. Bond, M. Fleischmann, and J. Robinson, *J. Electroanal. Chem.*, *168*: 299 (1984).

170. J. B. Cooper, A. M. Bond, and K. B. Oldham, *J. Electroanal. Chem.*, *331*: 877 (1992).
171. J. D. Norton, S. A. Anderson, and H. S. White, *J. Phys. Chem.*, *96*: 3 (1992).
172. M. I. Montenegro and D. Pletcher, *J. Electroanal Chem.*, *248*: 229 (1988).
173. R. A. Malmsten, C. P. Smith, and H. S. White, *J. Electroanal. Chem.*, *215*: 223 (1986).
174. J. O. Howell, W. G. Kuhr, R. Ensman, and R. M. Wightman, *J. Electroanal. Chem.*, *209*: 77 (1986).
175. D. O. Wipf, E. W. Kristensen, M. R. Deakin, and R. M. Wightman, *Anal. Chem.*, *60*: 306 (1988).
176. C. Amatore, C. Lefrou, and F. Pflüger, *J. Electroanal. Chem.*, *270*: 43 (1989).
177. D. Garreau, P. Hapiot, and J.-M. Savéant, *J. Electroanal. Chem.*, *281*: 73 (1990).
178. C. Amatore and C. Lefrou, *J. Electroanal. Chem.*, *324*: 33 (1992).
179. J. Osteryoung and M. M. Murphy, Normal and reverse pulse voltammetry at small electrodes, in Ref. 30, Section 3, pp. 123–138, and references therein.
180. J. Osteryoung, Square-wave and staircase voltammetry at small electrodes, in Ref. 30, Section 3, pp. 139–175, and references therein.
181. R. M. Wightman and D. O. Wipf, *Acc. Chem. Res.*, *23*: 64 (1990).
182. R. M. Wightman, The use of microelectrode for very rapid cyclic voltammetry, in Ref. 30, Section 3, pp. 177–186, and references therein.
183. J.-M. Savéant, Application of direct and indirect electrochemical techniques to the investigation of fast kinetics, in Ref. 30, Section 3, pp. 307–340, and references therein.
184. D. O. Wipf and R. M. Wightman, *Anal. Chem.*, *60*: 2460 (1988).
185. C. P. Andrieux, D. Garreau, P. Hapiot, J. Pinson, and J.-M. Savéant, *J. Electroanal. Chem.*, *243*: 321 (1988).
186. C. A. Amatore, A. Jutand, and F. Pflüger, *J. Electroanal. Chem.*, *218*: 361 (1987).
187. C. Amatore, A. Jutand, F. Pflüger, C. Jallabert, H. Strzelecka, and M. Veber, *Tetrahedron Lett.*, *30*: 1383 (1989).
188. H. Kawata, Y. Suzuki, and S. Nuzuma, *Tetrahedron Lett.*, *27*: 4489 (1986).
189. J. T. Maloy, Digital simulation of electrochemical problems, in Ref. 1, Chap. 16, pp. 417–444, and references therein.
190. W. J. Bowyer, E. E. Engelman, and D. H. Evans, *J. Electroanal. Chem.*, *262*: 67 (1989).
191. R. A. Marcus, *Discuss. Faraday Soc.*, *129*: 21 (1960).
192. Z. Smoluchowsky, *Physik. Chem.*, *92*: 129 (1917).

5

Scanning Electrochemical Microscopy

ALLEN J. BARD AND FU-REN FAN
The University of Texas at Austin
Austin, Texas

MICHAEL MIRKIN
Queens College–CUNY
Flushing, New York

I. INTRODUCTION

Scanning electrochemical microscopy (SECM) is a technique that is useful as both a form of scanning probe microscopy (SPM) for the imaging of surfaces and as an electrochemical tool. In the latter application, the SECM is considered as a dual-electrode electrochemical system like the rotating ring-disk electrode or closely spaced microband electrodes. The theory, instrumentation, and applications of SECM have been reviewed recently [1], so in this chapter we stress the basic principles of the technique and more recent applications to electrochemical systems.

A. Principles of SECM Imaging

As in other types of SPM, SECM is based on the movement of a very small electrode (the *tip*) near the surface of a conductive or insulating sample (the *substrate*). In amperometric SECM experiments, the tip is usually a conventional ultramicroelectrode (UME) fabricated as a conductive disk of metal or carbon in an insulating sheath of glass or polymer, although in some applications, especially where nanometer-sized electrodes are desired, conically shaped electrodes can be used. Potentiometric SECM experiments with ion-selective tips are also possible.

   In amperometric experiments, one notes how the current at the UME tip is perturbed by the presence of the substrate (Fig. 1). When the tip is far (i.e., greater than several tip diameters) from the substrate, the steady-state current, $i_{T,\infty}$, is given by

$$i_{T,\infty} = 4nFDca \tag{1}$$

where F is the Faraday constant, n the number of electrons transferred in the tip reaction (O + ne → R), D the diffusion coefficient of O, c is the concentration of O, and a is the tip radius. While equation (1) holds for a disk-shaped tip, similar equations hold for small tips of other geometry. When the tip is

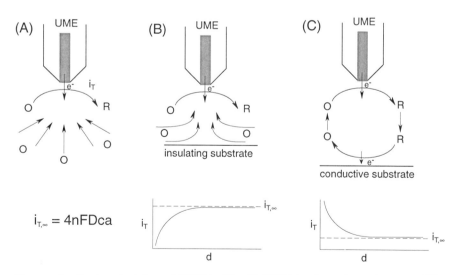

FIGURE 1   Basic principles of SECM: (A) with UME far from the substrate, diffusion of O leads to a steady-state current, $i_{T,\infty}$; (B) with UME near an insulating substrate, hindered diffusion of O leads to $i_T < i_{T,\infty}$; (C) with UME near a conductive substrate, positive feedback of O to the tip leads to $i_T > i_{T,\infty}$.

moved toward the surface of an insulating substrate, the tip current, i_T, decreases because the insulator blocks diffusion of species O to the tip from the bulk solution. The nearer the tip gets to the substrate, the smaller i_T becomes, approaching zero as the tip-to-substrate distance, d, approaches zero (Fig. 1B). On the other hand, with a conductive substrate, species R can be oxidized back to O. This produces an additional flux of O to the tip and hence an increase in i_T (Fig. 1C). In this case, the smaller the value of d, the larger is i_T, with $i_T \rightarrow \infty$ as $d \rightarrow 0$, when the oxidation of R on the substrate is diffusion controlled. The substrate can be maintained at a given potential to drive the desired reaction by a potentiostat, but frequently the substrate, which is usually much larger than the tip, is in contact with a solution of O in the regions away from the tip and will be poised at a sufficiently positive potential without an external bias. These simple principles form the basis for imaging with the SECM. Usually, the tip is rastered (in the xy plane) above the substrate, and variations in the tip current represent changes in surface topography and conductivity (or reactivity). One can separate topographic effects from conductivity effects by noting $i_{T,\infty}$ and the fact that over an insulator i_T is always less than $i_{T,\infty}$, while over a conductor i_T is always greater than $i_{T,\infty}$. An alternative approach, based on tip-position modulation, is discussed in Section II.C. Constant-current imaging, where d is varied to maintain i_T constant, is also possible (see Section IV.A).

B. Principles of Electrochemical Experiments

In electrochemical applications, the SECM combines many of the features of ultramicroelectrodes, thin-layer cells, and twin-electrode systems on a size scale not easily attainable in larger electrochemical cells. For example, consider the effective mass-transfer rate to a UME. For a UME far from a substrate, the mass-transfer coefficient [2], $m \sim D/a$, while for the tip near a conductive substrate ($d < a$), $m \sim D/d$. For example, with $D = 5 \times 10^{-6}$ cm^2/s, a tip 0.1 μm from a substrate would produce an m value of 0.5 cm/s. To attain an equivalent mass-transfer rate at a rotating disk electrode, rotation rates in excess of 10^6 rotations per minute would be required. The large effective m obtainable with the SECM implies that reactions with fast heterogeneous reaction rates can be measured, as discussed in Section IV.B.

The SECM is also useful for studying short-lived electrogenerated intermediates and determining the rate of homogeneous reactions coupled to an electron-transfer reaction at an electrode. In this case the small volume between the tip and substrate can be considered as a tiny electrochemical cell. The transit time by diffusion between tip and substrate is $d^2/2D$. Thus, with $d = 0.1$ μm, this time is about 10 μs. This represents the approximate half-life of electrogenerated species that can be detected in this arrangement. For a first-order homogeneous reaction following electron transfer, this allows measurement of rate constants on the order of 10^5 s^{-1}, while for second-order reactions with reactant concentrations of 1 mM, rate constants on the order of 10^8 M^{-1} s^{-1} are attainable (see Section IV.C.). Note that the characteristic time given above is also about that needed for the attainment of steady-state conditions at the SECM tip, indicating that under the usual scanning conditions (i.e., rastering in the xy plane above a substrate or moving the tip in the z direction toward the substrate) rapid scans can be employed while maintaining essentially steady-state conditions at the tip.

Finally, the SECM can be used to fabricate structures on electrode surfaces [3–11]. In this application, the tip is used to electrodeposit materials (e.g., metals) in a pattern determined by the xy-scan conditions. Alternatively, the conditions can be adjusted so that material is etched away from the substrate in a form of microelectrochemical machining. Such an approach has been employed to produce etched structures on metals and semiconductors.

II. INSTRUMENTATION

A. Basic Apparatus

A block diagram of the SECM instrument is shown in Figure 2. The basic apparatus consists of four major components: position controller for tip and substrate, electrochemical cell (including tip, substrate, counterelectrode, and reference electrode), bipotentiostat/programmer, and data/image acquisition and display system. The microelectrode tip is held on a piezoelectric pusher or a piezoelectric tube scanner, which is mounted on an inchworm-translator-driven x-y-z three-axis stage. The substrate, which is frequently mounted on a support by cementing it with a conductive cement (e.g., Ag epoxy) to a metal contact, is attached to the electrochemical cell with an O-ring seal. The edges of the substrate are sealed with epoxy to prevent leakage of solution to the electrical contact. The substrate can also be attached to a polymeric support (e.g., Teflon). In experiments involving biological samples (e.g., cells, enzymes), the samples are immobilized on insulating substrates which are then mounted on the electrochemical cell. The electrochemical cell is mounted on a steady platform. A programmable controller is used to control the movement of all three inchworm translators. A bipotentiostat equipped with a potential programmer is used to control the potentials of the tip and substrate vs. the reference electrode. The SECM instrument is controlled with a microcomputer equipped with an interface board. The interface board is used to acquire the electrochemical signals via the bipotentiostat, to supply the voltage for the piezoelectric pusher or tube scanner via high-voltage operational amplifiers, and to send the signals for control of the x-y-z micropositioning device via the inchworm movement controller. Details on the construction of SECM instruments can be found elsewhere [1,12].

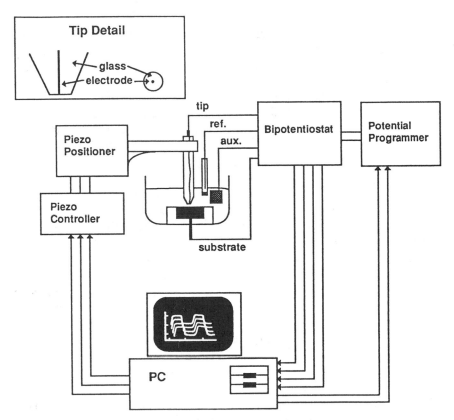

FIGURE 2 Block diagram of the SECM apparatus. (From Ref. 39, © 1990, American Chemical Society.)

B. Tip Preparation

Detailed information on tip preparation can be found in Ref. 1. The most popular amperometric tip used in SECM experiments is a disk-in-glass microelectrode. Construction of this type of tip is based largely on the fabrication techniques for the disk-shaped microelectrodes [13–15]. The tip is fabricated by placing a fine Pt or Au wire or a carbon fiber of the desired radius in a 10-cm-long 1-mm-ID Pyrex tube which is sealed at one end. After desorbing any impurities on the wire, by connecting the open end of the tube to a vacuum line and heating with a helix coil for about 1 h, the wire is sealed in the glass by increasing the heating coil temperature. The tip is formed by polishing this sealed end of the tube with sandpaper until the wire cross section is exposed and then by polishing with diamond paste (6, 1, and 0.25 μm, successively). Electrical connection to the fine wire is then made with silver paint to a copper wire. Finally, the glass wall surrounding the Pt, Au, or carbon disk is conically sharpened with emery paper and diamond paste to decrease the possibility of contact between the glass and substrate as the tip is moved close to the substrate.

Another type of amperometric tip is a cone-shaped microelectrode which is useful for directly probing soft interfacial films and for imaging. This type of tip is constructed by sharpening a 125- or 250-μm-diameter Pt–Ir (80–20%) rod by electrochemically etching it in a solution of saturated $CaCl_2$ (60% by volume), H_2O (36%), and HCl (4%) at about 20 V rms ac applied with a Variac transformer. A carbon rod or plate serves as the counterelectrode in the two-electrode etching cell. After etching, the tip is washed with Millipore reagent water and ethanol and then dried in air prior to insulation. The tip is insulated with molten Apiezon wax [16]. Several coatings of wax are usually required to insulate the tip. The very end of the tip is exposed by placing it in an electro-chemical scanning tunneling microscope (STM) immersed in a solution containing a redox electrolyte. The amount of exposed area of the tip can be controlled by the potential bias between the tip and substrate, the reference current setting of the feedback circuit of the STM, and the approach speed of the tip to the substrate surface. The exposed area of the tip can be estimated in situ from the steady-state tip current when the tip is far away from the substrate surface [equation (1)].

While most of the SECM work has been carried out with amperometric measurements, potentiometric probing is also possible [17,18]. Horrocks et al. [18] prepared antimony tips as pH sensors to image local pH changes around a corroding disk of AgI in aqueous potassium cyanide, a disk of immobilized urease hydrolyzing urea, and a disk of immobilized yeast cells in glucose solution. The antimony tips were prepared by drawing down Sb-filled glass capillaries to yield glass-coated Sb filaments with the diameter of Sb down to about 3 μm. Detailed procedures for preparing this kind of tip can be found in Ref. 18. Other types of SECM probes are also possible. For example, an enzyme electrode tip has been constructed to probe a particular species (i.e., peroxide) with high selectivity [19].

C. Tip-Position Modulation

In tip-position modulation SECM (TPM SECM) [20], the tip position (i.e., the distance between tip and substrate) is modulated with a small-amplitude sinusoidal motion normal to the substrate surface. This modulation causes a modulation in the tip current at the same frequency. Since the movement of the tip away from the substrate causes a decrease in the tip current over a conductor and an increase over an insulator, the tip current over a conductor is 180° out of phase compared with the tip current over an insulator. Thus, by detecting the in-phase component of the modulated tip current with a lock-in amplifier, the conductive nature of the substrate can be identified. Lock-in detection of the modulated current also offers the advantage of noise reduction and hence improved image resolution.

Figure 3 shows the SECM apparatus with the additional components that are required to provide for tip-position modulation [20]. The tip is mounted onto a spring-loaded linear translation stage that is driven by a piezoelectric pusher. The modulation voltage for the pusher is derived from the sine-wave reference oscillator output of a lock-in amplifier and is amplified to the desired value by a high-voltage operational amplifier. The dc response is measured with the lock-in amplifier to generate the phase-resolved rms ac response, and the dc and ac signals are acquired simultaneously through the analog-to-digital (A/D) conversion card.

As noted above, the TPM technique can be used to distinguish the conductive and insulating regions of a substrate and to filter the SECM signal, which provides an improved signal-to-noise ratio and greater sensitivity. Because the

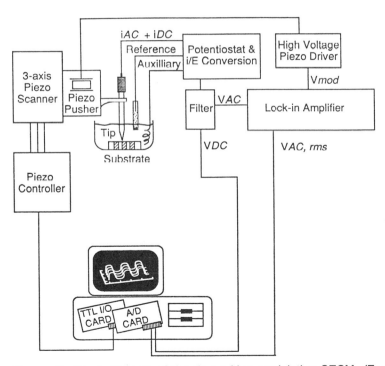

FIGURE 3 Block diagram of the tip-position modulation SECM. (From Ref. 20, © 1992, American Chemical Society.)

modulated tip current for both insulators and conductors is measured from a zero baseline, better sensitivity in imaging can be obtained, especially for insulators. A feedback circuit capable of performing constant-current imaging over a mixed insulator and conductor surface has been constructed based on this modulation technique [21].

III. THEORY

SECM theory has been reviewed recently [1,22] with some emphasis on the description of an uncomplicated quasi-reversible electron-transfer process at a tip electrode. Here we survey those results only briefly and focus on the processes, including homogeneous reactions in the gap and heterogeneous processes with finite substrate kinetics (these are related to studies at the semiconductor and chemically modified substrates discussed below). While fairly rigorous SECM theory has been developed based on the numerical solution of fairly complicated diffusion problems (either in terms of partial differential equations or multidimensional integral equations), in many cases analytical approximations allow much easier generation of theoretical dependencies and an analysis of experimental data. This section contains such approximations derived for different mechanisms. All theoretical results presented below concern the SECM with a disk-shaped tip electrode. Applications of non-disk-shaped tips (e.g., shaped as a cone or spherical cap) and the tip shape characterization are discussed in Refs. 1 and 23.

A. Uncomplicated Electron Transfer Under Steady-State and Non-Steady-State Conditions

1. Non-Steady-State Conditions

The general solution of the diffusion problem for an uncomplicated quasi-reversible non-steady-state process in SECM was obtained as a system of two-dimensional integral equations [24,25]. Two limiting cases, a diffusion-controlled process and one with totally irreversible kinetics, were treated numerically [25–27]. The non-steady-state SECM response depends on too many parameters to allow presentation of a complete set of working curves that would cover all experimental possibilities. The complexity of data analysis along with the short-time (or high-frequency) distortions of the measured current (probably owing to the capacitive coupling between the tip and the substrate electrodes [28,29]) have limited the applicability of non-steady-state (transient) SECM measurements, and we will not discuss this mode here further. Two recent SECM studies [30,31] employed transient measurements to examine diffusion inside a thin layer of AgBr and ion exchange at the polypyrrole–solvent interface; these processes could not be studied under steady-state conditions.

2. Steady-State Feedback Mode, Diffusion-Controlled Substrate Processes

The dimensionless current–distance curves for steady-state (time-independent) processes were obtained numerically for both insulating and conductive substrates, and several values of RG ($RG = Rg/a$, where Rg is the radius of the insulating glass surrounding the tip of radius a), assuming a diffusion-controlled mediator turnover, equal diffusion coefficients, and an infinitely large substrate [32]. An analytical expression [23] for a conductive substrate can be fit to the numerical results to yield the equation

$$I_T(L) = \frac{i_T}{i_{T,\infty}} = \frac{0.78377}{L} + 0.3315 \ \exp\left(\frac{-1.0672}{L}\right) + 0.68 \tag{2}$$

where $L = d/a$ is the normalized distance between the conductive substrate and the tip of radius a; this fits the $i_T\text{–}d$ curve over an L interval from 0.05 to 20 to within 0.7% (Fig. 4, curve 1). For an insulating substrate, a similar equation [equation (3a)] is slightly less accurate (to within 1.2%); however, the longer

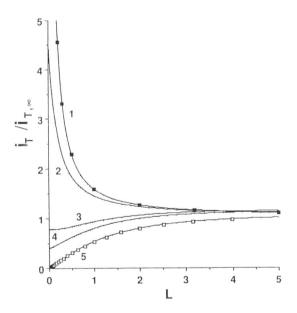

FIGURE 4 Steady-state tip current as a function of tip–substrate separation. 1, Diffusion-controlled process at a conductive substrate; 2–4, finite irreversible heterogeneous kinetics at substrate, $ka/D =$ (2) 5, (3) 1, and (4) 0.5; 5, substrate is an insulator. Solid lines are computed from (1) equation (2); (2) equation (7); (3, 4) equation (9); (5) equation (3b). Filled and open squares are data from Ref. 32.

expression [equation (3b)] is accurate to within 0.5% over the same L interval (Fig. 4, curve 5).

$$I_T^{ins}(L) = \frac{1}{0.292 + 1.5151/L + 0.6553 \exp(-2.4035/L)} \quad (3a)$$

$$I_T^{ins}(L) = \frac{1}{\begin{array}{c} 0.15 + 1.5385/L + 0.58 \exp(-1.14/L) + \\ 0.0908 \exp[(L - 6.3)/(1.017L)] \end{array}} \quad (3b)$$

By fitting an experimental current–distance curve to theory, one can determine the zero tip–substrate separation point ($L = 0$), which in turn allows the determination of L values essential for any quantitative SECM measurement.

From equation (2) one can derive the approximate equations for the tip at any potential, E, for either a nernstian tip process [equation (4)] or one under mixed diffusion–kinetic control [equation (5)][33]:

$$I_T(E, L) = \frac{0.68 + 0.78377/L + 0.3315 \exp(-1.0672/L)}{\theta} \quad (4)$$

$$I_T(E, L) = \frac{0.68 + 0.78377/L + 0.3315 \exp(-1.0672/L)}{\theta + 1/\kappa} \quad (5)$$

where $\theta = 1 + \exp[nf(E - E^{\circ\prime})]D_{Ox}/D_{Red}$, $\kappa = k^{\circ} \exp[-\alpha nf(E - E^{\circ\prime})/m_{Ox}$, k° is the standard rate constant, E is the electrode potential, $E^{\circ\prime}$ is the formal potential, α is the transfer coefficient, n is the number of electrons transferred per redox event, and $f = F/RT$ (where F is the Faraday constant, R is the gas constant, and T is temperature), and the effective mass-transfer coefficient for SECM is

$$m_{Ox} = \frac{4D_{Ox}[0.68 + 0.78377/L + 0.3315 \exp(-1.0672/L)]}{\pi a}$$

$$= \frac{I_T(L)}{\pi a^2 nFc} \quad (6)$$

One can see from (6) that at $L \gg 1$, $m_{Ox} \sim D/a$ (as for a microdisk electrode alone), but at $L \ll 1$, $m_{Ox} \sim D/d$, a thin-layer cell (TLC) type of behavior. This suggests that the SECM should be useful for studying rapid heterogeneous electron-transfer kinetics, since it should be easier to obtain very small (and variable) tip–substrate spacings than to produce microdisks with equally small radii. By decreasing the tip–substrate distance, the mass-transport rate can be increased sufficiently for quantitative characterization of the electron-transfer kinetics, preserving the advantages of steady-state methods (i.e., the absence of problems associated with ohmic drop, adsorption, and charging current).

At constant L, equations (4) and (5) describe reversible and quasi-reversible steady-state tip voltammograms, respectively. These curves can be obtained by scanning the potential of the tip while the substrate potential is held constant and the substrate process is diffusion controlled. Equation (5) was obtained assuming uniform accessibility of the tip surface (i.e., a uniform surface concentration of the electroactive species at the tip electrode). It is most suitable for analysis of steady-state voltammograms obtained with a liquid (i.e., mercury pool) substrate [34], where this assumption is more realistic. A somewhat longer equation [equation (25) in Ref. 35] was proposed for the SECM with a conventional, solid substrate electrode. Both equations were used in studies of fast heterogeneous kinetics [34–36], yielding essentially the same values of transfer coefficient and formal potential. The differences between extracted $k°$ values were about 20%.

3. Steady-State Feedback Mode, Finite Kinetics at a Substrate

The theory discussed above allows one to study kinetics of heterogeneous electron transfer at a tip electrode. There are, however, a number of interesting materials (e.g., single-crystal semiconductors, conductive polymers) unsuitable for tip preparation. These can be used in SECM experiments only as substrates, so the possibility of studying substrate kinetics is desirable. The calculations in Ref. 35 yielded an expression for the radius of the portion of the substrate surface participating in the SECM feedback loop, $r^\infty \cong 1 + 1.5d$. Thus at small tip–substrate separations (e.g., $L \leq 2$), a large substrate behaves as a virtual UME of a size comparable to that of the tip electrode. Therefore, SECM should allow one to study kinetics at large substrates with all of the advantages of microelectrode measurements [13]. Equations analogous to equations (5) and (6) are also available for an SECM process governed by finite substrate kinetics [31,37]. The situation is slightly different when the regeneration of the redox mediator (i.e., the feedback pictured schematically in Fig. 1C) proceeds via a heterogeneous chemical reaction rather than an electrochemical reaction on the substrate surface. The electrochemical rate constant is a function of electrode potential, while for a chemical process ($R \overset{k}{\rightarrow} O$) k is potential independent. The rate constant in the latter case can be extracted by fitting an experimental current–distance curve to the theoretical expression [30]

$$I_T^k = \frac{I_T}{1 + 1/\Lambda_c} + \frac{I'}{1 + \Lambda_c} \tag{7}$$

where I_T^k is the steady-state tip current under kinetic control normalized by $i_{T,\infty}$, and corresponding to the given L,

$$I' = 0.68 + 0.3315 \exp\left(-\frac{1.0672}{L}\right) \tag{8}$$

I_T is given by equation (2), and $\Lambda_c = kd/D$ is the dimensionless kinetic parameter. Equation (7) is applicable when the chemical reaction is sufficiently fast (i.e., $I_T^k \gg 1$ at $L \to 0$). In the opposite situation (i.e., $I_T^k < 1$ at $L \to 0$), the following equation is more accurate:

$$I_T^k = \frac{\pi}{4L(1 + 1/\Lambda_c)} + I_T^{\text{ins}} \tag{9}$$

where I_T^{ins} is given by equation (3). Figure 4 contains current–distance curves calculated for various values of k from 0 (i.e., an insulating substrate, curve 5) to ∞ (i.e., diffusion-controlled substrate process, curve 1). Unlike the diffusion-controlled process, when finite kinetics are present (curves 2 to 4), the current does not tend to infinity as the tip approaches the substrate but reaches some limiting finite value instead. This allows one to detect the kinetic behavior qualitatively and to get a rough estimate for k. One should avoid quantitative analysis of the approach curves with the highest value of $I_T^k \cong 1$ because in this case the accuracy of equations (7) and (9) is not high. The use of either a bigger or a smaller tip electrode is recommended in this situation.

B. Processes with Coupled Homogeneous Reactions in the Solution Gap

A homogeneous chemical reaction occurring in the gap between the tip and substrate electrodes causes a change in i_T; thus its rate can be determined from SECM measurements. If both heterogeneous processes at the tip and substrate electrodes are rapid (at extreme potentials of both working electrodes) and the chemical reaction (rate constant, k_c) is irreversible, the SECM response is a function of a single kinetic parameter $K = \text{const} \times k_c/D$, and its value can be extracted from I_T vs. L dependencies.

SECM theory has been developed for two mechanisms with homogeneous chemical reactions following electron transfer: a first-order irreversible reaction (E_rC_i mechanism) [27] and a second-order irreversible dimerization (E_rC_{2i} mechanism) [38]. (The solution obtained for a E_qC_r mechanism in terms of multidimensional integral equations [24] has not been utilized in any calculations.) In both cases the system of differential equations was solved numerically using the alternating-direction implicit (ADI) finite-difference method. Three approaches to kinetic analysis were proposed: (1) steady-state measurements in a feedback mode, (2) generation–collection experiments, and (3) analysis of the chronoamperometric SECM response. Unlike the feedback mode, the generation–collection measurements included simultaneous analysis of both I_T–L and I_S–L curves or the use of the collection efficiency parameter (I_S/I_T when the tip is a generator and the substrate is a collector). The chronoamperometric measurements were found to be less reliable [27].

The theory for both first- and second-order mechanisms was presented in the form of two-parameter families of working curves [27,38]. These curves were obtained by numerical solution of partial differential equations and represent steady-state tip current or collection efficiency as functions of K and L. We present here some generalizations of the theory, along with analytical approximations for the working curves. To understand this approach, it is useful first to consider a positive feedback situation with a simple redox mediator (i.e., without homogeneous chemistry involved) and with both tip and substrate processes under diffusion control. The normalized steady-state tip current can be presented as the sum of two terms,

$$I_T = I_f + I_T^{\text{ins}} \tag{10}$$

where I_f is the feedback current coming from the substrate and I_T^{ins} is the current due to the hindered diffusion of the electroactive species to the tip from the bulk of solution given by equation (3); all variables are normalized by $i_{T,\infty}$. The substrate current is

$$I_S = I_f + I_d \tag{11}$$

where I_f is the same quantity as in Eq. (10), representing the oxidized species which eventually arrives at the tip as a feedback current, and I_d is the dissipation current (i.e., the flux of species not reaching the tip). It was shown in Ref. 38 that I_S/I_T is more than 0.99 at $0 < L \leq 2$ (i.e., for any L within this interval the tip and the substrate currents are essentially equal to each other). Thus from (10) and (11),

$$I_d = I_T^{\text{ins}} \tag{12}$$

or

$$\frac{I_d}{I_S} = \frac{I_T^{\text{ins}}}{I_T} = f(L) \tag{13}$$

where $f(L)$ can be computed for any L as the ratio of the right-hand side of (3) to that of (2).

Analogously, for an electrochemical process followed by an irreversible homogeneous reaction of any order

$$\text{Tip:} \quad O + e^- \rightarrow R \tag{14}$$
$$\text{Gap:} \quad R + \cdots \xrightarrow{k_c} \text{products} \tag{15}$$
$$\text{Substrate:} \quad R - e^- \rightarrow O \tag{16}$$

one can write

$$I_T' = I_f' + I_T^{\text{ins}} \tag{17}$$
$$I_S' = I_f' + I_d' \tag{18}$$

where the variables labeled with the prime are analogous to unlabeled variables in (10) and (11), and I_T^{ins} is unaffected by the occurrence of the homogeneous reaction in (15). Since the species O are stable, the fraction of these species arriving at the tip from the substrate should also be unaffected by the reaction in (15) [i.e., the relation $I_d'/I_S' = f(L)$ holds true]. Consequently,

$$I_T' = I_T^{ins} + I_S' - I_d'$$
$$= I_T^{ins} + I_S'[1 - f(L)] \tag{19}$$

that is, for an SECM process with a following homogeneous chemical reaction of any order, the dependence I_T' vs. I_S' at any given L should be linear with a slope equal to $1 - f(L)$ and an intercept equal to $I_T^{ins}(L)$. Thus the generation–collection mode of the SECM (with the tip electrode serving as a generator) for these mechanisms is completely equivalent to the feedback mode, and any quantity, I_T', I_S', or I_S'/I_T', can be calculated from (19) for a given L if any other of these quantities is known. To check (19), we used the data simulated for an $E_r C_{2i}$ mechanism [38] and plotted I_T' vs. I_S' for different values of L. The slope and intercept values found from these linear dependencies were in excellent agreement with equations (19), (2), and (3).

For mechanisms with following irreversible reactions, one can expect the collection efficiency, I_S'/I_T', to be a function of a single kinetic parameter κ. If this parameter is known, the SECM theory for this mechanism can be reduced to a single working curve. After the function $κ = F(I_S'/I_T')$ is specified, one can immediately evaluate the rate constant from I_S'/I_T' vs. L or I_T' vs. L experimental curves. If only tip current has been measured, the collection efficiency can be calculated as

$$\frac{I_S'}{I_T'} = \frac{1 - I_T^{ins}/I_T'}{1 - f(L)} \tag{20}$$

For the $E_r C_i$ mechanism, one can easily deduce that $κ = k_c d^2/D$. Figure 5A represents the working curve, κ vs. I_S'/I_T', along with the simulated data. (Minor deviations are probably due to the limited accuracy of simulated data [27] used to compute this curve and errors in digitizing Fig. 10 from this reference.) The numerical results fit the analytical approximation

$$κ = F(x) = 5.608 + 9.347 \exp(-7.527x)$$
$$- 7.616 \exp\left(-\frac{0.307}{x}\right) \tag{21}$$

(solid curve in Fig. 5A), where $x = I_S'/I_T'$, within about 1%. For the $E_r C_{2i}$ mechanism, the choice of κ is not so straightforward, but we found empirically that

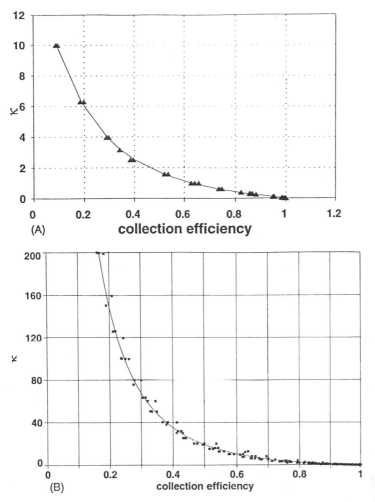

FIGURE 5 Kinetic parameter κ as a function of the collection efficiency (I_S^r/I_T^r). (A) E_rC_i mechanism; $κ = k_c d^2/D$, solid line was computed from equation (21), triangles are simulated data from Ref. 27. (B) E_rC_{2i} mechanism; $κ = c^*k_c d^3/aD$, solid line was computed from equation (22), squares are simulated data from Ref. 38.

using $κ = c^*k_c d^3/aD$, one can obtain an acceptable fit for all the data points computed in Ref. 38 (Fig. 5B). These data were fit to

$$κ = 104.87 - 9.948x - \frac{185.89}{\sqrt{x}} + \frac{90.199}{x} + \frac{0.389}{x^2} \qquad (22)$$

Although this approximation is less accurate, its use would not lead to an error of more than about 5 to 10%, which is within the usual range of experimental error. The invariability of k_c computed from different experimental points would assure the validity of the results.

IV. APPLICATIONS

A. Imaging

1. Constant-Height Imaging

In the constant-height mode, SECM images are obtained by rastering the tip across the substrate of interest in a constant reference plane above that of the substrate surface. Variations in the current or potential at the tip during the scanning produce the images. Surfaces imaged in the constant-height mode have included electrodes [28,39], minerals [40], membranes [41–43], and biological specimens [44–46]. Figure 6 is a constant-height image of a portion of a composite Kel-F/Au surface [21].

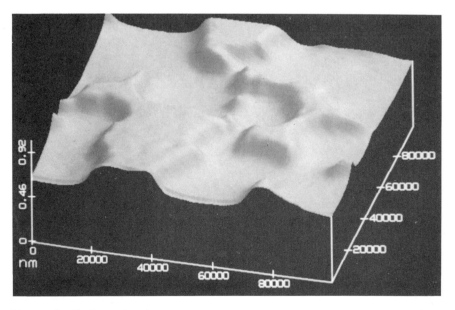

FIGURE 6 Surface plot of the tip current recorded during a constant-height scan. Scan size is 100 μm × 100 μm. Vertical axis is relative tip position in micrometers. This image was made with a 2-μm-diameter Pt tip using a 2.1 mM solution of $Ru(NH_3)_6^{3+}$ in a pH 4.0 buffer as the mediator species. The tip raster scan rate was 10 μm/s. (From Ref. 21, © 1993, American Chemical Society.)

2. Constant-Current Imaging

The resolution attainable with the SECM is largely governed by the size of the tip used and the distance between the tip and substrate. When the size of the tip used is very small (e.g., less than 100 nm in diameter) and the tip is scanned in close proximity to the substrate surface (e.g., about 50 nm above the surface), stray vibrations or surface irregularities can cause the tip to crash, so it is difficult to scan below these levels in the constant-height mode. Constant-current imaging, however, can be used to eliminate much of the problem of tip crashing. This mode of imaging is straightforward when the substrate surface consists of only insulating or only conducting material, because the piezofeedback can be set to counter a decrease in current by moving the tip either closer (in the case of a conductor) or farther away (in the case of an insulator). To image samples that contain both insulating and conducting regions, however, a method of recognizing the nature of the substrate and a feedback system capable of controlling the piezoelectric movement in the two regions are required. Wipf et al. have described a constant-current imaging device that uses an automatically switched servosystem in combination with the TPM signal (see Section II.C) to image surfaces containing both insulating and conducting regions [21]. Figure 7 shows the constant-current image of a portion of a Kel-F/Au composite surface used for the constant-height image in Figure 6. The constant-height image (Fig. 6) can be compared to the upper half of the constant-current image in Figure 7, since both are images of the same electrode region. The constant-current image shows more detail than the constant-height image, and it provides the true topography of the sample with better accuracy.

B. Heterogeneous Reactions and Reaction-Rate Imaging

1. Heterogeneous Electron Transfer at Metal Electrodes

The influence of finite heterogeneous kinetics on the SECM current–distance curves has been demonstrated experimentally by Wipf and Bard [47], who studied irreversible oxidation of Fe^{2+} on a glassy-carbon (GC) substrate. Later [25], these data along with those for a quasi-reversible reduction of $Ru(NH_3)_3^{3+}$ at a GC substrate were treated theoretically using the computational procedures of Refs. 24 and 27. The extracted kinetic parameters agreed well with the literature values; however, the calculations were time consuming and not routine. The procedure became simpler after the derivation of equations (5) and (6) and the development of a simplified approach to the analysis of steady-state voltammograms [33]. In this way, the kinetics of the fast oxidation of ferrocene at a Pt-tip electrode were measured [35]. The standard rate constant obtained (3.7 $\pm$ 0.6 cm/s) was about two to four times higher than that determined by fast-scan voltammetry [48–50], indicating that even very careful compensation of *iR* drop

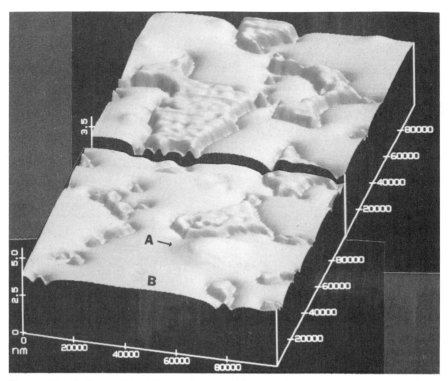

FIGURE 7 Surface plot of the *z*-piezo positioner voltage recorded during a constant-current imaging scan. The image shown is a composite of two consecutive scans. Scan size is 100 μm × 200 μm. Vertical axis is relative tip position in micrometers obtained from the piezo voltage. This image was made with a 2-μm-diameter Pt tip using a 2.1 mM solution of Ru(NH$_3$)$_6^{3+}$ in a pH 4.0 buffer as the mediator species. The tip position was modulated at a frequency of 160 Hz with a 100-nm peak-to-peak modulation amplitude. The tip raster scan was 10 μm/s. The negative and positive current reference levels were set at 500 and 780 pA, respectively, and $i_{T,\infty}$ was 80 pA, implying maintenance of about 2-μm tip–substrate spacing over both conductive and insulating regions. (From Ref. 21, © 1993, American Chemical Society.)

cannot guarantee the desired accuracy of measurements when the heterogeneous kinetics are rapid. Similar steady-state measurements with a liquid (mercury pool) substrate yielded kinetic parameters for the fast first stage of the reduction of C$_{60}$ in two highly resistive solvents, *o*-dichlorobenzene and benzonitrile [36]. The steady-state CVs were obtained in 0.2- to 0.4-μm-thick thin-layer cells formed inside the mercury pool, as described in Ref. 34, and two sets of kinetic parameters were calculated: $k° = 0.46 \pm 0.08$ cm/s, $\alpha = 0.43 \pm 0.05$ (*o*-dichlorobenzene) and $k° = 0.12 \pm 0.02$ cm/s, $\alpha = 0.52 \pm 0.05$ (benzonitrile). The kinetics of the reduction of $Ru(NH_3)_6^{3+}$ at a carbon fiber tip were also studied by this technique [34].

2. Heterogeneous Electron Transfer at Semiconductor Electrodes

In a recent review [51], Lewis noted that almost no reliable measurements of heterogeneous electron transfer to dissolved redox couples at nonilluminated semiconductor electrodes have been reported. This is due partly to the experimental difficulties encountered in making such measurements with conventional electrochemical techniques. Typical experimental problems include the presence of several parallel processes (e.g., semiconductor decomposition) which must be separated from the main redox reaction, surface heterogeneity and imperfection, and ohmic potential drop caused by the semiconductor resistance [52]. A recent study [37] involving several redox couples [e.g., $Ru(NH_3)_6^{3+/2+}$, $Fc^{+/0}$, and TMPD] and semiconductors (e.g., *n*- and *p*-type Si, *n*- and *p*-type WSe_2, and GaAs) demonstrated the possibility of eliminating some of these difficulties by using a semiconductor as an SECM substrate. In these experiments, the redox reaction of interest [e.g., reduction of $Ru(NH_3)_6^{3+}$] is driven at a diffusion-controlled rate at the tip. The rate of reaction at the semiconductor substrate is probed by measuring the feedback current as a function of substrate potential. The substrate was always held at a potential where no species present in bulk solution would react. Only species arriving from the tip reacted at the semiconductor surface. Thus the current at a large semiconductor electrode was very small (comparable to the tip current), eliminating problems associated with ohmic potential drop in solution and in the bulk of the semiconductor. Most irreversible parasitic processes at a semiconductor surface, such as corrosion, did not contribute to the tip current, and thus separation of the redox reaction of interest from parallel processes at the semiconductor was achieved (Fig. 8). Finally, apparently smooth low-defect areas most suitable for kinetic experiments could be found by scanning the tip over a semiconductor surface.

The most important conclusion made was that the low Tafel slope values obtained experimentally in many studies (in contradiction with the generally accepted model [53]) cannot be attributed to experimental artifacts, but probably reflect some fundamental properties of the semiconductor–electrolyte interface. Note that some systems, however, [e.g., $Ru(NH_3)_6^{3+/2+}$, at a *p*-WSe_2 in 0.5 *M* Na_2SO_4] displayed a transfer coefficient of 1, in agreement with the model proposed. The rate constant values for various semiconductor–redox couple combinations were extracted from steady-state voltammograms. Differences in reactivities of relatively defect-free portions of a semiconductor surface and those with apparently high defect density have also been demonstrated.

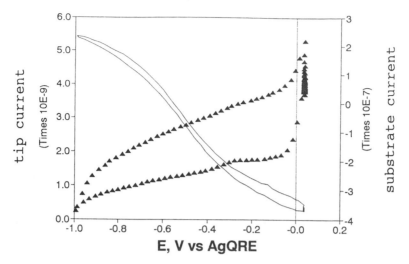

FIGURE 8 Tip (solid line) and substrate (symbols) currents as functions of the substrate potential for a reduction of $Ru(NH_3)_6^{3+}$ at an $n - WSe_2$ semiconductor [37]. The tip was a 10-μm-diameter Pt disk, the substrate was a single-crystal WSe_2. Note that the substrate current, although on a submicroampere range, was about two orders of magnitude higher than the tip current (i.e., some parallel processes at the semiconductor made a major contribution to i_s). A quasi-steady-state tip voltammogram was apparently free from this interference. Solution contained 1.3 mM $Ru(NH_3)_6^{3+}$ in 0.1 M KCl. Tip–substrate separation was about 2 μm. The substrate potential was swept at a rate of 100 mV/s.

3. Heterogeneous Reactions at Modified Electrodes

Unlike metal electrodes, most chemically modified electrodes (e.g., polymer-modified electrodes) are multiphase systems [54]. When an electrochemical reaction occurs at such an electrode, its response consists of different components: the faradaic currents associated with the interfacial reaction and with the changes in the modifying coating, as well as the charging current, which may be quite significant. The deconvolution of such a signal into its different components is not straightforward. Another typical problem pertains to the spatial localization of chemical or electrochemical reactions. In some cases the site of a reaction is clear; for other systems it is not clear where the reaction occurs on or within a film. The SECM may be helpful in solving both of these problems [30,31]: the signal of interest can be separated from the background processes, which do not contribute to the tip current (see Section IV.B.2), while the high spatial resolution offered by the SECM allows spatial localization of the process. Diagnostic criteria have been formulated for several different situations:

1. Regeneration of the mediator occurs at the film–solution interface (Fig. 9A). The position of the substrate obtained from the SECM approach curve (the zero-distance point) coincides with the z coordinate of the film–solution interface (this usually can be found as the point where the tip touches the substrate) as long as the mediator regeneration is fast. Here the shape of i_T–d curves is independent of the film thickness.

2. Regeneration of the mediator at the metal–film interface (Fig. 9B). If the tip does not penetrate the film, the maximum feedback current magnitude decreases with an increase in the film thickness, ℓ, and no positive feedback current is obtained when $\ell \gg a$ (where a is the tip radius). The substrate position obtained from the i_T–d curve with posi-

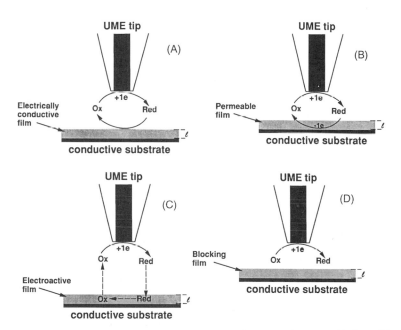

FIGURE 9 Schematic diagrams of the SECM experiments with four different types of mediator regeneration. (A) Regeneration of a mediator at a film–solution interface via heterogeneous chemical or electrochemical reaction (case 1 in the text). (B) Electrochemical regeneration of a mediator at a conductive substrate surface (case 2 in the text). (C) Regeneration proceeds by reaction between film and tip-generated species within the film (case 3 in the text). (D) Regeneration of a mediator at a substrate is blocked by resistive and impermeable film, resulting in negative feedback due to the hindered diffusion of redox species to the tip electrode (case 4 in the text). (From Ref. 30, © 1993, American Chemical Society.)

tive feedback relates to the metal–film interface rather than the film–solution interface.

3. Regeneration proceeds by reaction between the film and tip-generated species at the film–solution interface or within the film (Fig. 9C). In this case, the zero point on the i_T–d curve does not correspond to either the inner or outer boundary of the film and the approach curve shape deviates significantly from simple SECM theory. Unlike case 2, the positive feedback current does not vanish with an increase in ℓ as long as the charge transport in the film is rapid.

4. No regeneration occurs at the film–solution interface (Fig. 9D). When the film is impenetrable to the mediator (or so compact that mediator diffusion inside the film is slow), the substrate appears as an insulating one and the zero point on the i_T–d curve corresponds to the outer film boundary.

This methodology was used to analyze the mechanisms of the reactions of $Ru(NH_3)_6^{2+}$ and $Os(bpy)_3^{3+}$ with a Ag substrate covered with a AgBr film several micrometers thick [30]. The chemical reaction of AgBr with $Ru(NH_3)_6^{2+}$ was found to occur at the film–solution interface (case 1), but the electroreduction of $Os(bpy)_3^{3+}$ can take place only at the silver substrate surface, and thus the highly resistive AgBr layer blocks this process (case 4). The heterogeneous rate constant for the former chemical reaction (0.082 cm/s) was determined from current–distance curves, as described in Section III.A.3.

SECM can also be used to characterize the diffusion coefficient in thin solid films; this was demonstrated on the silver bromide thin film electrodeposited on a silver substrate [30]. The diffusion coefficient of bromide ion in the AgBr layer can be determined from the transit time $[t_x = \text{const}(d^2/D)]$ required for Br^- to diffuse from the substrate to the tip. Experimentally, a 10-μm-diameter Pt tip, biased initially at -0.5 V vs. SCE, is positioned in close proximity to the substrate by noting the i_T value obtained with a $Ru(NH_3)_6^{3+}$ mediator. After positioning, the tip potential is changed to 0.9 V vs. SCE, a value sufficiently positive for diffusion-limited Br^- oxidation. After application of a negative potential pulse to the Ag/AgBr electrode, AgBr is reduced to Ag, and the bromide ions released from the substrate will diffuse through the AgBr layer and the solution gap between the substrate and the tip and ultimately will be oxidized at the tip electrode. Figure 10 shows a fairly good linear relation between the transit time and the tip–substrate distance squared. The diffusion coefficient of Br^- determined from this linear relation is 5.6×10^{-7} cm^2/s, which is about 35 times lower than the diffusion coefficient for Br^- in aqueous solution and has been attributed to Br^- diffusion in AgBr.

Similarly, the oxidation of $Ru(NH_3)_6^{2+}$ at a polypyrrole-modified substrate is an electrochemical reaction whose rate is much lower at the resistive, reduced,

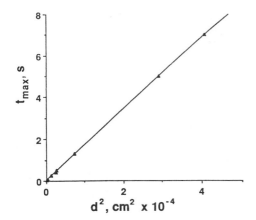

FIGURE 10 Time corresponding to the maximum tip current as a function of squared tip–substrate distance. Triangles were obtained by averaging experimental results with different pulse durations and filter frequencies. Current transients were obtained at a 10-μm-diameter Pt tip by application of a potential pulse of −0.5 V magnitude to the Ag/AgBr substrate while the tip potential was held at 0.9 V vs. SCE. Solution was 5 mM in $Ru(NH_3)_6^{3+}$ and 0.5 M in KNO_3. (From Ref. 30, © 1993, American Chemical Society.)

form of the polymer than at the conductive, oxidized, polypyrrole [31]. At the same time, the reduction of ferrocinium and $Os(bpy)_3^{3+}$ can proceed alternatively via an electrochemical mechanism (at oxidized polypyrrole) or as a chemical reaction (at reduced polypyrrole). In the latter case, the reaction rate does not decrease with a decrease in polymer conductivity and does not depend on the film thickness. The kinetics of the electrochemical electron transfer at oxidized polypyrrole was found to fit the metal electrode model rather than the redox polymer model.

The irreversible chemical regeneration of the SECM mediator at a substrate modified with immobilized glucose oxidase (GO; the enzyme from the mold *Aspergillus niger*) studied earlier [55] apparently corresponded to case 3. The catalytic oxidation of β-D-glucose to D-glucono-δ-lactone:

$$\beta\text{-D-glucose} + GO^{Ox} \xrightarrow{k_f} \delta\text{-gluconolactone} + GO^{Red} \qquad (23)$$

$$GO^{Red} + 2\ Med^{Ox} \rightarrow GO^{Ox} + 2\ Med^{Red} \qquad \text{(substrate)} \qquad (24)$$

$$Med^{Red} - e^- \rightarrow Med^{Ox} \qquad \text{(tip)} \qquad (25)$$

[where the mediators (Med) employed were FcCOOH (Fc = ferrocene), $Fe(CN)_6^{4-}$, or H_2Q (hydroquinone)] occurred inside a micrometer-thick porous enzyme layer. Although the fit between theory (assuming case 1) and experiment in this case was not as good as in experiments with flat metallic substrates, it was possible to establish zero-order enzyme-mediator kinetics and to evaluate apparent heterogeneous rate constants for several mediators.

4. Imaging Surface Reactivity

Because the SECM response is a function of the rate of the heterogeneous reaction at the substrate, it can be used to image the reactivity of surface features. Applications, including imaging mixed metal–insulator surfaces [56], substrates composed of two conductors with different reactivities (e.g., gold and glassy-carbon [57], or Pt and C [58]), as well as the inhomogeneities of thin polyelectrolyte coatings loaded with a redox mediator [42], have been reviewed recently [1]. Imaging of Prussian blue films in a ''direct mode'' of SECM, where the scanning tip slightly penetrates the film and its z-coordinate is recorded as a function of the (x, y) position while the tip potential and current are kept constant, has been described [59]. Here the SECM feedback occurred inside the solid phase. The resulting image depended strongly on the substrate potential, which in turn governed the film reactivity. A feedback detection scheme was also used to observe the localized reaction of glucose oxidase and mitochondria-bound NADH-cytochrome c reductase at the micrometer level (Fig. 11) [44]. The spatial resolution of such imaging is high for enzymatic systems with a fast catalytic turnover. For slower reactions, however, sensitivity constraints need to be considered and image resolution must be sacrificed in favor of adequate detection level. In these cases, collection-mode operation of the SECM could offer greater image resolution. A similar approach to the characterization of polymer films on electrodes is discussed in Section IV.D.

C. Studies of Homogeneous Chemical Reactions

A detailed review of processes with coupled first- and second-order homogeneous reactions studied by SECM can be found in Ref. 1. Among the processes with a first-order irreversible homogeneous reaction following electron transfer was the relatively slow oxidation of N,N-dimethyl-p-phenylenediamine (DMPPD) in aqueous solution (k_c values from 1.4 s^{-1} to 19 s^{-1} at pH 10.2 to 11.2). This was used as a model experimental system to test the developed theory [27]. The four-electron oxidation of epinephrine to adrenochrome (with adrenalinequinone as an intermediate) was also treated as a first-order process [29]. The SECM was also employed to prove the homogeneous character of the first chemical stage of the eight-electron oxidation of sodium borohydride [60]. The intermediate generated in the first two-electron wave was shown to diffuse into the solution and participate in a homogeneous reaction rather than react at the anode surface.

The quantitative theory developed for a process with product dimerization [38] was utilized in studies of the reductive coupling of both dimethyl fumarate

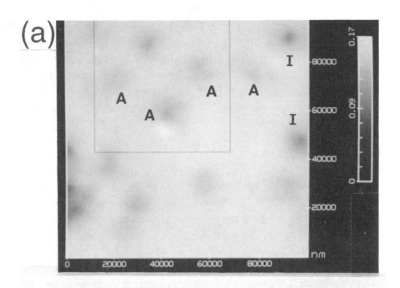

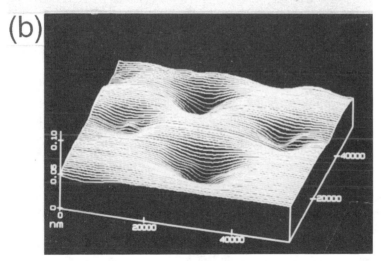

FIGURE 11 SECM image (a, 100- × 100-µm) of glass immobilized mitochondria with boxed region depicted as surface plot (b, 50- × 50-µm) for higher contrast. Image was taken with a 4-µm-radius carbon tip at +0.2 V vs. SCE in 50 m*M* phosphate buffer (pH 7.5) containing 0.1 *M* KCl, 0.25 *M* sucrose, 0.5 m*M* TMPD, and 50 m*M* NADH. Lightest image regions depict the greatest tip currents. A, Selected enzymatically active mitochondria; I, inactive mitochondria. (From Ref. 44, © 1993, American Chemical Society.)

(DF) and fumaronitrile (FN) in *N,N*-dimethylformamide. While the first dimerization reaction is relatively slow, the second one is quite fast and has been difficult to study with conventional electrochemical methods. Both steady-state tip and substrate current–distance experimental curves yielded a value of $k_c = 180 \ M^{-1} \ s^{-1}$ for DF, in good agreement with the rate constants measured by other methods [61]. From the I_T–L and I_S–L curves obtained at various FN concentrations (from 1.5 to 121 mM), a rate constant of $k_c = 2.0 \ (\pm \ 0.4) \times 10^5 \ M^{-1}$ was determined.

The main challenge in studying very fast kinetics and detection of short-lived intermediates with the SECM is obtaining very close tip–substrate spacing. Progress in this direction has made possible the study of reactions as fast as the hydrodimerization of acrylonitrile [62]. The competition between fast hydrodimerization and another (ECE) reaction complicated treatment of the kinetics of this process. The second-order rate constant for the hydrodimerization reaction determined by SECM was on the order of $5 \times 10^8 \ M^{-1} \ s^{-1}$ (i.e., much higher than the upper limit for rotating disk measurements or most other conventional electrochemical techniques). This also confirmed the earlier theoretical predictions [27,38] that the first-order rate constants in excess of $2 \times 10^4 \ s^{-1}$ and second-order kinetics with k_c up to $4 \times 10^8 \ M^{-1} \ s^{-1}$ should be accessible to SECM measurements under steady-state conditions.

D. Characterization of Polymer Films and Membranes

SECM has also been used to study thin films at interfaces. These studies have included polyelectrolytes, such as Nafion and protonated poly(4-vinylpyridine), and electronically conductive polymers, such as polypyrrole [42,63–68].

SECM studies of these films have made use of a unique type of cyclic voltammetry, tip–substrate cyclic voltammetry (T/S CV), which involves monitoring the tip current (i_T) vs. the substrate potential (E_S) while the tip potential (E_T) is maintained at a given value and the tip is held near the substrate. For example, $Os(bpy)_3^{2+/3+}$ incorporated in a Nafion film has been investigated [65]. As shown in Figure 12, the substrate cyclic voltammogram (i_S vs. E_S) of an $Os(bpy)_3^{2+}$-incorporated Nafion film in an aqueous solution containing $K_3Fe(CN)_6^{3-}$ shows only a wave for the $Os(bpy)_3^{2+/3+}$ couple, indicating that $Fe(CN)_6^{3-}$ does not diffuse into the Nafion coating. Thus any reaction between $Fe(CN)_6^{4-/3-}$ and $Os(bpy)_3^{3+/2+}$ is restricted to the solution–film interface.

Figure 12A shows typical T/S CV curves of the $Os(bpy)_3^{3+/2+}$–Nafion substrate in an aqueous solution in which the bulk species is $Fe(CN)_6^{3-}$. When the tip electrode is far from the substrate ($d = 500 \ \mu m$), $i_T = i_{T,\infty}$ and is essentially independent of E_S. When the tip electrode is close to the substrate ($d = 10 \ \mu m$), either negative or positive feedback effects arise, depending on the oxidation state of the $Os(bpy)_3^{2+/3+}$ couple in the Nafion. When E_S is swept positive of the

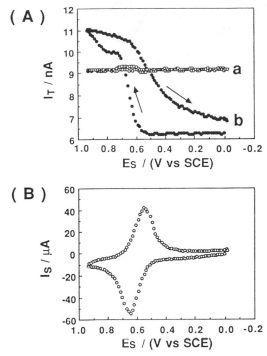

FIGURE 12 (A) T/S CVs: curve a, $d = 500$ μm; curve b, $d = 10$ μm; (B) S CV of a Nafion/Os(bpy)$_3^{3+/2+}$-covered Pt disk electrode (5-mm diameter) in 10 mM K$_3$Fe(CN)$_6$ and 0.1 M Na$_2$SO$_4$, scan rate $v = 50$ mV/s, $E_T = -0.4$ V vs. SCE. (From Ref. 65, © 1990, American Chemical Society.)

Os(bpy)$_3^{2+/3+}$ redox waves, $i_T > i_{T,\infty}$. This positive feedback of i_T is due to the regeneration of Fe(CN)$_6^{3-}$ in the solution gap region because of the oxidation of Fe(CN)$_6^{4-}$ by Os(bpy)$_3^{3+}$ at the solution–film interface. When E_S is negative of the redox wave, the film behaves as an insulator and $i_T < i_{T,\infty}$, since the Os(bpy)$_3^{2+}$ formed is unable to oxidize tip-generated Fe(CN)$_6^{4-}$ back to Fe(CN)$_6^{3-}$ and the film blocks diffusion of Fe(CN)$_6^{3-}$ to the tip.

Figure 13 shows the CVs of the Os(bpy)$_3^{3+/2+}$–Nafion film in an aqueous solution in which the bulk species is Fe(CN)$_6^{4-}$. The substrate CV in Figure 13B shows the catalytic oxidation of Fe(CN)$_6^{4-}$ by electrochemically generated Os(bpy)$_3^{3+}$. The substrate reaction in this solution causes the depletion of Fe(CN)$_6^{4-}$, rather than tip-generated Fe(CN)$_6^{3-}$, and the extent of this depletion is demonstrated in Figure 13A. When the tip is far from the substrate ($d = 220$ μm), and E_S is scanned to the potentials where the oxidation of Os(bpy)$_3^{2+}$ takes

(A)

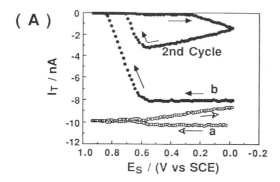

(B)

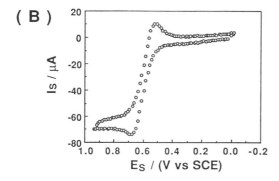

FIGURE 13 (A) T/S CVs: curve a, d = 220 μm; curve b, d = 10 μm; (B) S CV of a Nafion/Os(bpy)$_3^{3+/2+}$-covered Pt disk electrode (5-mm diameter) in 10 mM K$_4$Fe(CN)$_6$ and 0.1 M Na$_2$SO$_4$, scan rate v = 50 mV/s, E_T = 0.6 V vs. SCE. (From Ref. 65, © 1990, American Chemical Society.)

place, i_T decreases because of depletion of Fe(CN)$_6^{4-}$ in the solution region between substrate and tip (Fig. 13A, curve a). The time, t_d, for i_T to decrease below $i_{T,\infty}$ in the cyclic scan is about one-half of the time needed for the diffusion layers to grow out from the substrate to the tip (i.e., $t_d \sim d^2/2D$). This depletion effect can be seen for much larger tip distances as compared with the feedback mode if sufficient time is allowed for substrate diffusion layer growth. When the tip is moved closer to the film surface (e.g., d = 10 μm), more rapid and greater depletion of the Fe(CN)$_6^{4-}$ species at the substrate is detected at the tip (Fig. 13A, curve b). i_T decreases rapidly to zero immediately after Os(bpy)$_3^{2+}$ is oxidized to Os(bpy)$_3^{3+}$ because the large modified substrate electrode (Pt disk, 5 mm diameter) depletes Fe(CN)$_6^{4-}$ near the surface of the film. The almost total depletion of Fe(CN)$_6^{4-}$ is maintained until the substrate potential scan is reversed and the $Os(bpy)_3^{3+}$ species in the Nafion film is reduced. i_T then increases because $Fe(CN)_6^{4-}$ from the bulk solution diffuses into the gap region. This increase continues until i_T approaches the value characteristic of that over an insulator. However, when E_S is scanned back into the region where the oxidation of $Fe(CN)_6^{4-}$ takes place, i_T decreases again via the depletion effect.

Lee and Anson have applied SECM to examine $Os(bpy)_3^{2+/3+}$ ejection from Nafion coatings on electrodes [67]. The responses observed at a tip electrode during the recording of cyclic voltammograms at the substrate-coated electrode are shown in Figure 14. When the loading, X_{Os}, did not exceed 0.33, essentially no tip current was observed with the tip potential set to detect $Os(bpy)_3^{3+}$ or $Os(bpy)_3^{2+}$ (Fig. 14A), which demonstrated that neither $Os(bpy)_3^{3+}$ nor $Os(bpy)_3^{2+}$, but rather an electroinactive species, was ejected from the Nafion coating. However, when X_{Os} was increased to near 0.5, substantial cathodic tip currents were observed during the oxidation of $Os(bpy)_3^{2+}$ within the coating (Fig. 14B). The cathodic tip currents with the tip potential set at 0.3 V (curve III) were much larger than the anodic currents with the tip potential at 0.9 V (curve II), which demonstrated that the cations arriving at the tip consisted primarily of $Os(bpy)_3^{3+}$.

Direct measurements of electrochemical parameters have also been carried out by recording the tip current as it is moved from solution into a polymer phase and is ultimately brought into contact with the substrate [68]. In Figure 15 we describe direct measurements of film thickness and electrochemical parameters for a Nafion film containing $Os(bpy)_3^{2+}$ on an indium tin oxide (ITO) substrate immersed in an electrolyte solution containing only 40 mM NaClO$_4$. During the experiment, the tip is held at 0.80 V vs. SCE, where $Os(bpy)_3^{2+}$ oxidation is diffusion controlled. The Nafion-coated ITO substrate is biased at 0.20 V vs. SCE, where any $Os(bpy)_3^{3+}$ generated at the tip that reaches the substrate will be reduced back to $Os(bpy)_3^{2+}$ when the tip–ITO separation is small. When a conical tip (30-nm radius, 30-nm height) is outside the film in the solution, only a negligible current flows. As soon as the tip contacts the films, extra current begins to flow as the oxidation of $Os(bpy)_3^{2+}$ occurs at the tip. The onset of this extra current thus represents the film–solution boundary. The tip current increases as more of the tip enters the film, eventually leveling off when the whole active tip surface is within the film but the tip–ITO separation is still large. The tip current then increases steadily until the tip gets to within tunneling distance of the substrate, where a sharp increase in current takes place. The thickness of the film immersed in electrolyte, in this example 2200 Å, can be found directly from the approach curve as the difference in relative displacement between the film–solution interface coordinate and that for the onset of tunneling. Cyclic voltammetry can also be performed with the tip partially or completely within the film; a typical steady-state voltammogram is shown in Figure 16. Because the steady-state voltammogram is fairly insensitive to the exact

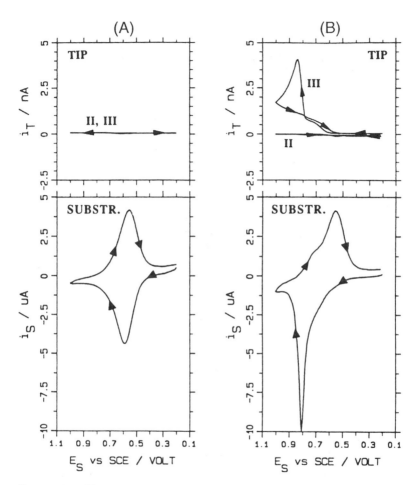

FIGURE 14 Tip currents measured during cyclic voltammetry of a Os(bpy)$_3^{3+/2+}$ -incorporated Nafion coating on a substrate electrode. The potential of the tip electrode was maintained at 0.3 or 0.9 V to monitor the concentration of Os(bpy)$_3^{3+}$ or Os(bpy)$_3^{3+}$, respectively. The tip was positioned about 3.5 μm above the substrate electrode. Cathodic tip currents are plotted upward and anodic, downward. (A) The substrate electrode was loaded with Os(bpy)$_3^{2+}$ and cycled several times to decrease X_{Os} to 0.33. (B) First voltammogram following exposure of the coated substrate electrode to 1 m*M* Os(bpy)$_3^{2+}$ for 25 h (X_{Os} ~ 0.5). Supporting electrolyte 0.2 *M* CH$_3$COONa adjusted to pH 4.6. Scan rate = 11.6 mV/s. The initial potential for all curves was 0.2 V vs. SCE. (From Ref. 67, © 1992, American Chemical Society.)

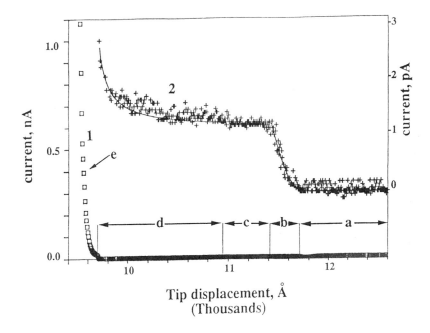

FIGURE 15 Dependence of the tip current versus distance: a, tip outside film; b, tip moving into film; c, tip inside film; d, onset of positive feedback; e, onset of tunneling. The displacement values are given with respect to an arbitrary zero point. The current observed during stages a to d is much smaller than the tunneling current and therefore cannot be seen on the scale of curve 1 (the left-hand current scale). Curve 2 is at higher current sensitivity to show the current–distant curve corresponding to stages a to d (the right-hand current scale). The solid line is computed for a conically shaped electrode with a height h = 30 nm and a radius r_0 = 30 nm for zones a to c and SECM theory for zone d. The tip was biased at 0.80 V vs. SCE, and the substrate was at 0.20 V vs. SCE. The tip moved at a rate of 30 Å/s. (From Ref. 68, © AAAS.)

shape of the tip, it can be approximated by the equivalent-size hemisphere. This allows estimation of the value of the diffusion coefficient of $Os(bpy)_3^{2+}$, $D = 1.2 \times 10^{-9}$ cm^2/s. The kinetic parameters of the electrode reaction have also been evaluated by the three-point method: the half-wave potential, $E_{1/2}$, and two quartile potentials, $E_{1/4}$ and $E_{3/4}$ [33]. A rate constant ($k°$) of 1.6×10^{-4} cm/s was obtained [68].

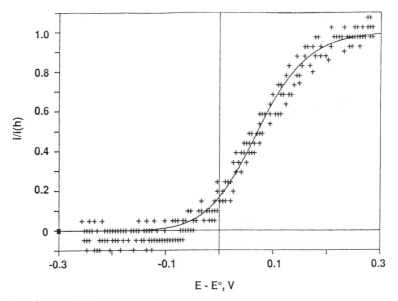

Figure 16 Voltammogram at a microtip electrode partially penetrating a Nafion film containing 0.57 M $Os(bpy)_3^{2+}$. Scan rate v = 5 mV/s. The substrate was biased at 0.2 V vs. SCE. The solid line is the theoretical response with D = 1.2 × 10^{-9} cm^2/s and $k°$ = 1.6 × 10^{-4} cm/s. (From Ref. 68, © AAAS.)

ACKNOWLEDGMENTS

The support of our research in SECM by grants from the Robert A. Welch Foundation, the National Science Foundation, and the Texas Advanced Research Program is gratefully acknowledged. Our thanks to Rose Buettner for her assistance with the preparation of this chapter.

REFERENCES

1. A. J. Bard, F.-R. F. Fan, and M. V. Mirkin, in *Electroanalytical Chemistry*, Vol. 18 (A. J. Bard, ed.), Marcel Dekker, New York, 1993, p. 243.
2. A. J. Bard and L. R. Faulkner, *Electrochemical Methods*, Wiley, New York, 1980, p. 27.
3. A. J. Bard, in *Microchemistry: Spectroscopy and Chemistry in Small Domains: Proceedings of the JRDC-KUL Joint International Symposium* (H. Masuhara, ed.), Elsevier, Amsterdam, 1994, pp. 507–520.
4. D. H. Craston, C. W. Lin, and A. J. Bard, *J. Electrochem. Soc., 135*: 785 (1988).
5. D. Mandler and A. J. Bard, *J. Electrochem. Soc., 136*: 3143 (1989).

6. O. E. Hüsser, D. H. Craston, and A. J. Bard, *J. Vacuum Sci. Technol. B*, *6*: 1873 (1988).
7. O. E. Hüsser, D. H. Craston, and A. J. Bard, *J. Electrochem. Soc.*, *136*: 3222 (1989).
8. D. Mandler and A. J. Bard, *J. Electrochem. Soc.*, *137*: 1079 (1990).
9. R. L. McCarley, S. A. Hendricks, and A. J. Bard, *J. Phys. Chem.*, *96*: 10,089 (1992).
10. D. Mandler and A. J. Bard, *J. Electrochem. Soc.*, *137*: 2468 (1990).
11. D. Mandler and A. J. Bard, *Langmuir*, *6*: 1489 (1990).
12. J. Kwak and A. J. Bard, *Anal. Chem.*, *61*: 1794 (1989).
13. R. M. Wightman and D. O. Wipf, in *Electroanalytical Chemistry*, Vol. 15 (A. J. Bard, ed.), Marcel Dekker, New York, 1988, p. 267.
14. M. Fleischmann, S. Pons, D. R. Rolison, and P. P. Schmidt, *Ultramicroelectrodes*, Datatech Systems, Morgantown, N.C., 1987.
15. M. I. Montenegro, M. A. Queirós, and J. L. Daschback, eds., *Microelectrodes: Theory and Applications*, NATO ASI Ser. Appl. Sci., Vol. 197, Kluwer Academic Publishers, Dordrecht, The Netherlands, 1991.
16. L. A. Nagahara, T. Thundat, and S. M. Lindsay, *Rev. Sci. Instrum.*, *60*: 3128 (1989).
17. G. Denuault, M. H. Troise Frank, and L. M. Peter, *Faraday Discuss.*, *94*: 23 (1992).
18. B. R. Horrocks, M. V. Mirkin, D. T. Pierce, A. J. Bard, G. Nagy, and K. Toth, *Anal. Chem.*, *65*: 1213 (1993).
19. B. R. Horrocks, D. Schmidtke, A. Heller, and A. J. Bard, *Anal. Chem.*, *65*: 3605 (1993).
20. D. O. Wipf and A. J. Bard, *Anal. Chem.*, *64*: 1362 (1992).
21. D. O. Wipf, A. J. Bard, and D. E. Tallman, *Anal. Chem.*, *65*: 1373 (1993).
22. A. J. Bard, F.-R. F. Fan, and M. V. Mirkin, in *Handbook of Surface Imaging and Visualization* (A. T. Hubbard, ed.), CRC Press, Boca Raton, Fla., in press.
23. M. V. Mirkin, F.-R. Fan, and A. J. Bard, *J. Electroanal. Chem.*, *328*: 47 (1992).
24. M. V. Mirkin and A. J. Bard, *J. Electroanal. Chem.*, *323*: 1, 29 (1992).
25. A. J. Bard, M. V. Mirkin, P. R. Unwin, and D. O. Wipf, *J. Phys. Chem.*, *96*: 1861 (1992).
26. A. J. Bard, G. Denuault, R. A. Friesner, B. C. Dornblaser, and L. S. Tuckerman, *Anal. Chem.*, *63*: 1282 (1991).
27. P. R. Unwin and A. J. Bard, *J. Phys. Chem.*, *95*: 7814 (1991).
28. A. J. Bard, F.-R. F. Fan, J. Kwak, and O. Lev, *Anal. Chem.*, *61*: 132 (1989).
29. R. C. Engstrom, T. Meaney, R. Tople, and R. M. Wightman, *Anal. Chem.*, *59*: 2005 (1987).
30. M. V. Mirkin, M. Arca, and A. J. Bard, *J. Phys. Chem.*, *97*: 10,790 (1993).
31. M. V. Mirkin, M. Arca, and A. J. Bard, *J. Phys. Chem.*, Submitted.
32. J. Kwak and A. J. Bard, *Anal. Chem.*, *61*: 1221 (1989).
33. M. V. Mirkin and A. J. Bard, *Anal. Chem.*, *64*: 2293 (1992).
34. M. V. Mirkin and A. J. Bard, *J. Electrochem. Soc.*, *139*: 3535 (1992).
35. M. V. Mirkin, T. C. Richards, and A. J. Bard, *J. Phys. Chem.*, *97*: 7672 (1993).
36. M. V. Mirkin, L. O. S. Bulhões, and A. J. Bard, *J. Am. Chem. Soc.*, *115*: 201 (1993).
37. B. R. Horrocks, M. V. Mirkin, and A. J. Bard, *J. Phys. Chem.*, *98*: 9106 (1994).
38. F. Zhou, P. R. Unwin, and A. J. Bard, *J. Phys. Chem.*, *96*: 4917 (1992).
39. A. J. Bard, G. Denuault, C. Lee, D. Mandler, and D. O. Wipf, *Acc. Chem. Res.*, *23*: 357 (1990).

40. P. R. Unwin and A. J. Bard, *J. Phys. Chem., 96*: 5035 (1992).

41. E. R. Scott, H. S. White, and J. B. Phillips, *J. Membr. Sci., 58*: 71 (1991).

42. I. C. Jeon and F. C. Anson, *Anal. Chem., 64*: 2021 (1992).

43. E. R. Scott, H. S. White, and J. B. Phillips, *Solid State Ionics, 53*: 176 (1992).

44. D. T. Pierce and A. J. Bard, *Anal. Chem., 65*: 3598 (1993).

45. A. J. Bard, F.-R. F. Fan, D. T. Pierce, P. R. Unwin, D. O. Wipf, and F. Zhou, *Science, 254*: 68 (1991).

46. C. Lee, J. Kwak, and A. J. Bard, *Proc. Natl. Acad. Sci. USA, 87*: 1740 (1990).

47. D. O. Wipf and A. J. Bard, *J. Electrochem. Soc., 138*: 469 (1991).

48. M. I. Montenegro and D. Pletcher, *J. Electroanal. Chem., 200*: 371 (1986).

49. D. O. Wipf, E. W. Kristensen, M. R. Deakin, and R. M. Wightman, *Anal. Chem., 60*: 306 (1988).

50. A. M. Bond, T. L. E. Henderson, D. R. Mann, T. F. Mann, W. Thormann, and C. G. Zoski, *Anal. Chem., 60*: 1878 (1988).

51. N. S. Lewis, *Annu. Rev. Phys. Chem., 42*: 593 (1991).

52. C. A. Koval and J. Olson, *Chem. Rev., 92*: 411 (1992).

53. H. Gerischer, in *Physical Chemistry: An Advanced Treatise*, Vol. 9A (H. Eyring, ed.), Academic Press, New York, 1970, p. 463.

54. R. W. Murray, in *Electroanalytical Chemistry*, Vol. 13 (A. J. Bard, ed.), Marcel Dekker, New York, 1984, p. 191.

55. D. T. Pierce, P. R. Unwin, and A. J. Bard, *Anal. Chem., 64*: 1795 (1992).

56. R. C. Engstrom, M. Weber, D. J. Wunder, R. Burgess, and S. Winquist, *Anal. Chem., 58*: 844 (1986).

57. D. O. Wipf and A. J. Bard, *J. Electrochem. Soc., 138*: L4 (1991).

58. R. C. Engstrom, B. Small, and L. Kattan, *Anal. Chem., 64*: 241 (1992).

59. H. Sugimura, N. Shimo, N. Kitamura, and H. Masuhara, *J. Electroanal. Chem., 346*: 147 (1993).

60. M. V. Mirkin, H. Yang, and A. J. Bard, *J. Electrochem. Soc., 139*: 2212 (1992).

61. I. B. Goldberg, D. Boyd, R. Hirasama, and A. J. Bard, *J. Phys. Chem., 78*: 295 (1974).

62. F. Zhou and A. J. Bard, *J. Am. Chem. Soc., 116*: 393 (1994).

63. C. Lee, J. Kwak, and F. C. Anson, *Anal. Chem., 63*: 1501 (1991).

64. J. Kwak, C. Lee, and A. J. Bard, *J. Electrochem. Soc., 137*: 1481 (1990).

65. C. Lee and A. J. Bard, *Anal. Chem., 62*: 1906 (1990).

66. J. Kwak and F. C. Anson, *Anal. Chem., 64*: 250 (1992).

67. C. Lee and F. C. Anson, *Anal. Chem., 64*: 528 (1992).

68. M. V. Mirkin, F.-R. F. Fan, and A. J. Bard, *Science, 257*: 364 (1992).

6

Electrochemical Impedance Spectroscopy: Principles, Instrumentation, and Applications

Claude Gabrielli
UPR15 CNRS, Université Pierre et Marie Curie
Paris, France

I. INTRODUCTION

For electrochemical processes of academic interest (i.e., simplified compared with the real world in order to describe the latter in fundamental terms), or for more applied purposes, the electrochemist aims at analyzing a reaction mechanism by chemical identification and kinetic characterization of reaction intermediates, or at estimating a distinctive parameter of a process (e.g., corrosion or deposition rate) from the measurement of a well-defined quantity. Therefore, to disentangle the coupling between mass transport and interfacial reactions, or to perform a test, the experimenter needs a technique capable of extracting informations during the evolution of an electrochemical process.

    Some of the techniques used to characterize the surface condition or adsorbed species at the interface require a vacuum chamber [low-energy electron diffraction (LEED), Auger electron spectroscopy, etc.]. Those that utilize electromagnetic radiation (optical: ellipsometry, or x-rays: EXAFS) are now being used for investigating the electrochemical interface, but they have some major difficulties when an alteration (dissolution, deposit, etc.) of the surface occurs. Therefore, electrical techniques are often the only ones usable for an in situ investigation of the electrochemical interface.

By controlling electrochemical reactions, the use of electrical quantities permits a kinetic study that allows couplings between various elementary phenomena to be dissociated. In this way the monoelectronic steps of the reaction mechanisms can be distinguished and the intermediates, often unstable, involved in the reactions can be counted. Although these techniques do not lead to an identification of the chemical bonds or of the intermediates in the chemical sense, they give information on the reaction rates occurring at the electrochemical interface and provide a certain characterization of the intermediates [1–9].

In addition to the steady-state techniques that allow the simplest processes to be investigated, non-steady-state techniques are necessary for studying more complex electrochemical systems. Use of these techniques is based on principles analogous to those that justify use of the relaxation methods in chemical kinetics at equilibrium. Perturbation of the electrochemical system shifts the reactions from their steady state. The rate at which they proceed toward a new steady state depends on their characteristic parameters (e.g., reaction-rate constants, diffusion coefficients, etc.). Analysis of the transient regime provides information on these parameters and on the phenomenological equations that relate them. As the various elementary processes evolve at different rates, the response of the system can be analyzed to dissect the global electrochemical process.

The choice of technique depends on the goal: either to establish a reaction mechanism (i.e., to test a model) or to determine the kinetic parameters of a known or assumed mechanism. Some transient techniques are commonly used because they are well adapted to extract some kinetic parameters when mass transport is the rate-determining step of the process [10].

Due to nonlinearities related to electron transfer, the calculation and handling of the interface response to a large-amplitude signal are often inextricable [11]. Therefore, in general, low-amplitude signals are used. Even in this case, the time response is often very complicated. On the other hand, the expression of the frequency response is generally much simpler; this allows easy handling of the experimental results for obtaining the parameters that are sought.

In some very favorable situations, several techniques can have comparable effectiveness. However, when complex heterogeneous reactions interact with mass transport, analysis of the current or potential time transients lead to poor results if a reaction mechanism has to be resolved. A frequency analysis is then more efficient. Therefore, measurement of the impedance by means of a sine wave in a wide frequency range has been largely developed.

Concepts that lead to the definition and to the validity conditions of electrochemical impedance are reviewed in Section II. Development of a measurement procedure and elaboration of the models used to compare to the experimental data are described in Section III. The instrumentation necessary for measuring the impedance is described in Section IV. In Section V we report some applications of impedance spectroscopy in various domains.

II. BASIC PRINCIPLES

An impedance is a quantity defined for a linear system. In this section, after defining the notion of transfer function, of which the impedance is a particular case, use of this quantity is extrapolated to nonlinear systems such as the electrochemical interface.

A. Linear Systems, Transfer Functions, and Impedances

In the sine-wave perturbation regime, the relationship between the current that flows through a circuit and the voltage applied to its ports can be characterized by the ratio $|Z|$ between the amplitudes of the current and the voltage and the phase shift, ϕ, between the rotating vectors, which at any time represent the instantaneous voltage $V(t)$ and current $I(t)$. These quantities are the modulus and phase shift of a vector Z, which can also be represented as a complex number $R + jX$. If the circuit is linear, this quantity is independent of both the amplitude of V and I and depends only on their ratio. As an example, the behavior of the simple circuits depicted in Figure 1 is examined briefly.

1. Series Circuit

A sine-wave current I at a frequency f, with pulsation $\omega = 2\pi f$, assumed as the phase reference flows through the series circuit. By definition, the ratio

$$Z = \frac{V}{I} = R_S - jX_S \tag{1}$$

where R_S is the resistance and $X_S = 1/C_S\omega$, the reactance of the circuit, is called the impedance. The vector $V_1 = R_S I$ is in phase with the current I, and the vector $V_2 = jX_S I$ has a $-\pi/2$ phase shift. Electrochemists are in the habit of considering the vector $jX_S I$ as positive. Therefore, in the complex plane (Re[Z], Im[Z]) where Re[Z] and Im[Z] mean, respectively, the real and imaginary parts of Z (i.e., the Nyquist diagram), capacitive reactances are plotted positively on the ordinate, contrary to the rules applied in electrotechniques. The impedance of the series circuit is plotted in Figure 1 in the Nyquist plane.

2. Parallel Circuit

A sine-wave voltage V with pulsation ω, which is assumed as the phase refer-

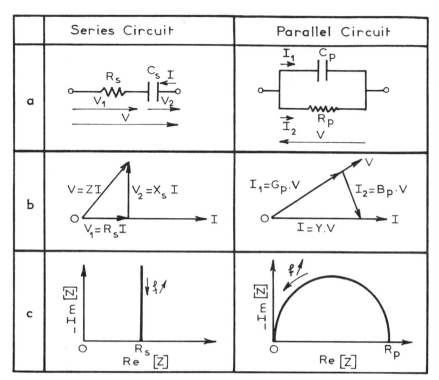

FIGURE 1 Frequency behavior of the series and parallel circuits: (a) scheme of the circuit; (b) vectorial representation; (c) plot in the Nyquist plane.

ence, is applied to the circuit. By definition, the ratio

$$Y = \frac{I}{V} = G_P - jB_P \tag{2}$$

where $G_P = 1/R_P$ is the conductance and $B_P = 1/X_P$ is the subceptance of the circuit, is called the admittance. The vector $I_1 = G_P V$ is in phase with the voltage V, whereas the vector $I_2 = jB_P V$ has a $-\pi/2$ phase shift. The impedance $Z = 1/Y$ of the parallel circuit is plotted in Figure 1 in the Nyquist plane.

3. Transfer Function and Impedance

In a more general way, a perturbation signal, current or voltage, $x(t)$, and its response, voltage or current, $y(t)$, are related in the frequency domain by a transfer function $H(\omega)$ such as

$$Y(\omega) = H(\omega)X(\omega) \tag{3}$$

where $Y(\omega)$ and $X(\omega)$ are the Fourier transforms of $y(t)$ and $x(t)$, such as

$$Y(\omega) = \int_{-\infty}^{+\infty} y(t)e^{-j\omega t} \, dt \tag{4}$$

If $x(t)$ is a current and $y(t)$ a voltage, $H(\omega)$ is an impedance; if $x(t)$ is a voltage and $y(t)$ a current, $H(\omega)$ is an admittance.

From a theoretical point of view, any kind of perturbing signal $x(t)$—white noise, step, pulse, or sine wave—can be used to measure the transfer function $H(\omega)$ and then the impedance. From an experimental point of view, the accuracy and the convenience of the measurement determine the more suitable perturbing signal. Given the instrumentation available on the market, the analysis by means of a sine-wave signal is often the most appropriate for the electrochemical investigations.

The impedance of the electrochemical interface $Z(\omega)$ is a complex number that can be represented either in polar coordinates or in cartesian coordinates:

$$Z(\omega) = |Z| \exp(j\phi) \tag{5}$$
$$Z(\omega) = \text{Re}[Z] + j \, \text{Im}[Z] \tag{6}$$

These quantities are related by the following relationships:

$$|Z|^2 = (\text{Re}[Z])^2 + (\text{Im}[Z])^2 \quad \text{and} \quad \phi = \arctan \frac{\text{Im}[Z]}{\text{Re}[Z]} \tag{7}$$

on the one hand, and

$$\text{Re}[Z] = |Z| \cos \phi \quad \text{and} \quad \text{Im}[Z] = |Z| \sin \phi \tag{8}$$

on the other hand.

Two types of plotting are used to describe these relationships; they are illustrated in Figure 2 in the case of the high-frequency equivalent circuit of an electrochemical cell (Fig. 2a), of which the impedance is equal to

$$Z(\omega) = R_e + \frac{1}{1/R_t + jC_d\omega} \tag{9}$$

where R_e is the electrolyte resistance, R_t the charge-transfer resistance, and C_d the double-layer capacitance. $Z(\omega)$ is plotted in Figure 2b in the complex plane with the negative imaginary part plotted above the real axis as is usual in electrochemistry. In Figure 2c a plot in the Bode plane shows the changes in $|Z|$ and ϕ with frequency; $\text{Re}[Z]$ and $\text{Im}[Z]$ can be plotted similarly. The admittance $Y(\omega) = Z^{-1}(\omega)$ can be analyzed in the same way.

B. Linearization of Nonlinear Systems

The response, $y(t)$, of a linear system to a perturbation $x(t)$ is determined by an

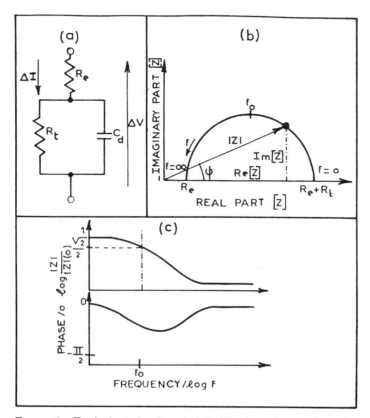

FIGURE 2 Equivalent circuit and plot of the impedance of an electrochemical cell, where R_e is the electrolyte resistance, R_t the charge-transfer resistance, and C_d the double-layer capacity. (a) Equivalent circuit; (b) plot in the complex plane (Nyquist plane); (c) plot in the Bode plane.

nth-order differential equation such as

$$b_0 \frac{d^n y(t)}{dt^n} + b_1 \frac{d^{n-1} y(t)}{dt^{n-1}} + \cdots + b_n y(t)$$

$$= a_0 \frac{d^m x(t)}{dt^m} + a_1 \frac{d^{m-1} x(t)}{dt^{m-1}} + \cdots + a_m x(t) \tag{10}$$

or by an ensemble of n first-order differential equations. Some more complex linear systems can also be described by partial differential equations. As an example, the behavior of an electrochemical interface, whose current is controlled both by the electrochemical reaction rates and by mass transport, is described by both ordinary differential equations for the reactions and partial differential equations for the transport.

When the interface is perturbed from equilibrium by means of an external energy source, a permanent flux of charge and matter is established. This is due to the electrochemical reactions, which allow electronic charge transfer between the electrode and the electrolyte. On the other hand, chemical and electric potential gradients promote transport of reactive species between the bulk of the electrolytic solution and the reaction interfacial zone. The elementary laws that govern mass transport and electrochemical reactions, and the complex couplings between these elementary phenomena, impose on the electrochemical systems a nonlinear behavior which can be rather pronounced. This feature results in, for example, potential dependent a_i and b_i coefficients in equation (10).

As an example, it is well known that charge transfer obeys an exponential activation law (Tafel law) such as

$$I_F = I_0 \exp\left(-\frac{\alpha nFE}{RT}\right) \tag{11}$$

where I_F is the faradaic current, E the potential, α the transfer coefficient, n the number of electrons involved in the transfer, F the Faraday constant (96,484 $C \cdot mol^{-1}$), R the gas molar constant (8.31 J/K $\cdot$ mol), T the absolute temperature (K), and I_0 the exchange current (A).

From the experimental point of view, there are many examples of nonlinear behavior in electrochemistry:

1. N-shaped $I-V$ curve for iron passivation in neutral or slightly acid medium [12]
2. S-shaped curve for zinc electrocrystallization in Leclanché medium [13] (2.67M NH_4Cl, 0.72M $ZnCl_2$, pH 5.2 with NH_4OH)
3. Z-shaped curve for iron passivation in molar sulfuric acid medium (pH 0) [14]

It can be shown that the behavior of a nonlinear system can be defined entirely in linear terms if the equivalent linear equations are known in each point of the steady-state $I-V$ curve [15,16]. Therefore, the local analysis of a nonlinear system can be reduced to linear system theory. From the experimental point of view it is sufficient to measure the impedance of an electrochemical cell by using a small-amplitude perturbing signal all along the $I-V$ curve. The interpretation is then simplified because the equations of the model that describes the interface behavior can be solved in the linear regime and therefore compared with experiment, whereas it is impossible or very difficult to do in the nonlinear regime.

As shown in Figure 3, a small-amplitude sine-wave voltage, $\Delta V(t) = |\Delta V| \sin \omega t$, is superposed on the dc polarization voltage, V_S. Consequently, a

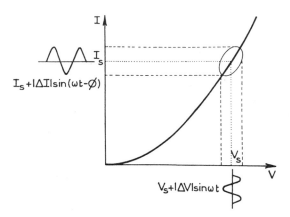

Figure 3 Small-signal analysis of a nonlinear system. Lissajous figure observed on an X-Y device when a sine-wave voltage $|\Delta V|$ sin ωt is superposed on the dc polarization voltage V_S.

small-amplitude sine-wave current, $\Delta I(t) = |\Delta I|$ sin$(\omega t - \phi)$, is superposed on the dc current I_S. If the two sine-wave signals are plotted on a X-Y recorder, a Lissajous ellipse, similar to the one shown in Figure 3, can be observed.

C. Linearity Domain

A Taylor expansion of the current $I(t)$ in the vicinity of a steady-state polarization point (V_S, I_S) gives

$$\Delta I = \left(\frac{dI}{dV}\right)_{V_S, I_S} \Delta V + \frac{1}{2}\left(\frac{d^2I}{dV^2}\right)_{V_S, I_S} \Delta V^2 + \cdots \tag{12}$$

To use a linear approximation, the amplitude of the perturbing signal ΔV has to be such that the sum of the higher-order terms $[\frac{1}{2}(d^2I/dV^2)_{V_S, I_S} + \cdots]$ is negligible compared to the first term; the latter, in a sine-wave analysis, is the fundamental term at the perturbing frequency responsible for the impedance. This allows a maximum amplitude $|\Delta V_{max}|$ to be defined beyond which a nonlinear distortion appears. To this end, the modulus, or the phase shift, which is more sensitive, is plotted vs. ΔV at a given frequency (Fig. 4a). It is established that above some value $|\Delta V_{max}|$, the quantity tested differs by a given $\epsilon\%$ from the value obtained at a very low level. The plot of $|\Delta V_{max}|$ vs. frequency defines a linearity domain (Fig. 4b) where the impedance measurement is valid. The "floor" value (crosshatched in Fig. 4b) of this domain $|\Delta V_{min}|$ is determined by the parasitic noise of the measurement arrangement. Therefore, under these conditions, small-signal analysis of a nonlinear system can be considered to be in

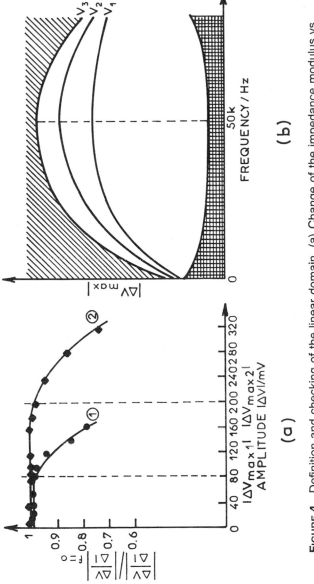

FIGURE 4 Definition and checking of the linear domain. (a) Change of the impedance modulus vs. perturbing signal amplitude. 1, Iron in sulfuric medium at 10 Hz; 2, passive nickel in sulfuric medium at 40 Hz. (b) Linearity domain (the amplitude of the measuring signal has to be in the non-cross-hatched region at any frequency) for various polarization points V_1, V_2, and V_3.

the linear regime [17]. For the frequencies generally used in electrochemistry (10^{-3} Hz to 50 kHz), the maximum amplitude is larger for the higher than for the lower frequencies, where faradaic processes impose nonlinearities.

III. MODELS

A. Generalities

The investigation of a physicochemical process often leads to the elaboration of a model. A rational representation, often in differential equation form, of a phenomenon that involves only the essential features of the real situation is sought. Such a model has two purposes: first, it has to take into account all the facts discovered experimentally; and then it has to predict system behavior in various situations. A model is based on the actual structure of the system, but in some cases it can also be useful, or sufficient, to elaborate a model of the "input–output" type (e.g., equivalent circuit), which describes the behavior of the system in some environment.

The spatial distribution of the variables $E(M, t)$ and $C_i(M, t)$—potential and concentration of species i at point M and time t—is determined by an ensemble of integrodifferential equations (charge and mass balance equations) subject to initial and boundary conditions (concentration at time zero, at the electrode surface, outside a diffusion layer, flux at the electrode surface, etc.). From this starting point, a model of the interface is obtained by making some hypotheses that allow the initial equations to be simplified. The interface is schematized in Figure 5, the electrochemical reactions evolve at the interface

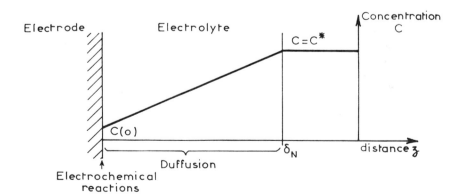

FIGURE 5 Scheme of the electrochemical interface in the z direction perpendicular to the electrode surface. δ_N is the Nernst layer thickness.

between the electrode and the electrolyte, and the reacting species diffuse in a layer of thickness δ_N (Nernst layer), which can be infinite or finite.

One of the major approximations assumes that the faradaic current is independent of the double-layer charging current and that the electrolyte resistivity is taken into account by considering an electrolyte resistance, R_e. Therefore, the equivalent circuit of the interface is generally considered as in Figure 6, where the faradaic impedance, Z_F, takes into account the electrochemical reactions and mass transport processes.

It is often assumed that dilute solutions are used and that they are electroneutral. Concerning transport, when the majority of ionic species, which are supposed not to take part in the reaction mechanism (supporting electrolyte), are present in the solution, the migration process can be neglected [18]. Generally, derivation of the concentration under convection control is difficult unless particular hydrodynamic conditions prevail. The well-known example of the rotating-disk electrode leads to a constant concentration gradient at the electrode surface (uniformly accessible electrode) [19]. It is located in a layer, called the Nernst layer, of thickness δ_N, in which the liquid is practically motionless. The equation of the transport is reduced to the Fick equation with good approximation:

$$\frac{\partial C_i}{\partial t} = D_i \nabla^2 C_i \tag{13}$$

The convection process is taken into account implicitly by applying the boundary condition for C_i ($C_i = C_i^*$) outside the Nernst layer ($z \geq \delta_N$).

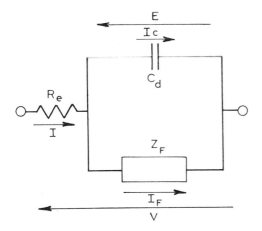

FIGURE 6 Equivalent circuit of an electrochemical cell. R_e, electrolyte resistance; C_d, double-layer capacity; Z_F, faradaic impedance.

Heterogeneous kinetic laws applied to adsorbed species involved in the reaction mechanism at the electrode surface:

$$\frac{dC_{S_i}}{dt} = \xi_i \tag{14}$$

where C_{S_i} is the surface concentration of species i and ξ_i is a particle source, which can be positive for a production or negative for a consumption of species i by a superficial reaction. It is often written as a function of the surface coverage θ_i such that

$$C_{S_i} = \beta_i \theta_i \tag{15}$$

where β_i is the maximum superficial concentration.

B. Examples of Models

Two typical processes are modeled below to show how an interfacial process can be expressed in terms of impedance: first, a mass-transport-limited charge transfer with an infinite and finite diffusion layer, and second, a two-step reaction mechanism with one adsorbed intermediate are considered.

1. Mass-Transport-Limited Charge Transfer

A quasi-reversible charge transfer is considered. The redox reaction at the interface is

$$\text{Ox} + ne \underset{k_b}{\overset{k_f}{\rightleftharpoons}} \text{Red} \tag{16}$$

where k_f and k_b are the forward and backward reactions and are equal to

$$k_f = k_f^{\circ} \exp\left(-\frac{\alpha nFE}{RT}\right) \quad \text{and} \quad k_b = k_b^{\circ} \exp\left[\frac{(1-\alpha)nFE}{RT}\right]$$

Species Ox and Red with concentrations C_O and C_R diffuse in the solution in the direction z perpendicular to the electrode surface:

$$\frac{\partial C_O}{\partial t} = D_O \frac{\partial^2 C_O}{\partial z^2} \quad \text{and} \quad \frac{\partial C_R}{\partial t} = D_R \frac{\partial^2 C_R}{\partial z^2} \tag{17}$$

where D_O and D_R are the diffusion coefficients of species Ox and Red. The boundary conditions are as follows:

1. The initial conditions assume that the solution is homogeneous (i.e., the concentrations are equal to C_O^* and C_R^*) before the current establishment:

$$t = 0, z \geq 0: \quad C_O(z, 0) = C_O^* \quad \text{and} \quad C_R(z,0) = C_R^* \tag{18}$$

2. The boundary conditions assume that outside the Nernst layer, the concentrations are equal to those in the solution bulk:

$$t \geq 0, z \geq \delta_N: \qquad C_O(z,t) = C_O^* \quad \text{and} \quad C_R(z, t) = C_R^* \qquad (19)$$

3. The boundary conditions at the electrode surface are imposed by the charge and mass balances. The Ox and Red fluxes are identical and equal to the faradaic current, I_F:

$$t \geq 0, z = 0:$$

$$D_O \frac{\partial C_O}{\partial t}(0,t) = -D_R \frac{\partial C_R}{\partial t}(0,t) = \frac{I_F(t)}{nFA} \qquad (20)$$

where I_F is given by heterogeneous kinetics:

$$I_F(t) = nFA(k_f C_O(0,t) - k_b C_R(0,t)) \qquad (21)$$

where the surface concentrations are supposed to be equal to the concentrations in the solution close to the electrode surface ($z = 0$).

By assuming that a small-amplitude perturbation $\Delta E \exp(j\omega t)$ ($\omega = 2\pi f$) is applied to the electrochemical system, the corresponding response of the concentrations $\Delta C_i \exp(j\omega t)$ and of the current $\Delta I_F \exp(j\omega t)$ is obtained by differentiating equations (17) and (21) and by removing the $\exp(j\omega t)$ terms:

$$j\omega \, \Delta C_i(z) = D_i \frac{\partial^2 \Delta C_i(z)}{\partial z^2} \qquad i = \text{O, R} \qquad (22)$$

$$\frac{\Delta I_F}{nFA} = -\frac{\alpha nF}{RT} k_f \overline{C}_O(0) \, \Delta E + k_f \, \Delta C_O(0) \qquad (23)$$
$$-\frac{(1 - \alpha)nF}{RT} k_b \overline{C}_R(0) \, \Delta E - k_b \, \Delta C_R(0)$$

where $\overline{C}_i(z)$ is the value of the species concentration for a dc polarization, such that

$$C_i(z,t) = \overline{C}_i(z) + \Delta C_i(z) \, \exp(j\omega t) \qquad (24)$$

Equation (22) and (23) lead to two important concepts: the charge-transfer resistance R_t and the Warburg impedance W. It has to be noticed that even if for calculation convenience the $\exp(j\omega t)$ terms are omitted, the preexponential terms ΔI_F and $\Delta C_i(z)$ also depend on ω.

a. Charge-Transfer Resistance

The charge-transfer resistance is defined by

$$\frac{1}{R_t} = \left(\frac{\partial I_F}{\partial E}\right)_{C_i} \qquad (25)$$

Equation (23) leads to

$$\frac{1}{R_t} = A \frac{n^2F^2}{RT} [-\alpha k_f \overline{C}_O - (1 - \alpha)k_b \overline{C}_R] \tag{26}$$

where $\overline{C}_i = C_i(0, t \to \infty)$.

b. Warburg Impedance

The resulting concentration perturbation $\Delta C_i(0, t)$ is deduced from the general solution of (22):

$$\Delta C_i(z) = M_i \exp\left(x\sqrt{\frac{j\omega}{D_i}}\right) + N_i \exp\left(-z\sqrt{\frac{j\omega}{D_i}}\right) \qquad i = O, R \tag{27}$$

The integration constants M_i and N_i are calculated from the boundary conditions and depend on the thickness of the Nernst layer.

Infinite-thickness diffusion layer In this case $M_i = 0$ and

$$\Delta C_i(z) = N_i \exp\left(-z\sqrt{\frac{j\omega}{D_i}}\right) \tag{28}$$

By substituting this value in the boundary conditions, it is shown that

$$\frac{\Delta C_O(0)}{\Delta I_F} = -\frac{1}{nFA\sqrt{j\omega D_O}} \qquad \text{and}$$

$$\frac{\Delta C_R(0)}{\Delta I_F} = -\frac{1}{nFA\sqrt{j\omega D_R}} \tag{29}$$

By combining (23), (26), and (29), the faradaic current response is obtained:

$$\Delta I_F = \frac{1}{R_t} \Delta E - \left(\frac{k_f}{\sqrt{D_O}} + \frac{k_b}{\sqrt{D_R}}\right) \frac{\Delta I_F}{\sqrt{j\omega}} \tag{30}$$

Therefore, the faradaic impedance is equal to

$$Z_F(\omega) = \frac{\Delta E}{\Delta I_F} = R_t\left(1 + \frac{\lambda}{\sqrt{j\omega}}\right) \tag{31}$$

where

$$\lambda = \frac{k_f}{\sqrt{D_O}} + \frac{k_b}{\sqrt{D_R}} \tag{32}$$

In (31), the term $R_t\lambda/\sqrt{j\omega}$ represents the Warburg impedance. The high-frequency limit of $Z_F(\omega)$ is equal to R_t.

By considering the double-layer capacitance C_d, the Randles equivalent circuit is obtained (Fig. 7a) [20], where the faradaic impedance Z_F in Figure 6

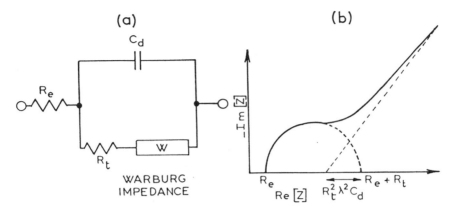

FIGURE 7 (a) Randles equivalent circuit; (b) electrochemical impedance for an infinite thickness diffusion layer: scheme of the impedance of the Randles equivalent circuit in the complex plane (Nyquist plane).

is replaced by R_t in series with the Warburg impedance, W. It can be shown that the high-frequency part of the impedance diagram is a semicircle similar to the one depicted in Figure 1, and the low-frequency part is a Warburg impedance. As shown in Figure 7b, the extrapolation of the 45° straight line, which represents the Warburg impedance in the complex plane, intersects the real axis at a value

$$R_0 = R_e + R_t - R_t^2 \lambda^2 C_d \tag{33}$$

Finite-thickness diffusion layer The Nernst hypothesis is now applied. It is assumed that the diffusing species concentration varies linearly in a layer of thickness δ_N:

$$C_i(z,t) = C_i(0, t) + (C_i^* - C_i(0, t)) \frac{z}{\delta_N} \qquad \text{for } z < \delta_N$$

$$C_i(z,t) = C_i^* \qquad \text{for } z \geq \delta_N \tag{34}$$

The application of the boundary conditions for $z = \delta_N$ leads to

$$\Delta C_i(z) = -2N_i \exp\left(-\delta_N \sqrt{\frac{j\omega}{D_i}}\right) sh\left[(z - \delta_N) \sqrt{\frac{j\omega}{D_i}}\right] \qquad \text{for } z < \delta_N$$

$$\Delta C_i(z) = 0 \qquad \text{for } z \geq \delta_N \tag{35}$$

A derivation similar to that in an infinite-thickness diffusion layer leads to

$$Z_F(\omega) = R_t\left(1 + \frac{k_f th\delta_N \sqrt{j\omega/D_O}}{\sqrt{j\omega D_O}} + \frac{k_b th\delta_N \sqrt{j\omega/D_R}}{\sqrt{j\omega D_R}}\right) \tag{36}$$

This impedance is plotted in Figure 8. Limiting cases show that:

1. If $\omega \to \infty$, the Warburg impedance is found again, because

$$\frac{th\delta_N \sqrt{j\omega/D}}{\sqrt{j\omega/D}} = \frac{1}{\sqrt{j\omega/D}} \tag{37}$$

2. If $\delta_N \to \infty$, the Warburg impedance is obviously found again.

2. Electrochemical Reactions

This approach was first used by Gerischer and Mehl [22] in their study of the impedance related to adsorbed intermediates in the hydrogen evolution reaction and was used intensively later by Epelboin et al. for various processes [23–27]. Species B arrives at the electrode surface and adsorbs onto the surface to produce species P through a two-step reaction mechanism involving an adsorbed intermediate, X:

$$
\begin{aligned}
B &\xrightarrow{k_1} X + e \\
X &\xrightarrow{k_2} P + e
\end{aligned} \tag{38}
$$

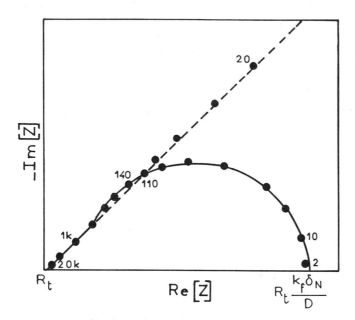

FIGURE 8 Solid line, calculated electrochemical impedance for a finite-thickness diffusion layer (Nernst hypothesis); dashed line, for comparison, the case of an infinite-thickness diffusion layer (frequency in hertz). (From Ref. 21.)

The model is based on the following hypotheses:

1. Adsorption and charge-transfer processes occur on a two-dimensional interface.
2. The reaction intermediate adsorption obeys a Langmuir isotherm with surface coverage θ.
3. The electrochemical reactions obey heterogeneous kinetics laws.
4. The reaction rates are exponentially potential activated (Tafel law):

$$k_i = k_i^\circ \exp(b_i E) \tag{39}$$

5. The maximum number of sites per surface unit that can be occupied by the adsorbate X is characterized by a coefficient β.

Charge and mass balances lead to the following equations:

$$\beta \frac{d\theta}{dt} = k_1 C_B (1 - \theta) - k_2 \theta \tag{40}$$

$$I_F = FA[k_1 C_B (1 - \theta) + k_2 \theta] \tag{41}$$

The solution of these equations for $d\theta/dt = 0$ gives, after elimination of θ, the equation for the steady-state I–E curve:

$$I_F = \frac{2FAk_1 k_2 C_B}{k_1 C_B + k_2} \tag{42}$$

When a small-amplitude sine-wave perturbation $\Delta E \exp(j\omega t)$ is superposed on the dc polarization voltage, the sine-wave responses $\Delta\theta$ and ΔI_F can be obtained by linearizing (40) and (41):

$$j\omega\beta \; \Delta\theta = [b_1 k_1 (1 - \theta)C_B - b_2 k_2 \theta] \; \Delta E - (k_1 C_B + k_2) \; \Delta\theta \tag{43}$$

and

$$\Delta I_F = FA\{[b_1 k_1 (1 - \theta)C_B - b_2 k_2 \theta] \; \Delta E + (-k_1 C_B + k_2) \; \Delta\theta\} \tag{44}$$

that is,

$$\Delta I_F = \frac{1}{R_t} \; \Delta E + FA(-k_1 C_B + k_2) \; \Delta\theta \tag{45}$$

where the charge-transfer resistance, R_t, is equal to

$$R_t = \left(\frac{\partial E}{\partial I_F}\right)_{\theta, C_B} = \frac{k_1 C_B + k_2}{FAk_1 k_2 C_B (b_1 + b_2)} \tag{46}$$

Elimination of $\Delta\theta$ between (43) and (44) and replacement of θ by its steady-state value lead to

$$Z_F(\omega)$$

$$= \frac{1}{FA} \frac{k_1 C_B + k_2}{k_1 k_2 C_B \{b_1 + b_2 + [(b_1 - b_2)(-k_1 C_B + k_2)]/(j\omega\beta + k_1 C_B + k_2)\}}$$

$$(47)$$

When the frequency tends to infinity, $Z_F(\omega)$ tends to R_t and the low-frequency limit is equal to the polarization resistance $R_P = (dE/dI_F)_{\text{steady state}}$. Z_F is plotted in Figure 9. For some values of the parameters, Z_F has a capacitive behavior or an inductive one.

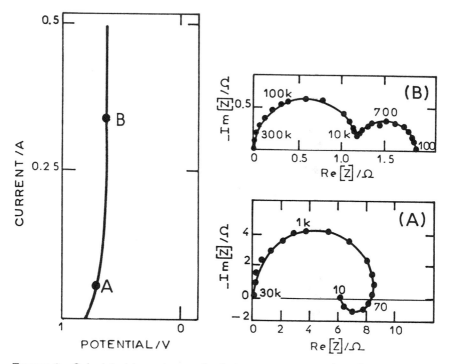

FIGURE 9 Calculated impedances for heterogeneous reactions [equation (47)]. Kinetic parameters are equal to $k_1 = 4 \times 10^8 \exp(36E)$; $k_2 = 10^{-3} \exp(10E)$; $\beta = 2 \times 10^{-9}$, $C_B = 10^{-4}$, and $C_d = 20$ μF. Impedances A and B are calculated at corresponding polarization points A and B on the I–E curve (frequency in hertz). (From Ref. 28.)

IV. INSTRUMENTATION

Prior to any impedance measurement, it is necessary to control the dc polarization of the interface. This function is essential for two reasons: first, it is necessary for plotting the steady-state $I-V$ curve, and second, it has to be used, on the one hand, to polarize the interface at the potential where the impedance has to be measured, and on the other hand, to impose the perturbing signal. The polarization control has to impose the desired potential and maintain the electrochemical system in the investigated state, variable or not. The main problem is the accessibility of this state, that is, the adaptation of the control device to the steady-state $I-V$ curve of the investigated electrochemical system.

A. Principles of the Polarization Control of an Interface

The first galvanostatic power supplies that imposed a dc current through a high-value resistor allowed the electrochemical systems to be studied out of equilibrium. However, electrochemists encountered metal–electrolyte interfaces whose $I-V$ curves had negative slopes (e.g., metal passivation). These domains were inaccessible to these control devices. A significant improvement was achieved when potentiostats were introduced because they provided access to domains forbidden to previous experiments.

1. Adaptation of the Control Device to the Interface

A system leaves its spontaneous state under the effect of a constraint. This is applied, in electrochemistry, by means of a power supply with voltage source $\mathscr{E}$ and internal resistance R (Fig. 10a). The polarization point imposed on the interface is determined by plotting in the $I-V$ plane (Fig. 10b):

1. On the one hand, the $I-V$ curve of the interface $I = f(V)$.
2. On the other hand, the $I-V$ curve imposed by the power supply $V = \mathscr{E} - RI$, called the "load line" d.

The operating or polarization point is given by the intersection M of these two curves (Fig. 10b).

To plot the $I-V$ curves experimentally, $\mathscr{E}$ is changed by keeping R constant. From the initial state O, by increasing $\mathscr{E}$, a succession of steady states is described; their plots give the $I-V$ curve (Fig. 10c). The OA part is described first, then when the load line becomes tangent to the considered $I-V$ curve at point A, the imposed polarization point jumps to A', which is another solution of the problem for the same $\mathscr{E}$ and R values.

$A'Q$ is then described. If $\mathscr{E}$ is decreased, the QB part is followed and the polarization point jumps from B to B'. Then $B'O$ is followed. So two jumps $A-A'$ and $B-B'$ are observed, forming a hysteresis cycle $A-A'-B-B'$. The states of the system characterized by the AB domain are consequently inaccessible to

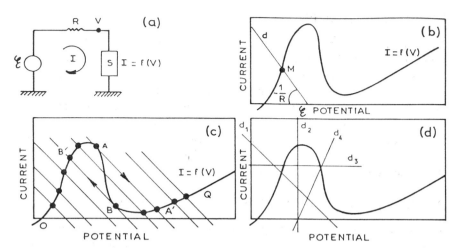

FIGURE 10 Principle of the polarization control of an electrochemical system: (a) scheme of a control device ($\mathscr{E}$, R) loaded by a system S; (b) determination of the polarization point M; (c) plot of the I–V curve; (d) typical positions of the load line: d_1, potentiometric control; d_2, potentiostatic control; d_3, galvanostatic control; d_4, negative impedance control.

the experimenter, when the previous ($\mathscr{E}$, R) power supply is used, because this device is not adapted to the investigated system.

2. Types of Control Devices

The slope of the load line, given by the $1/R$ value, can change from $-\infty$ to $+\infty$. So this straight line can have four typical positions, which define four different control devices, depicted in Fig. 10d:

d_1 If $R > 0$: potentiometric supply (previous case)
d_2 If $R = 0$: potentiostat
d_3 If R infinite: galvanostat
d_4 If $R < 0$: negative internal impedance supply

As an example, the I–V curves of a rotating disk electrode made of pure iron (Johnson–Matthey) immersed in an aqueous molar sulfuric acid solution obtained by means of a galvanostat (Fig. 11a) and a potentiostat (Fig. 11b) are given. The stabilization of the intermediate states located between A and B needs the use of a negative internal impedance supply whose description is beyond the scope of this chapter [14].

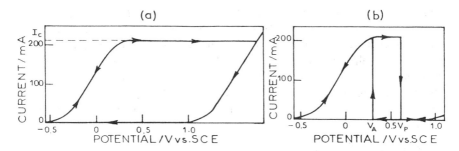

FIGURE 11 Current–voltage curve of a 5-mm-diameter Johnson–Matthey iron electrode rotating at 750 rpm in an aqueous 2 N H_2SO_4 solution, plotted by means of (a) a galvanostatic control device and (b) a potentiostatic control device, where V_p is the passivation potential curve and V_A is the Flade reactivation potential.

B. Impedance Measurements

If the values of the parameters of an already known mechanism are sought, it is possible to use a measurement signal or a data treatment which particularly sensitizes the parameters studied. On the other hand, if one is looking for a model of an unknown interface, it is necessary to use signals whose ferqucncy explores a broad domain to excite all the natural frequencies of the system. Impedance measurements in a wide frequency range will be the recourse.

1. Data Plotting

An impedance is usually characterized, as any complex number, either by a real part–imaginary part couple or a modulus–phase couple. However, the choice of one form rather than the other is not without importance. In fact, the real part–imaginary part relationship is 1:1 for a physicochemical system from the Kramers–Kronig relationship [29–31]. Therefore, it is sufficient, in principle, to measure only one of these components in the entire spectrum. On the contrary, the modulus–phase relationship is not always 1:1 (e.g., for nonminimal phase shift systems the Bode relationships are not valid). However, the nature of the measured impedance is not known, a priori, and an electrochemical impedance can have a nonminimal phase shift (e.g., where a negative real part is observed on the impedance diagram). Therefore, it is necessary to measure the two components of the modulus–phase couple. Although the simultaneous measurement of the real and imaginary parts is redundant, it is nevertheless better to perform the measurement in this way, as this redundancy contributes to an improvement of the measurement accuracy.

Data presentation can be carried out either in the complex plane ($Re[Z]$, $-Im[Z]$) with the frequency as a parameter, or in the Bode plane (i.e., modulus and phase shift versus logarithm of the frequency). The first representation is convenient on both these grounds: first, it accentuates the separation of the time constants of the electrochemical system studied, and second, it allows the experimental results and the diagrams depicted by the linearized tested models to be compared easily. The second presentation is essential when the impedance modulus changes by several orders of magnitude. The constraints imposed on the measurement arrangement by the frequency range explored are reviewed below.

2. Frequency Range

The time constants involved in electrochemical processes are often much greater than 1 s. Taking the double-layer capacity into account, the impedance of the interface has to be measured in a very wide frequency domain, ranging from several tens of kilohertz down to 10^{-2} or 10^{-3} Hz. In this vast domain the impedance changes between its high-frequency limit, equal to the electrolyte resistance, and its low-frequency limit, called polarization resistance, equal to the inverse of the slope of the I–V curve (dI/dV) at the polarization point investigated.

3. Transfer Function Analysis

The most convenient way to measure the impedance is to use a frequency response analyzer (FRA), whose principle is depicted in Figure 12. The signal $S(t)$ [i.e., the response of the system investigated to the signal delivered by the sine-wave generator $x(t) = X_0 \sin \omega t$] is correlated with two synchronous reference signals in phase and in $\pi/2$ phase shift (i.e., $\sin \omega t$ and $\cos \omega t$) in order to calculate

$$A = \frac{1}{NT} \int_0^{NT} S(t) \cos \omega t \, dt$$

$$B = \frac{1}{NT} \int_0^{NT} S(t) \sin \omega t \, dt \tag{48}$$

where

$$S(t) = X_0 \, |K(\omega)| \, \sin[\omega t - \phi(\omega)] + \sum_n A_n \sin(n\omega t - \phi_n) + b(t)$$

is the response (sum of the fundamental, harmonics, and noise) of a transfer function $K(\omega) = |K(\omega)| \exp[j\phi(\omega)]$. The measurement time is practically equal to an integral number, N, of the period $T = 1/f$.

    If the measurement time, NT (integration time), were infinite, all the components of $S(t)$, in particular the noise, at pulsation different from $\omega = 2\pi f$, would give a zero integral, and only the fundamental component would give a

FREQUENCY RESPONSE ANALYSER

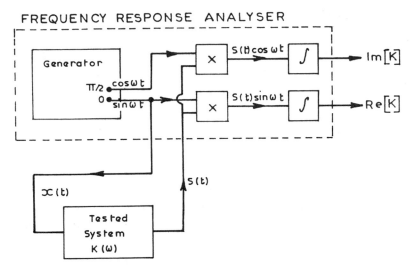

FIGURE 12 Working principle of a frequency response analyzer (FRA). Re[K], real part of the measured transfer function; Im[K], imaginary part of the measured transfer function; $x(t)$, perturbing signal; $S(t)$, cell response signal.

nonzero integral. The equivalent filter would have an infinitely narrow bandwidth, which would let pass only the component at the generator frequency. In this case

$$A = \lim_{N \to \infty} \frac{1}{NT} \int_0^{NT} S(t) \sin \omega t \, dt = X_0 \, |K(\omega)| \cos \phi(\omega)$$

$$B = \lim_{N \to \infty} \frac{1}{NT} \int_0^{NT} S(t) \cos \omega t \, dt = X_0 \, |K(\omega)| \sin \phi(\omega) \tag{49}$$

Therefore,

$$Re[K(\omega)] = \frac{A}{X_0} \quad \text{and} \quad Im[K(\omega)] = \frac{B}{X_0} \tag{50}$$

and two quantities proportional to the real and imaginary parts of the signal would be obtained. From equations (49) and (50), the modulus and phase can be obtained through (7). However, the measurement time, NT, cannot be infinite, and therefore the equivalent filter has a finite bandwidth whose value depends on the integration time. This leads to an estimation error in the measurement.

4. Measurement Error

The presence of parasitic noises in the measurement channels limits the measurement accuracy and introduces an error in the transfer function $K(\omega)$. If $\hat{K}(\omega)$ is the measured quantity, the errors in the modulus and the phase are defined by [32,33]

$$\epsilon^2[|\hat{K}|] = \frac{\text{var}(|\hat{K}|)}{(E[|\hat{K}|])^2}$$

$$\text{var } \hat{\epsilon} = \epsilon^2[|\hat{K}|] \qquad \text{rad} \tag{51}$$

where $\text{var}[\cdot]$ and $E[\cdot]$ mean variance and mean value.

It can be shown that for a potentiostatic arrangement, the impedance measurement in the presence of noise of a power spectral density ϕ_n is performed with an error ϵ at frequency f such that

$$\epsilon^2(|\hat{Z}|) = \frac{2}{N} \frac{\phi_n f}{a^2} \left(1 + \frac{|Z|^2}{R^2}\right) \tag{52}$$

where N is the number of periods, a the amplitude of the measuring sine wave, $|Z|$ the modulus of the measured impedance, and R the value of the resistance used to measure the current.

It has to be noted that the error would have been very low for a linear system, as a large-amplitude signal could have been used. However, for nonlinear systems, such as electrochemical systems, the perturbation level cannot be greater than some value ($a = |\Delta V_{\text{max}}|$) in order to stay in the linearity domain (Section II.C).

5. Experimental Arrangement Based on the FRA

Two typical experimental arrangements are employed: potentiostatic and galvanostatic arrangements (Fig. 13) [34]. The FRA performs two measurements simultaneously. First, the real part and the imaginary part of the transfer function between channel 1 and the output of the generator are measured: A_1 and B_1. Second, those between channel 2 and the generator, A_2 and B_2, are obtained. Finally, the FRA calculates the real and imaginary parts of the transfer function between channels 1 and 2 from A_1, B_1, A_2, and B_2, which are related directly to the impedance of the electrochemical cell without any influence of the adding amplifier (Σ) or of the control device. Only amplifiers G_I and G_V affect the measurement by their possible phase shifts and gain change in the high-frequency domain. However, if the two channels are balanced (i.e., if the two amplifiers have approximately identical phase shifts), the useful frequency range can be extended.

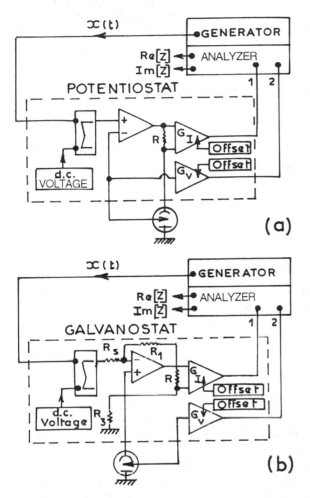

FIGURE 13 Schemes of the impedance measurement arrangement based on a FRA in potentiostatic (a) and galvanostatic (b) configurations. Input dc voltage is added to the sine-wave signal to impose V_S or I_S; the offsets are used to eliminate the dc voltage in the voltage and current responses.

State-of-the-art instrumentation allows automatic measurement and plotting of the interfacial impedance to be carried out now from 10^{-5} Hz to the MHz range. It can be based on a two- or four-channel FRA (e.g., Schlumberger-Solartron 1250, 1254, 1255, or 1260). The sine wave delivered by the generator of these analyzers can be programmed as a logarithmic frequency sweep by imposing the minimum and maximum frequencies of the range analyzed and the number of frequency points per decade. Table 1 gives the measurement time for various frequency ranges with five measurements per frequency decade.

V.   APPLICATIONS TO ELECTROCHEMICAL PROCESSES

The measurement of the impedance of an electrochemical interface has been applied to many domains. If studies of corrosion and corrosion protection represent 70% of the applications, energy sources, electrocrystallization, and material characterization are other examples where the use of this technique gave fruitful results. Various applications that illustrate the use of the electrochemical impedance measurement are described below; while some examples were intended to elucidate a reaction mechanism, others had only the purpose of finding a good working test.

A.   Anodic Behavior of Metals

The anodic behavior of numerous metals has been investigated by impedance spectroscopy. Some aspects of the dissolution of iron in a sulfuric medium were chosen to illustrate the course to be followed when dealing with an electrochemical study.

1.   Anodic Dissolution

The electrochemical impedance was measured during dissolution of an iron electrode in $1\,M\,H_2SO_4 + 1\,M\,Na_2SO_4$. These experiments were carried out in wide pH (0 to 5), current density (up to $0.1\,A\cdot cm^{-2}$), and frequency (10^{-3} to 10^5 Hz) ranges [35]. In addition to the high-frequency loop ascribed to the double-layer capacity and transfer resistance, three time constants were observed at potentials lower than that of the onset of the passivation process (Fig. 14).

To interpret the experimental data, it was necessary to take into account three adsorbed reaction intermediates and two ions in solution. The procedure consists of writing the kinetic equations that describe the concentration changes, c_i, of the five species [Fe(I), Fe*(I), Fe*(II), Fe(II), and Fe(II)$_s$] involved in the

TABLE 1   Measurement Times Necessary for Various Frequency Ranges Logarithmically Swept

Frequency range	Measurement time
50 kHz–0.1 Hz	1 min 15 s
50 kHz–0.01 Hz	6 min
50 kHz–0.001 Hz	40 min

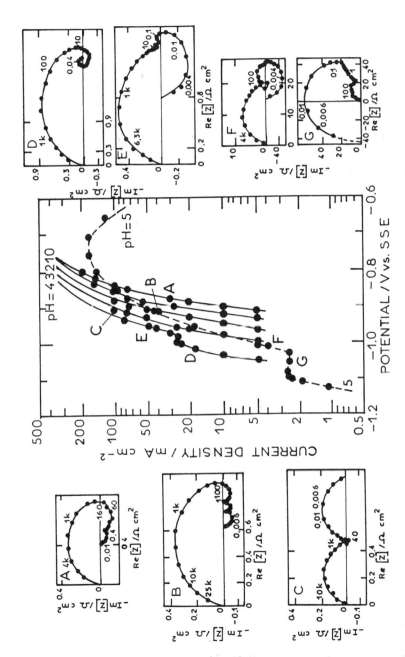

FIGURE 14 Measured steady-state *I–E* curves and impedances for a Johnson–Matthey iron electrode rotating at 1600 rpm in a 1 *M* Na₂SO₄ + 1 *M* H₂SO₄ aqueous solution deoxygenated by Ar bubbling at various pH. Impedances are measured at points marked on the *I–E* curves (frequency in hertz). (From Ref. 35).

reaction mechanism [similar to equation (40)]:

$$\frac{dc_i}{dt} = \sum_i k_i c_i \qquad 1 \le i \le 5 \tag{53}$$

The I–E curve and the impedance [similar to equations (42) and (47)] calculated from (53) were compared to the experimental results.

Forty possible reaction schemes were considered. The appearance at pH 5 of a prepassivation peak at -1.1 V, then of a passivation peak at -0.75 V, allow 10 of these models to be eliminated. On the other hand, the appearance of three simultaneous inductive loops is a criterion that allows 18 others to be eliminated: the change in shape from diagram F to diagram G is also very selective, and four other models were eliminated. The impedance of the eight possible remaining reaction mechanisms was calculated in detail. Only one mechanism involving the adsorbed intermediates Fe(I), Fe*(I), and Fe*(II) with the departure of two types of Fe(II) ions in the solution bulk leads to an impedance that varies with pH and potential, in good agreement with the experimental results in all experimental conditions explored. This mechanism is:

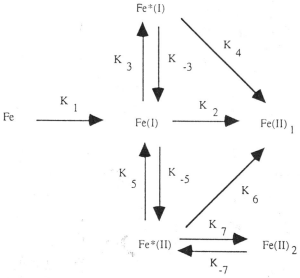

2. Iron Passivity

Numerous investigations on iron passivity were developed both in acidic and basic media, especially for understanding the stainless steel protection mechanism. Very slow processes are involved; they lead to a very low frequency loop whose low-frequency limit is very difficult to reach because impedance measurements at frequencies lower than 10^{-3} Hz are often necessary. In addition, electrolyte leakages between the iron electrode and the insulator that surrounds it often lead to experimental artifacts that impede the very low frequency part of the impedance diagram to be measured accurately.

In a sulfuric medium it was shown that the adsorbed intermediates Fe(II) and Fe(III) produce a thin passive layer which becomes more and more organized as a crystal when the potential becomes more anodic. The following reaction sequence was tested by an impedance technique (Fig. 15) [36,37]:

$$\text{Fe} \rightarrow \text{Fe(I)}_{ads} \rightleftharpoons \text{Fe(II)}_{ads} \rightleftharpoons \text{Fe(III)}_{ads} \rightleftharpoons \text{oxide}$$

$$\downarrow \qquad\qquad \downarrow$$

$$\text{Fe(II)}_{sol} \qquad \text{Fe (III)}_{sol}$$

B. Corrosion-Rate Estimation

Corrosion of a metal in an aqueous medium is an electrochemical process. Therefore, electrochemical techniques, particularly impedance techniques, are suitable for estimation of the corrosion rate x. The latter is equal to

$$x = \frac{dW}{dt} = \frac{MI_a}{zF} \tag{54}$$

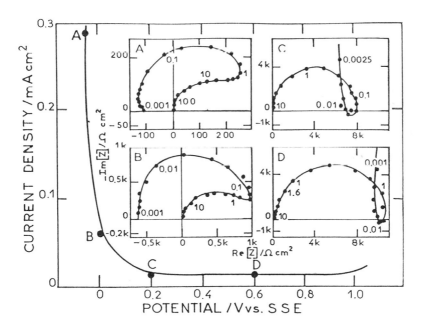

FIGURE 15 Measured *I–E* curve and impedances for passive iron in 2 *N* H$_2$SO$_4$ (1600-rpm rotating disk electrode). A, B, C, and D diagrams are measured at points marked on the *I–E* curves (frequency in hertz). (From Ref. 36.)

where W is the weight loss of the metal, M the atomic mass, z the dissolution valency, and I_a the anodic dissolution current. In the past, I_a was often obtained from the polarization resistance (impedance at $f = 0$) [38]. However, it was shown that impedance measurements over a large frequency range can provide more reliable results [39–41]. Various corrosion processes were investigated by means of these techniques. Most of these were carried out on uniformly corroded surfaces, but localized and stress corrosions are starting to be analyzed, too.

1. Principle of Corrosion-Rate Estimation

For uniform corrosion, the metal dissolution occurs on the entire surface in contact with the electrolyte. The total current I that flows through the interface is equal to

$$I = I_a + I_c \tag{55}$$

where I_c is the cathodic current. For spontaneous corrosion, the measurable total current is $I = 0$; therefore, it is impossible to determine directly the dissolution current I_a in order to calculate x from (54).

Two cases can be considered according to the instantaneous dependence, or lack thereof, of I_a and I_c on the potential. Differentiation of (55) leads in the first situation to

$$\frac{\Delta I}{\Delta E} = \frac{1}{Z_F} = \frac{\Delta I_a}{\Delta E} + \frac{\Delta I_c}{\Delta E} = \frac{1}{R_p} \tag{56}$$

where E is the electrode potential corrected for ohmic drop. If I_a and I_b obey Tafel laws with coefficients b_a and b_c such that

$$I_a = I_a^\circ \exp(b_a E) \quad \text{and} \quad I_c = I_c^\circ \exp(b_c E) \tag{57}$$

to spontaneous potential ($I = 0$), the impedance is equal to

$$\frac{1}{Z_F} = \frac{1}{R_p} = b_a I_a - b_c I_c \tag{58}$$

In this situation ($I = 0$),

$$I_a = -I_c = I_{\text{corr}} \tag{59}$$

and the relationship between the impedance and the corrosion current I_{corr} is then

$$\frac{1}{R_p} = I_{\text{corr}}(b_a + b_c) \tag{60}$$

In this case the polarization resistance is identical to the transfer resistance, and the faradaic impedance is not frequency dependent.

When I_a and I_c do not instantaneously follow the potential (e.g., if the current I_a depends on the surface coverage θ of an intermediate), the impedance is equal to

$$\frac{\Delta I}{\Delta E} = \frac{1}{Z_F} = \left(\frac{\Delta I_a}{\Delta E}\right)_\theta + \left(\frac{\Delta I_c}{\Delta E}\right)_\theta + \left(\frac{\Delta I_a}{\Delta\theta}\right)_E \left(\frac{\Delta\theta}{\Delta I}\right) \tag{61}$$

Contrary to equation (56), only the first two terms on the right-hand side of (61) give a frequency-independent contribution to Z_F. The third term depends on the dynamic behavior of θ versus E. It is time dependent and obeys an equation such as [similar to equation (40)]

$$\frac{d\Delta\theta}{dt} = \lambda\,\Delta a + \mu\,\Delta E \tag{62}$$

where λ and μ are coefficients. If ΔE and consequently $\Delta\theta$ vary sinusoidally with time, the frequency behavior of $\Delta\theta$ is obtained from (62) and is equal to

$$\frac{\Delta\theta(\omega)}{\Delta E} = \frac{\mu}{j\omega - \lambda} \tag{63}$$

In the high-frequency range, θ is "frozen"; that is, it cannot follow the potential changes as it does at lower frequencies $[\Delta\theta(\omega \to \infty) = 0]$.

Equation (61) and (63) lead to

$$\frac{1}{Z_F} = \frac{1}{R_t} + \left(\frac{\Delta I_a}{\Delta\theta}\right)_E \frac{\mu}{j\omega - \lambda} \tag{64}$$

Equation (64) shows the frequency dependence of Z_F. The limiting values of Z_F are equal to

$$\lim_{\omega \to \infty} Z_F = R_t$$

and

$$\lim_{\omega \to 0} Z_F^{-1} = R_t^{-1} - \left(\frac{\Delta I_a}{\Delta\theta}\right)_E \frac{\mu}{\lambda} = \frac{1}{R_p} \tag{65}$$

Therefore, the faradaic impedance changes between R_t and R_p. The corrosion-rate estimation is now considered in various situations. In can be shown that (64) leads to a relationship that gives I_a vs. R_t instead of R_p, as in the first situation. For various hypotheses about anodic and cathodic reactions, Table 2 presents the relationships between R_p and I_{corr}, on the one hand, and between R_t and I_{corr}, on the other hand. It is noticeable that I_{corr} is related to R_t in two attractive cases (partial limitation by diffusion and multistep electron transfer)

TABLE 2 Theoretical Relationships Between the Corrosion Current and the Charge Transfer and Polarization Resistances, Under Various Anodic and Cathodic Kinetics

Case	Kinetic control of:		Relation to corrosion current, I_{corr}	
	Anodic reaction	Cathodic reaction	Polarization resistance, R_p^{-1}	Charge-transfer resistance, R_t^{-1}
1	One step, Tafelian electron transfer	One step, Tafelian electron transfer	$(b_a + b_c)I_{corr}$	$(b_a + b_c)I_{corr}$
2	One step, Tafelian electron transfer	Purely diffusional	$b_a I_{corr}$	$(b_a + b_c)I_{corr}$
3	One step, Tafelian electron transfer	Mixed control, partly diffusional	Complicated equation, depends on the degree of control by diffusion	$(b_a + b_c)I_{corr}$
4	n_a irreversible Tafelian, consecutive steps (b_a^i)	n_c irreversible Tafelian, consecutive steps (b_c^i)	Complicated equation, depends on the whole set of rate constants	$\left(\dfrac{1}{n_a}\sum_{i=1}^{n_a} b_a^i + \dfrac{1}{n_c}\sum_{j=1}^{n_c} b_c^i\right)I_{corr}$
5	Passive dissolution	One irreversible Tafelian transfer on the passive area	$b_c I_{corr}$	Complicated equation, depends on the kinetics of dissolution and passivation

where the classical technique using R_p would fail. However, R_p has to be preferably used in the case of passive electrodes (stainless steel, anodized aluminum). Therefore, the use of a characteristic value of the impedance—R_t or R_p—has as its purpose indicating the corrosion-rate changes versus corrosive environment changes. Some examples follow.

2. Examples

a. Active or Inhibited Corrosion of Iron

Both anodic and cathodic reactions occur in two consecutive irreversible steps. Faradaic impedance measurements provide the values of $\Sigma_i b_a^i$ and $\Sigma_j b_c^j$ which are involved in the relation between R_t and I_{corr} (relation 4, Table 2). The impedance diagram shows two inductive loops at low frequencies. They were proven to originate from the potential dependence of, respectively, the surface coverages by the inhibiting species (hydrogen or hydrogen-bonded organic compound) at lower frequencies, and by the anodic intermediate species FeOH at higher frequencies. Good agreement was found between weight-loss experiments, and the absolute value of the corrosion rate deduced from R_t for either a high-purity (Johnson–Matthey) iron (0.8 mg cm^{-2} h^{-1}) or a pure iron of industrial origin (Holzer type) (2 mg cm^{-2} h^{-1}) (Fig. 16a).

Corrosion-rate estimations based on R_t and R_p measurements were compared with weight-loss data for a Holzer-type iron in 0.5 M H_2SO_4 as a function of an inhibitor (propargylic alcohol) concentration (Fig. 16b). The impedance diagram at the corrosion potential with inhibitor is similar to the one obtained for uninhibited iron corrosion, at least for low inhibitor concentration (Fig. 16a) [39]. Relative corrosion rates with respect to the uninhibited conditions can be expressed in terms of the dimensionless inhibiting efficiency:

$$H = \frac{100(x_0 - x)}{x_0} \tag{66}$$

where x_0 is the corrosion rate in the absence of an inhibitor and x the corrosion rate in the presence of an inhibitor. x is either measured directly by weight loss (H_{dir}) or calculated from R_t or R_p, which are assumed to be inversely proportional to x.

As shown in Figure 16b, the values of H obtained from R_t are in agreement with those of H_{dir}, except at very high concentrations, where predictions become pessimistic. The values of H_{R_p} can be considered as acceptable for low inhibitor concentration, but beyond an inhibitor concentration of 2 mM, they predict a negative inhibition (i.e., an acceleration of the corrosion), in complete disagreement with direct measurements. According to the theoretical derivation given above, this discrepancy is related to the increasing size of the low-frequency inductive arc as the inhibitor concentration is increased, resulting in a corresponding difference between R_p and R_t.

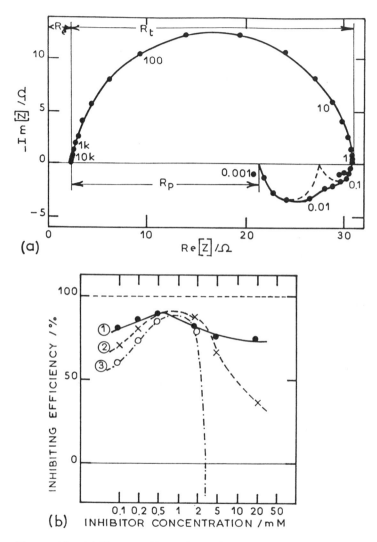

FIGURE 16 (a) Measured impedance for Holzer iron in deoxygenated 1 M H$_2$SO$_4$ aqueous solution polarized at the corrosion potential (1600-rpm rotating disk electrode; frequency in hertz). (b) Inhibiting efficiency vs. propargylic alcohol concentration determined from (1) weight loss, (2) R_t measurement, (3) R_p measurement (a negative inhibiting efficiency would indicate a corrosion acceleration). (From Ref. 39.)

b. Corrosion of Coated Metal

Faradaic impedance measurements on painted iron (see Fig. 17, inset) showed that the metal is corroded in the same way as when unprotected, but in a much smaller area at the spot where the protective layer presents porelike flaws. The isolating part of the paint gives rise to a distinct capacitive behavior at high frequencies [41].

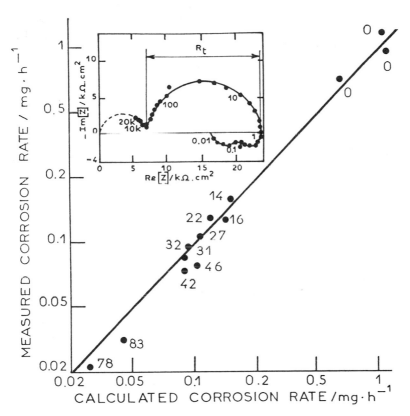

FIGURE 17 Comparison of the corrosion rate calculated from R_t and measured by weight loss for pure iron (Goodfellow) coated with various thicknesses of epoxy paint (paint thickness in micrometers is the parameter; 0 corresponds to the bare specimen). Inset: Measured impedance for an Armco iron electrode coated with epoxy paint 40 μm thick at corrosion potential. R_t is the charge-transfer resistance (frequency in hertz). (From Ref. 42.)

Corrosion rates calculated from R_t data were compared with weight loss for a number of specimens protected with epoxy paint, from 0 to 80 μm in thickness (Fig. 17). Coated thin foils of high-purity Goodfellow iron were exposed to attack in a $0.5M$ H_2SO_4 solution for 2 to 4 days. The rate of corrosion increased continuously during the exposure and R_t was measured as a function of time. The average calculated corrosion rate was compared with weight loss at different layer thicknesses. As shown in Figure 17, a very good correlation is found: the thicker the layer, the better the protection [42–44].

c. Anodized Aluminum Alloys

Impedance measurements can be used as a quality control test for the sealing and anodizing procedures as well as a corrosion resistance test for anodized Al alloys. The resistance against corrosion of aluminum-based alloys protected by an anodic oxide layer is usually evaluated by a salt spray test, which needs a long time exposure. The ac impedance method is faster and has the same degree of reliability [45–48].

As an example, the anodized sample to be tested is immersed in a 3% NaCl solution buffered at pH 4 and deoxygenated by nitrogen bubbling. The impedance is measured at its free corrosion potential. Figure 18 shows the impedance diagram. Two capacitive contributions can be resolved when the high-frequency behavior is enlarged (see Fig. 18, inset). The shorter time constant

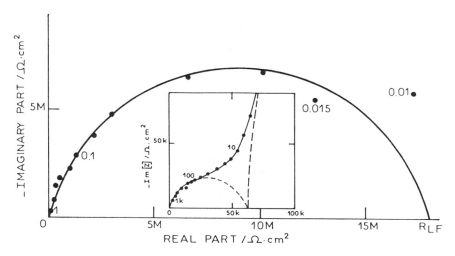

Figure 18 Corrosion of an anodized aluminum alloy. Measured impedance for a corroding 2024T4 alloy electrode anodized in chromic bath in $0.18\,M$ NaCl + $0.01\,M$ CH_3COOH + $0.02\,M$ CH_3COONH_4 (frequency in hertz). Inset: Enlarged diagram in the high-frequency range. (From Ref. 45.)

probably arises from the dielectric property of the oxide layer. A correlation between corrosion susceptibility and impedance behavior must reasonably be expected in the low-frequency domain. In fact, good correlation was found between the low-frequency resistance R_{LF} and the exposure time at which the first pit opened in the salt spray environment. Since the impedance technique is nondestructive, ac impedance and salt spray tests can be applied successively to the same sample so that a highly significant comparison can be made. No faradaic model of corrosion is available in the case of aluminum, and R_{LF} can hardly be dealt with in terms of either R_t or R_p. Therefore, the significance of the test must be founded on a large number of experiments. Table 3 gives, as an example, the selection made among 83 samples from the same batch on the basis of ac measurement or salt spray test. A "pass" in the salt spray test is recorded if no pits have appeared after 300 h of exposure to the spray. The "pass/fail" threshold value of R_{LF} ($5.5M \; \Omega \cdot cm^2$) was determined empirically. A similar correlation was found between the two techniques for different compositions of aluminum alloys and oxidizing baths.

ASTM B457 describes a procedure based on measurement of the impedance at 1000 Hz to determine the quality of sealing. However, some problems may arise when performing measurement at a single frequency. It is established that the high-frequency portion of the impedance spectra reflects the properties of the porous layer, while the properties of the barrier layer determine the low-frequency portion. Two time constants generally occur for a sealing procedure, but pronounced differences in the spectra are observed at high frequencies where the ASTM test applies, while very similar values may be obtained at low frequencies. This indicates a difference in the sealing mechanism; nevertheless, poor corrosion behavior may be observed for the sealing procedures even if the ASTM predictions are different. The present norm may result in an erroneous conclusion concerning the quality of the sealing procedure and the resulting corrosion resistance.

Considering all the experimental data, it appears that the corrosion resistance can be better determined from the low-frequency impedance data. How-

TABLE 3 Comparison of Impedance Measurement and Salt Spray Test[a]

Number of samples tested	83			
Impedance measurement $R_{LF} \leq 5.5 \; M\Omega \cdot cm^2$	Fail 76		Pass 7	
Salt spray test Pit occurs at 300 h	Fail 75	Pass 1	Fail 2	Pass 5

[a]Fail corresponds to correctly protected samples.

ever, for the "good" anodized Al alloys the low-frequency limit ($f = 0$) cannot often be determined in the practical frequency range that is experimentally accessible. Hence some authors [46] proposed to define a damage function D based on the value of the impedance at a higher frequency ($f = 0.1$ Hz) at time t, Z_t:

$$D = \log \frac{Z_0}{Z_t}$$

where Z_0 is the initial impedance value at 0.1 Hz.

As examples, Table 4 shows a summary of damage functions for three dichromate sealed and unsealed Al alloys for sulfuric (SAA) and chromic (CAA) anodizations exposed to 0.5 N NaCl solution. Some anodized Al alloys are so corrosion resistant ($D = 0$) that no corrosion occurs in 7 days.

d. Stainless Steel

The corrosion of stainless steel occurs in the passive range. As shown in Table 1, the quantity used for estimating the corrosion rate in the passive range is the polarization resistance R_p. This is only an assumption for stainless steel because the actual dissolution mechanism is not known. However, experimental results have shown the relevance of such a choice. As shown in Figure 19, the processes involved in the corrosion of stainless steel are so slow that the low-frequency limit, R_{LF}, can often be estimated only by extrapolation of a best-fitted half-circle to the experimental values [49].

C. Electrocrystallization of Metals

The electrocrystallization of a polyvalent metal, M, occurs in several steps, from the transport of the cations M^{z+} from the bulk of the electrolyte to incorporation

TABLE 4 Damage Functions for $t = 7$ Days (Dichromate Seal)

	Sealed	Unsealed
CAA Al 2024	1.51	1.37
SAA Al 2024	0.11	0.60
CAA Al 6061	0.04	0
SAA Al 6061	0.0	0
CAA Al 7075	0.18	1.91
SAA Al 7075	0.0	0.1

Source: Ref. 46.

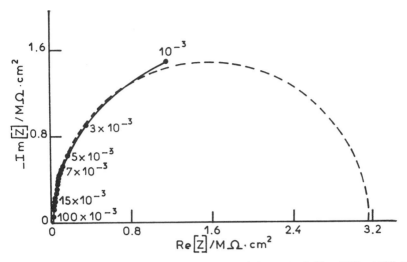

FIGURE 19 Measured impedance for a stainless steel (Fe–17Cr–16Ni–5.5Mo–2.7Cu–0.03C) corroding electrode in aerated 1.8M H_2SO_4. Solid line, experimental results; dashed line, best-fitted half circle. (From Ref. 49.)

of the metal in the crystal lattice. Therefore, many processes can limit the deposition rate and have to be taken into account in devising a model for the deposition: diffusion, surface coverage of the adsorbed reaction intermediates, and parameters related to the surface morphology of the deposit.

The adatom model assumes that the metal ion is discharged at the electrode and then, after diffusion along the electrode surface, the resultant adatom is incorporated into the metal at a growth site located at the edge of a nucleus. The crystallization impedance is therefore a function of diffusion in the solution, charge transfer, surface diffusion, and lattice incorporation. Simplified models, or more complex models, taking into account some or all of these four elementary processes, have been proposed. The zinc electrocrystallization is summarized below as an example.

The study of zinc electrocrystallization is generally considered as being of high interest, mainly because this metal is employed in electrochemical generators. Zinc deposits have been investigated in several acidic and basic media commonly used for industrial purposes (in particular, Leclanche media and alkaline electrolytes) [50–52].

To interpret the S-shaped polarization curve and the measured impedances (Fig. 20), the authors have suggested that an autocatalytic step (the third step

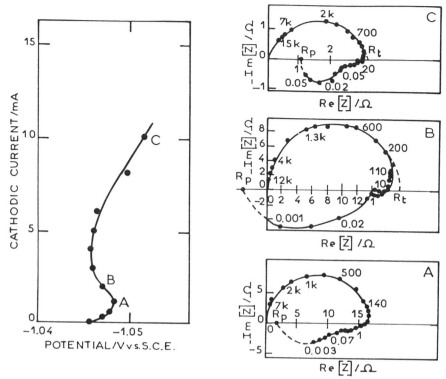

FIGURE 20 Zinc electrocrystallization in 1 *M* Na₂SO₄ + 1.5 *M* ZnSO₄, pH 4.3, 3000-rpm rotating disk electrode (*A* = 0.28 cm²): (a) polarization curve; (b) impedance measurements for points A (2 mA), B (3 mA), and C (10 mA). (From Ref. 50.)

in the following mechanism) is involved in the reaction mechanism.

$$H^+ + e \rightarrow H_{ads}$$

$$H^+ + H_{ads} + e \rightarrow H_2$$

$$Zn^{2+} + Zn_{ads}^+ + e \rightleftharpoons 2Zn_{ads}^+$$

$$Zn_{ads}^+ + H_{ads} \rightarrow H^+ + Zn$$

$$Zn_{ads}^+ + e \rightarrow Zn$$

$$Zn^{2+} + e \rightarrow \; + Zn_{ads}^+$$

In addition, when the current is increased, spongy, compact, and dendritic zinc electrodeposits can be observed successively. Improvement of the model, taking into account the localization of some of the interface reactions at particular sites distinguished with regard to the crystal growth, has allowed this change in deposit morphology to be explained. The interpretation is based on various couplings, between the interfacial reactions, the surface diffusion of Zn_{ads}^+, and the renewal of the growth sites. At low current, the electrode kinetics is governed by an equation such as

$$\beta \frac{d\theta}{dt} = f(\theta,E) + \beta D_S \frac{\partial^2 \theta}{\partial r^2}$$

where D_S is the surface diffusion coefficient, r the position on the surface, θ the surface coverage of Zn_{ads}^+, and $f(\theta,E)$ the source term given by the electrochemical reaction mass balance.

D. Solid Electrolytes

Studies of solid electrolytes have concentrated largely on the research of new materials with high electrical conductivity which are used, typically, in galvanic cells with a high energy density. The ionic conductivity in these materials is due to migration of negative or positive ions with high mobility. The materials are:

1. Isotropic anionic conductors with O^{2-} or F^- conduction
2. Cationic conductors, either isotropic with Ag^+ or Li^+ conduction or anisotropic with Na^+ conduction

Solid electrolytes have been studied by impedance measurements to better understand the various processes involved:

1. Transport numbers, concentrations of charge-carrying species, electronic and ionic defects, and their mobility.
2. Effect on conductivity of oxygen partial pressure, dopant concentration, impurities, grain boundaries, and second-phase precipitation.
3. Charge-transfer and polarization phenomena at the electrolyte–electrode interface (noble metals are often chosen for electrodes, as they are assumed to be chemically inert).

In addition, as solid electrolytes of practical importance are frequently polycrystalline, bulk and grain boundary conductivities have to be taken into account. Impedance diagrams of solid-state electrochemical cells are generally composed of capacitive arcs (semicircles and diffusion arcs) that can be assigned to the bulk of the ionic conductor and to the electrochemical reactions taking place at the interface [53–57]. Depending on the experimental conditions, only a few semicircles are readily observable. It should be noted that the frequency range of the impedance measurement is very wide (e.g., 10^{-3} to 10^8 Hz). As an example, some results concerning zirconia are given below.

At elevated temperatures, the conductivity of stabilized zirconia is due almost exclusively to the motion of oxygen vacancies, and hence it is used in galvanic cells. In particular, these cells are employed as monitoring elements when analyzing oxygen (in the range 10^{-20} to 1 atm) or when extracting oxygen from a gas mixture by electrolysis.

Impedance measurements have been used to select materials for electrochemical applications and for fundamental investigations. It has been shown that polarization phenomena at the electrode–electrolyte interface have a significant effect on cell behavior. This is demonstrated in Figure 21, where faradaic reaction (right), space charge (middle), and bulk conductivity (left) can be distinguished.

E. Prediction of the State of Charge of Batteries

The electrochemical power cells deliver energy as a consequence of electrochemical reactions, but they start to deteriorate from the moment that manufacturing is complete. This self-discharge may be quite significant and can be dangerous for starting equipment that has long periods of inactivity. Therefore, to assess the reliability of a power source, a cell status indicator would be of high interest if a high wastage rate of partially discharged but still usable batteries has to be avoided. On the other hand, continuous state-of-charge monitoring is also often needed, for example, in electrically powered cars (the state of charge of a battery is the ratio of residual capacity, at a given instant, to the maximum capacity available from the battery). Therefore, users need a simple, rapid, and reliable, nondestructive check of the energy available in the battery.

Tests based on impedance measurements of batteries have been investigated for various energy sources. The change in some impedance components was recorded when a prescribed amount of charge is withdrawn from the storage

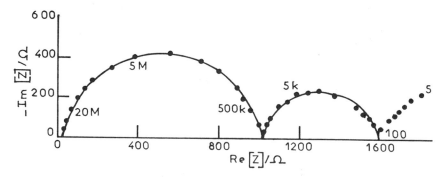

FIGURE 21   Measured impedance for a Pt/YSZ/Pt cell in oxygen at 457°C (frequency in hertz). (From Ref. 56.)

cell. A priori, it seems that the problem is less convoluted in the case of primary cells than for secondary cells. The reason for this is that primary cells are not subject to cycling as are secondary cells.

Usually, a two-electrode arrangement is used for impedance measurements. The inclusion of a third electrode inside a commercial cell is always a possibility—which is attractive for a fundamental kinetic investigation—but the incorporation is not very convenient for practical applications. A test of the state of charge of a cell based on parameters obtained from data on its impedance would be entirely satisfactory, reliable, and credible if sufficient kinetic interpretation could also be provided to justify the selection of those parameters for the test. The problem is to find the parameter that changes more markedly than others upon discharge.

It can be shown (Fig. 22) that the locus of the impedance generally follows a semicircular shape for the higher frequencies and becomes mainly resistive when the highest frequency is approached (about 10 kHz) [60]. The resistive

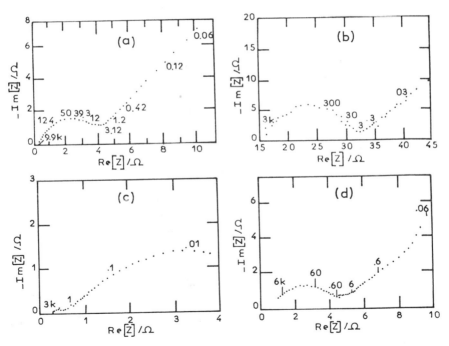

FIGURE 22 Measured impedances for fully charged batteries (frequency in hertz): (a) Leclanche (SP11, EverReady); (b) alkaline $Zn-MnO_2$ (MN150, Duracell); (c) alkaline $Zn-HgO$ (RM502R, Mallory); (d) Li-CuO (LC01, SAFT). (From Ref. 60.)

character observed as the frequency becomes infinite corresponds clearly to the ohmic resistance of the cell, electrolyte, electrodes, and current conductors. The high-frequency semicircular loop represents the faradaic process which controls the current and is generally identified with the electrode that forms the negative terminal of the cell, namely the anode, as the cathode is usually of larger area. The depolarizing reaction at the cathode is quite often the slowest of the two electrode reactions in the kinetic sense; however, the limitation imposed by the anode electrode area is generally sufficient to provide the major current control at higher frequencies. The change in diameter of the semicircle that occurs when a charge is delivered by the cell is clearly an attractive parameter for examination.

At least for the highest state of charge (between 100 and 90%) a linear relationship was found between the state of charge and the diameter of the high-frequency loop for the alkaline Zn/HgO and Li/SO_2 cells. With the Leclanche cell the low-frequency part of the semicircle is so merged with the Warburg line that the graphical determination of the diameter is quite difficult. Only a numerical fitting procedure is possible and has shown in some cases that the real part of the impedance measured at 31.2 Hz is a linear function of the state of charge in the range 100 to 90%.

For alkaline Zn/MnO_2 and Li/CuO cells, it seems rather difficult to devise a simple one- or two-frequency test. The cell impedance has to be analyzed over a wide frequency range to assess the charge status. For the $Li/SOCl_2$, in spite of a thorough kinetic investigation, a state-of-charge test has not yet been devised.

F. Characterization of Porous Electrodes

Porous metallic electrodes are used in various fields (power sources, electrochemical reactors, etc.), due to the large specific area that is necessary when a large reactive surface is needed in a small volume. This porosity leads to potential and concentration distributions in the electrolyte paths between the bulk of the solution and the metal surface. The simplest basic model is given by a narrow cylindrical pore of finite length in which the potential and concentration profiles are characterized in terms of the penetration depth of an ac signal. At high frequencies the electrode admittance is generally large and the ac signal is damped within a short distance. As the frequency decreases, the ac signal penetrates more and more deeply in the pore, and at low frequency the impedance is the same as for a flat surface having the same developed area. Therefore, the impedance is closely related to the geometry and size of the pores [62–65]. If in the gas phase these parameters can be determined by various methods (e.g., BET), in a liquid phase the characterization of the pore texture is poorly established.

For more intricate pore texture, an equivalent cylindrical pore electrode characterized by a radius, depth, and pore number can be estimated. Figure 23 shows the impedance diagram obtained with 28 mg of Raney nickel in 5 N KOH [64]. The apparent surface area is 1 cm^2. For frequencies greater than 1 Hz, the impedance diagram reveals a unit slope, whereas for frequencies lower than 0.6 Hz a capacitive loop plotted as a slightly depressed semicircle can be seen. Several resistive terms can be defined: R_t is the diameter of the low-frequency loop; R is the resistance of the electrolyte between the capillary tip of the reference electrode and the upper surface of the catalyst layer; R_p is the polarization resistance, equal to $R + \Omega + R_t$; R_0 indicates the electrolyte resistance measured in a cell without a catalyst.

To a first approximation, a Raney nickel electrode can be characterized, as previously, by an equivalent cylindrical pore. From the value of the capacity C, the specific surface area S is estimated ($S = 68$ m^2 g^{-1}). The charge-transfer resistance R_t is due to the oxidation of the adsorbed hydrogen at the Raney nickel electrode at open-circuit potential. This value allows the exchange current

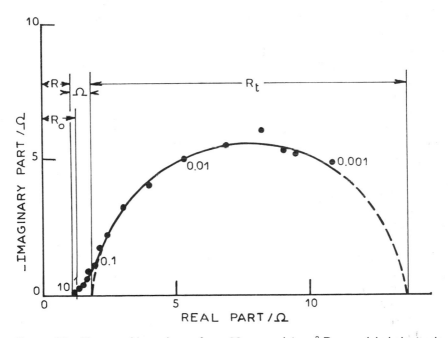

FIGURE 23 Measured impedance for a 28-mg and 1-cm^2 Raney nickel electrode in 5N KOH at 20°C. R is the electrolyte resistance outside the catalyst layer; R_0 is equal to R without the catalyst layer; R_t is the parallel resistance of C; $R_p = R + \Omega + R_t$ is the polarization resistance (frequency in hertz). (From Ref. 64.)

per unit catalyst mass J_0 to be evaluated ($J_0 = 0.145$ A g^{-1}), which shows a high reactivity of the electrode. The $R_0 - R$ value gives the global porosity, as the gas-phase measurement leads to the porosity inside a catalyst grain, and by difference the intergranular porosity is obtained.

However, if a single porosity is assumed, the estimated pores are too large compared with the value obtained from the gas-phase measurement. If micropores are supposed to exist in the catalyst grain and to have a length equal to the grain radius, the number of cylindrical pores per catalyst grain, n_p, can be calculated ($n_p = 2 \times 10^7$). Therefore, these results show that the complicated double-porosity texture of the Raney nickel electrode has been decrypted and described as a double family of pores by using impedance analysis.

G. Electrodes Modified with a Redox Polymer Film

The classical technique for these studies is linear sweep voltammetry coupled with chronoamperometry. The latter is used to determine the characteristic transport time of the charge within the film. Several models have been proposed to describe the kinetics of the modified electrodes under steady-state conditions with a substrate (i.e., a redox couple whose reaction is mediated by the redox polymer film) in solution. To interpret the simple case where there is no substrate in solution, the charge transfer at the electrode–polymer interface and the charge transport within the polymeric film have to be considered.

The charge transport in the film of thickness ϕ is assumed to be a diffusion-like process. Since the redox centers are attached to the polymer, they are not able to diffuse in a proper way, even though these centers may have a certain mobility. By mutual electron transfer between adjacent redox centers, however, a charge transport through the entire film is possible without significant displacement of the redox centers themselves. This charge transport obeys Fick's law, where the diffusion coefficient depends not only on the electron exchange rate (self-exchange rate) between the redox centers but also on the diffusion of these centers themselves.

In general, this charge-transport process is accompanied by counterion diffusion, which assures electroneutrality throughout the film. The charge transport within the film therefore depends on the counterion diffusion in the film, the self-exchange rate of the redox centers, and the mobility of the polymer "fixed" redox centers.

A simple model has been tested by impedance techniques [66–72]; it is based on the oxido-reduction of the redox active centers at the electrode–film interface and the diffusion of the redox couple within the polymer film. By assuming that the redox centers cannot leave the film to the solution, a faradaic impedance can be derived which agrees well with the experimental results.

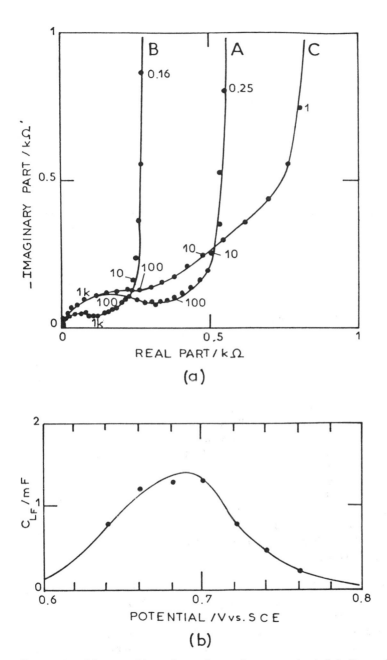

Figure 24 Measured impedance for a vitreous carbon disk (3 mm in diameter) electrode modified with a redox polymer film [Ru(bpy)$_2$[poly(4′-vinylpyridine)Cl]Cl] in 1 M HCl. (a) Impedance at (A) 0.64 V vs. SCE; (B) 0.7 V vs. SCE; (C) 0.76 V vs. SCE (frequency in hertz). (b) Change of the low-frequency capacity C_{LF} vs. potential. (From Ref. 66.)

For example, impedance measurements were performed at various potentials using redox polymer (Ru(bpy)$_2$[poly(4′-vinylpyridine)Cl]Cl)-coated glassy carbon electrodes (diameter: 3 mm) in 1 M HCl. The results are displayed in Figure 24. The low-frequency capacity C_{LF} has been determined by considering the low-frequency equivalent resistance-capacity circuit. The variation of C_{LF} was plotted versus potential [65]. The integration of this graph leads to a value of the surface concentration of the redox centers which is very close to the value obtained by cyclic voltammetry. By plotting the transfer resistance versus the potential, the Tafel coefficients and rate constants of the reaction rates can be attained.

The impedance analysis of a polymer-coated electrode allows all the accessible parameters to be evaluated by using only one technique. Hence this approach is very attractive for investigating modified electrodes.

REFERENCES

1. C. Gabrielli, *Identification of Electrochemical Processes by Freqeuncy Response Analysis*, Schlumberger-Solartron, Farnborough, Hampshire, England, 1980.
2. C. Gabrielli, *Use and Applications of Electrochemical Impedance Techniques*, Schlumberger-Solartron, Farnborough, Hampshire, England, 1990.
3. D. D. Macdonald, *Transient Techniques in Electrochemistry*, Plenum Press, New York, 1977.
4. E. Yeager, J. O'M. Bockris, B. E. Conway, and S. Sarangapani, eds., *Comprehensive Treatise of Electrochemistry*, Vol. 9, Plenum Press, New York, 1984.
5. J. R. Scully, D. C. Silverman, and M. W. Kendig, eds., *Electrochemical Impedance: Analysis and Interpretation*, STP 1188, ASTM, Philadelphia, 1993.
6. J. R. MacDonald, *Impedance Spectroscopy*, Wiley, New York, 1987.
7. C. Gabrielli, ed., Proc. First Int. Symp. Electrochemical Impedance Spectroscopy, *Electrochim. Acta, 35*(10) (1990).
8. D. D. Macdonald, ed., Proc. 2nd Int. Symp. Electrochemical Impedance Spectroscopy, *Electrochim. Acta, 38*(14) (1993).
9. R. Varma and J. R. Selman, eds., *Techniques for Characterization of Electrodes and Electrochemical Processes*, Wiley, New York, 1991.
10. R. S. Nicholson and I. Shain, *Anal. Chem., 36*:706 (1964).
11. C. Gabrielli, M. Keddam, and H. Takenouti, in *Materials Science Forum*, Vol. 8, *Electrochemical Methods in Corrosion Research* (M. Duprat, ed.), Trans Tech Publications, Aedermannsdorf, Switzerland, 1986, p. 417.
12. N. Sato and G. Okamoto, in *Comprehensive Treatise of Electrochemistry*, Vol. 4, *Electrochemical Materials Science* (J. O'M. Brockris, B. E. Conway, E. Yeager, and R. E. White, eds.), Plenum Press, New York, 1981, p. 193.
13. I. Epelboin, M. Ksouri, and R. Wiart, *J. Electrochem. Soc., 122*: 1206 (1975).
14. I. Epelboin, C. Gabrielli, M. Keddam, J. C. Lestrade, and H. Takenouti, *J. Electrochem. Soc., 119*: 1632 (1972).
15. J. W. White, *Proc. IEEE, 59*: 98 (1971).

16. J. W. White, *Proc. 6th Annual Princeton Conf. Information Sciences and Systems*, 1972, p. 173.
17. G. S. Popkirov and R. N. Schindler, *Electrochim. Acta, 38*: 861 (1993).
18. J. S. Newman, *Electrochemical Systems*, 2nd ed., Prentice Hall, Englewood Cliffs, N.J., 1991.
19. V. D. Levich, *Physicochemical Hydrodynamics*, Prentice Hall, Englewood Cliffs, N.J., 1962.
20. J. E. B. Randles, *Trans. Faraday Soc., 44*: 327 (1948).
21. M. Sluyters-Rembach and J. H. Sluyters, in *Electroanalytical Chemistry*, Vol. 4 (A. J. Bard, ed.), Marcel Dekker, New York, 1970.
22. H. Gerischer and W. Mehl, *Z. Elektrochem., 59*: 1049 (1955).
23. D. Schumann, *J. Electroanal. Chem., 17*: 45 (1968).
24. I. Epelboin and M. Keddam, *J. Electrochem. Soc., 117*: 1052 (1970).
25. I. Epelboin and R. Wiart, *J. Electrochem. Soc., 118*: 1577 (1971).
26. I. Epelboin, M. Keddam, and J. C. Lestrade, *Faraday Discussions Chem. Soc., 56*: 265 (1973).
27. L. Bai and B. E. Conway, *J. Electrochem. Soc., 138*: 2897 (1991).
28. I. Epelboin, C. Gabrielli, M. Keddam, and H. Takenouti, *Electrochim. Acta, 20*: 913 (1975).
29. D. D. Macdonald and M. Urquidi-Macdonald, *J. Electrochem. Soc., 132*: 2316 (1985).
30. M. Urquidi-Macdonald, S. Real, and D. D. Macdonald, *J. Electrochem. Soc., 133*: 2018 (1986).
31. J. M. Esteban and M. E. Orazem, *J. Electrochem. Soc., 138*: 67 (1991).
32. C. Gabrielli, F. Huet, and M. Keddam, *J. Electroanal. Chem., 335*: 33 (1992).
33. P. Agarwal, O. C. Moghissi, M. E. Orazem, and L. H. Garcia-Rubio, *Corrosion NACE*, 49: 278 (1993).
34. C. Gabrielli and M. Keddam, *Electrochim. Acta, 19*: 355 (1974).
35. M. Keddam, O. R. Mattos, and H. Takenouti, *J. Electrochem. Soc., 128*: 257, 266 (1981).
36. I. Epelboin, C. Gabrielli, M. Keddam, and H. Takenouti, in *Comprehensive Treatise on Electrochemistry*, Vol. 4 (E. Yeager, J. O'M. Bockris, B. E. Conway, and S. Sarangapani, eds.), Plenum Press, New York, 1984, p. 151.
37. M. Keddam, J. F. Lizee, C. Pallotta, and H. Takenouti, *J. Electrochem. Soc., 131*: 2016 (1984).
38. F. Mansfeld, in *Advances in Corrosion Science and Technology*, Vol. 6, Plenum Press, New York, 1976, p. 163.
39. I. Epelboin, C. Gabrielli, M. Keddam, and H. Takenouti, in *Proc. ASTM Symp. Electrochemical Corrosion Testing* (F. Mansfeld and U. Bertocci, eds.), STP 727, ASTM, Philadelphia, 1981, p. 150.
40. D. D. Macdonald and M. C. H. Mackubre, in *Proc. ASTM Symp. Electrochemical Corrosion Testing* (F. Mansfeld and U. Bertocci, eds.), STP 727, 1981, p. 110.
41. I. Epelboin, M. Keddam, and H. Takenouti, *J. Appl. Electrochem., 2*: 71 (1972).
42. L. Beaunier, I. Epelboin, J. C. Lestrade, and H. Takenouti, *Surface Technol., 4*: 237 (1976).
43. G. W. Walter, D. N. Nguyen, and M. A. D. Madurasinghe, *Electrochim. Acta, 37*: 245 (1992).

44. F. Mansfeld and M. W. Kendig in *Materials Science Forum*, Vol. 8, *Electrochemical methods in corrosion research* (M. Duprat, ed.), Trans Tech Publications, Aedermannsdorf, Switzerland, 1986, p. 337.
45. J. J. Bodu, M. Brunin, G. Sertour, I. Epelboin, M. Keddam, and H. Takenouti, *Aluminio, 46*: 277 (1977).
46. F. Mansfeld and M. W. Kendig, *Corrosion NACE, 41*: 490 (1985).
47. F. Mansfeld and M. W. Kendig, *J. Electrochem. Soc., 135*: 828 (1988).
48. J. Hitzig, K. Juttner, W. J. Lorenz, and W. Paatsch, *Corrosion Sci., 24*: 945 (1984).
49. C. Gabrielli, M. Keddam, H. Takenouti, Vu Quang Kinh, and F. Bourelier, *Electrochim. Acta, 24*: 61 (1979).
50. I. Epelboin, M. Ksouri, and R. Wiart, *J. Electrochem. Soc., 122*: 1206 (1975).
51. I. Epelboin, M. Ksouri, and R. Wiart, *Faraday Discussions Chem. Soc., 12*: 115 (1978).
52. M. Ksouri and R. Wiart, *Oberflache Surface, 18*: 61 (1977).
53. J. E. Bauerle, *J. Phys. Chem. Solids, 30*: 2657 (1969).
54. H. J. De Bruin and S. P. S. Badwal, *J. Australian Ceram. Soc., 14*: 20 (1978).
55. E. Schouler, M. Kleitz and C. Deportes, *J. Chim. Phys., 70*: 923 (1973).
56. S. P. S. Badwal and H. J. De Bruin, *Phys. Status Solidi. (a), 54*: 261 (1979).
57. N. Bonanos, R. K. Slotwinski, B. C. H. Steele, and E. P. Butler, *J. Mater. Sci., 19*: 785 (1984).
58. N. A. Hampson, S. A. G. R. Karunathilaka, and R. Leek, *J. Appl. Electrochem., 10*: 3 (1980).
59. S. A. G. R. Karunathilaka, R. Barton, M. Hugues, and N. A. Hampson, *J. Appl. Electrochem., 15*: 251 (1985).
60. S. A. G. R. Karunathilaka, N. A. Hampson, R. Leek, and T. J. Sinclair, *J. Appl. Electrochem., 10*: 357 (1980).
61. J. P. Randin, *J. Appl. Electrochem., 15*: 365 (1985).
62. R. De Levie, in *Advances in Electrochemistry and Electrochemical Engineering*, Vol. 6 (P. Delahay, ed.), Interscience, New York, 1967, p. 329.
63. J. P. Candy, P. Fouilloux, M. Keddam, and H. Takenouti, *Electrochim. Acta, 26*: 1029 (1981).
64. J. P. Candy, P. Fouilloux, M. Keddam, and H. Takenouti, *Electrochim. Acta, 27*: 1585 (1982).
65. C. Gabrielli, F. Huet, A. Sahar, and G. Valentin, *J. Appl. Electrochem., 22*: 801 (1992).
66. C. Gabrielli, O. Haas, and H. Takenouti, *J. Appl. Electrochem., 17*: 82 (1987).
67. C. Gabrielli, O. Haas, H. Takenouti, and A. Tsukada, *J. Electroanal. Chem., 302*: 59 (1991).
68. G. Lang, J. Bacskai, and G. Inzelt, *Electrochim. Acta, 38*: 773 (1993).
69. M. Sharp, B. Lindhom-Sethson, and E. L. Lind, *J. Electroanal. Chem., 345*: 223 (1993).
70. I. Rubinstein, J. Rishpon, and S. Gottesfeld, *J. Electrochem. Soc., 133*: 729 (1986).
71. I. Rubinstein, E. Sabatini, and J. Rishpon, *J. Electrochem. Soc., 134*: 3078 (1987).
72. T. B. Hunter, P. S. Tyler, W. H. Smyrl, and H. J. White, *J. Electrochem. Soc., 134*: 2198 (1987).

7

Principles and Applications of the Electrochemical Quartz Crystal Microbalance

MICHAEL D. WARD
University of Minnesota
Minneapolis, Minnesota

I. INTRODUCTION

Advances in electrochemistry in the last few decades can be attributed largely to the development of sophisticated instrumental techniques capable of probing electrode structures and interfaces in increasingly greater detail [1]. In the last decade, a new approach to examining electrodes and their interfaces, the electrochemical quartz-crystal microbalance (EQCM), has emerged as a powerful technique capable of detecting very small mass changes at the electrode surface that accompany electrochemical processes. This relatively simple technique only requires, in addition to conventional electrochemical equipment, an inexpensive radio-frequency oscillator, a frequency counter, and commercially available AT-cut quartz crystals. In this last decade the EQCM has evolved into a routine experimental method used in numerous electrochemical laboratories.

The EQCM has provided a rather simple and economical approach to the investigation of phenomena such as underpotential deposition, electrolyte adsorption, mass changes accompanying ion and solvent movement in redox polymer films, and electrochemically driven self-assembly. However, while the EQCM has been extremely useful in clarifying many electrochemical processes, the effects of liquids and thin films on the response of the EQCM are not yet fully understood, nor are the limitations imposed by these effects widely appreciated. The purpose of this review is to provide the reader with a fundamental understanding of the EQCM and several illustrative examples of applications that demonstrate its unique capabilities. It is hoped that this will encourage readers to consider adding the EQCM to their battery of electrochemical methods while informing them of some of the nonideal behavior that can affect interpretation of results. Other review articles can be consulted for different emphasis or greater detail [2–4].

II. BASIC PRINCIPLES OF OPERATION AND THEORY

A. Piezoelectric Effect

To understand the operation of the EQCM, a fundamental understanding of the piezoelectric effect is required. In 1880, Jacques and Pierre Curie discovered that mechanical stress applied to the surfaces of various crystals, including quartz, rochelle salt ($NaKC_4H_4O_6 \cdot 4H_2O$), and tourmaline, resulted in an electrical potential across the crystal whose magnitude was proportional to the applied stress [5]. This behavior is referred to as the piezoelectric effect, which is derived from the Greek word *piezin* meaning ''to press.'' This property exists only in materials that are acentric, that is, those that crystallize into noncentrosymmetric space groups. A single crystal of an acentric material will possess a polar axis due to dipoles associated with the arrangement of atoms in the crystalline lattice. The charge generated in a quartz crystal under mechanical stress is a manifestation of a change in the net dipole moment because of the physical displacement of the atoms and a corresponding change in the net dipole moment. This results in a net change in electrical charge on the crystal faces, the magnitude and direction of which depends on the relative orientation of the dipoles and the crystal faces. Following their initial discovery, the Curies discovered the converse piezoelectric effect, in which the application of a potential across these crystals resulted in a corresponding mechanical strain. It is this effect that is the operational basis of the EQCM.

The EQCM is actually the electrochemical version of the QCM, which has long been used for frequency control and mass sensing in vacuum and air. The QCM consists of a thin, AT-cut quartz crystal with very thin metal electrode ''pads'' on opposite sides of the crystal. The terminology ''AT'' simply refers to the orientation of the crystal with respect to its large faces; this particular crystal is fabricated by slicing through a quartz rod at an angle of approximately

35° with respect to the crystallographic x axis. The electrode pads overlap in the center of the crystal with tabs extending from each to the edge of the crystal where electrical contact is made. When an electrical potential is applied across the crystal using these electrodes, the AT-cut quartz crystal experiences a mechanical strain in the shear direction. Crystal symmetry dictates that the strain induced in a piezoelectric material by an applied potential of one polarity will be equal and opposite in direction to that resulting from the opposite polarity (Fig. 1). Therefore, an alternating potential across the crystal causes vibrational motion of the quartz crystal with the vibrational amplitude parallel to the crystal surface and the x direction. This oscillatory behavior and the electromechanical "motor generator" properties are the basis of a myriad of applications, including the QCM, sonar transducers, speakers, microphones, phonograph pickups, and quartz digital watches. It is important to note that the nature of the vibration is critical for liquid-phase applications; since the amplitude of vibration of the AT-cut crystal is *parallel* to the crystal–liquid interface, forces from the liquid that may dampen the vibration are minimized (see below).

The vibrational motion of the quartz crystal results in a transverse acoustic wave that propagates back and forth across the thickness of the crystal between the crystal faces. Accordingly, a standing-wave condition can be established in the quartz resonator when the acoustic wavelength is equal to twice the combined thickness of the crystal and electrodes. The frequency f_0 of the acoustic wave fundamental mode is given by equation (1), where v_{tr} is the transverse velocity of sound in AT-cut quartz (3.34×10^4 m s^{-1}) and t_Q is the resonator thickness. An assumption is made here that the velocity of sound in quartz and the electrodes is identical; while not rigorously true, for small electrode thicknesses the error introduced by this approximation is negligible. The acoustic velocity is dependent on the modulus and density of the crystal. The quartz crystal surface is at an antinode of the acoustic wave, and therefore the acoustic wave propagates across the interface between the crystal and a foreign layer on its surface. If one assumes that the acoustic velocity and density of the foreign layer are identical to those in quartz (cf. the assumption for the metal electrodes), a change in thickness of the foreign layer is tantamount to a change in the thickness of the quartz crystal. Under these conditions, a fractional change in thickness results in a fractional change in the resonant frequency; appropriate substitutions yielding the well-known Sauerbrey equation [equation (2)], where Δf is the measured frequency change, f_0 the frequency of the quartz resonator prior to a mass change, Δm the mass change, A the piezoelectrically active area, ρ_Q the density of quartz (2.648 g cm^{-3}), and μ_Q the shear modulus of AT-cut quartz (2.947×10^{11} dyn cm^{-2}). This equation is the primary basis of most QCM and EQCM measurements wherein mass changes occurring at the electrode interface are evaluated directly from the frequency changes of the quartz resonator. It is generally considered accurate as long as the thickness of the film

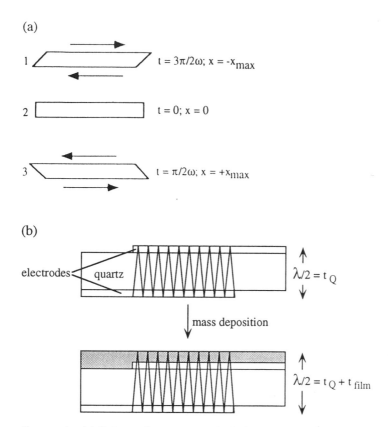

FIGURE 1 (a) Schematic representation of the shear vibration of an AT-cut quartz resonator. The time at which the crystal achieves maximum strain during oscillation is indicated. The crystal has maximum kinetic energy at $x = 0$, but maximum potential energy at $x = \pm x_{max}$, similar to classical oscillators. (b) Schematic representation of the transverse shear wave in a quartz crystal with excitation electrodes and a composite resonator comprising the quartz crystal, electrodes, and a thin layer of a foreign material. The acoustic wavelength is longer in the composite resonator due to the greater thickness, resulting in a lower frequency compared to the quartz crystal.

added to the QCM is less than 2% of the quartz crystal thickness. With this constraint, the errors resulting from the discrepancy between the acoustic propagation characteristics in quartz and the film are not severe. Deviations from equation (2) due to higher mass loadings may be compensated, however, by use of the *Z-match method* [6]. Although this method has been used for vacuum applications, it has yet to be employed in EQCM applications. Typical operating frequencies of the EQCM lie within the range 5 to 10 MHz, although recently we have determined that even 30-MHz quartz crystals can be operated successfully in EQCM applications [7]. These operating frequencies provide for mass detection limits approaching 1 ng cm^{-2}.

$$f_0 = \frac{v_{tr}}{2t_Q} \tag{1}$$

$$\Delta f = \frac{-2f_0^2 \, \Delta m}{A\sqrt{\mu_Q \rho_Q}} \tag{2}$$

The reader may better understand the mass-sensing properties of the QCM by comparing the motion of the quartz crystal to other oscillating systems, such as a vibrating string, a pendulum, or a mass on a spring. In all cases the amplitude is defined by the initial energy input and the resonant frequencies are defined by characteristics of mass and length. The quartz crystal motion can be described as moving about the $x = 0$ rest point between limits of $-x_{max}$ and $+x_{max}$. The magnitude of x_{max} will depend on the applied alternating voltage across the crystal. As with the more familiar oscillating systems, the potential energy of the crystal is at a maximum at $x = \pm x_{max}$, whereas the kinetic energy is at a maximum at $x = 0$. The effect of mass (or thickness) changes on the quartz resonant frequency can be understood by analogy to a classical system such as a vibrating string. Standing waves can exist in a vibrating string if their wavelengths are integral divisors of $2l$, where l is the length of the string. The fundamental frequency, f_0, is given by equation (3), where S is the tension on the string and m_l is the mass per unit length. An increase in the mass of the string or its length therefore results in a decrease in f_0. This is identical to increasing the thickness of a quartz crystal; in both cases the dimensional increase results in a longer distance over which the standing wave must propagate and a corresponding reduction in the frequency. Similarly, the stress applied to the vibrating string is analogous to the modulus of the quartz crystal; an increase in either of these quantities increases the velocity of the standing wave.

$$f_0 = \frac{(S/m_l)^{1/2}}{2l} \tag{3}$$

While the physics of quartz crystals and their mass-sensing properties can be understood from the aforementioned discussion, the key distinguishing feature is the negligible energy dissipation encountered by the quartz crystals during their oscillation. While a pendulum may lose considerable energy during oscillation because of friction, a quartz crystal loses only a minute amount due to phonon interactions that produce heat, vibrational damping by the mounting components, and acoustical losses to the environment. This property is generally characterized by the *quality factor*, **Q**, which is the ratio of the energy stored to energy lost during a single oscillation. For quartz crystals, this quantity can exceed 100,000. Low energy losses in oscillating systems are manifested as high accuracies. As a system loses more energy during oscillation, the period of the oscillation becomes less well defined. This is the basis for the widespread use of quartz crystals in timepieces and frequency control elements. In fact, the frequency of a typical quartz crystal can be determined to an accuracy of 1 part in 10^8. It is this precision that makes the QCM and EQCM so useful. In liquid applications, including the EQCM, **Q** will generally have values of 1000 to 3000, indicative of energy damping by the fluid. Nevertheless, the quartz crystals will still perform acceptably at these levels.

B.   Electromechanical Model of the EQCM

Quartz crystals are *electromechanical devices*, and therefore their mechanical vibrations can be described in terms of electrical equivalents [8]. This also serves to enhance understanding of the EQCM, particularly the conditions under which the Sauerbrey equation, (2), is valid. The quartz resonator can be described according to a mechanical model with elements of mass, compliance (the ability of an object to yield elastically under an applied force), and friction. The electrical equivalent of this system is an electrical circuit that has an inductor, a capacitor, and a resistor connected in series (Fig. 2). In this equivalent circuit, the inductor, L_1, represents the mass displaced during oscillation, C_1 the energy stored during oscillation (the compliance is the inverse of the elastic, or Hooke's, constant), and R_1 the energy dissipation due to losses that are tantamount to internal friction. To describe the quartz crystal behavior accurately, a parallel capacitance must also be included that represents the static capacitance of the quartz plate with its electrodes, and any stray parasitic capacitances. The complete circuit is commonly referred to as the *Butterworth–van Dyke circuit* [9]. The series branch of the circuit is referred to as the *motional branch* since it reflects the vibrational behavior of the crystal.

The relationship between this circuit and the quartz crystal is especially useful because the LCR branch is identical to a *tank circuit*, in which oscillations can be sustained by cycling of current between the capacitor and the inductor. When the capacitor in this circuit discharges through the inductor, a magnetic field is established around the inductor as it opposes the current. When the capacitor discharge is complete and the current falls to zero, the electromotive force in the inductor creates a current in the direction opposite to the original current and the capacitor recharges. Repetition of this cycle results in electrical oscillation, with the oscillations dampened by an amount proportional to *R*. In the case of quartz crystals, the *R* values are rather small and sustained oscillations are favored. As a result of the electromechanical relationship between

(a)

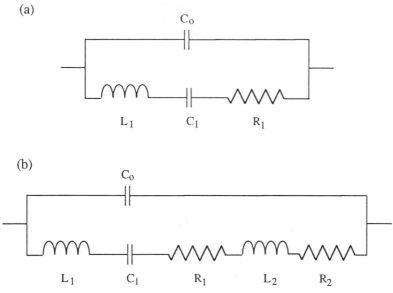

(b)

FIGURE 2 (a) Butterworth–van Dyke equivalent electrical circuit used to describe the mechanical properties of a quartz resonator. The components L_1, C_1, and R_1 in the motional branch of the circuit represent the inertial mass, compliance, and energy dissipation in the crystal, and C_0 represents the static capacitance of the quartz crystal. (b) Equivalent electrical circuit used to describe the mechanical properties of a quartz resonator immersed in a liquid. The inductance L_2 and resistance R_2 represent the mass and viscosity components of the liquid.

quartz crystals and electrical circuits, the equations of harmonic motion of the quartz crystals are closely related in form to the expressions describing the properties of the LCR tank circuit. This has been reviewed elsewhere and a detailed description is not given here. The equivalent electrical parameters in terms of crystal properties are given in equations (4) to (7), along with typical experimental values for these parameters. In these relationships, D_q is the dielectric constant of quartz, ϵ_0 the permittivity of free space, r a dissipation coefficient corresponding to the energy losses during oscillation, ϵ the piezoelectric stress constant, and c the elastic constant. Note that while L_1, C_1, and R_1 depend on ϵ, C_0 does not participate directly in piezoelectricity. It should also be noted that L_1 depends on the density; in fact, the quantity $t_Q^3 \rho/A$ is equivalent to the mass per unit area in the Sauerbrey equation. These equivalent representations provide a quantitative approach to examining the properties of the EQCM, the role of the liquid environment and thin films on the resonant frequency response, and the design of quartz resonators.

$$C_0 = \frac{D_Q \epsilon_0 A}{t_Q} \approx 10^{-12} \text{ F} \tag{4}$$

$$C_1 = \frac{8A\epsilon^2}{\pi^2 t_Q c} \approx 10^{-14} \text{ F} \tag{5}$$

$$R_1 = \frac{t_Q^3 r}{8A\epsilon^2} \approx 100 \ \Omega \tag{6}$$

$$L_1 = \frac{t_Q^3 \rho}{8A\epsilon^2} \approx 0.075 \text{ H} \tag{7}$$

C. Operational Considerations in the Liquid Phase

While the Butterworth–van Dyke circuit accurately describes the operation of the QCM in vacuum or the gas phase, it is not a sufficient description when the QCM or EQCM is used in liquids where the liquid density and viscosity alter the resonator characteristics. The density of the liquid effectively adds to the mass of the resonator, while its viscosity provides additional energy damping. The effect of a Newtonian liquid can be described by adding an additional inductance and resistance L_2 and R_2 in series with the motional branch of the Butterworth–van Dyke circuit, in which L_2 and R_2 are related to the extra mass of the liquid and its viscosity, respectively (Fig. 2b) [10]. The vibration of the quartz crystal parallel to the QCM–liquid interface results in the radiation of a shear wave into the liquid. The instantaneous shear wave velocity decays as an exponentially damped cosine function according to equation (8), where k is the propagation constant, z the distance from the resonator surface, A the maximum amplitude of the shear wave, and ω the angular frequency (Fig. 3). The inverse of the propagation constant k is the decay length, δ, which is given by equation (8), where ρ_L and η_L are the liquid density and viscosity, respectively. This leads to equation (10), which gives the dependence of the resonant frequency on $(\rho_L \eta_L)^{1/2}$ [11]. This dependence can be observed readily by the approximately 750 Hz decrease in frequency that occurs when a QCM is immersed in water. The observed frequency decrease effectively is attributed to the effective mass of liquid contained in this decay length. It should be noted that equations (8) and (9) rely on the assumption of a "no-slip" condition at the QCM–liquid interface; that is, the molecular layer of liquid directly in contact with the vibrating region of the QCM surface moves with the same velocity and amplitude as the vibrating region. Thus the shear wave then decays in the liquid because of the velocity gradient along the direction normal to the surface, that is, the viscosity. Although there are some reports claiming deviations from the no-slip

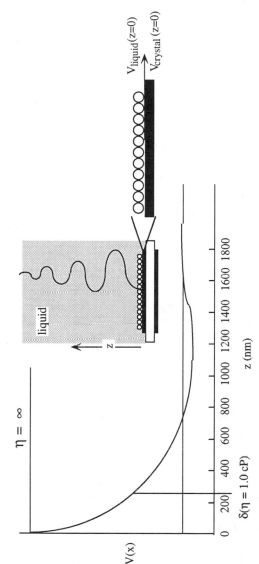

FIGURE 3 Description of the shear wave propagation in a newtonian fluid in terms of the shear velocity in the x direction (parallel to the resonator–liquid interface) as a function of distance from the resonator–liquid interface. If the viscosity of the film in contact with the resonator is infinitely large, the acoustic wave will propagate without loss. In water, the amplitude of the acoustic wave decays, with the decay length $\delta = 2500$ Å. The decay length represents the distance at which the amplitude of the shear wave decays to $1/e$ of its maximum value at the interface. The circles represent the layer of liquid molecules at the interface, at which a no-slip condition generally is assumed. That is, these molecules move with the same velocity and amplitude as the resonator ($V_{liquid} = V_{crystal}$).

condition, these are sparse and not yet conclusive [12].

$$V_x(z,t) = A e^{-kz} \cos(kz - \omega t) \tag{8}$$

$$\delta = \sqrt{\frac{\eta_L}{\pi f_0 \rho_L}} \tag{9}$$

$$\Delta f = -f_0^{3/2} \left(\frac{\rho_L \eta_L}{\pi \rho_Q \mu_Q}\right)^{1/2} \tag{10}$$

Inspection of (9) and (10) reveals that an increase in either the density or viscosity results in a decrease in the resonant frequency. This can be better understood by considering two extremes: in air, the quantity $(\rho_L \eta_L)^{1/2}$ is negligible, but when the liquid is actually a rigid solid, $(\rho_L \eta_L)^{1/2}$ is substantial. In the case of a rigid solid with $\eta = \infty$, the acoustic wave will propagate indefinitely and the frequency change would reflect the true thickness of the solid. If the thickness of this rigid film was large, crystal oscillation would be difficult because of the significant damping by the large mass. Therefore, it is the decay of the shear wave in liquids that enables operation of the QCM and EQCM in liquid media.

As the viscosity of the liquid becomes larger, the accuracy and performance of the resonator diminishes. These effects can be especially important when the EQCM is modified with polymer films, which may undergo changes in $(\rho_L \eta_L)^{1/2}$ during measurements. In addition, polymer films may be viscoelastic, which can further complicate interpretation of EQCM data (see below). Indeed, there are several examples of significant QCM responses to these changes under conditions where mass changes are not operative [13].

Impedance analysis is being used increasingly to evaluate the contributions from viscous and viscoelastic effects. There is not sufficient space in this chapter to give a detailed explanation of this approach, but it can be described briefly as a technique in which a voltage within a specified range of frequencies is broadcast across the crystal and the current measured. The impedance or admittance is measured, and the BVD equivalent circuit parameters determined by numerical fitting of the data. The mechanical properties corresponding to the electrical parameters can then be assessed. Commercially available impedance analyzers allow a complete set of parameters to be measured within 1 min, enabling dynamic measurements, albeit on a rather slow time scale.

There are other aspects of the EQCM that distinguishes their operation in liquids from that of the QCM in vacuum or the gas phase: namely, the effect of the liquid on the actual vibrating area of the quartz crystal. The Saurbrey equation assumes that the frequency shift resulting from a localized deposit is equivalent to the contribution of that deposit when it is a portion of a thin film of identical thickness distributed over the entire active QCM area. However, a general expression that accounts for localized or nonuniform mass deposits covering the QCM electrode to $r = r_e$ (where r_e is the ideal radius of the excitation electrodes, between which the electric field induces crystal motion) is given by equation (11), where $S(r,\theta)$ is the differential mass sensitivity (df/dm) and $m(r,\theta)$ is the mass distribution with respect to r, the distance of the deposit from the center of the crystal, and θ, the angle in the crystal plane with respect to the x axis. It has long been accepted, on the basis of theory and experiment, that $S(r,\theta)$ has a near-gaussian form in which the crystal vibration is maximum in the center and negligible at the edges [14]. Indeed, equation (10) and the Sauerbrey equation are based on the assumption that the vibrating area, and therefore the mass-sensitive area, is limited to inside the region where the excitation electrodes overlap. Any changes in $S(r,\theta)$ resulting from the liquid, or any crystal motion beyond the electrode edges, will cause deviations from the ideal response.

$$\Delta f = \frac{1}{\pi r_e^2} \int_0^{2\pi} \int_0^{r_e} S(r,\theta)m(r,\theta)r \, dr \, d\theta \tag{11}$$

Indeed, recently it has been shown in liquids that $S(r,\theta)$ and the actual vibrational area depend significantly on the liquid and its properties, including its viscosity [15]. QCM measurements performed in conjunction with scanning electrochemical microscopy demonstrated that indeed $S(r,\theta)$ was gaussian-like (Fig. 4). However, the shape and maximum value in the center were affected dramatically by the viscosity of the liquid in contact with the EQCM. Increasing the viscosity resulted in a suppression of the maximum, and also an increase in the mass sensitive area due to a phenomenon referred to as field fringing, in which the crystal vibrations extend *beyond* the electrode area. This results in an overall reduction in the sensitivity of the QCM. The details of these effects will depend on crystal contour, electrode geometry, and the liquid properties. Trapping the crystal vibrations within the excitation electrode boundaries is much more efficient for planoconvex crystals, in which one side of the crystal is contoured by polishing such that its shape resembles a convex lens. The difference in mass between the center and outer regions of the crystal results in a focusing of the acoustic energy toward the center of the resonator, as is evident from an increase in the value of $S(r,\theta)$ at the center and negligible mass sensitivity at $r > r_e$. Therefore, there appear to be distinct advantages to using planoconvex crystals. Understanding these effects is crucial if precise quantitative interpretations of EQCM data is required. At the very least, these effects should be appreciated in a qualitative way.

Another effect that needs to be considered during EQCM applications is the microscopic roughness of the resonator surface. The cavities on a rough surface can trap liquid, which will be manifested as an additional mass on the surface. The amount of trapped liquid will depend on the cavity geometry and size. This effect has been inferred in gold and copper surfaces on the QCM, in

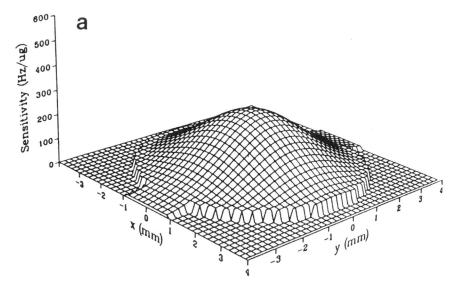

FIGURE 4 Sensitivity distributions $S(r, \theta)$ for 5-MHz AT-cut quartz crystals determined by a scanning electrochemical microscopy method in which 100-μm-diameter circular copper features are electrodeposited at different values of r, θ while simultaneously measuring electrochemical charge and frequency. The amplitude of these plots at a given value of r, θ represents the sensitivity of the EQCM to mass changes at that location, and the area under the plot represents the total, or integral, sensitivity C_f. The $S(r, \theta)$ plots shown were measured on (a) plano-plano and (b) plano-convex quartz crystals in aqueous 20 m*M* CuSO$_4$ (η_L = 0.8904 $\times$ 10^{-2} g cm^{-1} s^{-1}), and (c) plano-convex quartz crystals in aqueous 20 m*M* CuSO$_4$ containing 15% sucrose (η_L = 1.469 $\times$ 10^{-2} g cm^{-1} s^{-1}). (From Ref. 15a.)

which the electrode surfaces were roughened during electrochemical cycling through the oxide regions of the metal electrodes [16]. An extensive study of these gold surfaces involved comparison of the frequency shift with that expected for liquid trapped in surface cavities whose dimensions were measured with scanning electron microscopy. These studies indicated that the frequency changes were smaller than expected based on the SEM measurements, suggesting that the trapped liquid did not behave as a rigid mass. Therefore, surface roughness effects can be very difficult to quantify even when the exact roughness is known. These effects may be significant in many published EQCM studies because of the wide use of unpolished quartz crystals, which are commonly used because of their ready availability and low cost. These are widely sold for vacuum thickness monitor applications where good adhesion of metal films is required but liquid trapping is not a factor. Our laboratories have always used

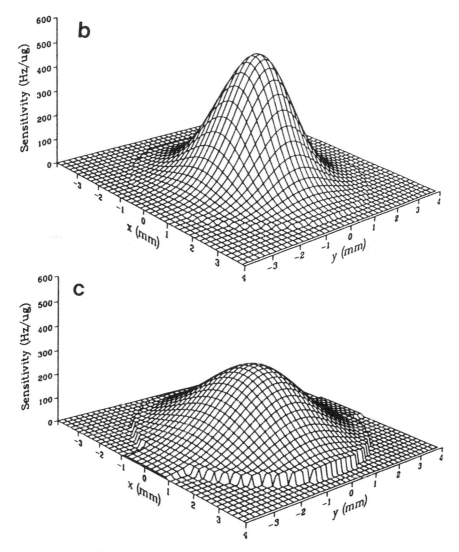

FIGURE 4 Continued

polished quartz crystals to minimize roughness effects. In any case, any published work employing the EQCM should contain a description of the quartz crystals and their roughness.

Generally, EQCM experiments are performed with one side of the quartz resonator immersed under a column of the electrolyte solution. Under these conditions, hydrostatic pressure results in a stress on the quartz crystal that affects the resonant frequency. Equation (12) describes a parabolic dependence of f_0 on the hydrostatic pressure [17,18]. The significance of this effect is probably minimal, as the hydrostatic pressure is generally constant (barring evaporation) for most experiments.

$$f_0 - f_0^{max} = A(p - p_{max})^2 \qquad (12)$$

III. ELECTROCHEMICAL APPLICATIONS OF THE EQCM

The EQCM has been used to examine a wide variety of electrochemical processes. Unfortunately, space does not permit a comprehensive review of the area, and the reader is referred to other reviews cited earlier. Rather, in this section we highlight some illustrative examples of EQCM applications that are meant to provide the reader with a general understanding of the scope of this method, starting with investigations of processes occurring directly at the electrode surface and moving on to investigations of thick films immobilized on the EQCM. These examples will be limited generally to those in which the data have been interpreted on the basis of mass changes only, that is, based on the Sauerbrey equation, (2). However, some examples are included that purposely demonstrate that this condition cannot always be assumed.

A. Experimental Apparatus and Operation

Having described the electromechanical and mass-sensing properties of the EQCM in liquids, it is appropriate to describe the experimental apparatus typically used in EQCM experiments (Fig. 5). Several versions of EQCM instrumentation has been described [19], differing mostly in details. The quartz crystals commonly have diameters of 0.5 and 1.0 in., with appropriately sized excitation electrodes. The crystals can be mounted to the bottom of a glass cylinder that assumes the role of the working electrode compartment of an otherwise conventional electrochemical cell. The crystal is mounted so that one of the excitation electrodes is facing the solution, the opposite electrode therefore facing air. Crystals have been mounted either between O-rings or with epoxy, the former being more convenient as the crystals can be easily demounted for reuse or further surface studies. The two excitation electrodes are electrically connected to an oscillator circuit that contains a broadband RF amplifier, so that the electrode facing solution is at hard ground. The circuit is designed so that the crystal is in a feedback loop, therefore driving the crystal at a frequency at which the maximum current can be sustained in a zero-phase angle condition. Several oscillator designs are available, although some require modification to supply sufficient gain to the crystal to sustain oscillation in liquids. The output of the oscillator is then connected to a conventional frequency meter for mea-

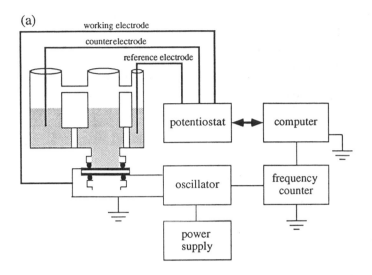

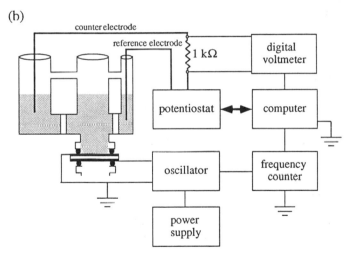

FIGURE 5 (a) Schematic representation of typical EQCM apparatus in which a Wenking potentiostat is employed with the working electrode at hard ground. (b) An alternative EQCM in which a conventional, commercially available potentiostat is used. In this arrangement the working electrode lead of the potentiostat is not connected to the EQCM working electrode. Rather, the current is measured by the voltage drop across a 1-kW resistor in series with the counterelectrode. This arrangement is required when using commercially available potentiostats because these potentiostats operate with the working electrode at virtual ground; connection of this virtual ground to the hard-grounded EQCM working electrode can result in oscillator instabilities.

surement. A critical feature of the EQCM is the potentiostat, which can be either a Wenking potentiostat or a more conventional potentiostat. The difference between these two potentiostats is in the working electrode: the Wenking potentiostat functions with the working electrode at hard ground, whereas current commercially available potentiostats generally function with the working electrode at virtual ground. Commercial potentiostats therefore can only be used if the working electrode is *not* connected to the potentiostat and the potential difference between the reference and hard ground used to control the working electrode potential [19e]. Because the current is generally measured at the working electrode side, this format requires that the current be measured by the voltage drop across a resistor in series with the counterelectrode connection. Because this equipment generally is available in most electrochemical labs, the EQCM practitioner need not build, nor buy, a custom-made Wenking potentiostat. Finally, a computer is used to collect frequency and electrochemical data simultaneously, as well as control the waveform applied to the working electrode. This arrangement allows simultaneous measurement of the electrochemical charge, current, voltage, and EQCM frequency. The time scale of the analysis is in a particularly useful domain for electrochemists. The time constant of a QCM resonator is fixed by $Q/\pi f_0$. For a 5-MHz QCM, $\mathfrak{Q} \approx 10^3$, and therefore the minimum sampling time is in the millisecond range. Frequency counters are capable of sampling the frequency output of the oscillator at 100-ms intervals. This capability enables analysis of the kinetics of a wide range of electrochemical processes, including electrodeposition and dissolution, nucleation, and growth and ion–solvent insertion in redox polymer films. It is important to stress that this equipment is not prohibitively expensive and is well within the reach, economically and technically, of most electrochemical investigators.

B. Data Interpretation

Interpretation of EQCM data is accomplished in a rather straightforward manner. Since the electrochemical charge represents the total number of electrons transferred in a given electrochemical process, it corresponds to mass changes occurring at the electrode surface. Accordingly, under ideal conditions, the frequency change measured with the EQCM will be proportional to the electrochemical charge and will be related to the *apparent* molar mass by eq. (13), where MW is the apparent molar mass (g mol^{-1}), Q the electrochemical charge, n the number of electrons involved in the electrochemical process, F the Faraday constant, and C_f(Hz g^{-1}) the sensitivity constant derived from the Sauerbrey relationship. Inspection of (13) reveals that plots of Δf vs. Q are particularly useful in the determination of MW/n, which represents the molar mass per electron transferred. This calculation, of course, depends on knowledge of C_f. Because this value can differ for different crystal contours, electrode geometries, and solution conditions (see above), it is important to calibrate the QCM with a well-behaved electrochemical reaction under conditions similar to those present during experiments of interest. This is generally accomplished with copper or silver electrodeposition, for which all terms on the right side of equation (13) are known except for C_f. The term *apparent* molar mass is stressed because in many cases the measured value of MW may not be that expected based on a simple stoichiometric relationship, but may involve solvent or coadsorbed species that can reveal considerable insight into the electrochemical behavior. Nonlinearities in plots of Δf vs. Q can be particularly useful for diagnosing nonideal behavior such as roughness and viscoelastic effects that may become evident over the range of frequency changes examined. An alternative approach to data analysis involves the relationship between the electrochemical *current* and the first derivative of the frequency change with respect to time, as given in equation (14), where v is the scan rate in units of $V\ s^{-1}$. This format is particularly useful for cyclic voltammetry experiments, as $d\Delta f/dt$ should appear similar in form to the voltammograms if the electrochemical events are accompanied by corresponding mass changes.

$$\Delta f = \frac{MW \cdot C_f Q}{nF} \tag{13}$$

$$i = \frac{d(\Delta f)}{dE}\frac{nvF}{MW \cdot C_f} \tag{14}$$

The utility of the EQCM method stems from its capability to measure electrochemical charge and current while simultaneously measuring mass changes with extraordinary sensitivity. A typical operating resonant frequency of 5 MHz provides a theoretical sensitivity of $0.057\ Hz\ cm^{-2}\ ng^{-1}$. Since the frequency can generally be measured to within an accuracy of 1 Hz, a 5 MHz EQCM can detect approximately $10\ ng\ cm^{-2}$. This translates roughly into 10% of a monolayer of Pb atoms. Much higher sensitivity can be realized with quartz crystals that operate at a higher fundamental frequency, as the sensitivity increases with f_0^2. Alternatively, the quartz crystal can be driven at one of its odd harmonic modes with appropriate circuitry, which provides an *n*-fold increase in sensitivity, where n is the harmonic number. While the third harmonic of a 5-MHz crystal (i.e., an operating frequency of 15 MHz) has been employed successfully in the examination of underpotential deposition on metal electrodes [20], general use of this approach can be limited by the lower stability of the harmonic modes relative to the fundamental mode. It is much more beneficial to employ crystals that can be operated at higher fundamental frequencies; however, above 10 MHz the quartz crystals generally become very fragile. Recently, we reported the successful operation of an EQCM that employed crystals with

f_0 = 30 MHz, in which the crystals were prepared by chemical etching of a AT-cut quartz disk. This treatment affords a very thin $\approx$ 50-μm-thick quartz "membrane" in the center of a thick outer quartz ring, the outer ring providing improved mechanical stability [7]. These crystals behaved ideally, with sensitivity constants in exact agreement with equation (2).

C. EQCM Investigations of Metal Electrodes and Films

As stated above, the most common technique for calibrating the EQCM involves simultaneous measurement of the electrochemical charge and resonant frequency during the electrochemical deposition or dissolution of a metal film. In this capability demonstrated by Bruckenstein and Shay, the deposition of 10 layers of Ag metal on a 10-MHz EQCM gave frequency shifts that were within 3% of the amount expected based on the Sauerbrey equation [21]. Therefore, the EQCM can be an effective method for examining the faradaic efficiencies of electroplating and dissolution processes such as corrosion [22].

The high sensitivity of the EQCM has enabled examination of underpotential deposition processes in situ. For example, examination of the underpotential deposition of Pb on the Au working electrode of the EQCM revealed mass and electrochemical charge changes in the UPD region corresponding to a hexagonally closest-packed monolayer. The ability to measure mass and charge simultaneously allowed determination of the electrovalency number for this process, which was found to be γ_{Pb} = 2.08 $\pm$ 0.10. Similar experiments revealed that other UPD processes occurred with γ values lower than expected for complete electron transfer, with γ_{Bi} = 2.7, γ_{Cu} = 1.4, and γ_{Cd} = 1.6 to 2.0 [23].

The EQCM has revealed several interesting features in UPD processes that otherwise would be difficult to detect. For example, in the aforementioned investigations with bismuth, it was discovered that a precipitous increase in electrochemical charge accompanied the third UPD peak, with a slight *decrease* in mass (Fig. 6). This behavior may be associated with the loss of electrolyte ions that adsorbed during the first two UPD events. More detailed studies of Pb UPD on Au and Ag revealed behavior that demonstrated the importance of the underlying substrate in UPD processes [24]. Whereas Pd UPD on Au electrodes in borate buffer was accompanied by a mass increase, UPD on Ag electrodes exhibited a mass *decrease* due to the desorption of BO_2^- ligands from a previously adsorbed anionic Pb(II) species. The adsorption of the latter was detected by the frequency decrease upon addition of Pb(II) to the electrolyte solution prior to electrochemical experiments. The data reflected the desorption of three BO_2^- ions per adsorbed Pb(II) species during the UPD process. These studies clearly reveal the value of the EQCM in probing electrode surface processes, providing information critical to complete understanding of rather complicated mechanistic schemes.

The EQCM has also enabled determination of the electrovalency numbers for anion adsorption. Simultaneous measurement of electrochemical charge and frequency during anodic adsorption of Br^- and I^- under conditions where monolayer coverage previously had been demonstrated revealed $\gamma_{Br}^- = -0.39 \pm 0.03$ and $\gamma_I^- = -1.01 \pm 0.03$ [25]. These studies indicate that while adsorption of iodide occurs with complete charge transfer, bromide ions partially retain their negative charge.

The capabilities of the EQCM provide a unique approach for examining electrode dissolution that are important in processes such as corrosion and electrochemical machining. While EQCM studies of dissolution have been rather limited, these reports indicate that important mechanistic information can be obtained readily with this method. For example, EQCM investigations of the anodic dissolution of nickel films revealed two maxima in the Δf vs. potential plots, indicating a potential-dependent dissolution of the α and β phases of NiH_x [26]. Analysis of the frequency changes and electrochemical charge in EQCM studies of the anodic dissolution of a nickel–phosphorus film revealed that two different Ni–P compositions were present in the film prior to dissolution. This was consistent with the known Ni–P phase diagram [27]. EQCM studies of the electrochemical dissolution of copper films in oxygenated sulfuric acid revealed that the dissolution rate was linearly dependent on $[O_2]$ and $[H^+]$, enabling the authors to conclude that a heterogeneous surface reaction was operative [28].

Several groups have demonstrated that the EQCM is also an ideal method for examining the electrochemically induced adsorption of hydrogen and deuterium in metal films such as palladium [29]. While these efforts were undoubtedly encouraged by reports of electrochemical cold fusion, such studies can have significant impact on understanding the commercially important isotopic separation of H and D. In addition to measurement of H/D adsorption, these EQCM studies have revealed the important role of stress in EQCM measurements. Cheek and O'Grady reported a rather novel approach to this issue by using a "double-resonator" technique that previously had been reported for measuring stresses in Si films resulting from ion implantation (in vacuum) [30]. This method involves the comparison of frequency responses from an AT-cut quartz crystals *and* a BT-cut quartz crystal. Both crystals have identical sensitivities to mass changes, but a compressive stress in a film on an AT-cut crystal results in a frequency decrease while an identical stress in a film on a BT-cut crystal results in an *increase* in frequency of similar magnitude. The amount of stress can be determined with equation (15), where K^{AT} and K^{BT} are the stress coefficients for the different resonators ($K^{AT} = 2.75 \times 10^{-12}$ cm^2 dyn^{-1} and $K^{BT} = -2.65 \times 10^{-12}$ cm^2 dyn^{-1}). Stress in thin films is manifested in a stress in the quartz crystal, which is tantamount to a change in the elastic modulus, μ_q. Because the

acoustic velocity depends upon this quantity, these stresses result in changes in
the resonant frequency (this effect is similar to tightening a violin string to
increase the frequency of vibration). Thus hydrogen absorption in a Pd film on
an AT-cut crystal gave a frequency decrease that was much larger than expected,
indicating compressive stresses in the Pd film upon H absorption: -351 Hz for

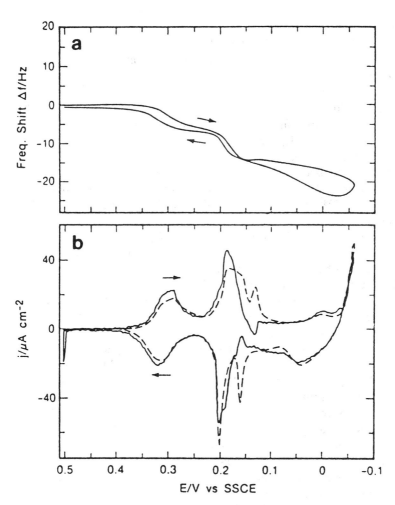

FIGURE 6 (a) Frequency shift of the EQCM during Bi underpotential deposition
on a Au electrode (1.0 mM Bi in 0.1 M HClO$_4$). (b) Current response during Bi
underpotential deposition on the Au EQCM electrode (- - -) and the current response expected from the observed frequency response (———). (From Ref. 23.)

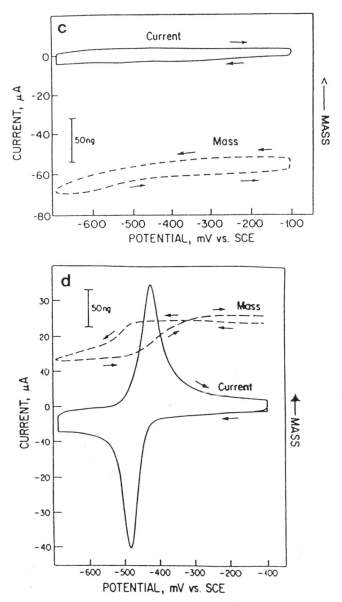

Figure 6 Continued. (c) Cyclic voltammetric scan showing the current and EQCM response obtained at a EQCM silver electrode in 0.1 M borate buffer at pH 9.15; scan rate, 50 mV s^{-1}. The electrode was conditioned at -100 mV (vs. SCE) for 30 s before initiating the potential scan. (d) Cyclic voltammetric scan showing the current and EQCM response obtained at a EQCM silver electrode in 0.1 M borate buffer at pH 9.15 containing $2.7 \times 10^{-5} M$ Pb(II); scan rate, 50 mV s^{-1}; starting potential, -100 mV. (From Ref. 24b.)

a 367-nm-thick Pd film, compared to -185 Hz expected for the mass change due to H absorption (Fig. 7). However, the same experiment on a BT-cut crystal resulted in a very small increase in frequency: $+20$ Hz for a 475-nm-thick Pd film.

$$S = (K^{AT} - K^{BT}) \left(\frac{t_Q^{AT} \Delta f^{AT}}{f_0^{AT}} - \frac{t_Q^{BT} \Delta f^{BT}}{f_0^{BT}} \right) \tag{15}$$

D. Thin-Film Growth

The EQCM provides an extremely useful approach to the in situ study of the nucleation and growth of a wide variety of thin films on electrodes, ranging from oxide films to molecular solids and monolayers. In principle, the ability to measure mass and electrochemical current simultaneously allows determination of chemical stoichiometries, faradaic efficiencies, and reaction kinetics. The EQCM therefore complements current transient methods typically used to study nucleation and growth, providing information that otherwise cannot be obtained readily. The time constant for EQCM measurements (see above) is within the time scale of many electrochemical processes.

The growth of metal oxide films has been investigated at copper, gold, and silver electrodes, the EQCM enabling determination of the stoichiometry of the oxides formed [16]. In addition, these studies revealed frequency responses that were consistent with morphological changes of the electrode surface accompanying oxide formation. This resulted in much larger changes in the frequency during oxide growth that was attributed to surface roughening and subsequent trapping of water in the cavities of the roughened surface (Fig. 8). The surface roughness was retained initially after reduction of the oxide back to the metal, but the frequency gradually returned to its original value, indicating dynamic changes in the morphology of the roughened metal electrodes. In addition to providing valuable mechanistic insight into oxide growth, these studies illustrate the sensitivity of the EQCM to surface-roughness effects.

EQCM studies of the electrochemical deposition of *molecular* films have also been studied. The adsorption of surfactants containing redox-active ferrocene (Fc) groups (referred to as C12 and C14, which designates the length of the alkyl chain) could be induced electrochemically, in that adsorption occurred when these species were in the reduced (Fc°) state (Fig. 9) [31]. Upon oxidation to the oxidized form (Fc$^+$), the surfactants desorbed rapidly. Interestingly, the desorption rate was dependent on the chain length of the surfactant. Whereas short-chain-length surfactants resulted in rapid desorption following electrochemical oxidation, the EQCM frequency changes indicated that long-chain surfactants desorbed much more slowly following oxidation. Mechanistic and thermodynamic information for these films was thereby attainable [32]. The nucleation, growth, and dissolution of thicker films of related species (i.e., di-

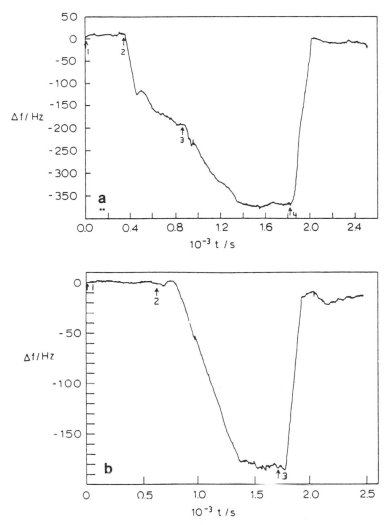

FIGURE 7 (a) Frequency shift vs. time for a 366.6-nm-thick Pd film on an AT-cut quartz resonator in 0.1 M LiOH/H_2O. The numbered arrows indicate the potential applied to the film (1) 0.00 V; (2) E scanned to and held at -1.14 V; (3) E scanned to and held at -1.27 V; (4) E scanned to and held at 0.0 V. The frequency decrease is larger than that expected based on the mass of the cathodic hydrogen absorption process and is attributed to compressive stress in the Pd film upon hydrogen absorption that decreases the frequency. (b) Frequency shift vs. time for a 475-nm-thick Pd film on an AT-cut quartz resonator in 0.1 M LiOH/D_2O. The numbered arrows indicate the potential applied to the film (1) 0.00 V; (2) E scanned to and held at -1.20 V; (3) E scanned to and held at 0.0 V. The frequency decrease is smaller than that expected based on the mass of the cathodic deuterium absorption process and is attributed to compressive stress in the Pd film upon deuterium absorption that increases the frequency. (From Ref. 29a.)

315

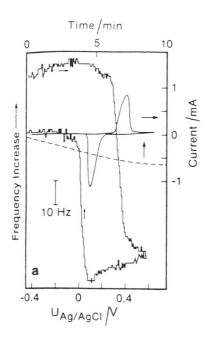

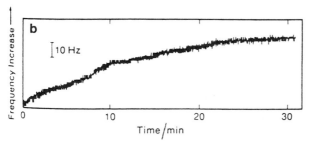

FIGURE 8 (a) Current vs. potential (———) and frequency vs. potential (jagged line) for a silver EQCM electrode during cyclic voltammetry from an initial potential of −0.4 V (vs. SCE). Scan rate: 20 mV s⁻¹, surface area: 0.33 cm². (- - -) Frequency vs. time with the electrode held at −0.4 V at the end of the cycle. Note that the frequency at the end of the cycle is approximately 30 Hz less than at the beginning, and this difference increases to 45 Hz after 10 min at −0.4 V. This indicates a significant change in the morphology of the electrode. (b) Frequency vs. time response for the silver electrode held at −0.4 V after the 10-min period depicted in (a). The frequency increase is attributed to a morphological relaxation of the silver electrode. (From Ref. 16b.)

heptyl viologen) was also examined with the EQCM. Examination of these processes at different potentials provided insight into the nucleation behavior of this system [33].

The EQCM has even been used to examine the adsorption of the proteins human IgG and anti-IgG on Ag electrodes [34]. Upon oxidation of the Ag electrodes, the frequency of the EQCM decreased due to formation of an insoluble layer of the proteins. These effects suggested that initial protein adsorption was related to interactions between oxidized Ag^+ ions on the electrode surface and functional groups on the protein.

The EQCM has also been useful in the investigation of the nucleation and growth of "electronic" materials. For example, the deposition of thin semiconductor films was examined, and the rather complicated mechanism involved in the deposition of thin films of Te was determined by comparison of the EQCM frequency changes and electrochemical charge [35]. Similarly, the electrodeposition of B-doped b-PbO_2 thin films was studied with the EQCM, and details of the film composition and catalytic activity toward oxygen atom transfer reactions were realized [36].

The nucleation and growth of *molecular charge-transfer salts* has also been examined, with an emphasis on optimizing the conditions for electrochemical crystal growth and elucidation of the morphology of the crystals [37]. It was found that by comparison of the electrochemical current and EQCM frequency, the growth of single crystals of $TTFBr_{0.7}$ (TTF = tetrathiafulvalene), an organic metal, could be distinguished from dendritic growth (Fig. 10a). The electrochemical crystallization of $TTFBr_{0.7}$ is accomplished by electrochemical oxidation of TTF in an electrolyte containing n-$Bu_4N^+Br^-$, which results in the formation of black crystals on the electrode surface. It was found that after a potential step to +0.3 V (vs. SCE), a brief current transient was observed with a concommitant decrease in frequency, indicating a correspondence between TTF oxidation and crystal growth. At longer times, however, the rate of mass increase at the electrode surface decreased significantly while the current increased. Since the latter was only consistent with an effectively increasing electrode area and therefore crystal growth, the lack of a corresponding frequency change in this region was attributed to the onset of dendritic growth. When dendritic growth is present, the crystalline mass was not attached rigidly to the vibrating EQCM surface. In a related study, electrochemical crystallization of $TTFBr_{0.7}$ upon continuous stepping between potentials at which growth (+0.13 V) and dissolution (−0.2 V) revealed that the frequency decrease during each growth cycle became larger, indicating a continuous increase in mass deposited in successive cycles (Fig. 10b). A continuous decrease in the frequency of the baseline at the dissolution potential was also observed, indicating that the dissolution was not complete at the negative potential. This behavior was attributed to the persistence of $TTFBr_{0.7}$ growth centers, which serve to facilitate crystal growth in

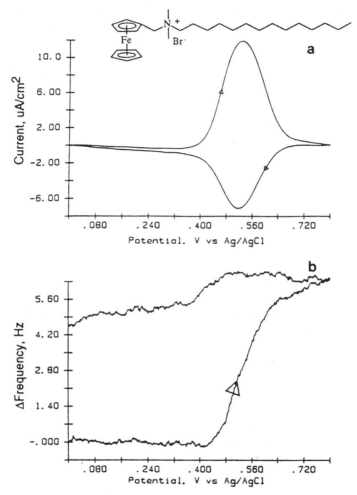

FIGURE 9 (a) Cyclic voltammetric scan for a 7 µ*M* solution of C14 from 0.0 to 0.80 V (vs. SCE) in 1 *M* H₃PO₄; 50 mV s⁻¹. The current was digitally smoothed and corrected for background. (b) The QCM frequency response for (a). (c) Cyclic voltammetric scan for a 22 µ*M* solution of C12 from 0.0 to 0.80 V in 1 *M* H₃PO₄; 50 mV s⁻¹. (d) The QCM frequency response for (c). Higher concentrations of the C12 compound were required to obtain the same coverage as for the C14. Note that the frequency response in (d) is less than that in (b), consistent with the lower molecular weight of C12. (From Ref. 31a.)

each potential cycle because crystal growth on existing nuclei is more favorable energetically than the formation of nuclei. The data were modeled on the basis of cylindrical crystals (the actual crystals are parallelipipeds with one long dimension and two nearly equivalent dimensions, resembling cylinders), and the relative growth kinetics and aspect ratios for these crystals were estimated. In related work, the behavior of thin films of TTF-tetracyanoquinodimethane immobilized on gold electrodes, and their catalytic activity toward mediated and direct electrochemical reactions of small molecules, were examined [38].

The growth and redox chemistry of Prussian blue and related films has been examined by EQCM, with particular attention paid to the degree of sol-

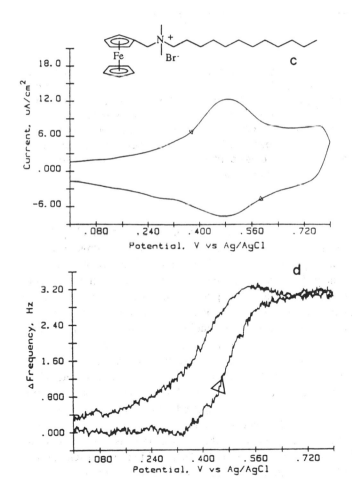

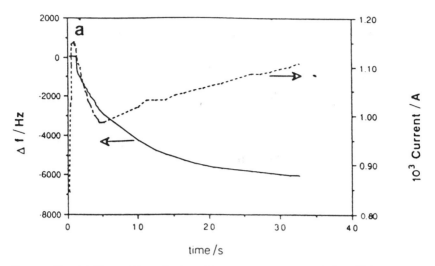

FIGURE 10   (a) Current transient and frequency response during TTFBr$_{0.7}$ electro-crystallization at +0.3 V (vs. SCE). (b) EQCM frequency response during repetitive double potential steps between −0.2 and +0.13 V. The electrode is held at each potential for 5 s. The baseline at −0.2 V steadily decreases due to the accumulation of persistent crystal growth centers, while the frequency excursions during the anodic potential step become increasingly larger because of the increasing surface area of these growth centers. (c) TTFBr$_{0.7}$ model used to describe the response in (b). (From Ref. 37a.)

vation of these films and the ion transport during electrochemical cycling. These studies would be expected to have an impact on the use of these films in sensors and electrochromic displays. The initial EQCM study of Prussian blue films found that the frequency decrease during electrochemical deposition was consistent with a high degree of hydration [39]. Subsequent EQCM studies of these films revealed that cation transport accompanied changes in the redox state during potential cycling, although at low-pH conditions proton transport was also involved. Ion and solvent transport in related nickel ferrocyanide films during potential cycling in 0.1 M CsCl solutions has also been examined. The results were consistent with transport of Cs$^+$ ions upon change in the redox state of the film [40]. The mass change calculated from equation (2), which assumes rigid-layer behavior, suggested that H$_2$O was expelled from the film as Cs$^+$ was incorporated during reduction. This was verified by a clever experiment in which the process was studied in H$_2$O and D$_2$O; the mass change associated with the solvent was found to increase by 10% in D$_2$O. This verified the participation of solvent and provided corroborative evidence of rigid-layer behavior.

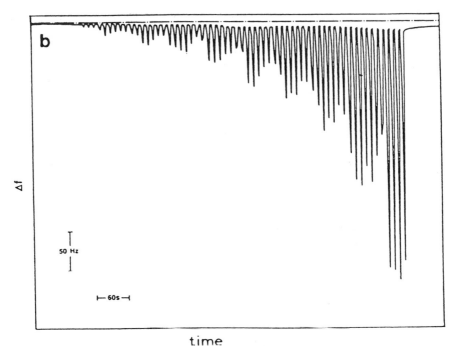

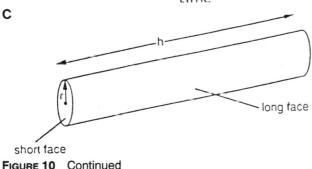

FIGURE 10 Continued

The prussian blue and nickel ferrocyanide films consist of a low-density pseudocrystalline lattice in which the metal atoms are organized by cyanide ligand bridges between octahedral metal centers. As a result, metal ion intercalation into the interstices in the lattice is affected by the size and hydration of the cation. The different amount of work required for intercalation of the different ions is manifested as a dependence of the formal potential of the films on the identity and concentration of the metal ion in the supporting electrolyte. Analysis of EQCM data acquired during redox cycling of Prussian blue films in propylene carbonate containing different relative amounts of $NaClO_4$ and $LiClO_4$ revealed that intercalation of Na^+ was favored over Li^+ by a factor of 15 [41]. This capability was demonstrated to be useful for the analytical identification of metal ions in a flow injection mode [42]. Simultaneous electrochemical and EQCM measurements were performed in which the Prussian blue film was held at a potential known to be selective for a given cation, and the charge and mass change associated with ion transport were measured. Comparison of the charge and mass change enabled determination of the molar mass, and therefore identification, of the cation.

The electrodeposition of polymer films has also been investigated with the EQCM, in particular the electrodeposition of poly(vinylferrocene) (PVF) films [43]. These studies indicated that in oxidative electrodeposition of PVF in CH_2Cl_2, initially more polymer was deposited than expected based on the deposition of one monomer unit per electron. It was suggested that during the initial stages, partially oxidized polymer deposited (i.e., large faradaic efficiencies), possibly with solvent or electrolyte trapped in the polymer. However, the results indicated that the total frequency change was smaller than that expected for the amount of charge passed during the deposition. This apparent loss of mass sensitivity suggests, as with some of the examples above, that the deposited film is not ideally rigid.

E. Electrochemically Active Polymer Films

Without a doubt, the EQCM has found its most extensive use in the investigation of electrochemically active polymer films, including redox and conducting polymers. Indeed, to a large extent it was the first report of an EQCM study of polypyrrole films by Kaufman et al. that heightened interest in the EQCM method [44]. The benefits of the EQCM were immediately obvious, as it became clear that the identity of the counterions exchanging between the films and electrolyte could be determined. In addition, the amount of solvent accompanying ion exchange could be determined by calculating the mass in excess of that expected from the electrochemical charge. Since these initial studies, numerous polymer films have been investigated with the EQCM, and many mechanistic and thermodynamic insights into their behavior have been realized. The practitioner should be cautious when using the EQCM with polymer films, as changes in the mechanical properties of the films are not uncommon when ion population and solvent swelling are involved. The corresponding changes in viscoelasticity can result in dramatic frequency changes that are unrelated to mass changes. In many cases, this warrants experimental verification that rigid-layer behavior is present over the thickness and composition ranges examined. This can be accomplished by performing experiments with different polymer thicknesses over a reasonable range, and by using impedance analysis to elucidate the contribution from changes in mechanical properties. If these factors are taken into account, the EQCM can be an especially valuable tool in examining electrochemically active polymer films.

1. Redox Polymer Films

It is reasonable to claim that poly(vinylferrocene) (PVF) is ubiquitous in the subdiscipline of polymer-modified electrodes. Similarly, it has been among polymer films examined most extensively with the EQCM. One of the initial investigations involved measurement of the frequency changes associated with ion transport required for electroneutrality and the amount of solvent accompanying ion transport in PVF films [45]. Based on a frequency decrease that accompanied oxidation of the PVF film, it was determined that ClO_4^- and PF_6^- inserted without accompanying solvent, with one equivalent of anions inserted for each equivalent of electrons (Fig. 11). The process was reversible, as evidenced by the frequency increase upon electrochemical reduction. Electrochemical oxidation of PVF films in the presence of other counterions, however, occurred with varying amounts of solvent incorporation, with the amount of incorporated water per ion decreasing in the order $Cl^- > IO_3^- \gg BrO_3^-$ (5) $> ClO_3^-$ (1) $\approx NO_3^-$ (1) $>$ $CH_3—C_6H_4—SO_3^-$(0.5) [46]. Indeed, cross-linking of the PVF film was necessary to prevent its dissolution upon oxidation. Unfortunately, determination of the amount of incorporated solvent is not feasible in the case of Cl^- or IO_3^- because of the dramatic changes in viscoelasticity that accompany the large degree of swelling when these ions are present. On the other hand, solvent incorporation for the other ions could be measured reliably (the amounts of solvent incorporated per counterion are indicated above within the parentheses).

Recently, the EQCM has been employed to investigate the transport in redox polymer films such as PVF and poly(thionine) of mobile neutral species such as water, particularly with regard to thermodynamic changes in solvent activity in the polymer film and transport kinetics [47,48]. These studies indicate that the number of solvent molecules per counterion need not be integral, and that it is feasible that neutral ion pairs are also involved in the transport. Comparison of EQCM frequency changes at different scan rates also revealed that mass changes and electrochemical charge changes do not always occur simultaneously [49]. That is, counterion motion to maintain electroneutrality during redox changes must always be established, but global equilibrium may lag behind. In particular, these studies revealed that electroneutrality would be achieved initially during redox by using counterions already present in the film, followed by transport of ions from the solution. Similar behavior was observed for redox-active poly(nitrostyrene) films [50]. The behavior has been attributed to potential gradients within the film that affect ion transport, and to the low dielectric constant of the films [51]. In the case of poly(thionine) experiments, comparison of current transients following a series of potential steps suggested that transport of different species occurs on different time scales (Fig. 12) [52].

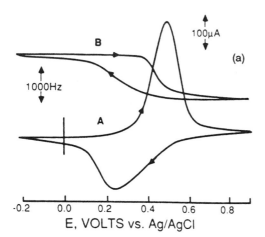

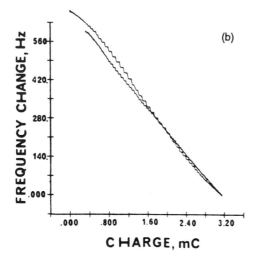

FIGURE 11   (a) A, Cyclic voltammogram of PVF on a gold EQCM electrode in 0.1 *M* KPF$_6$. Scan rate: 10 mV s^{-1}. B, EQCM frequency response obtained simultaneously with (A). (b) Plot of frequency vs. charge for a scan from 0.0 to 0.60 V and back for a PVF film in 0.1 *M* NaClO$_4$ + 0.1 *M* HClO$_4$. Scan rate: 25 mV s^{-1}. The linearity suggests ideal behavior and the slope gives the molar mass of the transport species. (From Ref. 45.)

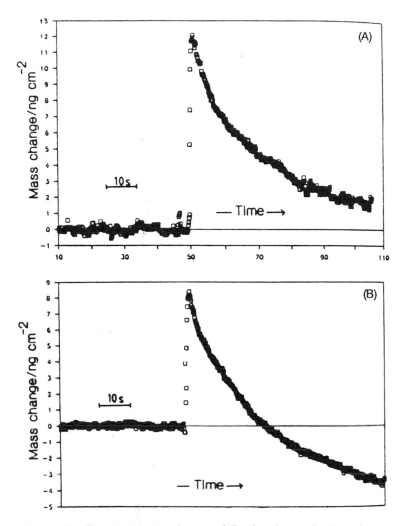

FIGURE 12 Transient mass changes following the application of a potential step from −0.1 to +0.5 V (vs. SCE) to a poly(thionine) film in $HClO_4$ solution: (A) solvent is H_2O, pH 1.6; (B) solvent is D_2O, pD 2.1. The potential step converts of fully reduced film to a fully oxidized film at long times. The pH and pD values were adjusted to achieve near-zero net mass change at long times after switching. Mass increases upward. The same polymer film was used in H_2O and D_2O. Note that the initial mass change is larger than that after global equilibrium is achieved at longer times. (From Ref. 52.)

It was surmised that proton transport necessary to achieve electroneutrality was rapid, but that global equilibrium occurred more slowly by transport of the neutral ion pair $H_3O^+ClO_4^-$ and H_2O. These conclusions were supported by experiments performed in D_2O, which exhibited trends consistent with the difference in mass of H and D. These results have important consequences in the design and synthesis of polymer films in applications where charging and discharging rates are critical, for example, in sensors or energy storage.

EQCM studies revealed that ion transport into PVF films upon oxidation was slower than their transport out of the film during reduction [53]. This is presumably because of the change in density accompanying the swelling of the oxidized film. It was estimated that in the first oxidation cycle of electrochemically deposited PVF films, water uptake began only after the film was approximately 40% oxidized, and that 50% of the water was retained after subsequent reduction [54]. Water transport in the following cycles was reversible, apparently with some water always retained in the polymer film. Transport in PVF films can also be severely affected by the nature of the counterion. For example, EQCM studies indicated that ion exchange of ferro- or ferricyanide into PVF during electrochemical cycling resulted in irreversible incorporation of $Fe(CN)_6^{3-/4-}$ in the film [55]. This process led to a slow decrease in the electroactivity of the film, presumably due to an electrostatic "cross-linking" by the multiply charged ion that inhibited further transport of $Fe(CN)_6^{3-/4-}$. It should be noted that variability in these results may be expected in different laboratories because of differences in film preparation.

The EQCM can also be used to examine the kinetics of chemical reactions of solution species with redox polymer films. The chemical oxidation of a PVF film by KI_3 was monitored by measurement of the frequency decrease associated with insertion of the I_3^- counterion following oxidation (Fig. 13). These experiments are possible because the PVF film can be held in its reduced form prior to the chemical reaction *in the presence of the oxidant* by holding the potential of the electrode negative of $E°$ (PVF/PVF$^+$). This established the initial conditions for this measurement. The results yielded the stoichiometry of the reaction (1 equiv. I_3^- per PVF$^+$ formed) as well as the pseudo-first-order rate constant for the reaction.

2. Conducting Polymers

As mentioned above, interest in the EQCM method heightened considerably after it was reported in the investigation of polypyrrole films [44]. Since that time, numerous reports have appeared describing attempts to deconvolute the typically complicated behavior involved in the preparation of conducting polymer films and their subsequent doping behavior. The value of the EQCM was immediately evident in those initial polypyrrole studies, in which it was determined that reduction

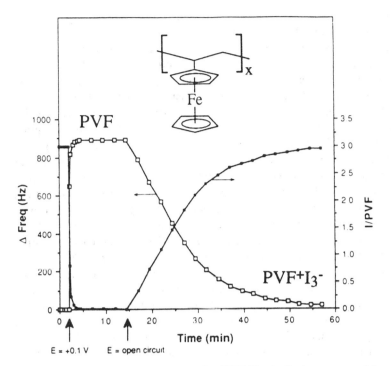

FIGURE 13 Frequency response of a PVF film in the presence of 5 m*M* KI$_3$ and 1.0 *M* KNO$_3$. The right-hand ordinate refers to the number of equivalents of I incorporated into the film normalized to the amount of PVF coverage on the piezo-electrically active area (0.28 cm^2). [Γ_{PVF}]$_0$ = 4.1 × 10^{-8} equiv. cm^{-2}. (From Ref. 19e.)

of electrochemically (by oxidation of pyrrole in LiClO$_4$) prepared polypyrrole films resulted in Li$^+$ insertion to maintain electroneutrality rather than expulsion of ClO$_4^-$. This behavior was attributed to strong ionic pairing between cationic sites on the polymer and ClO$_4^-$. It was also discovered that subsequent cycling of these films in the presence of *n*-Bu$_4$N$^+$*p*-CH$_3$C$_6$H$_4$—SO$_3^-$ resulted in frequency changes consistent with anion insertion during the oxidative doping step, with its expulsion during reduction. This behavior clearly revealed that the nature of ion transport in conducting polymers was strongly dependent on the nature of the counterions. The impact of this behavior on the design of energy storage systems with regard to weight, power density, and energy density prompted several other EQCM investigations of conducting polymers, including poly(pyrrole), poly(aniline), and poly(thiophene). Decreasing the mass of the transport species would lead to higher energy densities, whereas higher power densities would be realized by faster transport rates.

These investigations are exemplified by studies of poly(pyrrole) film growth in which it was concluded that film deposition involved the initial formation of soluble oligomers, which upon further polymerization precipitated on the electrode [56]. EQCM frequency changes that corresponded to the mass of deposited polypyrrole revealed a second-order dependence of the electropolymerization rate on pyrrole concentration, suggesting the bimolecular coupling of oxidatively formed radical cations was involved in the rate-determining step. Subsequent EQCM studies of poly(pyrrole) revealed more details of the ion transport and its dependence on film preparation. Electropolymerization of pyrrole in the presence of large, polymeric anions such as poly(4-styrenesulfonate) [57] and poly(vinylsulfonate) resulted in films whose transport properties differed significantly from films prepared in the presence of more conventional anions. In these studies conventionally prepared films exhibited frequency changes during doping–undoping cycles that indicated anion transport for ClO_4^-, BF_4^-, or PF_6^-. However, in electrolytes containing poly(4-styrenesulfonate), mixed transport of both poly(4-styrenesulfonate) anions and cations was observed in conventionally prepared films. If the films prepared in poly(4-styrenesulfonate) were used, transport during doping–undoping was completely dominated by the cation of the supporting electrolyte. Cation transport also dominated in the doping–undoping cycles of self-doped conducting copolymers of pyrrole and 3-(pyrrole-1-yl)propanesulfonate (Fig. 14).

Poly(thiophene)-based polymers, a conducting polymer closely related to poly(pyrrole), has also been examined with the EQCM [58]. These studies revealed the doping levels of poly(3-methylthiophene), as well as a nonlinear dependence of the charge on film thickness, the latter being inferred from the frequency change. Subsequent studies provided a thermodynamic model based on the EQCM response for transport of the ions, neutral ion pairs, and solvent based on the thermodynamic activity of the polymer film, similar to the model mentioned above. The transport of ions during doping–undoping cycles of poly(aniline), a conducting polymer film with doping behavior that is quite different from that of the aforementioned examples, has also been investigated with the EQCM [59]. These studies provided the faradaic efficiencies during oxidative film growth by comparison of the coulometric charge and the frequency decrease. Interestingly, it was discovered that the faradaic efficiencies for films prepared by cycling the potential were higher (ca. 40%) than those prepared at constant potential (ca. 10%). More important, mass changes during doping–undoping cycles revealed the extent of protonation of the polymer in both the doped and undoped states. Based on the EQCM results, it was concluded that each aniline ring contributed *one* electron to the conduction band of the polymer, in agreement with a proposed model [60]. Subsequent EQCM studies of poly(aniline) in nonaqueous solvents revealed that protons on the imine nitrogens of the polymer chains were retained during doping–undoping cycles, and that electroneutrality was maintained by transport of anions from the supporting electrolyte [61].

The format of the EQCM is amenable to combination with other in situ techniques. This was illustrated for the scanning electrochemical microscopy–EQCM experiments described earlier in this chapter. This capability is further demonstrated by ellipsometric studies of the nucleation and growth of poly(aniline) films performed in conjunction with the EQCM [62], which allowed complete characterization of the polymer films (Fig. 15). Whereas the EQCM provided the total mass of the deposit, ellipsometry provided the thickness. Accordingly, the combination of these techniques allows determination of the *density*, which is not realized by either technique alone. The major conclusions of these studies were that the kinetics for growth of the poly(aniline) films depended on the electrochemical conditions, and that self-assembled monolayers of aniline derivatives on the electrode surface promoted nucleation and films with higher densities.

IV. SUMMARY

The increasing use of the EQCM and the examples described in this chapter demonstrate the power of this method in elucidating the fundamental interfacial processes occurring at electrode surfaces. The rather simple concept and low cost of the equipment necessary to perform EQCM experiments will probably lead to further expansion into electrochemical laboratories as a routine and complementary tool for investigating electrode processes. However, many details of EQCM are still poorly understood, particularly those that involve the liquid phase. The contributions of interfacial slip, stress, surface roughness, viscosity, viscoelasticity, and influence of the liquid phase on propagation of the acoustic energy into unelectroded regions of the quartz crystal still require examination. The new practitioner (and even experienced ones) should expend some effort to grasp the basic principles of EQCM operation to avoid pitfalls and misinterpretations that can occur if these effects are ignored. Generally, verification of ideal behavior is possible through fairly simple experiments as discussed in this chapter or through more sophisticated impedance analysis techniques. It is anticipated that as the EQCM becomes more widely appreciated and used, many of these effects will be better understood. Nevertheless, the EQCM is well within the reach of most electrochemical laboratories, and properly used, can provide details of electrode processes that were previously unattainable.

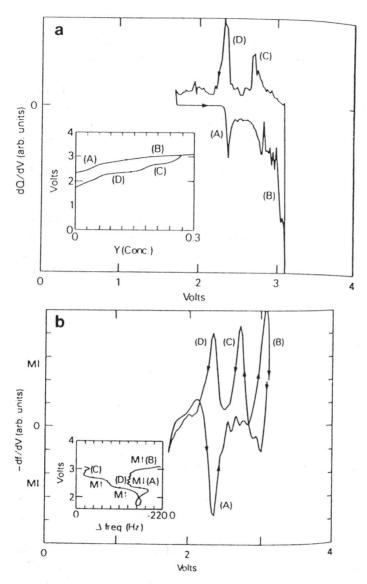

FIGURE 14 (a) Cyclic voltammetry data for a polypyrrole film immobilized on a QCM in 1.0 *M* LiClO$_4$/THF, presented as dQ/dV vs. V (V vs. Li). Oxidation occurs at peaks A and B while reduction occurs at C and D (inset shows V vs. Q). (b) EQCM data presented as *df/dV* vs. *V* showing differential mass changes associated with the oxidation and reduction features in (a). The up arrow (↑) indicates increasing mass, and the down arrow (↓) indicates decreasing mass. (From Ref. 44.)

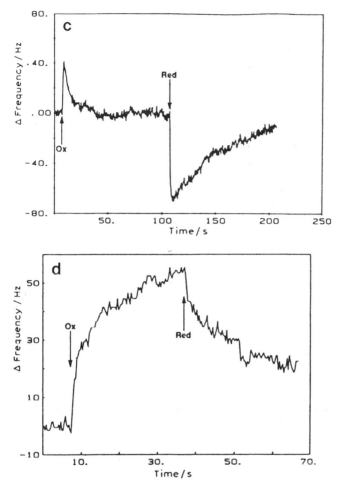

FIGURE 14 Continued. (c) EQCM response upon oxidation of a polypyrrole film followed by reduction in tetraethylammonium tosylate electrolyte. The polypyrrole film was prepared by electrochemical oxidation of pyrrole. (d) EQCM response upon oxidation of a polypyrrole-co-[3-(pyrrol-1-yl)propanesulfonate] film followed by reduction in tetraethylammonium tosylate electrolyte. In both cases the increase in frequency upon oxidation and decrease in frequency upon reduction are associated with transport of cations. (From Ref. 57d.)

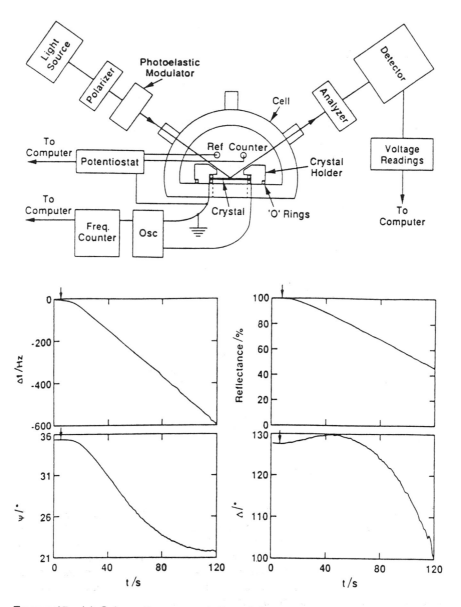

FIGURE 15 (a) Schematic representation of the experimental apparatus for simultaneous ellipsometry–EQCM measurements. (b) Optical and EQCM data for poly(aniline) growth on a Pt electrode in 1 *M* aniline containing 2 *M* HCl at 0.7 V (vs. SCE). Wavelength = 550 nm, angle of incidence = 65°. Δ and ψ are the ellipsometric parameters derived from the 50 kHz, 100 kHz, and dc components of the signal generated at the detector by the reflected beam. (From Ref. 62a.)

REFERENCES

1. H. Abruña, ed., *In Situ Studies of Electrochemical Interfaces*, VCH Chemical, New York, 1991.
2. M. D. Ward and D. A. Buttry, In situ interfacial mass detection with piezoelectric transducers, *Science, 249*: 1000 (1990).
3. D. A. Buttry and M. D. Ward, Measurement of interfacial processes at electrode surfaces with the electrochemical quartz crystal microbalance, *Chem. Rev., 92*: 1355 (1992).
4. D. A. Buttry, Applications of the quartz crystal microbalance to electrochemistry, in *Electroanalytical Chemistry*, Vol. 17 (A. J. Bard, ed.), Marcel Dekker, New York, 1991, p. 2.
5. P. Curie and J. Curie, Crystal physics: development of polar electricity in hemihedral crystals with inclined faces, *C.R. Acad. Sci., 91*: 294 (1880).
6. C. S. Lu and O. Lewis, Investigation of film-thickness determination by oscillating quartz resonators with large mass load, *J. Appl. Phys., 43*: 4385 (1972).
7. Z. Lin, C. Yip, I. S. Joseph, and M. D. Ward, Operation of an ultrasensitive 30 MHz quartz crystal microbalance in liquids, *Anal. Chem., 65*: 1646 (1993).
8. V. G. Bottom, *Introduction to Quartz Crystal Unit Design*, Van Nostrand Reinhold, New York, 1982.
9. (a) W. P. Mason, *Electromechanical Transducers and Wave Filters*, Van Nostrand, New York, 1948. (b) W. G. Cady, *Piezoelectricity*, Dover, New York, 1964.
10. S. J. Martin, V. E. Granstaff, and G. C. Frye, Characterization of a quartz microbalance with simultaneous mass and liquid loading, *Anal. Chem., 63*: 2272 (1991).
11. K. K. Kanazawa and J. G. Gordon III, Frequency of a quartz microbalance in contact with liquid sensors to the liquid phase by interfacial viscosity, *Anal. Chem., 57*: 1770 (1985).
12. (a) L. V. Rajakovic, B. A. Cavic-Vlasak, V. Ghaemmaghami, K. M. R. Kallury, A. L. Kipling, and M. Thompson, Mediation of acoustic energy transmission from acoustic wave, *Anal. Chem., 63*: 615 (1991). (b) A. L. Kipling and M. Thompson, Network analysis method applied to liquid phase acoustic sensors, *Anal. Chem., 62*: 1514 (1990). (c) W. Stockel and R. Schumacher, In situ microweighing at the metal/electrolyte junction, *Ber. Bunsenges. Phys. Chem., 345*: 91 (1987).
13. (a) C. E. Reed, K. K. Kanazawa, and J. H. Kaufman, Physical description of a viscoelastically loaded AT-cut quartz resonator, *J. Appl. Phys., 68*: 1993 (1990). (b) J. Wang, L. M. Frostman, and M. D. Ward, Self-assembled thio monolayers with carboxylic acid functionality: measuring pH-dependent phase transitions with the quartz crystal microbalance, *J. Phys. Chem., 96* (1992). (c) J. Wang, M. D. Ward, R. C. Ebersole, and R. A. Foss, *Anal. Chem., 65*: 2553 (1993).
14. (a) G. Sauerbrey, The use of a quartz crystal oscillator for weighing thin layers and for microweighing applications, *Z. Phys., 155*: 206 (1959). (b) D. M. Ullevig, J. F. Evans, and M. G. Albrecht, Effects of stressed materials on the radial sensitivity function of a quartz crystal microbalance, *Anal. Chem., 54*: 2341 (1982). (c) L. Koga, Y. Tsuzuki, S. N. Witt, and A. L. Bennet, Measurements of the vibrations of quartz plates, *Proc. Annual Frequency Control Symp., 14*: 53 (1960). (d) H. Fukuyo, A. Yokoyama, N. Ooura, and S. Nonaka, Vibration of biconvex circular AT-cut

plate, *Bull. Tokyo Inst. Technol., 72*: 1 (1965). (e) K. S. van Dyke, Strain patterns in thickness-shear quartz resonators, *Proc. Annual Frequency Control Symp., 111*: 1 (1957).

15. (a) A. C. Hillier and M. D. Ward, Scanning electrochemical mass sensitivity mapping of the quartz crystal microbalance in liquid media, *Anal. Chem., 64*: 2539 (1992). (b) M. D. Ward and E. J. Delawski, Radial sensitivity of the quartz crystal microbalance in liquid media, *Anal. Chem., 63*: 886 (1991). (c) C. Gabrielli, M. Keddam, and R. Torresi, Calibration of the electrochemical quartz crystal microbalance, *J. Electrochem. Soc., 138*: 2657 (1991). (d) B. A. Martin and H. E. Hager, Velocity profile on quartz crystals oscillating in liquids, *J. Appl. Phys., 65*: 2630 (1989).

16. (a) R. Schumacher, G. Borges, and K. K. Kanazawa, *Surface Sci., 163*: L621 (1985). (b) R. Schumacher, J. Gordon, and O. Melroy, Observation of morphological relaxation of copper and silver electrodes in solution using a quartz crystal microbalance, *J. Electroanal. Chem., 216*: 127 (1987). (c) A. Muller, M. Wicker, R. Schumacher, and R. N. Schindler, *Ber. Bunsenges. Phys. Chem., 92*: 1395 (1988).

17. K. E. Heusler, A. Grzegorzewski, L. Jackel, and J. Pietrucha, Measurement of a mass and surface stress at one electrode of a quartz oscillator, *Ber. Bunsenges. Phys. Chem., 92*: 1395 (1218).

18. E. P. EerNisse, in *Methods and Phenomena* (C. Lu and A. Czanderna, eds.), Vol. 4, Elsevier Sequoia, Lausanne, 1984.

19. (a) T. Nomura and M. Iijima, Electrolytic determination of nanomolar concentrations of silver in solution with a piezoelectric quartz crystal, *Anal. Chim. Acta, 131*: 97 (1981). (b) S. Bruckenstein and M. Shay, Experimental aspects of use of the quartz crystal microbalance in solution, *Electrochim. Acta, 30*: 1295 (1985). (c) D. A. Buttry, in *Electroanalytical Chemistry*, Vol. 17 (A. J. Bard, ed.), Marcel Dekker, New York, 1991, p. 1. (d) R. Schumacher, The quartz microbalance: a novel approach to the in-situ investigation of interfacial phenomena at the solid/liquid junction, *Angew. Chem. Int. Ed. Engl., 29*: 329 (1990). (e) M. D. Ward, Investigation of open circuit reactions of polymer films using the quartz crystal microbalance: reactions of polyvinylferrocene films, *J. Phys. Chem., 92*: 2049 (1988).

20. O. Melroy, K. K. Kanazawa, J. G. Gordon, and D. A. Buttry, Direct determination of the mass of an underpotentially deposited monolayer of lead on gold, *Langmuir, 2*: 697 (1987).

21. S. Bruckenstein and M. Shay, Experimental aspects of use of the quartz crystal microbalance in solution, *Electrochim. Acta, 30*: 1295 (1985).

22. H. E. Hager, R. D. Ruedisueli, and M. E. Buehler, The use of piezoelectric crystals as electrode substrates in iron corrosion studies: the real-time, in situ determination of dissolution and film formation reaction rates, *Corrosion, 42*: 345 (1986).

23. M. R. Deakin and O. Melroy, Underpotential metal deposition on gold, monitored in situ with a quartz crystal microbalance, *J. Electroanal. Chem., 239*: 321 (1988).

24. (a) M. Hepel and S. Bruckenstein, Tracking anion expulsion during underpotential of lead at silver using the quartz microbalance, *Electrochim. Acta, 34*: 1499 (1989). (b) M. Hepel, K. Kanige and S. Bruckenstein, In situ underpotential deposition study of lead on silver using the electrochemical quartz crystal microbalance: direct

evidence for lead(II) adsorption before spontaneous charge transfer, *J. Electroanal. Chem.*, *266*: 409 (1989). (c) M. Hepel, K. Kanige, and S. Bruckenstein, Expulsion of borate ions from the silver/solution interfacial region during underpotential deposition discharge of Pb(II) in borate buffers, *Langmuir*, *6*: 1063 (1990).

25. M. Deakin, T. Li, and O. Melroy, A study of the electrosorption of bromide and iodide ions on gold using the quartz crystal microbalance, *J. Electroanal. Chem.*, *243*: 343 (1988).

26. M. Benje, M. Eiermann, U. Pitterman, and K. G. Weil, An improved quartz crystal microbalance: applications to the electrocrystallization and -dissolution of nickel, *Ber. Bunsenges. Phys. Chem.*, *90*: 435 (1986).

27. M. Benje, U. Hofmann, U. Pitterman, and K. G. Weil, Anodic dissolution of thin nickel-phosphorus films, *Ber. Bunsenges. Phys. Chem.*, *92*: 1257 (1988).

28. R. Schumacher, A. Muller, and W. Stockel, An in situ study of the mechanism of the electrochemical dissolution of copper in oxygenated sulphuric acid: an application of the quartz microbalance, *J. Electroanal. Chem.*, *219*: 311 (1987).

29. (a) G. T. Cheek and W. E. O'Grady, Measurement of hydrogen uptake by palladium using a quartz crystal microbalance, *J. Electroanal. Chem.*, *277*: 341 (1990). (b) L. Grasjo and M. Seo, Measurement of absorption of hydrogen and deuterium into palladium during electrolysis by a quartz crystal microbalance, *J. Electroanal. Chem.*, *296*: 233 (1990). (c) N. Yamamoto, T. Ohsaka, T. Terashima, and N. Oyama, In situ electrochemical quartz crystal microbalance studies of water electrolysis at a palladium cathode in acidic aqueous media, *J. Electroanal. Chem.*, *274*: 313 (1989).

30. (a) E. P. EerNisse, Simultaneous thin film stress and mass-change measurements using quartz resonators, *J. Appl. Phys.*, *43*: 1330 (1972). (b) E. P. EerNisse, Extension of the double resonator technique, *J. Appl. Phys.*, *44*: 4482 (1973).

31. (a) J. J. Donahue and D. A. Buttry, Adsorption and micellization influence the electrochemistry of redox surfactants derived from ferrocene, *Langmuir*, *5*: 671 (1989). (b) I. Nordyke and D. A. Buttry, Redox surfactants are chemical probes of electrode surface functionalization derived from disulfide immobilization on gold, *Langmuir*, *7*: 380 (1991).

32. H. C. De Long, J. J. Donahue, and D. A. Buttry, Ionic interactions in electroactive self-assembled monolayers of ferrocene species, *Langmuir*, *7*: 2196 (1991).

33. G. S. Ostrom and D. A. Buttry, Quartz crystal microbalance studies of deposition and dissolution mechanisms of electrochromic films of diheptylviologen bromide, *J. Electroanal. Chem.*, *256*: 411 (1988).

34. E. S. Grabbe, R. P. Buck, and O. R. Melroy, Cyclic voltammetry and quartz microbalance electrogravimetry of IgG and anti-IgG reactions on silver, *J. Electroanal. Chem.*, *223*: 67 (1987).

35. E. Mori, C. K. Baker, J. R. Reynolds, and K. Rajeshwar, Aqueous electrochemistry of tellurium at glassy carbon and gold: a combined voltammetry-oscillating quartz crystal microgravimetry study, *J. Electroanal. Chem.*, *252*: 441 (1988).

36. L. A. Larew, J. S. Gordon, Y.-L. Hsiao, D. C. Johnson, and D. A. Buttry, Application of an electrochemical quartz crystal microbalance to a study of pure and bismuth-doped *b*-lead dioxide film electrodes, *J. Electrochem. Soc.*, *137*: 3071 (1990).

37. (a) M. D. Ward, Probing electrocrystallization of charge transfer salts with the quartz crystal microbalance, *J. Electroanal. Chem.*, *273*: 79 (1989). (b) M. D. Ward, Electrocrystallization of low dimensional solids: directed selectivity and investigation of crystal growth with the quartz crystal microbalance, *Synth. Metals*, *27*: B211 (1988).

38. M. S. Freund, A. Brajter-Toth, and M. D. Ward, Electrochemical and quartz crystal microbalance evidence for mediation and direct electrochemical reactions of small molecules at tetrathiafulvalene-tetracyanoquinodimethane (TTF-TCNQ) electrodes, *J. Electroanal. Chem.*, *289*: 127 (1990).

39. B. J. Feldman and O. R. Melroy, Ion flux during electrochemical charging of Prussian blue films, *J. Electroanal. Chem.*, *234*: 213 (1987).

40. S. J. Lasky and D. A. Buttry, Mass measurements using isotopically labeled solvents reveal the extent of solvent transport during redox in thin films on electrodes, *J. Am. Chem. Soc.*, *110*: 6258 (1988).

41. K. Aoki, T. Miyamoto, and Y. Ohsawa, The determination of the selectivity coefficient of Na^+ versus Li^+ on Prussian blue thin film in propylene carbonate by means of a quartz crystal microbalance, *Bull. Chem. Soc. Jpn.*, *62*: 1658 (1989).

42. M. R. Deakin and H. Byrd, Prussian blue coated quartz crystal microbalance as a detector for electroinactive cations in aqueous solution, *Anal. Chem.*, *61*: 290 (1989).

43. A. R. Hillman, D. C. Loveday, and S. Bruckenstein, Electrochemical quartz crystal microbalance monitoring of poly(vinylferrocene) electrodeposition, *Langmuir*, *7*: 191 (1991).

44. J. N. Kaufman, K. K. Kanazawa, and G. B. Street, Gravimetric electrochemical voltage spectroscopy: in situ mass measurements during electrochemical doping of the conducting of the conducting polymer polypyrrole, *Phys. Rev. Lett.*, *53*: 2461 (1984).

45. P. T. Varineau and D. A. Buttry, Applications of the quartz crystal microbalance to electrochemistry: measurement of ion and solvent populations in thin films of poly(vinylferrocene) as functions of redox state, *J. Phys. Chem.*, *91*: 1292 (1987).

46. P. T. Varineau, Ph.D. thesis, University of Wyoming, 1989.

47. S. Bruckenstein and A. R. Hillman, Consequences of thermodynamic restraints on solvent and ion transfer during redox switching of electroactive polymers, *J. Phys. Chem.*, *92*: 4837 (1988).

48. (a) A. R. Hillman, D. C. Loveday, M. J. Swann, S. Bruckenstein, and C. P. Wilde, Transport of neutral species in electroactive polymer films, *J. Chem. Soc. Faraday Trans.*, *87*: 2047 (1991). (b) A. R. Hillman, D. C. Loveday, and S. Bruckenstein, A general approach to the interpretation of electrochemical quartz crystal microbalance data. II. Chronoamperometry: temporal resolution of mobile species transport in polyvinylferrocene films, *J. Electroanal. Chem.*, *300*: 67 (1991).

49. (a) S. Bruckenstein, C. P. Wilde, M. Shay, A. R. Hillman, and D. C. Loveday, Observation of kinetic effects during interfacial transfer at redox polymer films using the quartz microbalance, *J. Electroanal. Chem.*, *258*: 457 (1989). (b) A. R. Hillman, D. C. Loveday, S. Bruckenstein, and C. P. Wilde, Criteria governing ion and solvent transport rates in electroactive polymers: the existence of kinetic permselectivity, *J. Chem. Soc. Faraday Trans.*, *86*: 437 (1990).

50. R. Borjas and D. A. Buttry, Solvent swelling influences the electrochemical behavior and stability of thin films of nitrated poly(styrene), *J. Electroanal. Chem.*, *280*: 73 (1990).

51. S. Bruckenstein, C. P. Wilde, M. Shay, and A. R. Hillman, Experimental observations on transport phenomena accompanying redox switching in polythionine films immersed in strong acid solutions, *J. Phys. Chem.*, *94*: 787 (1990). (b) A. R. Hillman, M. J. Swann, and S. Bruckenstein, General approach to the interpretation of electrochemical quartz crystal microbalance data. 1. Cyclic voltammetry: kinetic subtleties in the electrochemical doping of polybithiophene films, *J. Phys. Chem.*, *95*: 3271 (1991). (c) S. Bruckenstein, C. P. Wilde, and A. R. Hillman, Transport phenomena accompanying redox switching in polythionine films immersed in aqueous acetic acid solutions, *J. Phys. Chem.*, *94*: 6458 (1990).

52. S. Bruckenstein, A. R. Hillman, and M. J. Swann, Transient mass excursions during switching of polythionine films between equi-mass equilibrium redox states, *J. Electrochem. Soc.*, *137*: 1323 (1990).

53. A. R. Hillman, D. C. Loveday, M. J. Swann, R. M. Eales, A. Hamnett, S. J. Higgins, S. Bruckenstein, and C. P. Wilde, Charge transport in electroactive polymer films, *Faraday Discussions Chem. Soc.*, *88*: 151 (1989).

54. A. R. Hillman, N. A. Hughes, and S. Bruckenstein, Solvation phenomena in polyvinylferrocene films: effect of history and redox state, *J. Electrochem. Soc.*, *139*: 74 (1992).

55. M. D. Ward, Ion exchange of ferro(ferricyanide) into polyvinylferrocene films, *J. Electrochem. Soc.*, *135*: 2747 (1988).

56. C. K. Baker and J. R. Reynolds, Use of the quartz microbalance in the study of polyheterocycle electrosynthesis, *Synth. Metals*, *28*: C21 (1989). (b) C. K. Baker and J. R. Reynolds, A quartz microbalance study of the electrosynthesis of poly pyrrole, *J. Electroanal. Chem.*, *251*: 307 (1988).

57. (a) C. K. Baker, Y.-J. Qiu, and J. R. Reynolds, Electrochemically induced charge and mass transport in polypyrrole/poly(styrenesulfonate) molecular composites, *J. Phys. Chem.*, *95*: 4446 (1991). (b) K. Naoi, M. M. Lien, and W. H. Smyrl, Quartz crystal microbalance study: ionic motion across conducting polymers, *J. Electrochem. Soc.*, *138*: 440 (1991). (c) K. Naoi, M. M. Lien, and W. H. Smyrl, Quartz crystal microbalance analysis. 1. Evidence of anion or cation inversion into electropolymerized conducting polymers, *J. Electroanal. Chem.*, *272*: 273 (1989). (d) J. R. Reynolds, N. S. Sundaresan, M. Pomerantz, S. Basak, and C. K. Baker, Self-doped conducting copolymers: a charge and mass transport study of poly{pyrrole-co[3-(pyrrol-1-yl)propanesulfonate]}, *J. Electroanal. Chem.*, *250*: 355 (1988).

58. (a) R. Borjas and D. A. Buttry, EQCM studies of film growth, redox cycling and charge trapping of a *n*-doped and *p*-doped poly(thiophene), *Chem. Mater.*, *3*: 872 (1991). (b) S. Servagent and E. Vieil, In-situ quartz microbalance study of the electrosynthesis of poly(3-methylthiophene), *J. Electroanal. Chem.*, *280*: 227 (1990). (c) A. R. Hillman, M. J. Swann, and S. Bruckenstein, Ion and solvent transfer accompanying polybithiophene doping and undoping, *J. Electroanal. Chem.*, *291*: 147 (1990).

59. D. O. Orata and D. A. Buttry, Determination of ion populations and solvent content as functions of redox state and pH in polyaniline, *J. Am. Chem. Soc.*, *109*: 3574 (1987).

60. S. H. Glarum and J. H. Marshall, In situ potential dependence of poly(aniline) paramagnetism, *J. Phys. Chem.*, *90*: 6076 (1986).

61. H. Daifuku, T. Kawagoe, N. Yamamoto, T. Ohsaka, and N. Oyama, A study of the redox reaction mechanisms of polyaniline using a quartz crystal microbalance, *J. Electroanal. Chem.*, *274*: 313 (1989).

62. (a) J. Rishpon, A. Redondo, C. Derouin, and S. Gottesfeld, Simultaneous ellipsometric and microgravimetric measurements during the electrochemical growth of polyaniline, *J. Electroanal. Chem.*, *294*: 73 (1990). (b) I. Rubinstein, J. Rishpon, E. Sabatini, A. Redondo, and S. Gottesfeld, Morphology control in electrochemically grown conducting polymer films. 1. Precoating the metal substrate with an organic monolayer, *J. Am. Chem. Soc.*, *112*: 6135 (1990).

8

In Situ Synchrotron Techniques in Electrochemistry

JAMES MCBREEN

Brookhaven National Laboratory
Upton, New York

I. INTRODUCTION

Until the recent advances in x-ray and scanning microscopy techniques there were no in situ methods for verifying the structure of the electrode–electrolyte interface. X-ray techniques have the advantage if giving both structural and chemical information. Hard x-rays (with energies of $\approx$ 5 to 15 keV) have a large penetration depth in aqueous solution (≥ 1 mm), which permits in situ measurements. X-ray wavelengths are comparable to atomic dimensions, and hence x-ray scattering techniques probe structure at an atomic scale. The weak interaction of x-rays with matter permits in situ measurements and has also facilitated the development of the simple theories of diffraction based on the single scattering approximation. Electron diffraction, for instance, requires a much more complicated multiple scattering analysis. However, the weak interaction of x-rays yields small signals. In a typical in situ experiment a flat electrode surface contains about 10^{14} atoms, so the signal is small. The problem is exacerbated by scattering from the about 10^{20} atoms in the electrolyte. However, these difficulties have been overcome with use of the intense radiation from synchrotron sources. Early experiments on surfaces were performed parasititically at various physics accelerator facilities in the 1970s. The construction of dedicated storage rings, beginning at Daresbury, England in 1981, increased activity in the area. The application of these techniques to electrochemical systems began about 10 years ago and is still in its infancy. Another important development was the theory of extended x-ray absorption fine structure (EXAFS) in 1971 [1]. In EXAFS the signal-to-noise ratio is proportional to the square root of the beam intensity, so storage rings are necessary.

II. SYNCHROTRON RADIATION

A synchrotron produces electromagnetic-radiation-emitting relativistic charged particles that are undergoing acceleration transverse to their velocity. The particles are either electrons or positrons in storage rings. Electrons are used at the National Synchrotron Light Source (NSLS) at Brookhaven and positrons at the DCI facility at LURE in Orsay, France. Bending magnets on the storage ring confine the electrons to a closed orbit. The beam current is typically 100 to 250 mA (i.e., ca. 10^{12} electrons). The synchrotron radiation is produced by the centripetal acceleration of the high-energy electrons. This radiation has many desirable attributes; the most important is brightness, which accrues from the nature of synchrotron radiation. Brightness is the number of photons per second in unit bandwidth, per unit solid angle, per unit area of source. Brightness is the important parameter in in situ x-ray scattering experiments on flat electrodes. For a relativistic electron ($v \approx c$) the spatial distribution of the emitted radiation is confined to a narrow cone of opening angle $\theta \sim 1/\gamma$, where $\gamma = m_0 c^2/E$. E is the total energy of the electron and $m_0 c_2$ is its rest energy. The cone is directed along the tangent of the orbit, as shown in Figure 1. Since the particles in the storage ring trace out an arc, the synchrotron radiation sweeps a broad horizontal swath several degrees wide. However, vertical opening angle remains small, about 0.01°. Mirrors can be used to focus the beam to yield a very high flux on a small electrode sample. Apart from the natural collimation which gives high and stable intensity, other desirable features are the broad spectral band width and the 100% linear polarization in the plane of the electron orbit. The broad spectral range is crucially important for x-ray absorption spectroscopy (XAS). The power is radiated in a smooth featureless continuum without the spikes and structure associated with other sources. Characteristic spectral distribution curves are given for several facilities in Figure 2. The brightness of synchrotron sources is up to six orders of magnitude higher than conventional rotating anode sources.

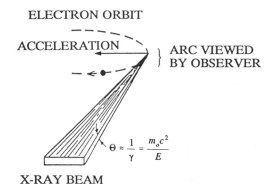

ELECTRON ORBIT

ACCELERATION

ARC VIEWED
BY OBSERVER

$$\Theta \approx \frac{1}{\gamma} = \frac{m_o c^2}{E}$$

X-RAY BEAM

FIGURE 1 Geometry of a synchrotron x-ray beam relative to motion of electrons in the storage ring.

III. X-RAYS: INTERACTION WITH MATTER AND DETECTION

A. Nature of X-Rays

1. Interaction of X-Rays with Matter

When x-rays impinge upon matter, they can either go right through it or be scattered or absorbed. X-rays can be elastically scattered, with the reflected x-rays having the same wavelength as the incident x-rays. This is known as *coherent scattering* and can explained by classical electromagnetic theory. The incident beam is considered to be an electromagnetic wave with an electric vector varying sinusoidally with time in a direction normal to the direction of the beam. The oscillating electric field accelerates the electrons back-and-forth producing radiation of the same frequency as the primary beam that radiates in all directions. Alternatively, the x-rays can transfer momentum to loosely bound electrons, and the scattered radiation has a longer wavelength. This phenomenon is known as the *Compton effect*. If the elastic or coherent scattered x-rays are the result of scattering from an ordered array of atoms, with a regular repeating pattern, the scattered rays can cooperate to build up diffracted rays. This is the basis of x-ray crystallography. Compton scattering is of no use in structural determinations.

2. X-Ray Absorption

On going through a sheet of material of thickness x, the attenuation of a x-ray beam of intensity I_0 can be expressed as

$$I = I_0 e^{-\mu x} \tag{1}$$

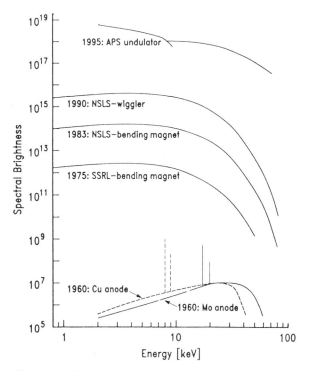

FIGURE 2 Spectral brightness of several synchrotron radiation sources compared to conventional rotating anode sources. The sharp spikes for the anode sources are the K_α and K_β lines. NSLS, National Synchrotron Light Source; SSRL, Stanford Synchrotron Radiation Laboratory; APS, Advanced Photon Source. The date indicates the year of implementation.

where μ is the linear absorption coefficient of the material and I is the intensity of the transmitted beam. The absorption can be put on a mass basis by replacing x by a term ρx and μ by μ/ρ, the mass absorption coefficient; ρ is the density. The equation then becomes

$$I = I_0 e^{-(\mu/\rho)\rho x} \tag{2}$$

Tables of μ/ρ as a function of x-ray energy are available for all the elements [2]. The mass absorption coefficient for a compound, an alloy, or a mixture can be calculated if the mass fractions g_i of the constituent elements i are known. Then

$$\frac{\mu}{\rho} = \sum_i g_i \left(\frac{\mu}{\rho}\right)_i \tag{3}$$

Except for discontinuities due to absorption edges, the absorption coefficient varies with wavelength approximately according to the relationship

$$\frac{\mu}{\rho} = kZ^4\lambda^3 \tag{4}$$

where Z is the atomic number, λ the wavelength, and k is a constant. This smooth variation of the absorption with the wavelength is due to the interaction of the x-rays with the outer-shell electrons. Absorption edges are due to the x-ray having sufficient energy to eject a core electron. All elements have K edges at unique energies due to the ejection of $1s$ electrons, and elements with $Z > 9$ have three L edges, due to the ejection of $2s$ or $2p$ electrons. The core hole can decay by two mechanisms. One is by the filling of the core hole by an electron from an outer shell. For K edges, this process is dominated by the production of K_α fluorescence radiation at an energy less than the original exciting radiation. The other process is nonradiative emission of Auger electrons. The two processes are competing and their relative strengths depend on the atomic number of the absorber. Auger emission is favored for light elements, whereas x-ray fluorescence is favored for heavier elements.

3. X-Ray Absorption by Water

Figure 3 shows the dependence of the penetration depth of x-rays in water on x-ray energy or wavelength. Because of the use of gold-coated mirrors in surface-scattering experiments, the cutoff for many beam lines is about 12 keV. So in the design of an in situ experiment the path length in the electrolyte should be no more than 1 or 2 mm.

B. Detection of X-Rays

The simplest way to detect the intensity of x-rays is by the use of ion chambers. These chambers contain two electrodes, biased with a voltage of about 300 V, and a suitable gas mixture that depends on the x-ray energy range of interest. Part of the x-rays that enter the chamber ionize the gas, producing electrons and positive ions that are collected at the electrodes; the resulting current is a measure of the x-ray intensity. When greater sensitivity or discrimination against background photons is required, solid-state detectors are used. These are based on intrinsic semiconductor diodes as detecting elements. Another scheme is the use of scintillation counters in conjunction with photomultiplier tubes [3].

IV. X-RAY SCATTERING

A. General Aspects

Prior to the availability of synchrotron x-ray sources, in situ x-ray scattering experiments, in electrochemistry, were confined to x-ray diffraction studies of

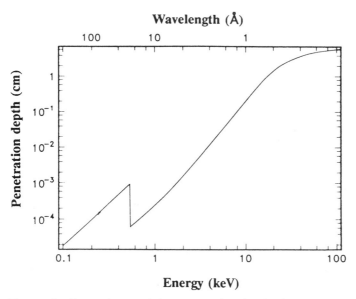

FIGURE 3 Dependence of the penetration depth of x-rays in water on x-ray energy. The drop at 543 eV is due to the oxygen *K* edge.

intercalation reactions and to electrode surfaces with high surface areas. In the early 1960s there were the pioneering studies of Salkind [4], Falk [5], and Briggs [6,7] on nickel hydroxide electrodes. About 20 years later similar studies were made on other intercalation reactions, such as the TiS_2 electrode [8]. More recently, Fleischman used a computer-controlled x-ray diffraction system to study several reactions, such as the oxidation of nickel hydroxide, iodine absorption on graphite, the adsorption of Pb on Ag, and high-surface-area platinized platinum electrodes [9,10]. There are also several other reports of in situ investigations of metal hydride formation [11,12], film formation on lithium electrodes [13], and lithium intercalation into metal oxides [14]. With the advent of synchrotron sources the emphasis is on x-ray scattering on well-defined crystal surfaces. The design of these experiments requires a knowledge of x-ray absorption, x-ray optics, crystal structure, and x-ray diffraction in two and three dimensions. There are several recent reviews on these topics [15–18]. Only a brief review is given here.

1. Experimental Aspects

In situ x-ray scattering experiments involve measurements of surface reflectivity or determinations of diffraction patterns from surface layers on the low-index planes of flat single crystals. Surface x-ray diffraction experiments use the grazing incidence x-ray geometry (GIXS). In surface diffraction experiments the x-rays enter the cell at a very low angle (<0.5°). Provisions have to be made in cell design to minimize the path length of the x-rays in the electrolyte at such low angles. Most published results have been done in cells of the design shown in Figure 4 [19]. The cell is machined from Kel-F and is approximately 3 cm in diameter and 2 cm high. The working electrode is a flat single crystal (1.25 cm in diameter) that fits in a hole at the center of the cell. The electrode extends slightly above the outer lip of the cell to accommodate incoming x-rays at grazing incidence. The electrode is covered with a thin polypropylene film that fits over the electrode and is sealed at the cell edges by an O-ring. The electrolyte is only a thin layer (~20 μm) between the polypropylene window and the electrolyte. The counterelectrode and the reference electrode capillary are in countersunk holes close to the periphery of the cell. When electrochemistry is performed, such as the deposition of an adsorbed monolayer, the polypropylene window is inflated by pumping electrolyte into the cell. This reduces *iR* drops, provides a current path, and yields adequate electrolyte volume. The potential is fixed and the electrolyte withdrawn. The thin electrolyte film is adequate for maintaining control of electrode potential in the absence of any significant faradaic process. In this type of experiment the scattering from the remaining electrolyte is on the same order as that from an adsorbed monolayer. In x-ray reflectivity experiments, where measurements are made at higher incidence angles, more conventional cells with better current distribution can be used [20]. There

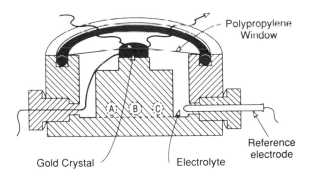

FIGURE 4 Cell design of x-ray scattering experiments on flat single-crystal surfaces. The Au working electrode is clamped in a Kel-F holder and is covered by a polypropylene window that is held under tension by the outer O-ring. A and C are electrolyte ports; B is the counterelectrode. The cell is in an outer chamber (not shown), with Kapton windows, filled with nitrogen to prevent ingress of oxygen through the polypropylene window.

is also a recent report of a cell design with a transmission geometry [21]. This is shown in Figure 5. The counter and reference electrodes are inside a funnel-shaped reservoir. The bottom of the funnel has a diameter similar to that of the working electrode (2.5 mm). The funnel is placed over the working electrode at a distance of about 3 mm, and electrochemical contact with the working electrode is via a concave drop that is held in place by surface tension. This design permits both GIXS measurements down to zero angle incidence and reflectance measurements at higher angles. The electrochemical cell is mounted on a Huber goniometer that can be mounted directly in a four-circle diffracto-meter. Figure 6 is a schematic of a four-circle diffractometer. This has four concentric motions, three of which, called ω, ϕ and χ, move the sample and one that moves the detector by an angle 2θ. The angular movements are indi-cated in Figure 6 and are explained in a paper by Busing and Levy [22]. The

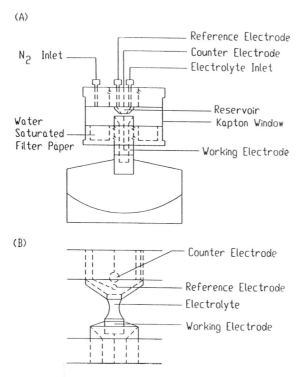

FIGURE 5 (A) Cell for transmission x-ray scattering experiments on flat single crystals; (B) magnified cutaway view of electrolyte reservoir with Pd bead refer-ence electrode. The cell is inside a helium-filled case with a Kapton window that permits measurements over 360°.

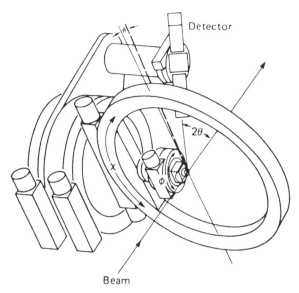

Beam

FIGURE 6 Schematic of a vertically mounted four-circle diffractometer. The meanings of the various angles are given in Ref. 22.

diffractometer is a large piece of equipment weighing as much as $\frac{1}{2}$ ton. In practice the orientation of the sample, the position of the detector, and the counting of the scattered photons are all computer controlled. Because of the polarization of the beam, the cell is usually mounted with the electrode surface plane vertical.

2. X-Ray Optics

X-rays obey the same rules of geometric and physical optics as visible light, and the laws of reflection and refraction hold. Reflection and refraction are described by Snell's law and the Fresnel equations, which are covered in textbooks on optics [23,24]. The only difference between light and x-rays is in the index of refraction and the absorption coefficient. An important consequence of this is that, for light, air has a smaller index of refraction than most dense medium, whereas for x-rays the refractive index of most materials is smaller than that of air. Because of this, total reflection of x-rays can occur at a small critical angle. This effect can be used to enhance the contribution of the surface to the signal and decrease penetration into the bulk. This is why the GIXS approach is used in surface diffraction experiments.

The reflection and refraction of x-rays at a flat interface between two materials with different indices of refraction n and n' are illustrated in

Figure 7. When the incident beam at an angle α falls on the surface it is split into a refractive and a refractive wave. The respective angles of reflection and refraction are α and α'. According to Snell's law,

$$n \cos \alpha = n' \cos \alpha' \tag{5}$$

The refractive index of a material at x-ray energies is written as

$$n = 1 - \delta - i\beta \tag{6}$$

where

$$\delta = \frac{\rho_e \lambda^2}{\pi} \frac{e^2}{mc^2} \tag{7}$$

and ρ_e is the electron density, e the electron charge, λ the x-ray wavelength, and m the mass of the electron. For typical materials δ is on the order of 10^{-5} to 10^{-6}. The imaginary term is

$$\beta = \frac{\lambda \mu}{4\pi} \tag{8}$$

where μ is the linear absorption; β is on the order of 10^{-6} to 10^{-9}. When δ is small, Snell's law can be reduced to

$$\alpha'^2 = \alpha^2 = 2(\delta' - \delta) \tag{9}$$

and the critical angle for total external reflection (α_c) is

$$\alpha_c = \sqrt{2 (\delta' - \delta)} \tag{10}$$

The respective refractive indices of air, water, and metal are such that the critical angle at the water–metal interface is much larger at the air–water interface. Thus a simple cell with a thin layer of electrolyte on top of the electrode can be

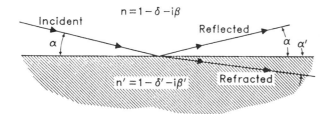

FIGURE 7 Geometric optics of x-rays, showing reflection and refraction of x-rays that are incident onto an interface between two materials with indices of refraction n and n'.

designed to operate in the total reflection mode at the electrode–electrolyte interface.

The intensities of the reflected and refracted waves are embodied in Fresnel's equations. These are derived from Maxwell's equations by applying the boundary conditions holding at a surface of discontinuity between two media of different refractive indices. From Snell's law and the Fresnel boundary conditions, the intensities of the reflected and refracted beams can be derived. Similarly, the depth (z) dependence of the intensity of the transmitted x-rays and the surface intensity of the x-rays can be calculated [15]. Figure 8 shows calculated reflectivity, penetration depth, and surface intensity for a water–silver interface at 8.04 and 17.44 keV. For $\alpha < \alpha_c$ the x-rays penetrate the metal to only a few tens of angstroms. These effects make surface diffraction experiments possible.

B. X-Ray Diffraction

Metal crystals consist of an ordered array of atoms, all of which are contained in sets of parallel lattice planes spaced a distance d apart. This is illustrated in Figure 9. If x-rays are incident on the planes at an angle θ, a small fraction (10^{-4} to 10^{-2}) of their incident radiation is specularly reflected at the same angle. A peak in the intensity of the scattered radiation will occur when reflections from successive planes interfere constructively. Since the path-length difference between planes is $2d \sin \theta$, the condition for constructive interference, known as *Bragg's law*, is

$$n\lambda = 2d \sin \theta \tag{11}$$

where n is an integer and λ is the x-ray wavelength.

1. Crystal Structure Lattice Planes and Miller Indices

Miller indices are used as a system of notation for the faces of a crystal and for specifying the orientation of planes within a crystal relative to the crystal axis. The Miller indices for a plane in a cubic crystal are derived by finding the intercept of the plane with the crystal axes, taking the reciprocal of the intercepts, and reducing the reciprocals to the three smallest integers having the same ratio. The reciprocal intercepts are written (hkl). Figure 10 shows examples of lattice planes and their Miller indices for a simple cubic lattice. Further details for other crystal systems can be found in textbooks on x-ray crystallography [25–28].

2. Reciprocal Lattice and Three-Dimensional Diffraction

When Bragg's equation is written in the form

$$\sin \theta = \frac{n\lambda}{2} \frac{1}{d} \tag{12}$$

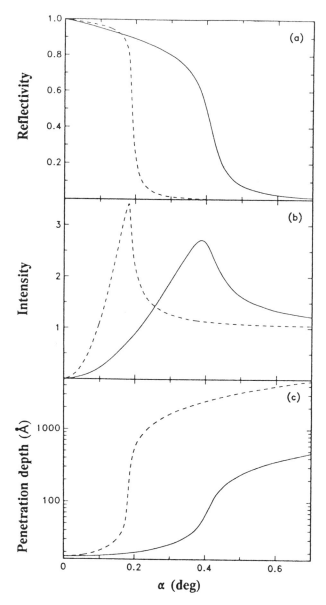

FIGURE 8 Relationship between x-ray incidence angle (α) and (a) reflectivity, (b) intensity, and (c) penetration depth in the metal for the Ag–water interface. Data are given for two x-ray energies, 17.44 keV (- - -) and 8.04 keV (—). The respective critical angles (α_c) for the two energies are 0.19° and 0.41°.

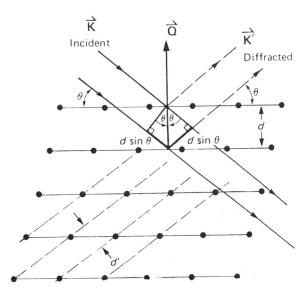

FIGURE 9 Bragg reflection from lattice planes with spacing *d*. The solid lines are the incident wave vector **k** and the dashed lines the reflected wave vector **k'**.

it is seen that sin θ is inversely proportional to *d*, the interplanar spacing in the crystal lattice. From the expression (12) it is clear that three-dimensional structures with large *d* will exhibit compressed diffraction patterns, and structures with small *d* will exhibit patterns with the peaks spread far apart. Interpretation of the x-ray patterns would be simplified if the relationship between sin θ and the lattice spacing could be replaced by a direct one. This is essentially done by constructing a reciprocal lattice based on 1/*d*, a quantity that varies directly as sin θ. Another problem that arises in x-ray diffraction is that the reflections come from an infinite number of planes with different slopes. The slope of a plane can be fixed by the geometry of the normal of the plane. Furthermore, the normal has one less dimension than the plane. In addition to tabulating the interplanar spacings, the reciprocal lattice also includes the slopes of the planes. The reciprocal lattice is built by constructing a normal from a common origin to every plane in the direct lattice and limiting its length so that it equals the reciprocal of the interplanar spacing of its plane. A point is placed at the end of each limited normal. The collection of such points is a lattice array that constitutes the reciprocal lattice. These positions in reciprocal space, where diffraction occurs, are referred to as *Bragg points*.

The reciprocal lattice and the scattering process are best formulated in terms of vectors. The direct lattice of a cubic unit cell can be defined by the

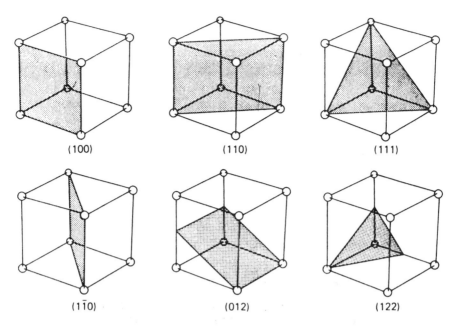

| (100) | (110) | (111) |

| (1$\bar{1}$0) | (012) | (122) |

Figure 10 Lattice planes and Miller indices for a simple cubic lattice.

orthogonal vectors **a**, **b**, and **c**. The corresponding reciprocal lattice has a unit cell defined by the vectors **a***, **b***, and **c***, where

$$\mathbf{a}^* = \frac{\mathbf{a} \times \mathbf{c}}{\mathbf{a} \cdot \mathbf{b} \times \mathbf{c}} \qquad \mathbf{b}^* = \frac{\mathbf{c} \times \mathbf{a}}{\mathbf{a} \cdot \mathbf{b} \times \mathbf{c}} \qquad \mathbf{c}^* = \frac{\mathbf{a} \times \mathbf{b}}{\mathbf{a} \cdot \mathbf{b} \times \mathbf{c}} \tag{13}$$

and $\mathbf{a} \cdot \mathbf{b} \times \mathbf{c} = V$, the volume of the crystal unit cell. The vectors **a**, **b**, and **c** have units of Å and the vectors **a***, **b***, and **c*** have units of Å^{-1}. Further details of the vector treatment of the reciprocal lattice can be found in books on x-ray diffraction [25–28].

3. Surface Structure

Quite often, to minimize the free energy, the equilibrium position of surface atoms can be different from that given by the lattice periodicity of the bulk. Often, the interatomic spacing between the top atomic layers differs from that deeper in the bulk, without any noticeable change in lateral symmetry. This is known as *surface relaxation*. A typical effect is a 5 to 10% contraction between the first and second layers and a smaller but measurable expansion between the second and third and the third and fourth layers. In a few cases, particularly with precious metals, the surfaces are reconstructed, and the lateral symmetry of the surface often differs from that of the bulk. This is called *surface reconstruction.*

The notation used to describe surface reconstructions and surface overlayers is described in texts on surface crystallography [29]. The lattice vectors **a′** and **b′** of an overlayer are described in terms of the substrate lattice vectors **a** and **b**. If the lengths $|\mathbf{a'}| = m|\mathbf{a}|$ and $|\mathbf{b'}| = n|\mathbf{b}|$, the overlayer is described as $(m \times n)$. Thus a commensurate layer in register with the underlying atoms is described as (1×1). The notation gives the dimension of the two-dimensional unit cell in terms of the dimensions of an ideally truncated surface unit cell.

4. Diffraction in Two Dimensions

Before considering surface layers it is instructive to consider diffraction from a free-standing two-dimensional solid. Such a solid has no diffracting planes normal to the surface. The effect of this can be visualized by taking the three-dimensional crystal and gradually increasing the d spacing along the z axis (d_z). This is illustrated in Figure 11 (top). The corresponding reciprocal space representation of the diffraction is shown below. The Bragg points are observed at equal spacings, corresponding to the multiple order of reflections normal to the surface. As the d spacing in the z direction increases, the Bragg points move

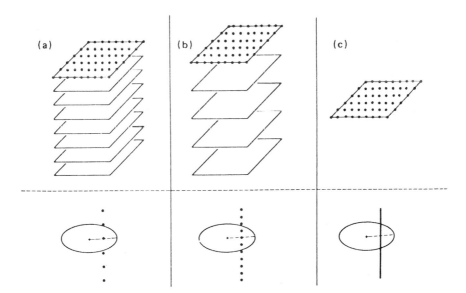

FIGURE 11 Effect of expansion of d_z, the real interplanar lattice spacing (top), on the reciprocal space representation (bottom). For simplicity only one reflection is shown. When $d_z = \infty$ the Bragg points coalesce into a Bragg rod.

closer together. As d_z approaches infinity, the Bragg spots are no longer discrete points but form a continuous streak of essentially constant intensity normal to the monolayer. These streaks are called *Bragg rods*.

5. Crystal Truncation Rods

Termination of the crystal lattice at a surface gives rise to two-dimensional-like diffraction features that have been termed *crystal truncation rods* (CTRs) or a condition of nonspecular reflectivity [30,31]. CTRs are more complicated than simple Bragg rods from two-dimensional solids in that they are not flat between the Bragg points. These are illustrated in Figure 13. The c direction of the crystal is taken as being parallel to the surface normal. The scattering intensity for the CTR has a $(\sin \pi Q_c)^{-2}$ shape, whereas the Bragg rod for a monolayer is flat (dashed line, Fig. 12). The peaks correspond to Bragg peaks from the bulk. The looping behavior between peaks contains valuable information about the surface structure. For instance, if the surface is reconstructed to a more compact structure, there will be an increase in the reflectivity from the CTR. Conversely, if the surface is roughened, there will be a decrease in the reflectivity. These deviations in reflectivity from the ideal truncation case are typically one to two orders of magnitude. Often, a cursory examination of these loops can reveal a lot of information about surface structure.

6. Surface X-Ray Diffraction

Details on surface diffraction are found in several excellent reviews [32–35]. Figure 13 shows a typical surface x-ray diffraction experiment. In the glancing angle (ca. 0.5°) geometry the x-ray beam impinges onto the surface at an angle ϕ and is detected at a combination of an in-plane grazing angle (2θ) and a second grazing angle ϕ'. The incident and exit angles are controlled by the sample tilt angle χ. Corrections for background are made by doing an ω scan

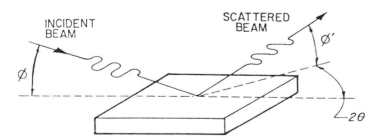

FIGURE 12   Grazing incidence x-ray scattering geometry. The x-rays are incident on the sample at an angle ϕ and are detected at a combination of two angles 2θ and $2\phi'$.

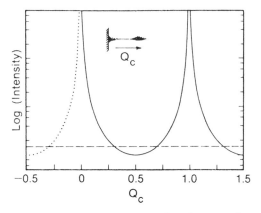

FIGURE 13 Scattering intensity of a crystal truncation rod for an ideally truncated surface. The intensity is plotted vs. the perpendicular scattering vector Q_c. The metal surface is at $Q_c = 0$ and the first Bragg peak is at $Q_c = 1$. The scattering intensity at the anti-Bragg position ($Q_c = 0.5$) is close to the scattering intensity for a free-standing monolayer, which is given by the dashed line.

in the same way as for bulk crystallography. The background measured nearby is subtracted. The contribution of the surface layer to the signal is enhanced by keeping the angle of incidence below the critical angle (ca. 0.5°). The low angle of incidence permits the beam to sample a large number of scatters. Thus the combination of the intense beam from the synchrotron and the grazing incidence geometry make surface diffraction experiments possible. The data are analyzed in a manner similar to that used for bulk diffraction. For instance, with changes in the unit cell on reconstruction, other truncation rods appear that are not associated with the Bragg points. By obtaining sufficient reflections, the unit cell of the surface layer can be determined. Similarly, if an adsorbed monolayer is not in register with the underlying atoms, extra truncation rods appear. This two-dimensional scattering is lower in intensity than that from three-dimensional peaks by about a factor of 10^6. Analysis of the truncation rods yields the in-plane structure of the surface layer and its relationship to the underlying bulk crystal structure.

C. X-Ray Reflectivity

Specular reflectivity is obtained by recording the reflectivity in the plane of incidence of the x-rays. When applied to single-crystal surfaces it is sometimes referred to as the specular rod. It gives structural information in the surface normal direction. Figure 14 shows reflectivity data for a Au(100) surface [36]. The reflectivity data are fitted to a model that sums up the interference caused

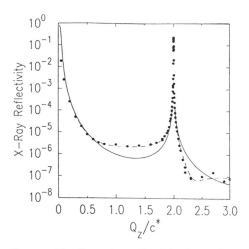

FIGURE 14 Specular reflectivity for a clean Au(100) surface in vacuum at 310 K (••••). The solid line is calculated for an ideally terminated lattice. The dashed line is a fit to the data with a reconstructed surface with a 25% increase in the surface density combined with a surface relaxation that increases the space between the top and next layers by 19%. In addition, the data indicate that the top layer is buckled or corrugated with a buckling amplitude of 20%.

by x-ray scattering among the surface layers. This can be done using continuum models based on the Fresnel equations [37,38] or on discrete models that include surface roughness, the surface density, and the deviation of the top layer spacing from the bulk spacing. Details are given in several publications by Ocko and Gibbs [36,39,40]. In a specular reflectivity experiment the incident beam strikes the surface at an angle θ and the intensity is recorded in the plane of incidence at an angle 2θ. Data are collected as a function of θ. The diffuse background scattering is determined from measurements slightly off the specular direction. After subtraction of the background the data are normalized by the incidence intensity. To determine electrode morphology it is necessary to perform rocking scans, where the detector angle 2θ is fixed and the incidence angle θ is varied. This yields information on off-specular scattering that is related to the electrode roughness [41]. So a combination of measurements of CTRs and x-ray reflectivity can provide information on atomic roughness, surface relaxations, surface reconstructions, the location of adsorbed atoms, and the density distribution of an electrolyte above an electrode. A convenient technique for following electrochemical reactions is to fix the sample at an anti-Bragg position and step or scan the electrode potential. This is useful for the study of surface reconstruction kinetics or the growth of surface layers.

Specular reflectivity can also be used to study thick surface layers up to 1000 Å. Since the specular reflectivity of x-rays from a surface depends on the variation of the index of refraction along the normal to the surface, the method can be used to probe the normal electron density profile. The reflectance technique has been used to study oxide growth on metals [38], and it yields information on oxide thickness, roughness, and stoichiometry. It is the only technique that can give information on the buried metal–oxide interface. It is also possible to get information on duplex or multiple-layer oxide films or oxide films consisting of layers with different porosity. Films with thicknesses of anywhere from 10 to 1000 Å can be studied. Thus the technique can be used to study surface layers all the way from monolayers to thick oxide films. The information derived from reflectivity studies accrues from the fact that whenever the index of refraction changes within a sample, part of the incident wave is reflected and part is transmitted. Interference between the beams reflected at different interfaces leads to an oscillatory pattern in the reflectivity as a function of incident angle. The effect is analogous to the optical interference fringes form oil films on water. Figure 15 shows reflectivity results for an electrochemically grown oxide film on a 2000-Å Ta film on a soda-lime glass slide [42]. For a single oxide layer on a metal the frequency of the oscillation is proportional to the oxide layer thickness and the amplitude depends on surface roughness. The roughness also determines the attenuation of the reflectivity at higher scattering angles. The

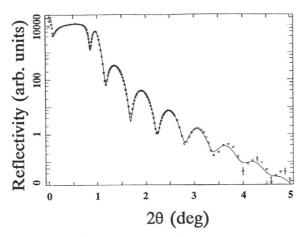

FIGURE 15 X-ray reflectivity data for an oxide film grown on a 2000-Å Ta film that was sputtered on to a soda-lime glass slide. The oxide films were grown by ramping the potential, in 0.05 M H_2SO_4, at 0.1 mV/s to 6.0 V SCE. The data indicate a growth of Ta_2O_5 with an anodizing ratio of 20.0 ± 1.0 Å/V.

oscillations can be interpreted to determine the electron density profile with depth below the surface. The models also account for the widths of the interfaces and can elucidate effects due to interface intermixing, interdiffusion, or roughness. Models for multiple layers have been developed [38]. The amplitude of the beam reflected at each interface is proportional to the change in electron density at that interface. This is directly related to changes in the refractive index profile in the z direction. Structural information is extracted by comparing the experimental data with that derived from a model. Details of various theoretical treatments of x-ray reflectivity are beyond the scope of this review and the reader is referred to several excellent papers in the literature [36–41,43–45].

D. X-Ray Standing Waves

There have been a few papers and reviews on the application of x-ray standing waves to in situ electrochemical studies [46–51]. X-ray standing waves (XSWs) are the result of interference between the coherently related and Bragg diffracted beams from a perfect crystal. The effect is illustrated in Figure 16. The standing wave extends from inside the crystal to about 1000 Å beyond the surface. The position of the nodal and antinodal planes of the standing wave can be varied by varying the incident angle θ. In practice, experiments are done in cells of the type illustrated in Figure 4. In the case of an adsorbed monolayer, if the x-ray energy is greater than an absorption edge, there will be a modulation in the fluorescence yield as the substrate is rocked in angle. The angular dependence of this modulation is a measure of normal distance for the adsorbed overlayer. XSWs from single crystals can be used to probe monolayers. Longer

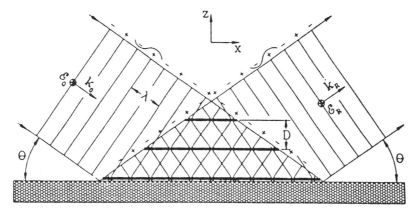

FIGURE 16 X-ray standing wave formed by the interference between the incident and Bragg diffracted beams. The result is a standing wave with a nodal wavelength D that corresponds to the d spacing of the diffracting planes.

distances (2 to 200 Å) can be probed using layered synthetic microstructures [46,50], and even longer distances can be probed using XSWs generated by total external reflection. So in principle the technique can also be used to probe the diffuse double layer. Despite this there are very few publications on the application of XSWs. Part of the problem is the stringent requirement for a perfect or near-perfect crystal to generate the standing wave. Indeed, if this requirement is met, the experiments can be done with a conventional x-ray source. The requirement for a perfect crystal is such that the equipment for handling crystals of soft metals such as Cu and the cell mount have to be specially designed so as not to strain the crystal. Theoretical treatments of the standing wave also require the more complex dynamical theory [51]. The other x-ray scattering techniques use the simple kinematical theory that treats scattering from each unit cell as being independent of that from other cells. The dynamical theory takes into account all wave interactions within the crystal.

V. X-RAY ABSORPTION SPECTROSCOPY

A. General Aspects

The application of x-ray absorption spectroscopy (XAS) to structural determinations is a relatively new technique. It is the subject of two books [52,53] and several reviews [54–58]. XAS is simply the accurate determination of the x-ray absorption coefficient of a material as a function of photon energy, in an energy range that is below and above the absorption edge of one of the elements in the material. Absorption measurements are usually done at K or L_3 edges, as absorption at the L_3 edge is much higher than at the other L edges. The K edge is due to absorption by a $1s$ and the L_3 to absorption by a $2p_{3/2}$ core states. Measurements at the K edge are most suitable for elements of low atomic number (low Z) and for first-row transition metal elements. In the case of platinum the K edge is at 78.4 keV, where the synchrotron photon flux is too low, whereas the L_3 edge for platinum is at 11.5 keV, which is close to the energy maximum for the photon flux from conventional synchrotron sources. The simplest method for XAS is a transmission XAS experiment. Figure 17 is a schematic representation of the experimental configuration. It consists of an x-ray source, a double-crystal monochromator, a thin sample of the material, ionization chamber detectors for monitoring the beam intensity before and after it passes through the material, and a data acquisition system. The data acquisition system is used for several purposes. This includes stepping the monochromator to pass the desired photon energies, alignment of the sample in the beam, and monitoring the signals from the detectors. This is the scheme used at the National Synchrotron Light Source (NSLS) at Brookhaven National Laboratory (BNL). The time for ob-

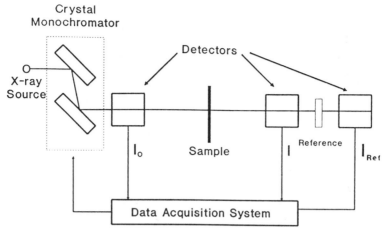

FIGURE 17 Experimental setup for a transmission x-ray absorption experiment.

taining a full spectrum is typically 20 min and is mainly limited by the dead time needed for rotation of the crystals in the monochromator.

B. X-Ray Absorption Spectra

1. XANES

Figure 18 shows a x-ray absorption spectrum for β-Ni(OH)$_2$ taken at the Ni K edge (8333 eV). The near-edge part of the spectrum is referred to as the x-ray absorption near-edge structure (XANES). The oscillations, starting about 50 eV above the edges, are the extended x-ray absorption fine structure (EXAFS). The fine structure in the XANES region can be explained in terms of (a) transition of the ejected photoelectron to unoccupied states in the vicinity of the Fermi level, and (b) to the long mean free path of the low-energy photoelectron, which results in multiple scattering around the excited atom. The shape of the edge yields information on both the type and symmetry of the ligands around the excited atom. Edge shifts due to core–hole interactions are indicative of changes in oxidation state. So XANES yields important chemical information about the absorbing atom.

The physical mechanism of the XANES can be understood from the Fermi golden rule, where the probability (w) of a transition from an initial state $|i>$ to a final state $<f|$ by the action of a perturbation H' is given by

$$w = \frac{2\pi}{\hbar} \; |<f|H'|i>|^2 \; \rho(E_i) \tag{14}$$

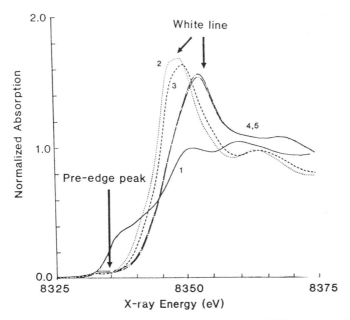

FIGURE 18 Nickel *K*-absorption edges for nickel foil and nickel oxide electrodes. 1, Nickel foil; 2, ex situ data for an uncharged electrode and in situ for electrodes, 3, after first discharge; 4, after first charge; 5, after second charge.

where $\rho(E_i)$ is the density of final states at the energy of the initial state E_i and $H' = Ae^{ikr}$, the x-ray photon. In the dipole approximation the transition is restricted by the selection rule $\Delta L = \pm 1$. Thus at the K edges of the first-row transition elements the transition is mostly from 1s to 4p states. Small preedge peaks due to the weaker quadrupole ($\Delta L = \pm 2$) are often observed in cases where there is an inversion of symmetry (e.g., the rock-salt structure). These correspond to 1s-to-3d transitions. These are illustrated in Figure 19 for MnO. For tetrahedral atoms such as MnO_4^- this is dipole allowed. The result is a greatly enhanced peak for the 1s-to-3d transition. This is shown in Figure 19. Crystal field splitting effects can introduce features in the XANES. In octahedral symmetry the p states are threefold degenerate. In going to tetragonal symmetry the p levels are split. This introduces shoulders in the XANES. Covalent bonding can result in mixing of p and d orbitals, which also introduces features and distortions in the XANES. It is clear that the XANES yields a lot of information on the chemistry and symmetry of the environment of the excited atom.

There is not yet a complete theory of the XANES. The present status has been reviewed by Bianconi [59]. With further developments it will be possible

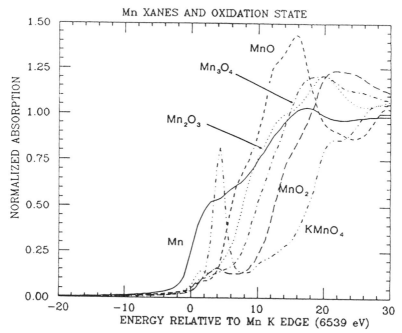

FIGURE 19 XANES data at the Mn *K* edge for various oxides of Mn and for KMnO₄.

to glean much more information from these spectra. Recently, Lytle et al. have shown the possibility of getting detailed structural information from XANES spectra [60].

2. EXAFS

In XANES region the low-energy photoelectron has a long mean free path and can undergo multiple scattering. In the EXAFS region the process can be explained by simple backscattering, and the theory has been worked out in detail [1]. A small fraction of the outgoing photoelectron wave, associated with the excited electron, is backscattered by surrounding atoms. The EXAFS is a final-state interference effect whereby, depending on energy (E), the backscattered wave interferes constructively or destructively with the outgoing wave. The EXAFS function $\chi(k)$ is

$$\chi(k) = \frac{\mu(E) - \mu_0(E)}{\mu_0(E)} \tag{15}$$

where μ and μ_0 are the x-ray absorption coefficients of the absorbing atom in the material of interest and in the free state, respectively. The difference $\mu - \mu_0$ depends on the local structure of the absorbing atom and represents the EXAFS. The division by μ_0 normalizes the EXAFS to a per atom basis.

The EXAFS is separated from the XAS spectrum by first subtracting the preedge background and then separating the EXAFS from the low-frequency oscillations of the background by use of a cubic spline function. An EXAFS spectrum for $Ni(OH)_2$ is shown in Figure 20. Structural information can only be extracted from the EXAFS after conversion from energy space to wave vector space using the relationship

$$k = \sqrt{\frac{2m}{\hbar^2}} (E - E_0) \tag{16}$$

where $\hbar = h/2\pi$, m is the mass of the electron, and E_0 is the threshold energy of the core shell of the excited atom. The EXAFS spectrum is the superimposition of contributions from different coordination shells to the backscattering process. The theoretical expression that relates the measured EXAFS to the structural parameters is given by

$$\chi(k) = \sum_j A_j(k) \sin[2kR_j + \phi_j(k)] \tag{17}$$

where j refers to the jth coordination shell and R_j is the average distance between the absorbing atom and the neighboring atoms in the jth shell. $\phi(k)$ is the phase shift suffered by the photoelectron in the scattering process. $A_j(k)$ is the amplitude function, which is expressed as

$$A_j(k) = \frac{N_j}{kR_j^2} S_0^2(k)F_j(k)e^{-2(R_j-\Delta)/\lambda} e^{-\sigma_j^2 k^2} \tag{18}$$

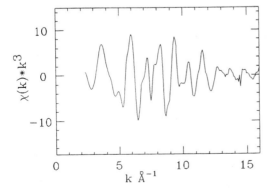

FIGURE 20 EXAFS spectrum for a nickel oxide electrode.

where N_j is the average coordination number and $F_j(k)$ is the backscattering amplitude of the atoms in the jth shell. σ_j^2 is a Debye–Waller term which accounts for the static and thermal disorder present in the materials, $S_0^2(k)$ is an amplitude reduction term which accounts for relaxation of the absorbing atom and multielectron excitations (shake up/off) at the absorbing atoms. λ is the mean free path of the photoelectron and is a correction to Δ since S_0^2, and $F_j(k)$ already accounts for the photoelectron losses in the first shell. Data analysis of the EXAFS involves the determination of R_j, N_j, and σ^2. The other parameters such as ϕ_j and F_j are obtained from standard compounds or theoretical calculations.

The contributions of individual shells to the EXAFS can be isolated by doing a Fourier transformation on the EXAFS:

$$\theta = \frac{1}{\sqrt{2\pi}} \int_{k_{\min}}^{k_{\max}} k^n \chi(k) e^{2ikr} \, dk \tag{19}$$

This is shown in Figure 21 for $Ni(OH)_2$ and it is called a radial structure function (RSF). The result are peaks in r space that correspond to interatomic distances between the central atom and the individual coordinating shells. The r values are shifted to lower values than the actual coordination distances because of the phase-shift term $[\phi_j(k)]$. The EXAFS spectrum is the sum of the χ values of the different coordination shells. The property of the Fourier transform is such that the Fourier transform of the EXAFS is the sum of the Fourier transforms of the χ values of the individual shells. Therefore, it is possible to isolate the data of one shell from the data for other shells by an inverse Fourier transformation of the corresponding peak in the radial structure function. Figure 22 shows the backtransformation of the first shell of $Ni(OH)_2$. The result is a sinusoidal EXAFS function. Structure parameters are determined by fitting the EXAFS for individual shells using either empirical or theoretical phase and amplitude functions. The determined parameters are N, R, $\Delta\sigma^2$, and ΔE_0. $\Delta\sigma^2$ is the Debye–Waller factor with respect to some standard. Since E_0 is not related to any recognizable feature in the XAS spectrum, it is treated as a sliding parameter. Both N and $\Delta\sigma^2$ are strongly correlated but can be distinguished. The same is true for R and ΔE_0. These effects can be decoupled by doing k^1 and k^3 weighted fits in both k and r space. The effects on N and R are the same for both weightings, whereas $\Delta\sigma^2$ affects predominantly the k^3 transform and ΔE_0 the k^1 transform. Furthermore, the effects of R and ΔE_0 on the magnitude and imaginary parts of the transform are reversed. Thus there are many checks to arrive at an unique set of structural parameters. From the discussion above it is clear that important structural information can be derived from the EXAFS. However, it is also apparent that the data analysis is tedious and has to be done

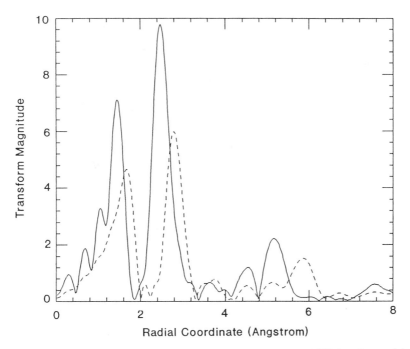

FIGURE 21 A comparison of Fourier transform of the EXAFS for charged (—) and discharged (- - -) nickel oxide electrodes (k^3 weighted, Δk = 2.6 to 14 Å^{-1}). The first peak at about 1.5 Å is the Ni–O contribution, the second peak at about 2.7 Å is the first Ni–Ni coordination shell, and the peak at about 5 Å is the third Ni–Ni coordination shell.

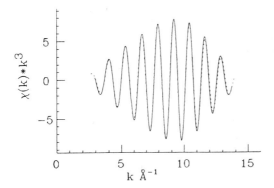

FIGURE 22 Backtransform of the Ni–O peak for the charged electrode (Δr = 0.00 to 2.10) (—). The dashed line is a two-shell fit to the data.

with care. There is a book [52], a review [61], and several papers on data analysis [62,63].

X-ray absorption spectroscopy (XAS) is an ideal method for in situ studies of electrochemical systems because both the probe and signal are penetrating x-rays. The great advantage is that XAS is element specific, and this permits investigation of the chemical environment of a constituent element in a composite material. It also permits studies of dilute systems. Since XAS probes only short-range order, it can provide structural information on amorphous materials, liquids, gases, adsorbed monolayers, and hydrated ions and complexes in solution. Because of this, XAS is finding application in an enormous variety of electrochemical systems. Examples are in situ electrode studies of UPD layers, passive films on iron, Cr species in passive films, carbon-supported fuel cell catalysts, redox polymer electrodes, battery electrode materials, and additives in battery electrodes. XAS has been invaluable in the study of ion-conducting polymers, where the ion-conducting phase is amorphous.

C. Time-Resolved XAS

Another technique of dispersive XAS permits time-resolved spectra acquisition with time scales as low as 5 ms [64]. This is the scheme used on the dispersive beam line at LURE-CNRS in Orsay, France. The x-ray optics consists of a bent triangular Si(311) monochromator crystal to focus and disperse the quasi-parallel polychromatic x-ray beam from a positron storage ring. The bent crystal yields a correlation between the photon energy and the direction of the photon beam. This is shown schematically in Figure 23. As a result, the beam converges to a focal spot where the sample is located. On passing through the sample the beam diverges toward a position-sensitive detector consisting of 1024 sensing elements spaced over 2.56 cm. The photon energy reflecting angle correlation yields a pixel number-energy correlation in the detector array. Since there is no movement in the monochromator data, acquisition times are limited only by the response of the detector. At present the facility at LURE has a bandpass of about a 700 eV. Thus XAS can only be recorded to about 600 eV above the absorption edge. However, it has the great advantage of time-resolved measurements and a resolution of 10 meV for energy shifts at the absorption edge [65]. This is due to the absence of mechanical movement in the monochromator. There have been recent reports of other facilities [66,67]. Dent and his co-workers [68] report on an interesting facility that can do both dispersive EXAFS and x-ray diffraction. Allen et al. [67] report on the use of a four-point crystal bender that yields a much larger bandpass (700 to 2000 eV) while maintaining good resolution (2 to 5 eV).

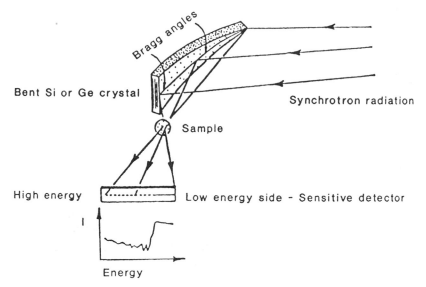

Bent Si or Ge crystal

Synchrotron radiation

Sample

High energy

Low energy side - Sensitive detector

I

Energy

FIGURE 23 Experimental setup for dispersive EXAFS.

D. Signal Detection

If the sample can be made thin enough, and if the concentration of the excited element is sufficient, the preferred method is a transmission measurement with ionization chamber detectors. The emitted fluorescence x-rays and the Auger electrons are proportional to the absorption and can be used as a signal. Fluorescence is the preferred method in dilute systems when the absorption by the excited element is only a few percent of the total absorption. Fluorescence cannot be used in concentrated samples because of self-absorption effects. The Auger electron yield is surface sensitive [69–72] and has been used in studies of emersed electrodes [73,74]. This is particularly true when used in conjunction with glancing incidence geometry [75]. Furthermore, if a helium atmosphere is used in the detector, the sensitivity is increased by a factor of 100.

Most XAS studies of adsorbed monolayers on electrode surfaces have been done in cells similar to that shown in Figure 4, with the x-rays diffracting on the surface at a glancing angle. The reflected beam cannot readily be used for detection of the EXAFS because of distortions in the signal. So the fluorescence signal is monitored with a solid-state detector. Because of the weak signal, early experiments required many repeating scans and took as long as 50 h [76]. Use of a scintillation detector in conjunction with a photomultiplier tube [3] reduces the data acquisition time to 5 h [74]. New multielement solid-state detectors should reduce the data acquisition time to less than 1 h [77].

VI. APPLICATIONS OF IN SITU X-RAY METHODS IN ELECTROCHEMISTRY

A. Surface Structure of Clean Metal Surfaces

Surface x-ray scattering (SXS) is now being used to determine the structure of metal electrode surfaces and its dependence on electrode potential and electrolyte composition. So far all of this work is on Au and Pt surfaces.

1. Au(111)

Ocko and his co-workers have carried out extensive SXS studies on Au(111) [78–80]. They found that the top layer of gold atoms undergoes a reversible transition between the (1×1) bulk termination and a $(p \times \sqrt{3})$ uniaxial discommensurate (striped) phase on changing the electrode potential in the negative direction. Below a critical potential in all electrolytes, $p = 23$, which is identical to that obtained in vacuum. On sweeping the potential in the negative direction the reconstruction starts at a critical potential that depends on the electrolyte composition. Analysis of this critical potential in NaF, NaCl, and NaBr electrolytes revealed that the critical driving force in the reconstruction was the surface charge. In all electrolytes the surface phase transition occurs at an induced surface charge density of 0.07 ± 0.02 electron per atom.

    Time-dependent SXS was used to follow the kinetics of the surface phase transition on Au(111) [80]. By stepping the potential from $-0.2V$ (vs. Ag/AgCl) to positive potentials, it was possible to follow the kinetics of the lifting of the reconstruction by studying the dependence of the reflected intensity of a reflection from the surface x-ray diffraction pattern. Also, by doing similar experiments while stepping the potential from 0.6 V to negative potentials it was possible to follow the kinetics of the nucleation and growth of the reconstructed phase. The studies revealed that the presence of chloride and bromide ions accelerate the kinetics of the reconstruction.

    X-ray reflectivity measurements revealed that in the $(23 \times \sqrt{3})$ phase, the interlayer spacing of the top gold layer is expanded by 3.3% relative to the bulk layer spacing. No interlayer relaxation was found for the top layer in the (1×1) phase. The data also showed that adsorbed chloride and bromide had enormous effects on the specular reflectivity profiles.

2. Au(001)

In situ x-ray specular reflectivity and glancing incident angle x-ray diffraction on Au(100) revealed that in 0.01 M $HClO_4$ that at negative potentials the surface undergoes a hexagonal reconstruction to form a layer with a mass density 21% greater than the underlying bulk layers [81]. The reconstruction could be completely lifted at 0.5 V. The excess atoms are segregated to the surface where they form a layer with a density corresponding to 22% of a bulk layer. The reconstruction fully recovers on sweeping the potential to −0.3 V.

3. Surface Roughening

You et al. [82] have recently used x-ray reflectivity to study oxidation-induced surface roughening of Pt(111) in 0.1 M HClO$_4$. Surface roughness can be induced simply by cycling the electrode into the oxide region. They found that there is a critical oxidation charge, of 1.6 electrons per Pt surface atom, below which no roughening takes place. If that charge is exeeded, a roughened surface is found on the subsequent reduction. They also found that the surface roughens progressively on cycling. They suggest that the dynamic scaling behavior proposed for interface growth is applicable to the cyclic oxidation–reduction roughening of Pt(111).

Robinson et al. [83] used SXS to study Au(100) in 0.1 M HClO$_4$. They found that potential excursions into the oxide region did not induce appreciable surface roughening. However, they observed roughening in the hydrogen region (−0.5 V SCE).

B. Adsorbed Monolayers on Metals

1. Pb on Ag(111)

Underpotential deposited (UPD) monolayers are ideal systems to study by either SXS or XAS. The adsorbed metal often has a large atomic number and consequently, has a large scattering cross section for SXS studies. If the layers are ordered, they can be studied by GIXS. The element-specific nature of XAS makes it an ideal candidate for studying UPD layers. XAS gives information on the chemical state of the adsorbed species as well as information of its interaction with the substrate and the electrolyte. Thus it is a powerful probe of double-layer structure.

The most extensive studies on UPD layers have been carried out on UPD Pb on Ag(111). X-ray scattering studies have been done in both perchlorate and acetate electrolytes [84–86]. The results indicate that in both electrolytes the Pb forms an incommensurate hexagonal monolayer compressed compared to bulk Pb. In addition, the Pb monolayer is rotated 4.5° from the Ag[011] direction. Figure 24 shows a representation of one domain of a Pb monolayer on Ag(111). The spacing between the Pb atoms decreases with decreasing electrode potential. This permitted calculation of the two-dimensional compressibility κ_{2D} = 1.25 ± 0.05 Å^2/eV. The compressibilities are identical in both electrolytes, but the near-neighbor spacing is slightly greater in the acetate solution.

Samant et al. [87] obtained XAS spectra on UPD Pb on Ag(111) in the fluorescence mode using a germanium solid-state detector in conjunction with

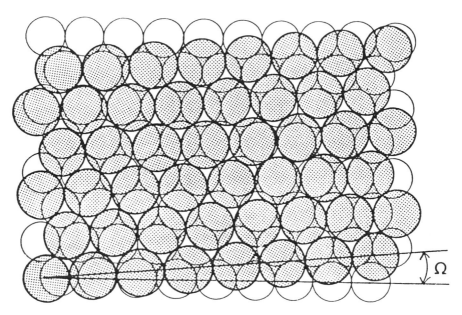

FIGURE 24 Schematic representation of one domain of a monolayer of Pb on Ag(111). The open circles are the Ag atoms and the shaded circles the Pb atoms. The rotational eptiaxy angle between the Ag and Pb lattices is $\Omega = 4.5°$.

a germanium filter. The data quality was exceptionally good. The XANES data showed that the UPD Pb was fully reduced to the zerovalent state. The EXAFS indicate that the Pb species are associated with oxygen species from the electrolyte. The Pb–O bond length changes with potential from 2.33 ± 0.02 Å at 0.53 V to 2.38 ± 0.02 Å at -1.0 V (vs. Ag/AgCl, 3 M KCl). There was no detectable contribution from the Pb–Ag interaction in the EXAFS. This was attributed to either a large thermal motion of the Pb atoms or to an incommensurate Pb layer on the Ag surface.

2. Tl on Ag(111)

SXS studies show that like Pb, Tl forms a two-dimensional incommensurate hexagonal layer that is compressed relative to the bulk and rotated by 4 to 5° with respect to the substrate [88,89]. The layer is also isotropically compressed at negative potentials with $\kappa_{2D} = 1.54 \pm 0.10$ Å^2/eV. An analysis of the intensities along the truncation rods revealed that the Tl layer was 3.05 Å above the Ag surface.

3. Bi on Ag(111)

SXS studies that Bi on Ag(111) forms a totally different structure to that found for Pb or Tl on Ag(111) [90]. The structure is unusual in that it has a rectangular lattice that is uniaxially commensurate with the hexagonal substrate. The structure is illustrated in Figure 25. There are two Bi adatoms per rectangular unit cell, and one adatom is displaced from the centered position by 0.35 Å. The commensurate Bi rows lie along the rows of threefold hollow sites on the Ag(111) surface. Unlike Pb or Tl, the Bi layer compresses uniaxially along the incommensurate direction [Ag(011)], thus preserving the uniaxially commensurate structure. The measured compressibility $\kappa_{2D} = 0.79 \pm 0.04$ Å^2/eV.

4. Ag on Au(111)

There are two reports of EXAFS studies of Ag on Au(111), one in 0.1 M HClO$_4$ [91], the other in 0.5 M NaClO$_4$ [92]. The data in 0.5 M NaClO$_4$ are superior. This is due to the use of a palladium filter and the averaging of many more scans (180 vs. 35). Also, data were obtained with the beam polarization parallel and perpendicular to the electrode surface. The 180 scans, at 30 min/scan, represent a data acquisition time of 90 h. The Ag XANES indicated that the UPD Ag was fully reduced. Analysis of the EXAFS indicated an Ag—Ag bond distance of 2.88 Å and an Ag—Au bond distance of 2.91 Å. The polarization

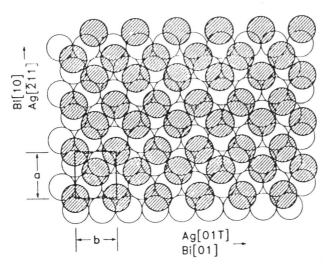

FIGURE 25 Structure of UPD Bi on Ag(111). Open circles are Ag atoms and shaded circles are the Bi atoms. The Bi layer is commensurate in the *a* direction and incommensurate in the *b* direction.

dependence of the EXAFS indicates that the Ag layer was commensurate with the surface, with the Ag atoms occupying threefold hollow sites. Analysis of the EXAFS reveals coordination of the Ag by oxygen species with an Ag—O bond distance of 2.21 Å. Unlike the case of Pb on Ag, the Ag—O bond length did not vary with applied potential. The EXAFS results in 0.1 M HClO$_4$ also indicated coordination of Ag with oxygen and the presence of Ag in threefold hollow sites on Au(111).

5. Bi on Au(111)

There is one report of an SXS study of Bi on Au [93]. The structure of the UPD monolayer, between 10 and 190 mV of the bulk Bi deposition potential, is essentially identical to that described above for Bi on Ag. However, the compressibility is larger, $\kappa_{2D} = 1.15 \pm 0.05$ Å^2/eV. A 25% coverage (2 × 2) hexagonal phase is stable at potentials between 200 and 280 mV positive of the bulk Bi deposition potential.

6. Cu on Au(111)

There are several reports of SXS studies of Cu on Au(111) [76,94,95]. Among these is the first report of the use of SXS to study UPD layers [94]. All the publications confirm the coordination of the adsorbed Cu with oxygen. They also agree that at monolayer coverage the Cu forms a (1 × 1) commensurate layer with the Cu atoms in threefold hollow sites. By using a detector based on a plastic scintillator and photomultiplier, Tadjeddine et al. were able to get XAS spectra with high resolution. The XANES revealed that the Cu was present as Cu$^+$, with charge transfer from the Cu to the Pt. At lower coverages they found evidence for ($\sqrt{3} \times \sqrt{3}$) and c(5 × 5) structures.

7. Cu on Au(100)

Tourillon et al. [96] reported on an XAS study that showed that at monolayer coverage of Cu on Au(100) the Cu exists as Cu$^+$. The Cu was coordinated with four oxygens with a Cu—O bond length of 1.97 Å. There were four Cu neighbors with a Cu—Cu bond distance of 2.66 Å, a value between that found for the Au—Au lattice spacing (2.88 Å) and the Cu—Cu lattice spacing (2.54 Å). This was interpreted as indicative of reconstruction of the Au underlayer. The results were consistent with the adsorbed O stabilizing Cu adsorption on the on top position on Au. On depositing a second Cu layer, the EXAFS clearly indicated that the Cu—O interaction had disappeared. The EXAFS also indicated that the second Cu layer had Cu—Cu spacing that was close to the Cu—Cu lattice spacing. The second layer caused the Cu—Cu spacing of the first layer to relax to a spacing that was commensurate with the Au(100) surface. The second layer also causes the atoms in the first layer to drop into fourfold hollow sites.

8. Cu on Pt

XAS has been used to study UPD Cu on Pt(100) [97], on polycrystalline Pt on float glass [98], and on carbon-supported Pt particles [99,100]. The use of a large polycrystalline electrode (55 cm^2) and carbon-supported Pt are devices to enhance the signal-to-noise ratio. In the case of the carbon-supported Pt it was possible to do measurements in the transmission mode and avoid complications from beam polarization effects. About 50% of the Pt atoms in that carbon-supported particles (diameter = 25 Å) are surface atoms. Thus electronic effects induced by the adsorbate can be detected in the Pt XANES.

XANES at the Cu *K* edge on Pt(100) and on carbon-supported Pt indicate that the adsorbed Cu is Cu$^+$. XANES at the Pt L_3 edge on carbon-supported Pt indicate a partial filling of the Pt *d* bands on the adsorption of Cu [99]. This confirms the transfer of charge to the Pt substrate. All the EXAFS data indicate the coordination of Cu with oxygen. The EXAFS study of UPD Cu on Pt by Durand et al. [97] in 1 *M* NaClO$_4$ + 10^{-3} *M* HClO$_4$ indicated that the UPD Cu atoms were associated with 0.6 oxygen atoms at a Cu—O distance of 1.97 Å. Since the study was done on a horizontal flat single-crystal surface the Cu—Pt interactions could not be seen because of the polarization of the x-ray beam in the plane of the synchrotron ring. On the basis of the absence of a Cu—Pt coordination shell they concluded that the Cu was adsorbed on the top position on the Pt and not in hollow sites. In the latter case there would be a Cu—Pt contribution since the Cu—Pt bonds would not be orthogonal to the plane of polarization of the beam. An interesting dual-layer model was used to account for the Cu—Cu interactions. In situ EXAFS studies of adsorbed copper on carbon-supported platinum were done in 0.5 M H$_2$SO$_4$ + 2 × 10^{-3} *M* Cu^{2+} at 0.05 V SCE. Data analysis of the nearest coordination shells around the copper is consistent with Cu adsorption on Pt on the top position with a Cu—Pt bond length of 2.67 Å. The results also indicate that copper is also coordinated with three oxygens at a distance of 2.06 Å and one sulfur at a distance of 2.37 Å. This indicates coadsorption of bisulfate ions. The bisulfate is probably adsorbed on top of the copper, where it facilitates charge transfer from the copper to the platinum and yields the adsorbed Cu$^+$ species that have been observed in XANES studies. A model for the adsorption is given in Figure 26 [100].

9. Pb on Pt

SXS has been used to study UPD Pb on Pt(111), Pt(110), and Pt(100), [101]. No ordered structures were observed on either Pt(110) or Pt(100), so no information could be obtained from SXS. In the case of Pt(111), ordered structures were found only at potentials close to the bulk Pb deposition potential. The adsorbed monolayer had a rectangular (3 × 3) cell, with each Pb having four

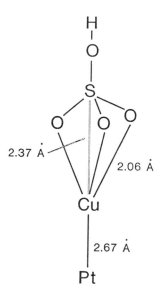

FIGURE 26   Model to fit EXAFS of UPD copper on carbon-supported Pt.

neighboring atoms at 3.18 Å and two others at 4.81 Å. X-ray reflectivity measurements indicated that the Pb adlayer was spaced 2.34 Å above the Pt substrate.

Recently, in situ x-ray absorption spectroscopy (XAS) was used to study the structure of adsorbed Pb on carbon-supported Pt in 1 M HClO$_4$ + 5 × 10^{-3} M Pb^{2+}, in the potential range of −0.24 to 1.15 V SCE [102]. The XAS measurements were done using a Canberra 13-element germanium detector. The XANES measurements indicated that in the UPD region, the Pb species are essentially neutral Pb atoms. At all potentials positive to the main UPD peak in the cyclic voltammogram, the Pb is in the Pb(II) state. Analysis of the EXAFS data for potentials more negative to 0.0 V SCE required a two-shell fit involving Pb—Pb and Pb—Pt interactions. At −0.24 V SCE the Pb—Pb bond distances were 3.05 Å, and the Pb—Pt bond distances were 2.86 Å. The results are in excellent agreement with the SXS results for Pb on Pt(111). At more positive potentials, in the UPD region, the data could be fitted to a single Pb—Pt shell, indicating a high degree of lateral disorder in the layer. There is no evidence of Pb interaction with oxygenated species in the UPD region. On stripping the UPD layer the Pb was present as hydrated Pb^{2+} ions. At more positive potentials there was clear evidence of incorporation of Pb into the platinum oxide layer. The EXAFS results indicate that repeated cycling changes the nature of the UPD layer. These XAS results illustrate the complementary information that can be obtained by doing the SXS and XAS measurements on UPD layers. Also, XAS does not have the limitation of requiring an ordered layer.

10. I on Au(111) and Pt(111)

Recent SXS studies of Au(111) in 0.01 M KI show that adsorbed iodine on Au(111) forms an electrocompressive ($p \times \sqrt{3}$) phase [103]. At a critical coverage of 0.409 the ($p \times \sqrt{3}$) phase transforms to a rotated hexagonal phase. X-ray reflectivity reveals that the Au—I interlayer spacing is only 2.3 Å, indicating strong chemisorption.

There is one brief report on an in situ XAS study of I on Pt(111) [104]. Analysis of the EXAFS gave a Pt—I bond length of 2.64 Å, a value in good agreement with ex situ determinations.

C. Chemically Modified Electrodes

Kim et al. [105] have reported a XANES study of (μ-oxo)bis[iron *meso*-tetrakis(4-methoxyphenyl)porphyrin] (FeTMPP)$_2$O, irreversibly adsorbed on high-surface-area carbon (Black Pearls) as a function of potential. The XANES indicate a change in coordination symmetry during the oxidation–reduction process. In the adsorbed, oxidized state the (FeTMPP)$_2$O retains its μ-oxo character. In the reduced state it exists as FeTMPP without axial ligation.

Elder et al. [106] used XAS to follow the copper redox reaction in a Nafion-coated colloidal graphite electrode. The copper was incorporated into the Nafion as the copper(I) complex bis(2,9-dimethyl-1,10-phenanthroline) copper(I)tetrafluoroborate (Cu^I(DMP)$_2$BF$_4$). XANES results showed conversion to Cu(II) after prolonged electrolysis. The Cu–N bond length decreased from 2.06 Å to 2.02 Å, and the coordination number increased from four to five.

Albarelli et al. [107] have done similar XANES studies on Pt electrodes modified with electropolymerized films of [M(v-bpy)$_3$]$^{2+}$, where v-bpy is 4-vinyl-4'-methyl-2,2'-bipyridene and M is Ru or Os. Changes in the oxidation state of both metal elements could be observed. In addition, the Os XANES indicated a change in coordination symmetry on oxidation.

D. Metal Deposition

GIXS has been used to study UPD Pb on Ag(111) and the initial stages of bulk Pb deposition [108]. As discussed above, the UPD layer was compressed 1.4% relative to bulk Pb. This compressive strain increases with applied potential, and at the onset of bulk deposition the ad-layer is compressed 2.8%. This large compressive strain inhibits expitaxial deposition and the Pb grows as islands with a (111) texture that are randomly oriented in the plane of the substrate. After deposition of about approximately five monolayers, the initial ad-layer appears to reconstruct. This behavior was also seen on the deposition od the second layer of Cu on Au(100) [96].

There has been a recent report of in situ x-ray reflectivity and x-ray diffraction studies of the electrodeposition of thick films of Ni and NiFe alloy [109]. The authors used an ingenious flow cell design that permitted simultaneous x-ray studies, and electrodeposition with good current distribution, under carefully controlled hydrodynamic conditions. In the case of Ni deposition from 0.2 M NiCl$_2$, diffraction peaks for nickel oxide–hydroxide were observed when either the current density or the bulk pH was high. This is consistent with the precipitation of nickel oxide/hydroxide due to the high-rate cathodic reaction, causing an increase in pH at the electrode surface. In situ reflectivity measurements were done during the electrodeposition of the equivalent of 80 nm of NiFe alloy, over a period of 2.5 h, on top of a smooth 110-nm sputtered NiFe film. No change in the oscillation period of the reflectivity was observed during the electrodeposition. This was interpreted as being due to the formation of island deposits rather than a smooth epitaxial layer.

E. Corrosion

The early work on the application of XAS to corrosion and the passive film on iron has been reviewed by Long and Kruger [110]. Several groups are applying XAS and x-ray reflectivity to the study of oxide films on metals. The experimental difficulties in applying XAS to the studies of passive films have been reviewed by Kerkar et al. [111]. These can be mitigated through the use of very thin layers of metal so that when they are passivated essentially all the metal is converted to the passive film. The technique was first proposed by Long et al. [112] and has been adapted by later workers [111,113–116]. The metal films of interest are sputtered on to Au- or Ta-coated thin Mylar film. The underlying Au or Ta film serves as a current collector. Kerkar et al. have obtained both XANES and EXAFS of passive films on iron. The EXAFS indicate the presence of edge-shared FeO$_6$ octahedra. Alloying with Cr increases the disorder in the passive film. The Cr EXAFS indicate that Cr is incorporated into the passive layer as a phase essentially identical to Cr(OH)$_3$. Bardwell et al. [114] used XANES studies at the Cr K edge to establish the formation of Cr(VI) in the transpassive region. Davenport et al. [113] have used XANES to study the passive film on sputtered AlCr thin films in a borate buffer electrolyte. The results show that the valence state of the chromium in the passive film can be changed reversibly from Cr(III) to Cr(IV) by stepping the potential between −1.5 and 2.0 V MSE. If the potential is stepped, there is no dissolution of the hexavalent chromium, normally a highly soluble species. If the potential is increased slowly rather than stepped, the Cr(IV) dissolves [115].

There are preliminary reports on the application of x-ray reflectivity to the study of oxide films on copper [117–119]. Nagy and co-workers have developed an excellent cell design for these studies. The technique offers great promise in the investigation of oxide films and in the elucidation of the buried metal–oxide interface.

F. Battery and Oxide Electrodes

1. Ni(OH)$_2$ Electrodes

McBreen and his co-workers have carried out an extensive study of the β-Ni(OH)$_2$ electrode [120–124] in 8.4 M KOH + 0.5 M LiOH. XAS spectra were obtained in the transmission mode using a spectroelectrochemical cell shown in Figure 27. Figure 21 shows a comparison of the Fourier transforms for charged and discharged hydroxide. Analysis of the EXAFS gave structural parameters for the discharged material that were in agreement with the best neutron diffraction data [120,123]. The structure of the charged material was unknown because it is amorphous. This work elucidated the structure of the charged material for the first time. The EXAFS showed that the Ni(OH)$_2$ layers in the brucite structure undergo a contraction during oxidation to NiOOH. The octahedral oxygen coordination around the Ni undergoes a tetragonal distortion yielding four short Ni—O bonds with a bond distance of 1.88 Å and two with bond distances of 2.07 Å. The changes in the XANES features were consistent with the results of the EXAFS analysis. The white line was lower in the oxidized material because of the distortion in the coordination symmetry (see Fig. 18). The investigation of the Ni(OH)$_2$ electrode also included a time-resolved XANES [121] and EXAFS [124] study, wherein time-resolved XAS was coupled with cyclic voltammetry.

Capehart et al. [125] have done an XAS study of the oxidation of α-Ni(OH)$_2$. They found that the Ni—O bond distance contracted from 2.05 Å to 1.86 Å on oxidation. They argued that the shorter bond distance was consistent with a 3.67 valence state of the oxidized phase. The valence state had been determined from electrochemical and chemical measurements. No XANES data were presented to support the higher valence state.

2. V$_2$O$_5$ Electrodes

Cartier et al. [126] have used XAS to study the reduction of V$_2$O$_5$ in a LiAsF$_6$/propylene carbonate-based electrolyte, as the reduction proceeds from 0 to 2.4 lithiums per mole of V$_2$O$_5$. The XANES data showed the conversion of V(V) to V(IV) during the reduction. The EXAFS data showed that the reduction process introduced disorder in the oxide structure. This begins at 0.9 Li per mole of V$_2$O$_5$ and is more pronounced when 2.4 Li are intercalated. The results indicated that the square pyramid V$_2$O$_5$ units are gradually modified as the reduction proceeds and are probably broken up at deep depths of discharge. This has implications regarding the reversibility of the system.

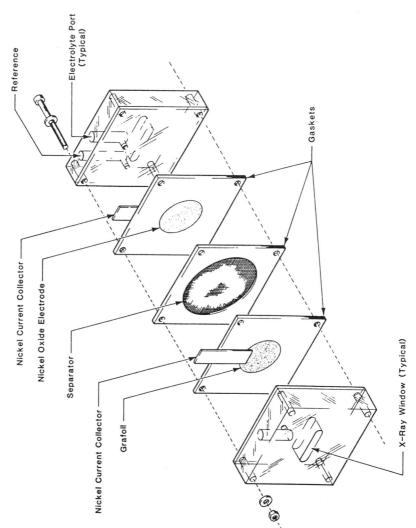

FIGURE 27 Spectroelectrochemical cell for in situ studies of nickel oxide electrodes.

3. Deposition of Mossy Zinc

McBreen [127] used XAS to study the deposition of mossy zinc in 7.4 M KOH + 0.74 M ZnO. The zinc was deposited in a thin cavity at either −45 mV or −65 mV vs. a zinc wire reference in the same electrolyte. The EXAFS for the deposit at −65 mV showed that the electrolyte within the pores of the mossy deposit was pure KOH. All the zincate was exhausted. At −45 mV there was some zincate in the electrode pores. The first peak of the Fourier transform of the EXAFS was considerably reduced in comparison to that found for Zn foil. This strongly suggests that mossy zinc may be a nanophase material. There was also evidence, from shifts in the peaks of the Fourier transform, that at −65 mV the deposit was oriented with the c axis parallel to the direction of the current.

4. Chemically Modified MnO_2

Conway et al. [128] have used XAS at the Mn K edge and the Bi L_3 edge to study Bi modified MnO_2 in 9 M KOH. The Bi modification renders the normally irreversible MnO_2 electrode rechargeable. The XAS data clearly showed reduction to Bi metal on deep discharge. On charge the Bi is oxidized prior to the onset of oxidation of the manganese active material. In partially charged and fully charged electrodes the Bi is Bi(III) but is not Bi_2O_3. There is some evidence that in the initial stages of charge the Bi(III) species are on the surface of the manganese oxide particles. A complete set of experiments were run at the Mn K edge at various stages of charge and discharge. The Mn results indicate that at the end of the one-electron discharge the manganese oxide is highly disordered. However, on completion of the two-electron discharge the product reverts to an ordered phase. The EXAFS of the recharged electrode was identical to that for the undischarged electrode. These results elegantly confirm the reversibility of the system, particularly with respect to structure changes.

5. Cu_2O Electrodes

Druska and Strehblow [129] have done XAS studies on Cu_2O layers formed by the electrochemical reduction of saturated CuO_2^{2-} in 5 M KOH. The XAS studied were done in a cell similar to that used by McBreen [120]. The electrolyte was a borate buffer. The EXAFS results showed that the Cu_2O layers have a near-order structure similar to that of crystalline Cu_2O.

G. Fuel Cell Electrocatalysts

1. Pt on Carbon

Weber et al. [130] have done in situ XAS studies during oxygen reduction on a carbon-supported Pt fuel cell electrode in 0.5 M H_2SO_4. By observing the intensity of the XANES white line at the Pt L_3 edge, they could follow the reduction of the surface oxide with potential. They concluded that the surface stoichiometry on the particles was closer to $PtO_{0.5}$ rather than PtO, which is proposed for the surface of bulk electrodes. In their paper they stress that XAS is an averaging technique. This can present problems in investigating gas diffusion electrodes. If only a small portion of the catalyst is utilized, the signal from the unutilized catalyst that is transited by the x-ray beam will mask the true results from the active catalyst. To avoid this problem, Robinson and co-workers [131,132] have worked with carbon-supported platinum catalysts in the flooded mode, in 1 M H_2SO_4. They obtained XAS spectra in the potential range 0.1 to 1.2 V SCE. They observed considerable enhancement of the white line on going to positive potentials. Analysis of the EXAFS confirmed that the oxide had a short-range structure similar to α-PtO_2.

McBreen and co-workers [133] also use flooded electrodes in XAS studies of carbon-supported Pt and Pt alloy electrocatalysts. By obtaining data at the L_2 and L_3 edges they could determine the effect of potential, anion absorption, and alloying on the occupancy of the Pt d bands. They studied the effect of potential on the Pt XANES at the L_3 edge in 1 M $HClO_4$. They found an enhancement of the white line at positive potentials similar to the results of Robinson [131]. They also found a broadening of the white line in the hydrogen adsorption region similar to that observed by Boudart for gas-phase adsorption [134]. Boudart attributes it to electronic transitions into Pt—H antibonding orbitals. Figure 28 illustrates the effect of anion adsorption in the double-layer region (0.3 V SCE). Sulfate adsorption increases the white line slightly. Perchlorate, a non-adsorbing anion, yields spectra that are identical to those found for Pt foil. By studying the effect of alloying additions of first-row transition metals at 0.3 V SCE in 1 M $HClO_4$, it was possible to elucidate the effect of the alloying addition on the Pt d-band occupancy. Analysis of the EXAFS at the Pt L_3 edge gave information on the effect of the alloying addition on the Pt—Pt bond distances. There was a 1:1 correlation between the d-band occupancy and Pt—Pt bond length. A comparison of results at 0.3 V and 0.6 V SCE revealed that the alloying additions inhibited the formation of surface oxides on Pt at 0.6 V. This may be the source of the enhanced catalysis of oxygen reduction by the alloys.

H. Conducting Polymer Electrodes

Tourillon and co-workers found that the conductivity of poly(3-methyl thiophene), having $SO_3CF_3^-$ as the counterion, could be improved by a factor of 3 by cathodic polarization in an aqueous solution of $CuCl_2$ or $FeCl_3$ [135]. Prolonged polarization in $CuCl_2$ resulted in the formation of metallic Cu clusters in the polymer. In a series of studies they used time-resolved XAS studies to

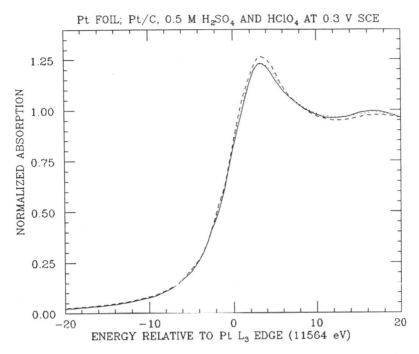

Pt FOIL; Pt/C, 0.5 M H₂SO₄ AND HClO₄ AT 0.3 V SCE

ENERGY RELATIVE TO Pt L₃ EDGE (11564 eV)

NORMALIZED ABSORPTION

FIGURE 28 Comparison of the XANES for Pt foil (—) and a Prototech carbon-supported Pt catalyst in 0.5 *M* H₂SO₄ (- - -) and 1 *M* HClO₄ (••••) and at 0.3 V SCE.

elucidate the mechanism of the processes occurring in the polymer during cathodic polarization in aqueous $CuCl_2$ and $FeCl_3$ electrolytes [135–137]. The results of the XAS studies indicated that in the case of $CuCl_2$ the Cu^{2+} is first reduced to Cu^+. Initially, the Cu^+ is associated with oxygen due to ion pairing with $SO_3CF_3^-$. Further polarization results in coordination of the Cu^+ with S from the polymer. The bridging of the polymer chains through the Cu—S linkages increases the conductivity. Prolonged polarization results in the nucleation of Cu clusters on the polymer fibers by a disproportionation mechanism. In the case of $FeCl_3$ it was found that the interaction of the Fe^{3+} with the S was immediate upon polarization. The Fe^{3+} was found to be coordinated with both S and O, indicating partial hydration of the Fe^{3+}. Further polarization resulted in the reduction of Fe^{3+} to hexa-aquo Fe^{2+} ions and interaction with the S is lost. No formation of Fe clusters occurs due to dedoping of the polymer. These studies demonstrate the power of time-resolved XAS.

I. Far-Infrared Studies

Recently, far-infrared radiation has also become available at NSLS [138]. This radiation is between 100 and 1000 times higher than that available from standard blackbody sources. There has been one report of the use of this beamline for in situ electrochemical studies of the Ag(111) surface in 0.1 M HClO$_4$ [139]. The authors confirmed the decomposition of ClO$_4^-$ to form adsorbed chloride.

VI. CONCLUSIONS AND FUTURE DIRECTIONS

The application of in situ synchrotron radiation techniques to electrochemistry is still in its infancy. The work reviewed here shows the great versatility of the techniques. This ability to study electrode processes at the molecular level brings electrochemistry once again to the forefront of surface science. It is also attracting many physicists to the field. The element-specific nature of XAS permits fundamental investigations of the complex electrode systems in batteries and fuel cells. Synchrotron sources with higher intensity will benefit x-ray scattering studies on dilute systems. Also, advances in solid-state detectors will benefit studies of surface layers and time-resolved measurements. High-intensity sources will have less impact on XAS, though they may permit photoemission studies. The present trend in increased application of spectroscopic methods to electrochemistry will continue. Synchrotron techniques will be important since they provide both chemical and structural information.

ACKNOWLEDGMENT

This work was supported by the U.S. Department of Energy Division of Conservation and Renewable Energy under Contract DE-AC02-76CH00016.

REFERENCES

1. D. E. Sayers, E. A. Stern, and F. W. Lytle, New technique for investigating non-crystalline structures: Fourier analysis of the extended x-ray absorption fine structure, *Phys. Rev. Lett.*, 27: 1204 (1971).
2. W. H. McMaster, N. Kerr Del Grande, J. H. Mallet, and J. H. Hubbell, *Compilation of X-Ray Cross Sections*, Lawrence Radiation Laboratory, Livermore, Calif., 1969.
3. G. Tourillon, D. Guay, F. Lemonnier, F. Bartol, and M. Badeyan, X-ray absorption spectroscopy: a fluorescence detection system based on a plastic scintillator, *Nucl. Inst. Methods*, A294: 382 (1990).
4. A. J. Salkind and P. F. Bruins, Nickel cadmium cells. I. Thermodynamics and x-ray studies, *J. Electrochem. Soc.*, 109: 256 (1962).
5. S. U. Falk, Investigations on reaction mechanism of the nickel cadmium cell, *J. Electrochem. Soc.*, 107: 661 (1960).

6. G. W. D. Briggs, An instrument for x-ray diffraction of electrodes, *Electrochim. Acta, 1*: 297 (1959).
7. G. W. D. Briggs and W. F. K. Wynne-Jones, Nickel hydroxide electrode; the effects of ageing. I. X-ray diffraction study of the electrode process, *Electrochim. Acta, 7*: 241 (1962).
8. R. R. Chianelli, J. C. Scanlon, and B. M. L. Rao, Dynamic x-ray diffraction, *J. Electrochem. Soc., 125*: 1563 (1978).
9. M. Fleischman, A. Oliver, and J. Robinson, In situ x-ray diffraction studies of electrode solution interfaces, *Electrochim. Acta, 31*: 899 (1986).
10. M. Fleischman and B. W. Mao, In situ x-ray diffraction studies of Pt electrode/solution interfaces, *J. Electroanal. Chem., 229*: 125 (1987).
11. J. N. Andrews and A. R. Ubbelhode, Overvoltage and diffusion of hydrogen through iron and palladium, *Proc. Roy. Soc., A253*: 6 (1959).
12. K. Machida and M. Enyo, In-situ x-ray diffraction study of metal hydride formation in palladium and palladium-based alloy electrodes under cathodic charging, *Bull. Chem. Soc. Jpn., 62*: 415 (1989).
13. G. Nazri and R. H. Muller, In situ x-ray diffraction of surface layers on lithium in nonaqueous electrolyte, *J. Electrochem. Soc., 132*: 1385 (1985).
14. N. G. Sudorgin and V. B. Nalbandyan, Quantitative x-ray monitoring of electrode process in sealed cells reduction of zirconium β-molybdate by lithium, *Elektrokhimiya, 28*: 122 (1992).
15. M. F. Toney and O. R. Melroy, Surface x-ray scattering, in *Electrochemical Interfaces: Modern Techniques for In-Situ Interface Characterization* (H. D. Abruña, ed.), VCH Publishers, New York, 1991, p. 55.
16. H. D. Abruña, X-ray probes of electrochemical interfaces, in *Modern Aspects of Electrochemistry*, Vol. 20 (J. O'M. Bockris, R. E. White, and B. E. Conway, eds.), Plenum Press, New York, 1989, p. 265.
17. H. D. Abruña, Probing electrochemical interfaces with x-rays, *Adv. Chem. Phys., 77*: 255 (1990).
18. I. K. Robinson, Structure factor determination in surface x-ray diffraction, *Aust. J. Phys., 41*: 359 (1988).
19. M. G. Samant, M. F. Toney, G. L. Borges, L. Blum, and O. R. Melroy, Grazing incidence x-ray diffraction of lead monolayers at a silver(111) and gold(111) electrode/electrolyte interface, *J. Phys. Chem., 92*: 220 (1988).
20. Z. Nagy, H. You, R. M. Yonco, C. A. Melandres, W. Yun, and V. A. Maroni, Cell design for in situ x-ray scattering study of electrodes in transmission geometry, *Electrochim. Acta, 36*: 209 (1991).
21. K. M. Robinson and W. E. O'Grady, A transmission geometry electrochemical cell for in situ x-ray diffraction, *Rev. Sci. Instrum., 64*: 1061 (1993).
22. W. R. Busing and H. A. Levy, Angle calculations for 3- and 4-circle x-ray and neutron diffractometers, *Acta Cryst., 22*: 457 (1976).
23. M. Born and W. Wolf, *Principles of Optics*, Pergamon Press, Oxford, 1970.
24. F. A. Jenkins and H. E. White, *Fundamentals of Optics*, McGraw-Hill, New York, 1957.
25. B. D. Cullity, *Elements of X-Ray Diffraction*, Addison-Wesley, Reading, Mass., 1956.

26. C. S. Barrett and T. B. Massalski, *Structure of Metals: Crystallographic Methods, Principles and Data*, McGraw-Hill, London, 1966.

27. G. H. Stout and L. H. Jensen, *X-Ray Structure Determination*, Wiley, New York, 1989.

28. M. J. Buerger, *X-Ray Crystallography*, Wiley, New York, 1942.

29. L. J. Clarke, *Surface Crystallography*, Wiley, New York, 1985.

30. I. K. Robinson, Crystal truncation rods and surface roughness, *Phys. Rev.*, *B33*: 3830 (1986).

31. K. M. Robinson, I. K. Robinson, and W. E. O'Grady, Structure of Au(100) and Au(111) single crystals surfaces prepared by flame annealing, *Surface Sci.*, *262*: 387 (1992).

32. I. K. Robinson, Surface crystallography, in *Handbook of Synchrotron Radiation*, Vol. 3 (G. Brown and D. E. Moncton, eds.), Elsevier Science Publishers, Amsterdam, 1991, p. 221.

33. I. K. Robinson and D. J. Tweet, Surface x-ray diffraction, *Rep. Prog. Phys.*, *55*: 599 (1992).

34. P. H. Fuoss and S. Brennan, Surface sensitive x-ray scattering, *Annu. Rev. Mater. Sci.*, *20*: 365 (1990).

35. R. Fridenhans'l, Surface structure determination by x-ray diffraction, *Surface Sci. Rep.*, *10*: 105 (1989).

36. D. Gibbs, B. M. Ocko, D. M. Zehner, and S. G. J. Mocherie, Absolute x-ray reflectivity study of the Au(100) surface, *Phys. Rev.*, *B38*: 7303 (1988).

37. L. G. Parratt, Surface studies of solids by total reflectance of x-rays, *Phys. Rev.*, *95*: 350 (1954).

38. H. Yoo, C. A. Melandres, Z. Nagy, V. A. Maroni, W. Yun, and R. M. Yoncoo, X-ray reflectivity study of the copper–water interface in a transmission geometry under in situ electrochemical control, *Phys. Rev.*, *B45*: 11,288 (1992).

39. B. M. Ocko and S. G. J. Mocherie, Reversible faceting of the copper(110) surface: x-ray Fresnel reflectivity, *Phys. Rev.*, *B38*: 7378 (1988).

40. B. M. Ocko, D. Gibbs, K. G. Huang, D. M. Zehner, and S. G. J. Mocherie, Structure and phases of the Au(001) surface: absolute x-ray reflectivity, *Phys. Rev.*, *B44*: 6429 (1991).

41. S. K. Sinha, E. B. Sirota, S. Garoff, and H. B. Stanley, X-ray and neutron scattering from rough surfaces, *Phys. Rev.*, *B38*: 2297 (1988).

42. D. G. Wiesler, M. F. Toney, C. S. McMillan, and W. H. Smyrl, Interfacial density profiles of anodic oxides of tantalum and niobium measured by x-ray reflectivity, in *The Applications of Surface Analysis Methods to Environmental/Materials Interactions* (D. R. Baer, C. R. Clayton, and G. D. Davis, eds.), The Electrochemical Society, Pennington, N.J., 1991, p. 440.

43. J. J. Benattar, J. Daillant, L. Bosio, and L. Leger, Physical properties of ultra thin films studied by x-ray optical techniques: Langmuir–Blodgett multilayer structures, organic monolayers on water and the spreading of polymer micro-droplets, *J. Phys.*, *C7*: 39 (1989).

44. J. Als-Nielsen, X-ray reflectivity studies of liquid surfaces, in *Handbook of Synchrotron Radiation*, Vol. 3 (G. Brown and D. E. Moncton, eds.), Elsevier Science Publishers, Amsterdam, 1991, p. 471.

45. J. Als-Nielsen and H. Möhwald, Synchrotron x-ray scattering studies of Langmuir films, in *Handbook of Synchrotron Radiation*, Vol. 4 (S. Ebashi, M. Koch, and E. Rubenstein, eds.), Elsevier Science Publishers, Amsterdam, 1991, p. 1.
46. M. J. Bedzyk, D. Bilderback, J. White, H. D. Abruña, and M. G. Bommarito, Probing electrochemical interfaces with x-ray standing waves, *J. Phys. Chem.*, *90*: 4926 (1986).
47. G. Materlik, M. Schmäh, J. Zegenhagen, and W. Uelhoff, Structure determination on single crystal electrodes with x-ray standing waves, *Ber. Bunsenges. Phys. Chem.*, *91*: 292 (1987).
48. H. D. Abruña, M. G. Bommarito, and D. Acevedo, The study of solid liquid interfaces with x-ray standing waves, *Science, 250*: 69 (1990).
49. R. W. James, *The Optical Principles of the Diffraction of X-Rays*, Oxbow Press, Woodbridge, Conn., 1982.
50. T. W. Barbee, Jr., and W. K. Warbuton, X-ray evanescent- and standing-wave fluorescence studies using a layered synthetic microstructure, *Mater. Lett., 3*: 17 (1984).
51. B. W. Batterman and H. Cole, Dynamical diffraction of x-rays by perfect crystals, *Rev. Mod. Phys., 36*: 681 (1964).
52. B. K. Teo, *EXAFS: Basic principles and Data Analysis*, Springer-Verlag, Berlin, 1986.
53. D. C. Koningsberger and R. Prins, *X-Ray Absorption: Principles, Applications, Techniques of EXAFS, SEXAFS and XANES*, Wiley, New York, 1988.
54. P. A. Lee, P. H. Citrin, P. Eisenberger, and B. M. Kincaid, Extended x-ray absorption fine structure: its strengths and limitations as a structural tool, *Rev. Mod. Phys., 53*: 769 (1981).
55. T. M. Hayes and J. B. Boyce, Extended x-ray absorption fine structure spectroscopy, *Solid State Phys., 37*: 173 (1982).
56. E. A. Stern and S. M. Heald, *Handbook of Synchrotron Radiation*, Vol. 1 (E. E. Koch, ed.), Elsevier Science Publishers, Amsterdam, 1983, p. 955.
57. H. Winick and S. Doniach, *Synchrotron Radiation Research*, Plenum Press, New York, 1980.
58. L. R. Sharpe, W. R. Heineman, and R. C. Elder, EXAFS spectroelectrochemistry, *Chem. Rev., 90*: 705 (1990).
59. A. Bianconi, XANES spectroscopy, in *X-Ray Absorption: Principles, Applications, Techniques of EXAFS, SEXAFS and XANES* (D. C. Koningsberger and R. Prins, eds.), Wiley, New York, 1988, p. 573.
60. F. W. Lytle, R. B. Greegor, and A. J. Panson, Discussion of x-ray absorption near-edge structure: application to Cu in the high-T_c superconductors $La_{1.8}Sr_{0.2}CuO_4$ and $YBa_2Cu_3O_7$, *Phys. Rev., B37*: 1550 (1988).
61. D. E. Sayers and B. A. Bunker, Data analysis, in *X-Ray Absorption: Principles, Applications, Techniques of EXAFS, SEXAFS and XANES* (D. C. Koningsberger and R. Prins, eds.), Wiley, New York, 1988, p. 211.
62. J. B. A. D. vanZon, D. C. Koningsberger, H. F. J. van't Blik, and D. E. Sayers, An EXAFS study of the structure of the metal–support interface of highly dispersed Rh/Al_2O_3 catalysts, *J. Chem. Phys., 82*: 5742 (1985).

63. F. B. M. Duivenvoorden, D. C. Koningsberger, Y. S. Uh, and B. C. Gates, Structures of alumina supported osmium clusters $(HOs_3(CO)_{10}\{OAl\})$ and complexes $(Os^{II}(CO)_{n=2or3}\{OAl\}_3)$ determined by extended x-ray absorption fine structure spectroscopy, *J. Am. Chem. Soc.*, *108*: 6254 (1986).

64. E. Dartyge, L. Depautex, J. M. Dubuisson, A. Fontaine, A. Jucha, P. Leboucher, and G. Tourillon, X-ray absorption in dispersive mode: a new spectrometer and a data acquisition system for fast kinetics, *Nucl. Instrum. Methods*, *A246*: 452 (1986).

65. H. Tolentino, E. Dartyge, A. Fontaine, and G. Tourillon, X-ray absorption spectroscopy in the dispersive mode with synchrotron radiation: optical considerations, *J. Appl. Cryst.*, *21*: 15 (1988).

66. F. D'Acapito, F. Boschenerini, A. Marcelli, and S. Mobilio, Dispersive EXAFS apparatus at Frascati, *Rev. Sci. Instrum.*, *63*: 899 (1992).

67. P. G. Allen, S. D. Conradson, and J. E. Penner-Hahn, A 4-point crystal bender for dispersive x-ray absorption spectroscopy, *J. Appl. Cryst.*, *26*: 172 (1992).

68. A. J. Dent, M. P. Wells, R. C. Farrow, C. A. Ramsdale, G. E. Derbyshire, G. N. Greaves, J. W. Couves, and J. M. Thomas, Combined energy dispersive EXAFS and x-ray diffraction, *Rev. Sci. Instrum.*, *63*: 903 (1992).

69. M. E. Kordesch and R. W. Hoffman, Electron yield extended x-ray absorption fine structure with the use of a gas-flow electron detector, *Phys. Rev.*, *B29*: 491 (1984).

70. G. Tourillon, E. Dartyge, A. Fontaine, M. Lemonnier, and F. Bartol, Electron yield x-ray absorption spectroscopy at atmospheric pressure, *Phys. Lett.*, *A121*: 251 (1987).

71. T. Elam, J. P. Kirkland, R. A. Neiser, and P. D. Wolf, Depth dependence for extended x-ray absorption fine-structure spectroscopy detected via electron yield in He and in vacuum, *Phys. Rev.*, *B38*: 26 (1988).

72. T. Gireaudeau, J. Mimault, M. Jaouen, P. Chartier, and G. Tourillon, Sampling depth in conversion-electron detection used for x-ay absorption, *Phys. Rev.*, *B46*: 7144 (1992).

73. M. E. Kordesch and R. W. Hoffman, Electrochemical cells for in situ EXAFS, *Nucl. Instrum. Methods*, *222*: 347 (1984).

74. K. I. Pandya, K. Yang, R. W. Hoffman, W. E. O'Grady, and D. E. Sayers, Electron yield detectors for near surface EXAFS at atmospheric pressure, *J. Phys.*, *C8*: 159 (1986).

75. G. J. Hansen and W. E. O'Grady, A cell for x-ray absorption studies of the emersed electrochemical interface, *Rev. Sci. Instrum.*, *61*: 2127 (1990).

76. O. R. Melroy, M. G. Samant, G. L. Borges, J. G. Gordon II, L. Blum, J. H. White, M. J. Albarelli, M. McMillan, and H. D. Abruña, In plane structure of underpotentially deposited copper on gold(111) determined by surface EXAFS, *Langmuir*, *4*: 728 (1988).

77. L. R. Furenlid, J. Beren, R. H. Beuttenmuller, H. W. Kraner, L. C. Rogers, S. Rescica, D. Stephani, and S. P. Cramer, Fluoresence XAFS detectors: new tools for electrochemical interface characterization, *Proc. Symp. X-Ray Methods in Corrosion and Interfacial Electrochemistry* (A. Davenport and J. G. Gordon II, eds.), The Electrochemical Society, Pennington, N.J., 1992, p. 364.

78. J. Wang, A. J. Davenport, H. S. Isaacs, and B. M. Ocko, Surface charge induced ordering of the Au(111) surface, *Science*, *255*: 1416 (1992).

79. B. M. Ocko, A. Gibaud, and J. Wang, Surface x-ray diffraction study of the Au(111) electrode in 0.01 *M* NaCl: electrochemically induced surface reconstruction, *J. Vacuum Sci. Technol.*, *A10*: 3019 (1992).

80. J. Wang, B. M. Ocko, A. J. Davenport, and H. S. Isaacs, In situ x-ray diffraction and -reflectivity studies of the Au(111)/electrolyte interface: reconstruction and anion absorption, *Phys. Rev.*, *B46*: 10,321 (1992).

81. B. M. Ocko, J. Wang, A. Davenport, and H. Isaacs, In situ X-ray reflectivity and diffraction studies of the Au(100) reconstruction in an electrochemical cell, *Phys. Rev. Lett.*, *65*: 1466 (1990).

82. H. You, D. J. Zurawski, Z. Nagy, and R. M. Yonco, In situ x-ray reflectivity study of incipient oxidation of Pt(111) surface in electrolyte solutions, submitted.

83. K. M. Robinson, I. K. Robinson, and W. E. O'Grady, Electrochemically induced surface roughness on Au(100) studied by surface x-ray diffraction, *Electrochim. Acta*, *37*: 2169 (1992).

84. M. G. Samant, M. F. Toney, G. L. Borges, L. Blum, and O. R. Melroy, In situ grazing incidence x-ray diffraction study of electrochemically deposited Pb monolayers on Ag(111), *Surface Sci.*, *193*: L29 (1988).

85. M. G. Samant, M. F. Toney, G. L. Borges, L. Blum, and O. R. Melroy, *J. Phys. Chem.*, *92*: 220 (1988).

86. J. B. Kortright, P. N. Ross, O. R. Melroy, M. F. Toney, G. L. Borges, and M. G. Samant, Electrochemically adsorbed Pb on Ag(111) studied with grazing incidence x-ray scattering, *J. Phys.*, *C7*: 153 (1989).

87. M. G. Samant, G. L. Borges, J. G. Gordon II, O. R. Melroy, and L. Blum, In situ surface extended x-ray absorption fine structure spectroscopy of a lead monolayer at a silver(111) electrode/electrolyte interface, *J. Am. Chem. Soc.*, *109*: 5970 (1987).

88. M. F. Toney, J. G. Gordon, M. G. Samant, G. L. Borges, O. R. Melroy, L. S. Kau, D. G. Wiesler, D. Yee, and L. B. Sorrenson, Surface x-ray scattering measurements of the substrate-induced spatial modulation of an incommensurate adsorbed monolayer, *Phys. Rev.*, *B42*: 5594 (1990).

89. M. F. Toney, J. G. Gordon, M. G. Samant, G. L. Borges, and O. R. Melroy, Underpotentially deposited thallium on silver(111) by in situ surface x-ray scattering, *Phys. Rev.*, *B45*: 9362 (1992).

90. M. F. Toney, J. G. Gordon, M. G. Samant, G. L. Borges, D. G. Wiesler, D. Yee, and L. B. Sorrenson, In situ surface x-ray scattering measurements of electrochemically deposited Bi on Ag(111): structure, compressibility, and comparison with ex situ low-energy electron diffraction measurements, *Langmuir*, 7: 796 (1991).

91. J. H. White, M. J. Albarelli, H. D. Abruña, L. Blum, O. R. Melroy, M. G. Samant, G. L. Borges, and J. G. Gordon II, Surface extended x-ray absorption fine structure of underpotentially deposited silver on Au(111) electrodes, *J. Phys. Chem.*, *92*: 4432 (1988).

92. M. G. Samant, G. L. Borges, and O. R. Melroy, In situ surface EXAFS of an underpotentially deposited silver monolayer on gold(111), *J. Electrochem. Soc.*, *140*: 421 (1993).

93. C. H. Chen, K. D. Kapler, A. A. Gewirth, B. M. Ocko, and J. Wang, Electrode-posited bismuth monolayers on Au(111) electrodes: comparison of surface x-ray scattering, scanning tunneling microscopy, and atomic force microscopy lattice structures, *J. Phys. Chem., 97*: 7290 (1993).

94. L. Blum, H. D. Abruña, J. White, J. G. Gordon II, G. L. Borges, M. G. Samant, and O. R. Melroy, Study of underpotentially deposited copper on gold by fluorescence detected EXAFS, *J. Chem. Phys., 85*: 6732 (1986).

95. A. Tadjeddine, D. Guay, M. Ladouceur, and G. Tourillon, Electronic and structural characterization of underpotentially deposited submonolayers and monolayer of copper on gold(111) studied by in situ x-ray absorption spectroscopy, *Phys. Rev. Lett., 66*: 2235 (1991).

96. G. Tourillon, D. Guay, and A. Tadjeddine, In-plane structural and electronic characteristics of underpotentially deposited copper on gold(100) probed by in situ x-ray absorption spectroscopy, *J. Electroanal. Chem., 289*: 263 (1990).

97. R. Durand, R. Faure, D. Aberdam, C. Salem, G. Tourillon, D. Guay, and M. Ladoucier, In situ x-ray absorption study of underpotential deposition of copper on platinum(100), *Electrochim. Acta, 37*: 1977 (1992).

98. T. E. Furtak, L. Wang, J. Pant, P. Pansewicz, and T. M. Hayes, Structure of the copper monolayer/platinum–electrode interface as measured by in situ x-ray absorption spectroscopy, *Proc. Symp. X-Ray Methods in Corrosion and Interfacial Electrochemistry* (A. Davenport and J. G. Gordon II, eds.), The Electrochemical Society, Pennington, N.J., 1992, p. 146.

99. J. McBreen, W. E. O'Grady, G. Tourillon, E. Dartyge, and A. Fontaine, XANES study of underpotential deposited copper on carbon supported platinum. *J. Electroanal. Chem., 307*: 229 (1991).

100. J. McBreen, EXAFS studies of absorbed copper on carbon supported platinum, *J. Electroanal. Chem., 257*: 373 (1993).

101. R. R. Adzic, J. Wang, C. M. Vitus, and B. M. Ocko, The electrodeposition of Pb monolayer on low index Pt surfaces: an x-ray scattering and scanning tunneling microscopy study, *Surface Sci., 293*: 1876 (1993).

102. J. McBreen and M. Sansone, In situ XAS study of adsorbed Pb on carbon supported Pt, *J. Electroanal. Chem., 373*: 227 (1994).

103. B. M. Ocko and J. Wang, Surface structure of the Au(111) electrode, *Proc. NATO Workshop on Synchrotron Techniques in Interfacial Electrochemistry* (C. A. Melandres and A. Tadjeddine, eds.), Kluwer, Amsterdam, 1994.

104. J. D. Gordon II, O. R. Melroy, G. L. Borges, D. L. Reisner, H. D. Abruña, P. Chandrasekhar, and L. Blum, Surface EXAFS at an electrochemical interface iodine on platinum(111), *J. Electroanal. Chem., 210*: 311 (1986).

105. S. Kim, I. T. Bae, M. Sandifer, P. N. Ross, R. Carr, J. Woicik, M. Antonio, and D. A. Scherson, In situ XANES of an iron porphyrin irreversibly adsorbed on an electrode surface, *J. Am. Chem. Soc., 113*: 9063 (1991).

106. R. C. Elder, C. E. Lunte, A. F. M. M. Rahman, J. R. Kirschoff, H. D. Dewald, and W. R. Heineman, In situ observation of the copper redox in a polymer modified electrode using EXAFS spectroelectrochemistry, *J. Electroanal. Chem., 240*: 361 (1988).

107. M. J. Albarelli, J. H. White, G. M. Bommarito, M. McMillan, and H. D. Abruña, In situ surface EXAFS at chemically modified electrodes, *J. Electroanal. Chem.*, *248*: 77 (1988).

108. O. R. Melroy, M. F. Toney, G. L. Borges, M. G. Samant, J. B. Kortright, P. N. Ross, and L. Blum, An in situ grazing incidence x-ray scattering study of the initial stages of electrochemical growth of lead on silver(111), *J. Electroanal. Chem.*, *258*: 403 (1989).

109. M. J. Armstrong, G. M. Whitney, and M. F. Toney, In-situ x-ray scattering during electrodeposition under controlled hydrodynamics, *Proc. Symp. X-Ray Methods in Corrosion and Interfacial Electrochemistry* (A. Davenport and J. G. Gordon II, eds.), The Electrochemical Society, Pennington, N.J., 1992, p. 62.

110. G. G. Long and J. Kruger, Surface x-ray absorption spectroscopy, EXAFS and NEXAFS, for the in situ and ex situ study of electrodes, in *Techniques for Characterization of Electrodes and Electrochemical Processes*, (R. Varma and J. R. Selman, eds.), Wiley, New York, 1991, p. 167.

111. M. Kerkar, J. Robinson, and A. J. Forty, In situ studies of the passive film on iron and iron/chromium alloys using x-ray absorption spectroscopy, *Faraday Discussions Chem. Soc.*, *89*: 31 (1990).

112. G. C. Long, J. Kruger, D. R. Black, and M. Kuriyama, EXAFS study of the passive film on iron, *J. Electrochem. Soc.*, *129*: 240 (1982).

113. A. J. Davenport, H. S. Isaacs, G. S. Frankel, A. G. Schrott, C. V. Jahnes, and M. A. Russak, In situ x-ray absorption study of chromium valency changes in passive oxides on sputtered AlCr thin films under electrochemical control, *J. Electrochem. Soc.*, *138*: 337 (1991).

114. J. A. Bardwell, G. I. Sproule, B. MacDougall, M. J. Graham, A. J. Davenport, and H. S. Isaacs, In situ XANES detection of Cr(VI) in the passive film on Fe–26Cr, *J. Electrochem. Soc.*, *139*: 371 (1992).

115. G. S. Frankel, A. J. Davenport, H. S. Isaacs, A. G. Schrott, C. V. Jahnes, and M. A. Russak, X-ray absorption study of electrochemically grown oxide fims on Al–Cr sputtered alloys, *J. Electrochem. Soc.*, *139*: 1812 (1992).

116. A. J. Davenport, H. S. Isaacs, J. A. Bardwell, B. MacDougall, G. S. Frankel, and A. G. Schrott, In situ studies of passive film chemistry using x-ray absorption spectroscopy, *Corrosion Sci. 35*: 19 (1993).

117. Z. Nagy, H. You, R. M. Yonco, C. A. Melandres, W. Yun, and V. A. Maroni, Cell design for in situ x-ray scattering study of electrodes in transmission geometry, *Electrochim. Acta, 36*: 209 (1991).

118. C. A. Melandres, H. You, V. A. Maroni, and Z. Nagy, Specular x-ray reflection for the "in situ" study of electrode surfaces, *J. Electroanal. Chem.*, *297*: 549 (1991).

119. H. You, Z. Nagy, C. A. Melandres, D. J. Zurawski, R. P. Chiarello, R. M. Yonco, H. K. Kim, and V. A. Maroni, X-ray reflectivity studies of the metal/solution interphase, *Proc. Symp. X-Ray Methods in Corrosion and Interfacial Electrochemistry* (A. Davenport and J. G. Gordon II, eds.), The Electrochemical Society, Pennington, N.J., 1992, p. 73.

120. J. McBreen, W. E. O'Grady, K. I. Pandya, R. W. Hoffman, and D. E. Sayers, EXAFS study of the nickel hydroxide electrode, *Langmuir, 3*: 428 (1987).

121. J. McBreen, W. E. O'Grady, G. Tourillon, E. Dartyge, A. Fontaine, and K. I. Pandya, In situ time resolved x-ray absorption near edge structure of the nickel oxide electrode, *J. Phys. Chem.*, *93*: 6308 (1989).

122. K. I. Pandya, W. E. O'Grady, D. A. Corrigan, J. McBreen, and R. W. Hoffman, Extended x-ray absorption fine structure investigations of nickel hydroxides, *J. Phys. Chem.*, *94*: 21 (1990).

123. K. I. Pandya, R. W. Hoffman, J. McBreen, and W. E. O'Grady, In situ x-ray absorption spectroscopic studies of nickel oxide electrodes, *J. Electrochem. Soc.*, *137*: 383 (1990).

124. D. Guay, G. Tourillon, E. Dartyge, A. Fontaine, J. McBreen, K. I. Pandya, and W. E. O'Grady, In situ time resolved EXAFS study of the structural modifications occurring in nickel oxide electrodes betwen their fully oxidized and reduced states, *J. Electroanal. Chem.*, *305*: 83 (1991).

125. T. W. Capehart, D. A. Corrigan, R. S. Conell, K. I. Pandya, and R. W. Hoffman, In situ extended x-ray absorption fine structure spectroscopy of thin film nickel hydroxide electrodes, *Appl. Phys. Lett.*, *58*: 865 (1991).

126. C. Cartier, A. Tranchant, M. Verdaguer, R. Messina, and H. Dexpert, X-ray diffraction and x-ray absorption studies of the structural modifications induced by electrochemical lithium intercalation into V_2O_5, *Electrochim. Acta*, *35*: 889 (1990).

127. J. McBreen, EXAFS studies of battery materials, *Proc. Symp. X-Ray Methods in Corrosion and Interfacial Electrochemistry*, (A. Davenport and J. G. Gordon II, eds.), The Electrochemical Society, Pennington, N.J., 1992, p. 214.

128. B. E. Conway, D. Y. Qu, and J. McBreen, In situ and ex situ spectroelectrochemical and x-ray absorption studies on rechargeable, chemically modified and other MnO_2 materials, *Proc. NATO Workshop on Synchrotron Techniques in Interfacial Electrochemistry* (C. A. Melandres and A. Tadjeddine, eds.), Kluwer, Dordrecht, 1994, p. 311.

129. P. Druska and H. H. Strehblow, In situ examination of electrochemically formed Cu_2O by EXAFS in transmission, *J. Electroanal. Chem.*, *335*: 55 (1992).

130. R. S. Weber, M. Peuckert, R. A. Dalla Betta, and M. Boudart, Oxygen reduction on small supported platinum particles. II. Characterization by x-ray absorption spectroscopy, *J. Electrochem. Soc.*, *135*: 2535 (1988).

131. M. E. Herron, S. E. Doyle, S. Pizzini, K. J. Roberts, J. Robinson, G. Hards, and F. C. Walsh, In situ studies of a dispersed platinum on carbon electrode using x-ray absorption spectroscopy, *J. Electroanal. Chem.*, *324*: 243 (1992).

132. M. E. Herron, S. E. Doyle, K. J. Roberts, J. Robinson, and F. C. Walsh, Instrumentation and cell design for in situ studies of electrode surfaces using x-ray synchrotron radiation, *Rev. Sci. Instrum.*, *63*: 950 (1992).

133. S. Mukerjee, S. Srinivasan, M. P. Soriaga, and J. McBreen, Oxygen reduction electrocatalysis on binary alloys at interfaces with proton exchange membrane fuel cells, *J. Electrochem. Soc.*, in press.

134. M. G. Samant and M. Boudart, Support effects on electronic structure of platinum clusters on Y zeolite, *J. Phys. Chem.*, *95*: 4070 (1991).

135. D. Guay, G. Tourillon, and A. Fontaine, Electrochemical inclusion of copper and iron species in a conducting polymer observed in situ using time resolved x-ray absorption spectroscopy, *Faraday Discussions Chem. Soc.*, *89*: 41 (1990).

136. G. Tourtillon, E. Dartyge, A. Fontaine, and A. Jucha, Dispersive x-ray spectroscopy for time resolved in situ observation of electrochemical inclusion of metallic clusters within a conducting polymer, *Phys. Rev. Lett., 57*: 603 (1986).

137. D. Guay, G. Tourillon, E. Dartyge, A. Fontaine, and H. Tolentino, In situ observations of electrochemical inclusion of copper and iron species in a conducting polymer by using time-resolved x-ray absorption spectroscopy, *J. Electrochem. Soc., 138*: 399 (1991).

138. G. P. Williams, The initial scientific program at the NSLS infrared beamline, *Nucl. Instrum. Methods, A291*: 8 (1990).

139. A. E. Russell, G. P. Williams, A. S. Lin, and W. E. O'Grady, Far-infrared synchrotron radiation for in situ studies of the electrochemical interface, *J. Electroanal. Chem., 356*: 309 (1993).

9

Recent Applications of Ellipsometry and Spectroellipsometry in Electrochemical Systems

SHIMSHON GOTTESFELD, YEON-TAIK KIM*, AND ANTONIO REDONDO
Los Alamos National Laboratory
Los Alamos, New Mexico

I. INTRODUCTION

Ellipsometry is an experimental technique employed primarily for measurement of optical spectra of highly absorbing solids and of surface films on solid or liquid substrates. It is used mostly in a reflection mode, that is, the incident probing beam is reflected from the interface examined and information on the probed interface is obtained from analysis of the reflected beam. In the study of interfacial systems in electrochemistry, the main use of ellipsometry has been for measurements of optical properties and thicknesses of surface films formed on bulk electrode surfaces, as well as for measurements of electrochemical conversion processes in such films. Such optical information complements the electrochemical information on the relevant interfacial process (e.g., the electric charge associated with film growth, the change in interfacial capacitance following the growth of a surface film, and/or the charge and impedance associated with a film conversion process). In electrochemical systems, ellipsometry can be applied in situ in the spectral domain between the near ultraviolet (UV) and

---

*Present affiliation: University of Alabama, Tuscaloosa, Alabama.

the near infrared (IR). Therefore, it is an effective tool for the measurement of optical spectra and thicknesses of surface films on bulk metal or semiconductor electrodes. Recording of such spectra can be performed during the electrochemical growth of the surface film and under conditions of ordinary electrochemical experimentation (e.g., galvanostatic or potentiostatic experiments).

Ellipsometry is based on the simultaneous measurement of the amplitude and phase parameters of a collimated, monochromatic beam of light reflected from the probed sample. It can effectively follow changes in interfacial optical properties brought about by surface films of thicknesses ranging from subnanometer to thousands of nanometers. Advanced spectroellipsometers enable recording of a spectrum with good signal-to-noise ratio within 1 s or less. A lateral resolution of 10 μm in the ellipsometric determination of local surface film thickness has also been demonstrated. In ordinary ellipsometric measurements, only the specularly reflected beam is probed. Further optical information may be obtained from examination of off-specular reflection, but this option has been used only in very few studies of electrochemical systems. Off-specular reflection measurements are discussed in Section III.

The ellipsometric amplitude and phase parameters for a specularly reflected beam are designated by ψ and Δ, repsectively, and are defined as

$$\tan \psi \, \exp(i\Delta) = \frac{r_p}{r_s} \tag{1}$$

where r_p and r_s represent the (complex) ratio of the Fresnel reflection coefficients for parallel and perpendicular polarizations, respectively. An ellipsometer measures simultaneously ψ and Δ at a discrete wavelength, λ, whereas a spectroellipsometer measures this pair of parameters as function of λ in a certain spectral range by employing a dispersive or a spectrographic optical system.

Detailed equations in Section VI and the appendix to this chapter describe (a) the dependence of the Fresnel reflection coefficients, r_p and r_s, on the optical properties of electrode–solution and electrode–film–solution interfaces, and (b) the resulting dependencies of the ellipsometric parameters, ψ and Δ, on the optical properties of these two types of interfaces, as derived from equation (1). Computer programs written according to such equations calculate the ellipsometric parameters $\{(\psi_0, \Delta_0)(\lambda)\}$ and $\{(\psi,\Delta)(\lambda)\}$ for electrode–solution and electrode–film–solution interfaces, respectively. This requires a set of optical parameters that fully define the interface. Such sets of optical parameters can be written in general form as[*] $\{\hat{n}_{el}(\lambda), n_{sol}(\lambda), \phi\}$ and $\{\hat{n}_{el}(\lambda), \hat{n}_{sol}(\lambda), \hat{n}_{film}(\lambda), d_{film}, \phi\}$

---

[*]The complex index of refraction, $\hat{n}$ (where the caret is used to denote that the quantity is complex), is normally written in terms of its real, n, and imaginary, k, parts as $\hat{n} = n - ik$. An equivalent description employs the complex dielectric constant, $\epsilon = \epsilon' - i\epsilon''$, where $\epsilon = \hat{n}^2$. k is related to the absorption coefficient, $\alpha(cm^{-1})$, according to $\alpha = 4\pi k/\lambda$.

for electrode–solution and electrode–film–solution interfaces, respectively. $\hat{n}_{el}$ is the complex refractive index of the metal or semiconductor electrode, n_{sol} the refractive index of the electrolyte, ϕ the angle of incidence employed in the experiment, $\hat{n}_{film}(\lambda)$ the complex refractive index of the surface film, and d_{film} is its thickness.

The goal of a spectroellipsometric measurement is to derive $\hat{n}_{film}(\lambda)$ and d_{film} from the ellipsometric measurements $\{(\psi_0,\Delta_0)(\lambda)\}$ and $\{(\psi,\Delta)(\lambda)\}$, taken on the reference "bare" electrode and on the filmed electrode in a certain λ range. The reference measurement of the bare electrode–solution interface is required to evaluate $\hat{n}_{el}(\lambda)$. Evaluation of the optical spectrum and thickness of the surface film involves fitting to experiment of computer-calculated spectra $\{(\psi_{calc},\Delta_{calc})(\lambda)\}$. Such spectra are generated for a range of optical properties and thicknesses of the surface film, searching for the best fit to the measured ellipsometric spectrum. $\{(\psi,\Delta)(\lambda)\}$. This derivation, described in detail in Section VI, makes it possible to solve the optical spectrum and thickness of a surface film formed on an electrode surface requiring no additional information beyond that obtained from the ellipsometric measurement itself.

Table 1 summarizes the four most common types of computer-aided calculations and evaluations in ellipsometry, with the top line describing calculations for the reference state of a bare electrode–solution interface, and the bottom line showing calculations for the state of a filmed electrode.

Applications of ellipsometry in electrochemistry have been summarized recently in a review published in 1989 [1]. In this chapter we have included mainly those developments reported following the publication of that review. The reader should consult Refs. 1 to 4 for more detailed background and more information on the less recent applications of ellipsometry in electrochemistry.

Some of the more important recent developments in the application of ellipsometry to electrochemical systems have been:

TABLE 1 Basic Types of Calculations and Evaluations in Ellipsometry of Electrochemical Systems

Calculation of ellipsometric readings from known or assumed optical properties		Calculation of optical properties from ellipsometric measurements	
Given	Calculated	Measured/given	Calculated/fitted
$\{\hat{n}_{el}(\lambda),\ n_{sol}(\lambda),\ \phi\}$	$\{(\psi_0,\Delta_0)(\lambda)\}$	$\{(\psi_0,\Delta_0)(\lambda),\ n_{sol}(\lambda),\phi\}$	$\hat{n}_{el}(\lambda)$
$\{(\psi_0,\Delta_0)(\lambda),\ \hat{n}_{film}(\lambda),d_{film},\phi\}$	$\{(\psi,\Delta)(\lambda)\}$	$\{(\psi,\Delta)(\lambda),\ \hat{n}_{el}(\lambda),\phi\}$	$\hat{n}_{film}(\lambda),d_{film}$

1. The development of fast automatic spectroscopic ellipsometers and their use in real time, in situ spectroscopic studies of film growth on electrode surfaces
2. The development of spatially resolved ellipsometry, which provides information on the spatial distribution of film thickness with a resolution of 10 μm
3. The application of ellipsometry simultaneously with the quartz-crystal microbalance (QCM) technique for the derivation of thickness/optical properties (ellipsometry) and mass buildup (QCM) during the growth of a film on an electrode surface

In this chapter we describe these new developments in detail.

The chapter is organized according to subjects of interfacial electrochemistry studied most recently with ellipsometry and spectroellipsometry. Following a section on instrumentation, meant to introduce the reader to the capabilities of modern ellipsometers and spectroellipsometers, we discuss applications of ellipsometry in the fields of electrochemical film growth, electrochemical film conversion processes, and self-assembled and LB films at various solid–liquid and liquid–air interfaces. Although not strictly considered ellipsometric techniques, we also discuss here some off-specular reflection measurements recently applied to the study of electrochemical interfaces. We follow our survey of experimental ellipsometric studies by a chapter on computer-aided analysis of ellipsometric and spectroellipsometric data, describing models and computer codes developed for such analysis.

We hope to show in this chapter that the recent activity at the interface of ellipsometry and electrochemistry has demonstrated the important capabilities of this optical technique, including real-time in situ measurements of film spectra and film thickness and high lateral resolution—all achieved in measurements of surface films on ordinary bulk electrodes. We hope that the recent applications of modern ellipsometry and spectroellipsometry described in this chapter and the clarification offered here for key issues in ellipsometric data analysis will further increase the interest of electrochemists in this technique.

II. INSTRUMENTATION

Instrumentation in ellipsometry has advanced within the last 20 years from the manual null ellipsometer to a real-time spectroscopic ellipsometer, with a concomitant five orders of magnitude increase in the rate of data acquisition, an increase of two orders of magnitude in precision and an order of magnitude in accuracy. The ellipsometric setup always consists of three parts: polarized light generation, the probed sample, and polarized light detection. A variety of configurations exist, corresponding to different modes for the evaluation of the ellipsometric parameters of the probed interface. One relatively simple way to determine the light-polarization change brought about by reflection from the probed sample is null-point detection [5]. This classical ellipsometric technique is accurate and requires a very simple instrumental setup but is limited by its slowness, requiring a significant fraction of a minute to complete one reading of ψ, Δ at a single wavelength. The null technique has been described in detail in a number of review articles [6]. Here we focus on faster, radiometric methods of detection, where the polarization state of the incident beam is continuously varied in a periodic mode. The optical information on the sample can thus be obtained by synchronously detecting and analyzing the reflected beam-intensity modulation. Two such methods that have been in use during the last two decades are rotating element ellipsometry and polarization modulation ellipsometry (PME). Both methods generally employ photomultiplier-tube (PMT) detectors. In addition, two new extensions of rotating element ellipsometry to be described here employ an optical multichannel analyzer (OMA) [7] or a charge-coupled device (CCD) [8] to follow the periodic intensity modulation of the reflected beam. The OMA detection scheme enables real-time spectroscopic monitoring of film growth, whereas the CCD detection scheme has been employed for recording of surface ellipsometric images with a resolution of 10 to 20 μm.

The most popular radiometric scheme is the rotating element ellipsometer with PMT detection. This scheme is the first to be discussed. Then we turn to the rotating element spectroellipsometer with OMA detection, followed by a discussion of the dynamic imaging microellipsometer. Finally, polarization modulation ellipsometry employing a photoelastic modulator is described.

A. Rotating Element Ellipsometry

In rotating element ellipsometry, the readings of Δ and ψ are derived from the waveform of the reflected intensity as determined by the settings of fixed optical elements in the optical train, the optical parameters of the reflecting surface and the rotating optical element. In the simplest case, either the polarizer or the analyzer is rotated, keeping the other element fixed. We consider here the case of a rotating polarizer/fixed analyzer, presented schematically in Figure 1. The reverse case of a rotating analyzer/fixed polarizer was considered in detail in Ref. 1. By passing through a rotating polarizer, assembled in the hollow shaft of a synchronous motor operating typically at 50 to 100 Hz [9], the unpolarized collimated incident light (usually obtained from a Xe arc source followed by a monochromator and collimator) is transformed into plane-polarized light of azimuth varying at angular frequency 2ω. Such a plane-polarized beam will be, in general, elliptically polarized after being reflected by the surface of the sample. At each point in time, the elliptically polarized light reflected from the sample passes through the analyzer, which samples a component of the intensity along

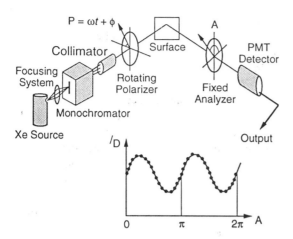

FIGURE 1 Optical configuration and detector irradiance, I_D, as function of time, t, for a rotating polarizer ellipsometer. (Reprinted with permission from the American Chemical Society.)

its fixed angle A. Optical encoder signals from the rotating element trigger and clock the collection of the intensity waveform incident on the detector, typically with a resolution of 30 readings per cycle, as shown schematically in Figure 1. The variation of the intensity with time measured at the detector with such resolution (Fig. 1) is sufficient to construct the waveform accurately and evaluate the ellipsometric parameters (Δ, ψ) at the wavelength of the measurement as described in the following. Scanning of the wavelength of the monochromator enables collection of (Δ, ψ) as a function of photon energy. This obviously requires that the state of the surface examined does not vary significantly during the scan of λ.

As shown in Figure 1, at a rotational velocity ω of the rotating element, the intensity signal at the detector contains dc and 2ω components. The ellipsometric parameters ψ and Δ of the reflecting surface can be computed directly from these intensity components. The equations of rotating element ellipsometry are described briefly here for ideal conditions (i.e., perfect polarizers and perfect calibration of the angular readings). The electric field of the reflected beam at the detector can be represented using the Jones formalism [2]:

$$\mathbf{E} = \begin{pmatrix} 1 & 0 \\ 0 & 0 \end{pmatrix} \begin{pmatrix} \cos A & \sin A \\ -\sin A & \cos A \end{pmatrix} \begin{pmatrix} r_p & 0 \\ 0 & r_s \end{pmatrix} \begin{pmatrix} \cos P & -\sin P \\ \sin P & \cos P \end{pmatrix} \begin{pmatrix} 1 & 0 \\ 0 & 0 \end{pmatrix} \begin{pmatrix} E_0 \\ E_0 \end{pmatrix}$$

$$(2)$$

where A is the fixed angle of the analyzer and P is the periodically varying angle of the polarizer with respect to the plane of incidence. The rotating element azimuthal angle can be represented as $(\omega t + \phi)$. The intensity at the detector is the product of the electric field in the reflected beam and its conjugate. For the case of a rotating polarizer/fixed analyzer considered here, the intensity at the detector can be represented by the first two terms of a Fourier series:

$$I = |E|^2 = I_0 |\alpha \cos(2\omega t + \phi) + \beta \sin(2\omega t + \phi) + 1| \tag{3}$$

where

$$I_0 = \frac{1}{2} |r_s|^2 |E_0|^2 \cos^2 A (\tan^2 \psi + \tan^2 A)$$

From the definition of the ellipsometric parameters in equation (1), α and β can be shown to be related to ψ and Δ, according to

$$\alpha = \frac{\tan^2 \psi - \tan^2 A}{\tan^2 \psi + \tan^2 A} \qquad \beta = \frac{2 \cos \Delta \tan \psi \tan A}{\tan^2 \psi + \tan^2 A}$$

The desired ellipsometric parameters can thus be derived from

$$\cos \Delta = \frac{\beta}{\sqrt{1 - \alpha^2}} \tag{4}$$

$$\tan \psi = \sqrt{\frac{1 + \alpha}{1 - \alpha}} \tan A \tag{5}$$

A similar set of equations for the case of a rotating analyzer/fixed polarizer are given in Ref. 1.

To perform the Fourier transform required to derive α and β [equation (3)], the PMT signal is typically digitized by an analog-to-digital converter initiated and synchronized during the cycle by pulses from the optical encoder mounted around the shaft of the rotating polarizer. Normally, an average of 10 to 20 complete optical cycles is collected to enhance the signal-to-noise ratio. The digitized intensities, I_q, $q = 1, 2, 3, \ldots, N$ (typically, $N = 30$), at each optical trigger pulse are Fourier analyzed to determine the dc term, I_0, and the Fourier coefficients, α and β, in equation (3) according to the following equations [9]:

$$I_0 = \frac{1}{N} \sum_{q=1}^{N} I_q$$

$$\alpha = \frac{(2/N)\sum_{q=1}^{N} I_q \cos[2\pi(q - 1)/N]}{I_0} \tag{6}$$

$$\beta = \frac{(2/N)\sum_{q=1}^{N} I_q \sin[2\pi(q - 1)/N]}{I_0}$$

Rotating element ellipsometers with a mechanically scanned monochromator, of the general type shown in Figure 1, are available commercially, for example, from Rudolph (New Jersey).

B. Rotating Element Fast Spectroscopic Ellipsometer

Figure 2 shows a spectroellipsometer based on a rotating element that is capable of measuring a complete $\{(\psi, \Delta)(\lambda)\}$ spectrum at high speed. To achieve this end, a spectrographic optical system replaces the mechanically scanned monochromator, and a multichannel analyzer based on a Si-photodiode array replaces the PMT. The optical multichannel analyzer consists of two parts: the array detector with 1024 pixels and a controller unit. Each pixel integrates all photons collected over the exposure time. To analyze the integrated signal, Fourier analysis of the raw data must be replaced by Hadamard analysis. The resulting Hadamard summations can be described as

$$S_1 = \int_0^{T/4} S(t)\omega \, dt = \frac{I_0}{2\omega} \left(\alpha + \beta + \frac{\pi}{2} \right)$$

$$S_2 = \int_{T/4}^{T/2} S(t)\omega \, dt = \frac{I_0}{2\omega} \left(-\alpha + \beta + \frac{\pi}{2} \right)$$

$$S_3 = \int_{T/2}^{3T/4} S(t)\omega \, dt = \frac{I_0}{2\omega} \left(-\alpha - \beta + \frac{\pi}{2} \right) \qquad (7)$$

$$S_4 = \int_{3T/4}^{T} S(t)\omega \, dt = \frac{I_0}{2\omega} \left(\alpha - \beta + \frac{\pi}{2} \right)$$

where $S(t) = I_0 (\alpha \cos 2\omega t + \beta \sin 2\omega t + 1)$.

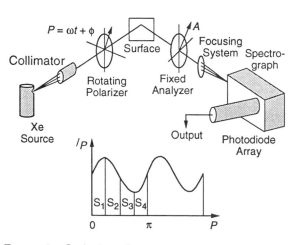

FIGURE 2 Optical configuration and pixel output waveform, I_p, for the rotating-polarizer spectroscopic ellipsometer (see Ref. 3). (Reprinted with permission from the American Chemical Society.)

By rearranging the relations above, we obtain

$$\alpha = \frac{\omega}{2I_0} (S_1 - S_2 - S_3 + S_4)$$

$$\beta = \frac{\omega}{2I_0} (S_1 + S_2 - S_3 - S_4) \qquad (8)$$

$$I_0 = \frac{\omega}{\pi} (S_1 + S_2 + S_3 + S_4)$$

From the relations above, the ellipsometric parameters, tan ψ and cos Δ can be obtained according to equations (4) and (5). Measurements of the waveform $S(t)$ can be performed simultaneously at many pixels, each corresponding to a photon energy within the range covered by the photodiode array. Consequently, the spectral variation of tan ψ and cos Δ [i.e., the complete information $\{(\psi,\Delta)(\lambda)\}$] is obtained on a time scale of the order of 10 ms [3,7]. Normally, improved signal-to-noise ratio requires averaging about 100 wavelength scans and thus about 1 s is required for the collection of a (ψ,Δ) spectrum of good signal-to-noise quality.

The recent introduction of this instrumental approach, which enables fast collection of surface ellipsometric parameters in a wide wavelength range, has converted ellipsometry into a real-time spectroscopic tool. Fast spectroellipsometric probing of surface films during their formation on electrode surfaces—an attractive capability in the research of electrochemical interfacial systems—has been made possible. A fast spectroscopic ellipsometer of the type described schematically in Figure 2 is commercially available from Sopra (France).

C. Dynamic Imaging Microellipsometry

The operating principle of dynamic imaging microellipsometry (DIM) is essentially identical to that of rotating element ellipsometry, but DIM has a two-dimensional array detector which enables mapping of the surface ellipsometric parameters with a spatial resolution of approximately 20 μm. Sugimoto and co-workers have reported two-dimensional images of surface films using a single detector and achieving a spatial resolution of approximately 10 μm [10]. However, the temporal resolution achieved was very poor because of the need to scan the probed area of the sample. Recently, Cohn and co-workers developed rapid full-field dynamic imaging microellipsometry (DIM) [8]. Good spatial resolution (approximately 10 to 20 μm) at a temporal resolution on the order of a minute per image was achieved, enabling surface mapping of passive films and localized corrosion processes on electrode surfaces [11]. The instrument is configured in the standard polarizer, sample, compensator, analyzer (PSCA) scheme, in combination with an imaging lens and a charged-coupled device (CCD) video camera. The radiometric approach enables one to determine the light intensities at the elements of the two-dimensional array detector. The intensities are then converted to a full-field ellipsometric parameter map (an "ellipsogram") with a lateral resolution of approximately 20 μm. A block diagram of the DIM system is illustrated in Figure 3. The mathematical form of the reflected intensity that Cohn and co-workers used is

$$I(p) = \frac{G^2}{2} \{a \cos^2 P \tan^2 \psi + b \sin^2 P$$

$$+ (c \cos \Delta - d \sin \Delta) \sin 2P \tan \psi\}$$

$$a = 1 + \cos 2C \cos 2(A - C) \tag{9}$$

$$b = 1 - \cos 2C \cos 2(A - C)$$

$$c = \sin 2C \cos 2(A - C)$$

$$d = \sin 2(A - C)$$

This equation can be obtained using the Jones matrix and vector formalism [2]. Here A and C are analyzer and compensator angles, respectively, and are fixed during measurement. The polarizer angle P is variable and should be set to at least three different readings at which the intensity is recorded, to determine the three unknowns in equation (9), G (the system gain), Δ, and ψ. In practice, the intensities at four specific polarizer positions (0°, 45°, −45°, and 90°) are mea-

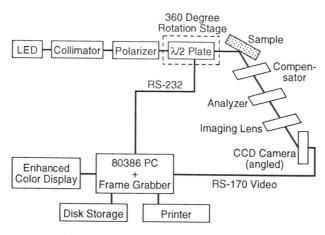

FIGURE 3 Schematic illustration of the dynamic imaging ellipsometer (DIM) setup. The backbone of the setup is similar to the traditional polarizer–specimen–compensator–analyzer configuration, with the extra half-wave plate added for fast change of the state of incident polarization (see Ref. 11). (Reprinted by permission from the Electrochemical Society.)

sured and digitized, and the four-image algorithm leads to simple calculations of the ellipsometric parameters from equation (9), according to

$$\psi = \tan^{-1} \sqrt{\frac{I(0°)}{I(90°)}} \sqrt{\frac{b}{a}} \tag{10}$$

$$\Delta = \cos^{-1} \frac{I(45°) - I(-45°)}{2\sqrt{I(0°)I(90°)}} - \tan^{-1} \frac{d}{c}$$

An effective way to achieve four intensity images has been implemented by Cohn and co-workers [8] using a fixed polarizer followed by a half-wave plate (see Figure 3). The half-wave plate was mounted on a computer-controlled rotation stage and could thus be used to rotate the plane of incidence of the linearly polarized incident beam, as required for the solution of ψ and Δ according to equation (10). The ellipsogram-to-ellipsogram temporal resolution achieved was 20 s and was limited by the rotation speed of the half-wave plate [11].

D. Polarization Modulation Ellipsometry

Polarization modulation ellipsometry employs modulation of the state of polarization at high frequency (50 kHz), as achieved by a standing strain wave applied to a fused quartz crystal placed in the path of the incident beam (Fig. 4). Unlike the automatic radiometric ellipsometers described above, this type of automatic photoelastic modulation ellipsometer does not have any moving parts. The modulated beam is reflected from the sample and passes through a fixed analyzer before reaching the detector. A scheme of a polarization modulation ellipsometer is given in Figure 4. The intensity waveform which reaches the PMT has ω, 2ω, and dc components, where ω is the imposed photoelastic modulation frequency. The ellipsometric parameters are obtained from the expressions

$$\frac{I_\omega/I \, dc}{R_\omega \, (\text{cal})} = \sin 2\psi \sin \Delta \tag{11}$$

$$\frac{I_{2\omega}/I \, dc}{R_{2\omega} \, (\text{cal})} = \cos 2\psi$$

In equation (11), I_ω, $I_{2\omega}$, and I_{dc} are the intensity components of ω, 2ω, and dc, respectively, measured by two lock-in amplifiers and a dc amplifier. $R_\omega(\text{cal})$ and $R_{2\omega}(\text{cal})$ are intensity ratios obtained under well-defined calibration conditions, typically measured once a day. Wavelength scanning to collect the spectra of ψ and Δ can readily be performed with the photoelastic modulation ellipsometer, requiring only a linear scan of the amplitude of the modulation to adjust for the varying wavelength. This is easily done with a linear voltage ramp

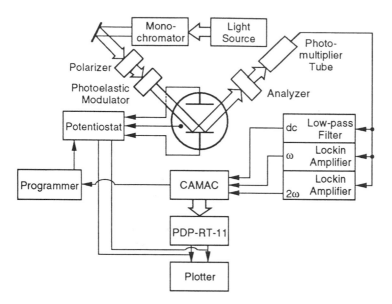

FIGURE 4 Schematic diagram of an automatic ellipsometer based on polarization modulation. Modulation at a frequency ω results in an intensity waveform at the photodetector with ω, 2ω, and dc components from which the ellipsometric parameters can be evaluated. (Reprinted by permission from Elsevier).

input to the power supply of the modulator. A spectroellipsometer based on photoelastic modulation has been employed in recent investigations of several electrochemical systems [12]. Its performance in terms of precision, accuracy, and speed is very comparable to that of a rotating element spectroellipsometer employing an ordinary monochromator (Fig. 1), with the advantage of no moving parts, which helps minimize mechanically and electrically induced noises. For more information on automatic ellipsometers based on polarization modulation, see Ref. 13.

E. Combined Ellipsometry/EQCM

Rishpon and co-workers have demonstrated a system that enables simultaneous ellipsometric and electrochemical quartz-crystal microbalance (EQCM) measurements during electrochemical film growth [12b]. The scheme of the system is shown in Figure 5. In this particular case, an automatic ellipsometer with photoelastic modulation was employed. Ellipsometers with a rotating element could also be used, in principle, for such a combined measurement, although the minimized mechanical and electronic noise of the photoelastically modulated

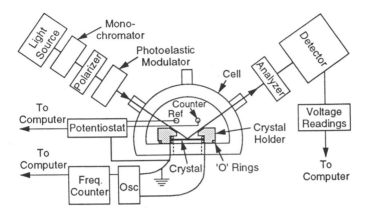

FIGURE 5 System for combined spectroellipsometry/quartz crystal microbalance (QCM) measurements. The probed electrode is a metal film deposited on the quartz crystal, located on the optical center of the ellipsometric system. When a surface film is grown, both film mass and film thickness/optical properties can be determined simultaneously (see Ref. 16). (Reprinted by permission from Elsevier.)

ellipsometer may be an important advantage when an EQCM measurement is performed simultaneously. Antisymmetric Pt keyhole patterns sputtered onto the quartz crystal served as the pair of contacts for probing the mass changes brought about by an electrochemical process. The electrochemical process takes place on one of the keyhole-shaped Pt film which faccs thc electrolyte solution and serves as the (grounded) working electrode in the electrochemical cell. The crystal is pressed between a pair of O-rings and onto the planar wall of the semicylinder. This arrangement enables easy positioning of the crystal center (i.e., the center of the electrode) on the optical center of the ellipsometric system. The voltage or current, the resonance frequency of the EQCM, and the three components of the ellipsometric signal (dc, ω, and 2ω) were digitized and stored. Evaluation of the ellipsometric parameters according to equation (11) was also performed subsequently on the same computer.

Ellipsometry and EQCM yield complementary information during film growth. Ellipsometry yields optical properties (or optical spectra) and film thickness, whereas the EQCM measures the overall change of the mass of the film during film growth or during cycling. The combined information of film thickness and film mass enables monitoring of the mass density of films during their electrochemical growth. The system shown in Figure 5 was used in several investigations of film growth and film conversion processes to be described in Sections III and IV.

III. ELLIPSOMETRIC STUDIES OF FILM GROWTH ON ELECTRODE SURFACES

Studies of film growth on electrode surfaces have always been a central subject in ellipsometric investigations of electrochemical systems. We describe in this section primarily ellipsometric studies of the growth of conducting polymer and oxide films on electrode surfaces. A significant number of recent ellipsometric studies of film growth in electrochemical systems have been performed at a single wavelength. However, the number of spectroellipsometric studies has increased continuously. In single-wavelength ellipsometry, the plot of ψ vs. Δ during film growth provides the basis for the evaluation of the film's optical parameters (real and imaginary components of $\hat{n}_{film}$ at a single λ) and film thickness as a function of time. This approach is mostly effective under conditions of uniform film growth, i.e., when $\hat{n}_{film}$ does not vary with film thickness (see Section VI). The form of the ψ vs. Δ plot obtained during film growth can also serve as a qualitative indicator for the optical properties of the film at the wavelength employed. These aspects are described in detail in Section IV of Ref. 1. In cases of more complex film growth processes, analysis of each growth stage, i, has to be performed independently to evaluate $(\hat{n}_{film})_i$ and $(d_{film})_i$. This can be achieved effectively with spectroellipsometry, as described in Section VI.

A. Ellipsometric and Spectroellipsometric Studies of the Electrochemical Growth of Conducting Polymer Films

The electrochemical synthesis and the electrochemical activity of conducting polymer films have attracted significant interest in recent years. The electrochemical growth of such films has been the subject of several ellipsometric investigations aimed primarily at correlations between film deposition conditions and film thickness and morphology. Another target of ellipsometric investigations of conducting polymer films has been the spatial propagation of the electrochemical film conversion process (see Section IV). We first describe here some recent real-time ellipsometric measurements at a single wavelength during the growth of conducting polymer films. Earlier work (before 1987) on this subject is described in Ref. 1. An ellipsometric study of the growth of films of polyaniline (PANI) was reported by Carlin et al. [15]. The polyaniline films were deposited by potential cycling and ellipsometric parameters were taken at a single wavelength during film growth. The data measured were evaluated by assuming growth of a single uniform film, to determine the refractive index (assumed constant) and the thickness of the film. The uniform film growth model resulted in large discrepancies for film thicknesses higher than 100 nm. In recent work, Rishpon et al. [12b], have observed ellipsometrically similar deviations from uniform growth for the galvanostatic deposition of polyaniline films. These authors showed that such deviations are more easily discernible when the film is measured ellipsometrically (at 550 nm) under cathodic applied potentials (i.e., in the nonabsorbing form). Their results are depicted in Figure 6. Rishpon et al. went on to elucidate the nature of this deviation from uniform film growth and demonstrated that the characteristic "looping in" of the experimentally observed ψ vs. Δ plots could be explained by a bilayer of PANI, consisting of a denser base layer covered by an open-structured external layer of graded density. The external layer was simulated with optical properties falling linearly from $\hat{n}_{film}$ = $1.50 - 0.01i$ to $\hat{n}_{film} = 1.38 - 0.003i$ as the overall thickness of the film increases from 140 to 580 nm. Such a trend reflects a gradual drop in the density of the external layer of the film. The form of the calculated ellipsometric growth curve evaluated from this bilayer model is demonstrated in Figure 7. This example shows that, even when collected at just a single wavelength, ellipsometric growth curves can provide insight into variations of film density with film thickness. This approach works out well provided that the film exhibits little absorption at the wavelength of the measurement. Trade-offs between optical properties and thickness could upset such attempts of analysis in single-wavelength ellipsometry in the case of highly absorbing films [12].

The electrochemical growth of films of polyaniline from various aqueous electrolytes was monitored by Robertson and co-workers [16] who employed

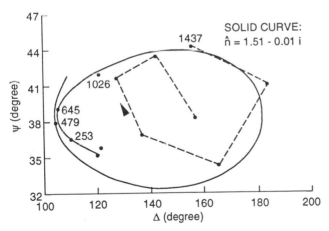

FIGURE 6 Experimental and fitted ψ vs. Δ plots for the galvanostatic growth of a PANI film (77 $\mu A/cm^2$) in aqueous HCl, with the ellipsometric readings (λ = 550 nm) taken at -0.2 V vs. SCE (i.e., for the reduced form of the film). Film thicknesses shown in angstrom units. The deviation from the fitted plot corresponding to single-film growth (solid curve) is apparent for film thicknesses above 140 nm, highlighted by dashed lines, (see Ref. 12b). (Reprinted by permission from Elsevier.)

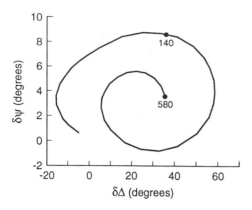

FIGURE 7 Qualitative simulation of the "looping-in" trend obtained experimentally (Fig. 6), based on the growth of a base film 140 nm thick, with subsequent growth of a layer of decreasing n_{film} changing linearly from $150 - 0.01i$ to $1.38 - 0.003i$ as the overall thickness increases from 140 to 580 nm (see Ref. 12b). (Reprinted by permission from Elsevier.)

automated null detection ellipsometry at the single, He–Ne laser wavelength of 632.8 nm. Deposition from 0.1 M aniline/0.1 M perchloric acid electrolyte produced the most uniform film. Figure 8 shows an ellipsometric growth curve for a film of PANI grown from perchloric acid at two different current densities. Here, the complex refractive indices of the two films are close to each other ($1.569 - 0.369i$ and $1.567 - 0.3296i$). However, these complex refractive indices are quite different than those reported by others for PANI films: $1.2 - 0.1i$ for galvanostatic growth from HCl [12b,17] and $1.1 - 0.3i$ for galvanostatic growth from sulfuric acid [18]. These lower values of $\hat{n}_{film}$ were measured for a PANI film at the same wavelength (633 nm) as that employed by Robertson and co-workers. Such differences in optical properties may be caused by the different acid solutions employed for the growth of the PANI film, which result in a different film composition and could bring about different polymer nanostructures. The results of Robertson et al. suggest at first glance a higher film density for PANI in the perchloric acid electrolyte. However, as predicted by the Kramers–Kronig relationship, the real part of $\hat{n}_{film}$ could be strongly affected not only by film density but also by the proximity to an absorption peak (see Ref. 1, sec. V). The reason the real part of $\hat{n}_{PANI}$ at 633 nm could be as small as, or even smaller than $n_{electrolyte}$ [12b,18], is the major absorption peak of PANI centered in the near IR (see Fig. 26 below). Thus differences in optical parameters of PANI measured at 633 nm (Ref. 16 vs. Refs. 12b and 18) could originate in part from different near-IR spectra of PANI films grown under dif-

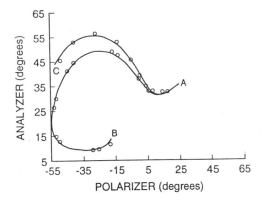

FIGURE 8 Ellipsometric film growth curves for the growth of a PANI film on a Pt electrode at 17 μA/cm² (curve A) and at 43 μA/cm² (curve B). The fitted curves correspond to n_{film} = 1.569 − 0.369i growing to a thickness of 280 nm in the first case (lower curve) and n_{film} = 1.567 − 0.3296i growing to a thickness of 165 nm in the second case (upper curve). (Reprinted by permission from the Electrochemical Society.)

ferent electrochemical conditions and in part from real differences in film density. These two physical properties could actually be, in turn, interconnected, because the IR spectrum of the film is a function of the long-range order in the polymer. Optical data at a single wavelength may thus be insufficient for effective comparative analysis, particularly in cases of such highly absorbing films as as-grown PANI, polypyrrole, or polythiophene.

The electrochemical growth of polyaniline has been studied extensively by a Los Alamos group. Part of their earlier ellipsometric studies were performed at single wavelengths [12b,19]. In most of these studies, polyaniline was galvanostatically deposited on Pt electrodes from aqueous HCl solutions. Following each stage of galvanostatic growth, the current was interrupted and single-wavelength ellipsometric readings were taken under a pair of cathodic and anodic biases, corresponding to the doped and undoped states of the polyaniline film. The results obtained from this analysis, which were described in detail in Refs. 1, 12b, and 19, showed a reasonable fit of the ellipsometric ψ vs. Δ plots to the uniform film growth approximation for both oxidized and reduced forms of the PANI film. However, during earlier stages of the growth of a conducting polymer film, significant variations in film properties usually take place as nucleation centers of increasing surface density form, and subsequently, densification of the new phase takes place simultaneously with the increase in total film thickness. What is required in such cases for a detailed description of the film growth process is to solve independently the unique optical properties of the film at a given stage of growth, together with the thickness of the film at this stage. If the optical measurements are limited to a single λ, the solution of the three unknowns, n_{film}, k_{film}, and d_{film}, at each point in time during film growth requires to provide a third measured parameter in addition to ψ and Δ. One option is to add a reflectometric measurement at the same λ and φ [1,20]. This approach was first applied for the optical analysis of conducting polymer film growth on electrodes by Hamnett, Hillman, and co-workers [21,22]. Redondo and co-workers later adopted this approach to the study of the growth of PANI films [12a], showing that combined ellipsometric/reflectometric measurements reveal variations in n_{film} and k_{film} during galvanostatic PANI film growth within the thickness range 0 to 150 nm. These results are shown in Figure 9. The variations in $\hat{n}_{film}$ are seen to correspond to a sequence of film densification steps (increase of n_{film} and k_{film}, sometime at constant film thickness) and film growth steps. The conclusion from these two different descriptions for the growth of PANI given in Refs. 19 and 12a, are (a) in the case of a complex film growth process, a uniform film growth model could provide, at least in some cases, a reasonable first approximation in terms of averaged film properties and film growth rate, hiding, however, some finer details of the growth sequence; and (b) reflectometric measurements could be very effective in revealing such finer details (Fig. 9) of the film growth process.

In situ optical measurements revealing simultaneous film densification and film growth during the anodic deposition of conducting polymers were first reported by Hamnett and co-workers [21–23]. They studied early stages in the

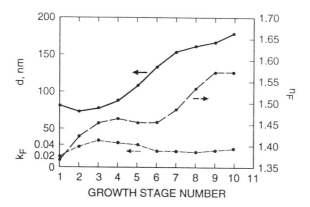

FIGURE 9 Variations in film thickness (solid), n_{film} (long-dashed) and k_{film} (short-dashed) during the growth of a PANI film as evaluated from combined ellipsometric-reflectometric measurements (see Ref. 19). (Reprinted by permission from Molecular Crystals and Liquid Crystals.)

growth of polypyrrole on a Pt electrode under potentiostatic conditions with single-wavelength ellipsometry and employed complementing reflectometric measurements to analyze their results. They presented their results as variations of n_{film}, k_{film}, and d_{film} with time [23]. Such results for potentiostatic growth of PPy from acetonitrile are shown in Figure 10. Two discontinuities in the variation of n_{film} and k_{film} are seen as the film thickness roughly increases linearly with time. The first discontinuity occurs around 10 s after initiation of the anodic potentiostatic step and is apparently associated with the initiation of polymer densification from highly dispersed nucleation sites. This rise in both components of $\hat{n}_{film}$ takes place at a measured film thickness of around 12 nm, interpreted by Hamnett and co-workers as the maximum length of a chain of PPy that can grow perpendicularly to the electrode surface from a nucleation site. The film thickness keeps increasing, simultaneously with film densification, up to d_{film} = 35 nm, reached after 30 s. At this point, another discontinuity in film optical parameters occurs, probably corresponding to the growth of a less dense external layer of the polymer, as discussed above for the case of polyaniline [12b]. Another, rather different example of complex film growth studied by combined ellipsometry/reflectometry at single λ, was described in Ref. 22 for the potentiostatic growth of poly(vinylferrocene-co-vinylpyrrolidone). In the latter case, long-chain oligomers are available in solution prior to the application of the anodic growth potential. In the oxidized form, these oligomers become insoluble and nucleate out onto the electrode surface. This results in a measured increase in n_{film} and in k_{film} beginning instantaneously at $t = 0$, and a much faster initial rise in film thickness, reaching 120 nm within 2 s, as seen in Figure 11.

Figures 9 to 11 demonstrate how interesting details of the complex process of polymer film growth can be obtained from analysis of single λ ellipsometry/reflectometry results. A possible difficulty with such an analysis, however, is that the reflectometric complementing measurements required are usually less reliable. Unlike the ellipsometric measurements, reflectometric measurements could be affected significantly by light absorption in the solution phase (e.g., when soluble, light-absorbing oligomers tend to form in parallel with film growth), as well as by drifts of light source intensity and detector signal amplification. Hence an important advantage of spectroscopic ellipsometry in the study of such complex growth processes is that it requires no additional, nonellipsometric measurements for the analysis of variations in the spectrum of the film during its growth (see Section VI). In the rest of this section we describe spectroellipsometric investigations of conducting polymer film growth processes on electrode surfaces.

Arwin and Aspnes [24] reported earlier ex situ spectroscopic ellipsometry measurements on films of polypyrrole and poly(n-methyl pyrrole). A more recent, in situ spectroellipsometric study of the growth of polypyrrole was performed by Kim and co-workers [7b]. In this study a rapid-scanning spectro-

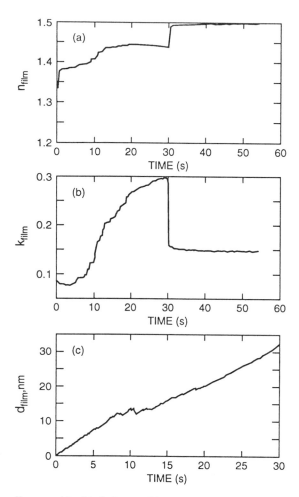

FIGURE 10 Variations with time of n_{film}, k_{film}, and d_{film} during the potentiostatic growth of polypyrrole from acetonitrile, evaluated from combined ellipsometric–reflectometric measurements (notice the different time scales). The increases in the components of $\hat{n}_{film}$ after 10 s correspond apparently to the onset of film densification from dispersed nucleation sites (see Ref. 23). (Reprinted by permission from Pergamon.)

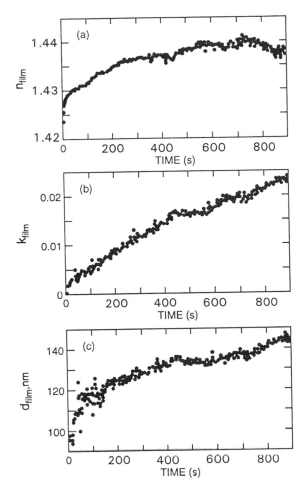

FIGURE 11 Combined ellipsometric/reflectometric measurements on a film of poly(vinylferrocene-co-vinylpyrrolidone), grown from a solution containing oligomers. In this case, the optical readings clearly show an instantaneous increase of the refractive index of the film and a fast initial rise in film thickness (see Ref. 22). (Reprinted by permission from Elsevier.)

scopic ellipsometer with an optical multichannel analyzer was employed for in situ real-time monitoring and analysis of pyrrole electropolymerization on a gold electrode. Spectra were collected as a function of time over the wavelength range 370 to 820 nm during potentiostatic deposition (0.6 V vs. SCE) from an aqueous 0.1 M KNO$_3$ solution. Film growth was monitored up to a thickness of approximately 50 nm. The experimental conditions were similar to those employed earlier by Hamnett and co-workers [23] in their single λ ellipsometric study of anodic PPy film growth. From the ellipsometric spectrum of the interface with the thickest PPy film grown, both a spectrum for this film and a film thickness (49 nm) were determined employing a single-film model, utilizing the analytical approach developed by Arwin and Aspnes [24]. The spectrum of the thickest PPy film grown is given in Figure 12. An alternative approach, utilizing a Lorentz harmonic oscillator model, provided a similar dielectric spectrum for the PPy film, as shown in the same figure, and a film thickness of 47.5 ± 2 nm [7b]. For details of these methods of evaluation of film spectra, see Section VI. The spectrum for this PPy film reveals an interband electronic transition at 3.8 eV and band-to-defect transitions at 1.65 and 2.3 eV. Using the spectrum evaluated for the thickest (49-nm) film, the ellipsometric spectra obtained as a function of time at earlier points in the growth of the film were analyzed based on effective medium theory, employing a linear regression analysis algorithm [7b]. Such analysis allowed the authors to characterize the evolution of film thickness and structure during the potentiostatic deposition process. A three-stage nucleation and growth process, similar to that described before by Hamnett and co-workers [23], could be revealed. The evaluated change in the thickness and density of the growing PPy film can be seen in Figure 13. Between 50 and 100 s following the potential step application, the evaluated thickness shows a linear

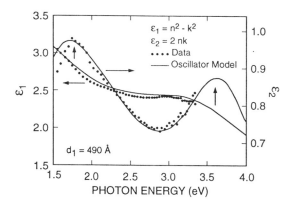

FIGURE 12 Real and imaginary components of the dielectric function of an electropolymerized film of polypyrrole, determined from the raw spectroellipsometric results for the thickest film grown assuming a single, uniform film model. The thickness (49 nm) was also derived from this analysis of the spectroellipsometric results. The vertical arrows denote resonance energies derived using a damped-harmonic oscillator model depicted by the solid lines (see Ref. 7b). (Reprinted by permission from the Electrochemical Society.)

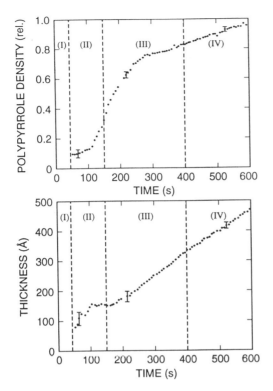

FIGURE 13 Four stages in the variations of the relative density (top) and thickness (bottom) of a film of PPy during the potentiostatic growth of the film from 0.1 *M* KNO$_3$. These variations were derived from dynamic spectroellipsometric measurements during film growth, based on effective-medium theory (see Ref. 7b). The densities are related to those of the final, 49-nm-thick film. (Reprinted by permission from the Electrochemical Society.)

increase from approximately 7.5 to 15 nm. Next, the thickness stays constant (at 15 nm) for 50 s and only at $t = 150$ s the thickness starts to grow again in a linear fashion. During the same period, the optical density increases significantly. This suggests [7b] nucleation followed by coalescence, with the latter occurring first at an almost constant film thickness of 15 nm. This is followed subsequently by combined densification and growth and a layer-by-layer growth at constant rate beyond a film thickness of 35 nm.

The complex growth mechanism and morphology of films of conducting polymers raise some questions on the reliability of information obtained from such optical measurements, even when performed in a spectroscopic mode.

In particular, trade-offs between film thickness and film spectra can occur. A recent contribution that addresses such concerns is the work of Tian and co-workers. They tried to correlate the electronic properties and film thicknesses evaluated from spectroellipsometric measurements during the growth of poly(3-methylthiophene), with images of the same film obtained by atomic force microscopy (AFM) [25]. The AFM images revealed stages of nucleation, first monolayer formation, thin fibrous layers, and thick granular layers. The ellipsometric thicknesses could not be properly correlated with the AFM images observed, especially at the earliest stages of film growth. This suggests that the optical model (a uniform single layer) is too simple to explain the earlier stages of nucleation. Variations of film density along the thickness dimension of the polymeric film, typically increasing monotonically from the film–solution interface toward the metal–film interface, have been noticed by neutron reflectometry measurements [26]. These measurements are based on a probe with a typical wavelength much shorter than that used in optical ellipsometry, resulting in much higher resolution along the film thickness dimension. It is not clear that spectroellipsometry could distinguish effectively between a film with a graded density and a film of uniform effective density that has an effective thickness different from that of the real graded film. We address such problems in Section VI, where a typical example of the challenges in the analysis of spectroellipsometric data is given for the growth of a film of PANI.

B. Combined Ellipsometry/EQCM Investigations of Conducting Polymer Film Growth

Rishpon et al. have suggested simultaneous ellipsometry/quartz crystal microbalance (QCM) measurements for the study of polymer film growth and built the cell and system shown in Figure 5 for the study of the growth of PANI [12b,c,17] and PPy/glucose oxidase [27] films. Ellipsometry yields optical film thickness, whereas the EQCM measures the overall film mass buildup. The combined information of film thickness (ellipsometry) and film mass (QCM) makes it possible to monitor the apparent mass density of films during their electrochemical growth. Examples for this combined measurement were given by Rishpon et al. in Refs. 12b,c and 17. A plot of the shift in QCM resonance frequency vs. optical thickness for the anodic, highly absorbing form of PANI, as measured during anodic galvanostatic growth of PANI, is shown in Figure 14. The apparent mass density obtained for film growth at 77 $\mu A/cm^2$ was 1.45 g/cm^3. A much smaller density, 1.1 g/cm^3, was derived in a similar way for the potentiostatic growth of PANI (at 0.7 V vs. SCE) from the same aqueous HCl solution [12b], demonstrating the capability of these combined measurements to detect variations in space filling in conducting polymer films as a function of the conditions of electrochemical growth. The smaller film density detected in

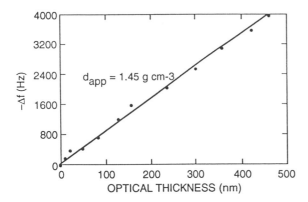

the case of potentiostatic growth of PANI was interpreted as the result of over-oxidation processes leading to soluble products. The authors suggested that such dissolution processes take place to a lesser extent during the galvanostatic growth of PANI, because the electrode potential is slowly drifting to less anodic values [12b].

Rubinstein et al. [12c] used an ellipsometer at single wavelength together with the quartz crystal microbalance to monitor the effect of a preadsorbed self-assembled monolayer on the morphology of polyaniline grown galvanostatically on a gold electrode. The polyaniline film was first grown, for reference, on a bare gold electrode, and next on a gold electrode with a surface pretreated by immersion in a solution of *p*-aminothiophenol (PATP). During such pretreatment, a monolayer of PATP self-assembles on the gold electrode surface. The same electrochemical conditions were used in galvanostatic growth experiments on the bare and on the PATP-covered gold electrodes. Figure 15 shows a comparison of the ellipsometric growth curves measured during the galvanostatic deposition of polyaniline on the two different surfaces. The polyaniline film grown on the bare electrode had a complex refractive index of $1.31 - 0.02i$ at 600 nm, whereas a complex refractive index of $1.18 - 0.21i$ was measured at the same wavelength for the film grown on a PATP-treated Au electrode. The PATP treatment thus leads to a substantial increase in the optical density of the polyaniline film at this wavelength. The authors interpreted this finding as the result of facilitated surface wetting, enabling enhanced coverage, and organization in the growing conducting polymer film. The facilitated wetting is ap-

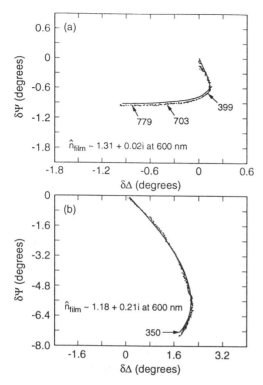

FIGURE 15 Raw ellipsometric data at 600 nm and fitted growth curves for uniform film growth in the cases of (a) galvanostatic growth of PANI on bare gold and (b) growth of PANI under identical conditions on PATP-treated gold (see Ref. 12c). The thicknesses designated on the curves are in angstrom units. (Reprinted by permission from the American Chemical Society.)

parently enabled by the self-assembled PATP monolayer. The PATP molecule bonds to the gold surface through its thiol group, leaving the para-amino group oriented toward the solution. Under anodic applied potential, this well-oriented amino group apparently forms a cation radical through which it can effectively bond to aniline monomers. This enables the buildup of a polymeric phase well bonded to the electrode surface. In this manner, the bifunctional PATP monolayer can serve as an effective "monomolecular glue" between the chemically dissimilar gold and PANI phases. The higher mass density of the PANI film on the PATP-treated Au electrode was confirmed in the same work by simultaneous QCM/ellipsometry measurements [12c]. This combined measurement showed that the mass density of the PANI film increased from 1.1 g/cm^3 on the bare electrode to 1.8 g/cm^3 on the PATP-treated electrode. An agreement was thus demonstrated between the optical evidence for film densification (increased k_{film}) and the evidence on increased mass density as obtained from combined ellipsometry/QCM. A later spectroellipsometric work by the same group addressed variations in PANI film density/morphology resulting from the application of a cathodic bias to a PANI film grown at constant anodic current [17]. These results are described in the next section, devoted to electrochemical film conversion processes.

Rishpon and Gottesfeld [27] reported the use of combined QCM/ellipsometric measurements for the study of polypyrrole/glucose oxidase (PPy-GO) composite films grown anodically on Pt film electrodes from phosphate buffer solutions. A significant increase in the optical density of the PPy film, from $1.5 - 0.13i$ to $1.8 - 0.25i$, was detected according to the single λ ellipsometric measurements as the GO enzyme concentration in solution was increased from zero to 1.4 mg/mL. The authors suggested that this indicates improved space filling in anodic PPy films, as a result of incorporation of the enzyme into the film. An increase in the apparent mass density of the PPy film was also demonstrated following GO incorporation according to combined ellipsometry/QCM measurements. However, the possibility in this case of significant changes in the viscoelastic properties of the PPy film as a result of enzyme incorporation was brought up as a complicating factor in the interpretation of QCM data. The authors argued that the effects measured suggest mutual stabilization of the PPy and the enzyme in the composite layer.

C. Spectroellipsometric Studies of the Growth of Anodic Oxide Films

Two recent spectroellipsometric investigations have looked into long-standing issues in interfacial electrochemistry: the nature of the passive film on iron and the nature of the thick oxide layer formed on Pt electrodes in acid solutions under high anodic polarizations. Chin and Cahan [28] described an extensive study of the passive film formed between 0.4 and 1.6 V vs. RHE on an iron electrode in borate buffer. The spectroellipsometric/reflectometric results have enabled evaluation of thicknesses and spectra (1.5 to 5.0 eV) of the passive films formed potentiostatically on iron under these conditions. The spectra for ϵ_1 and ϵ_2 of the passive film on iron at four different electrode potentials are shown in Figure 16. They are clearly very similar and, to a good approximation, this is a case of uniform film growth. The increased thickness of the film is the only significant difference observed between these three electrode potentials, increasing according to the analysis of the spectroellipsometric results from 1.0 nm at 0.4 V to 5.0 nm at 1.6 V. Chin and Cahan ascribe the broad peak centered around 3.5 eV, and additional weak spectral features at 2.2 and at 3.0

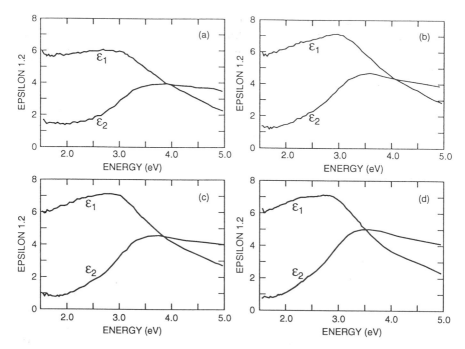

FIGURE 16 Complex dielectric functions for a passive film grown on iron in borate buffer at four different applied potentials: (a) +0.4 V; (b) +0.8 V; (c) +0.95 V; (d) +1.65 V vs. RHE; derived from spectroellipsometric/reflectometric measurements by assuming a single uniform film (see Ref. 28). (Reprinted by permission from the Electrochemical Society.)

eV (the latter could be detected only in second derivatives of the spectra), to polycrystalline Fe_2O_3 with spectral shifts caused by partial protonation of the oxide. The uniformity and homogeneity of the layer are further established by independence of the spectrum or thickness evaluated on the angle of incidence employed in the spectroellipsometric measurement [28].

Gottesfeld et al. [29] used a spectroellipsometer to study the growth of relatively thick anodic oxide layers on Pt electrodes during potentiostatic polarization within the oxygen evolution region in sulfuric acid solutions. A commercial automatic spectroscopic ellipsometer (S2000) fabricated by Rudolph, based on a fixed polarizer/rotating polarizer configuration, was employed. The overall interfacial process that takes place at the Pt electrode under these conditions involves the growth first of a thin, highly absorbing (and probably highly conducting) film which reaches a maximum thickness of around 3.2 to 3.5 nm (film α). This is followed by slow breakdown of the passivity of this film leading

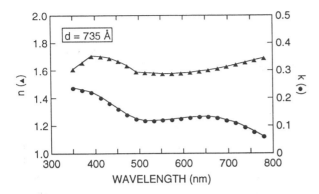

FIGURE 17 Spectrum of a thick oxide film (film β) which forms on a Pt electrode after 88 h of potentiostating at 2.1 V vs. RHE in aqueous sulfuric acid. Significant values of k_{film} in the visible and near IR suggest a spectrum of midgap states in this oxide film. The oxide is apparently associated with a band gap of about 4 eV (see Ref. 29). (Reprinted by permission from the Electrochemical Society.)

to growth on top of it of a much thicker oxide film (film β) which reaches thicknesses as high as 80 nm after 4 days of potentiostatic growth at 2.1 V vs. RHE. This sequence of growth and the specific thicknesses of the two oxide films obtained could be well established from the spectroellipsometric measurement, as described in detail in Ref. 29. The spectroellipsometric information on the general structure of the bilayer and the thicknesses of the two films was similar to that reported before from a single λ ellipsometric measurement of this interfacial process [30]. Additional valuable information from the recent spectroellipsometric measurements were the spectra of the two films, particularly that of the thicker oxide layer. The spectrum of the thick, external oxide layer film β, formed on a Pt electrode in sulfuric acid after 88 h at 2.1 V vs. RHE, is shown in Figure 17. From the spectrum evaluated, the major absorption peak is seen to be well above 3.5 eV. However, a distribution of midgap states of significant density is present in film β, suggesting electronic conductivity enabled by electron hopping mechanisms. These spectroscopic findings help in understanding of the electrochemical behavior of the thicker oxide film on Pt. Thanks, apparently, to the midgap states demonstrated by the spectroellipsometric measurements, film β can have beneficial effects in the process of oxygen evolution [29] while presenting no electronic conductivity limitations. The last work [30] also showed that film β grows linearly with time under a constant applied potential, implying a rate-determining step for film growth independent of transport through this layer.

In a recent spectroellipsometric investigation, Barbero and Kotz [31] showed that the electrochemical activation of glassy carbon electrodes produces a layer of highly porous and hydrated carbon. The spectrum for the surface layer formed by either potential multicycling or constant potentiostatic polarization could be well expressed by an effective-medium-theory expression, with the volume fraction of water and the thickness of the film being the only adjustable parameters. The authors found that during potentiostatic activation (1.86 V in 1 *m* sulfuric acid), the void fraction increases gradually with time up to 70% as the film thickens to 120 nm in 25 min. During multicycling, the porous film has a constant void fraction of 80% and reaches a thickness of 80 nm after 200 cycles between −0.5 and 1.86 V. The conclusions here are in accordance with a previous, single λ ellipsometric investigation of the same surface process by Kepley and Bard [32].

D. Dynamic Imaging Ellipsometry

Ellipsometry is by nature a surface averaging technique typically employing a probing light beam with a cross-sectional area of several mm^2. It is therefore limited in its ability to examine surface nonuniformities or localized surface phenomena caused by a nonuniform current distribution or by effects of microstructure in surfaces of polycrystalline metal samples. Several investigators have attempted to overcome this limitation by scanning single-point ellipsometric measurements. Sugimoto et al. [10] examined passive films on stainless steels and reported a spatial resolution of 10 μm. This work was described in an earlier review on ellipsometric investigations of electrochemical systems [1]. Such a scanning approach suffers from poor temporal resolution. The attractive alternative—a radiometric full-field imaging approach with good temporal resolution and a spatial resolution of 20 μm—was recently developed by Cohn and co-workers [8,11], who developed the dynamic imaging microellipsometer (DIM) described in Section II. Streinz and co-workers [11] have recently demonstrated the use of the DIM for examining passive film growth on surfaces with microstructural inhomogeneities. They reported results for passive film growth on a polycrystalline iron sample and on an AlTa alloy sample that contained intermetallic precipitates. Figure 18 shows an ellipsogram, a surface profile of the ellipsometric parameters Δ and ψ, recorded along the surface of a polycrystalline iron electrode passivated in a borate buffer solution. The sample has grains with areas on the order of 0.1 mm^2. The darker areas on the Δ map correspond to grains with lower values of Δ (i.e., grains covered by a thicker passive film). The uniformity of the ψ map suggests that the spatial variations of Δ are probably not caused by roughening phenomena [11]. From separate measurements of the growth kinetics on the grains designated 1 and 2 in the ellipsogram, Streinz and co-workers concluded the difference between oxide

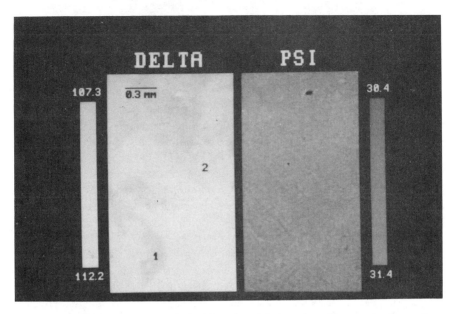

FIGURE 18 Ellipsogram recorded with a dynamic imaging ellipsometer for an iron sample passivated at 500 mV vs. SCE in a borate buffer solution. Grains are differentiated in the Δ map but not in the ψ map, indicating passive film thickness differences from grain to grain of the polycrystalline sample (see Ref. 11).

thicknesses on these grains to be as large as 0.5 nm. They ascribed this difference to the different orientations of grain surfaces—(111) in grain 1 and (110) in grain 2—as could be determined from subsequent SEM examination of the etched surface. Since the passive film on iron is crystalline, differences in properties of oxide films grown on such two different grain surfaces are quite expected [11]. In another part of this work, these authors have demonstrated for AlTa that oxide film growth on intermetallic precipitate sites is limited as compared with the growth of the same passive film along most of the metal alloy surface [11]. This was demonstrated for an Al–1% Ta alloy surface passivated in a borate buffer electrolyte. Implications of such microinhomogeneities of oxide film thickness are discussed in Ref. 11.

The DIM represents a significant development in ellipsometry, making it possible to achieve spatial resolution on the order of 10 μm in the determination of local surface film thickness and properties. Information on the degree of nonuniformity of passive films is very valuable in the analysis of passivity failure mechanisms. In addition, nonuniformity of film thickness/properties is of great interest in a number of other fields, including film deposition by sol-gel or CVD techniques. Such information can also be effectively provided with dynamic imaging ellipsometry.

E. Effects of Surface and Film Roughness Studied Optically in Electrochemical Systems

In all cases discussed hitherto, the assumption made was that the growth of a surface film can be fully described optically by the effects on specular reflection. Indeed, ellipsometric measurements have been almost always confined to one angle corresponding to specular reflection. However, this approach ignores further optical information which can be obtained from nonspecular scattering of the incident beam. The degree of nonspecular reflection becomes increasingly significant in electrochemical systems as the roughness of the surface layer formed on the electrode substrate increases. Such roughness could be the result of nonuniform growth of a patchy film along the electrode surface, combined processes of film formation and substrate corrosion, and roughness within a three-dimensional film in films with a microgranular structure. These problems have been treated recently theoretically by Urbach and co-workers [33]. To study surface roughness, the incident light beam is typically set at 90° and the reflected intensity is measured as function of angle of observation. Measurements of the angular dependence of diffuse light scattering are usually considered outside the scope of ellipsometry. We decided, however, to include a brief description of such measurements, because they use a complementary optical tool for study of the electrochemical interface and because some interesting examples have recently been reported.

Reference 34 gives an interesting example of measurement of the angular dependence of diffusely scattered light intensity performed to study an electrochemical surface process. The relatively simple instrumental system shown in Figure 19 was used to study the formation of an array of crystallites on the surface of an iron electrode in aqueous 1 M NaOH. As the iron sample is cycled in this medium between -1.2 and 0 V vs. SCE, discrete crystallites form from iron corrosion products and, as a result, the overall intensity of the diffusely scattered light increases with the number of cycles. A histogram of the one-dimensional angular-resolved light-scattering intensity, measured at one constant applied anodic potential, is shown in Figure 20 following given numbers of potential cycles. The angular dependence of the intensity provides valuable information on the topology of the iron surface during very early stages of this surface process: the number of maxima in such a histogram is correlated with the number of crystallites formed on the examined surface, the positions of the maxima are correlated with their x coordinate (the plane of the electrode is xy), and the height of the maxima is correlated with the radii of the crystallites and their relative refractive index, $m = n_{crystallite}/n_{electrolyte}$. Treatment of the intensity

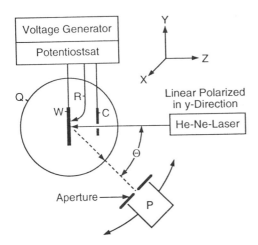

FIGURE 19 Experimental setup for angular resolved laser light scattering under potential control (Q is a cylindrical quartz cell). The angle of detection, θ, is varied in the experiment to record $I(\theta)$ (see Ref. 34). (Reprinted by permission from the Electrochemical Society.)

according to Rayleigh's scattering theory results in the following expression for the amplitude of light, a_k, scattered from an individual crystallite:

$$a_k = 4\pi^2 r_k^3 \lambda^{-2}(m^2 - 1)(m^2 + 2)R^{-1} \tag{12}$$

where r_k is the radius of the center and R is the distance of the probe from the scattering center. An array of such scattering centers serves essentially as a diffraction element, such that the total amplitude of the scattered light, dA, measured within an aperture $d\omega$ of the probe is given by

$$dA(\theta) = \Sigma a_k \exp(-i\Phi_k) \, d\omega \tag{13}$$

where the phase angle Φ_k is equal to $2\pi x_k \lambda^{-1} \sin \theta$, x_k being the typical interparticle distance.

Specifically, in the histogram shown in Figure 20 it is clear that the amplitudes of almost all the scattered intensity peaks increase with the number of voltammetric cycles, whereas their number and angular positions remain approximately constant. This means that most of the crystallites are formed on the surface in the first cycle (i.e., preferentially at nucleation sites that can be observed even before any electrochemical treatment). The angular dependence of light scattering thus provides valuable information on this very nonuniform film which consists of an array of crystallites. It could be shown [34] that crystallites as small as 10 nm are detectable by such a diffuse light-scattering measurement.

Observation of such a surface process by conventional ellipsometry (or spectroellipsometry) at the specular reflection angle could only provide infor-

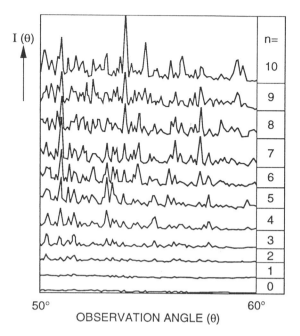

FIGURE 20 Histogram of the one-dimensional angular resolved light-scattering intensity distribution, $I(\theta)$, recorded for the corrosion layer on iron during consecutive oxidation cycles at constant potential; n is the number of cycles applied (see Ref. 34). (Reprinted by permission from the Electrochemical Society.)

mation on the thickness of the effective surface film (equal to the total height of the crystallites) and the volume fraction of the crystallites in the effective crystallite–solution composite layer. The additional diffuse scattering information, as generated by the relatively simple arrangement shown in Figure 19, provides much more structural detail in the lateral dimension. Finally, the mechanism for the formation of these crystallites on the iron electrode was identified by Oelkrug and co-workers [34] as crystallization of $Fe(OH)_2$ by a reductive dissolution–precipitation mechanism. This process starts from FeOOH that forms on the iron surface by direct anodization. The large increase in scattering intensity observed during the anodic half-cycles [34] was ascribed to conversion from $Fe(OH)_2$ to FeOOH within the crystallites, bringing about a significant increase in the value of m $(=n_{\text{crystallite}}/n_{\text{electrolyte}})$.

In another recent contribution, Kruijt and co-workers [35] reported on measurements of diffusely scattered light at a constant angle of reflection of 30° with respect to the incident (90°) beam, performed during the potentiodynamic passivation of nickel in acid solutions. It is rather expected that by slowly scanning the potential of the nickel electrode through the active–passive transition region, metal dissolution would take place to some extent in such an acid solution. Indeed, the intensity of scattered light has been found to increase during such an electrochemical routine. A complete theory for such findings is still missing [33,35]. However, such studies of roughening [35] clearly demonstrate that ellipsometric results for the potentiodynamic growth of a passive layer on a metal like nickel in acid solutions cannot be analyzed based simply on a single film model. The accompanying substrate surface roughening could lead to a significant apparent absorption coefficient (a nonzero k_{film}) in the single film assumed, even when the passive film is actually completely transparent. This point was explained in detail in Section Vb of Ref. 1. Finally, Foontokov and co-workers [36] have demonstrated effects of large-scale ordering in hydrogen absorbed on platinum using measurements of potential-modulated scattered-light intensity.

In summary, we like to draw attention to the additional information obtained from angular resolved measurements of diffusely scattered light. While the measurements show high sensitivity—the root mean square of the roughness can be determined accurately in the nanometer domain with an accuracy on the order of 0.1 nm and the resolution along the surface is around 20 nm—they await a detailed theory. No theory has yet been advanced for the effect of a granular film with micrometer or nanometer grain dimensions, a structure reported, for example, for conducting polymer films according to STM measurements [37]. The analysis of ellipsometric measurements usually assumes a single uniform surface film with effective optical properties. While effective-medium theory can well describe effective optical properties for composite films, a significant amount of information could be lost in this description. In cases where the height and surface density of crystallites, patches, or pits are of interest, scattered-light measurements could provide such information. This optical measurement maintains all the advantages of ellipsometry in the study of electrochemical systems, i.e., it is an in situ technique that can be applied to ordinary bulk electrodes.

IV. ELLIPSOMETRIC STUDIES OF ELECTROCHEMICAL FILM CONVERSION AND RESTRUCTURING

Conversion processes in electrochemically active films have been a central subject of interest in several fields of electrochemistry, including electrochemical power systems, electrochromic films, and electrochemical insulator/conductor switching phenomena. Such conversion processes involve, in general, variation of the average oxidation state through the complete volume of the electrochemically active film. This requires, in turn, injection–ejection of electronic and counter ionic charges throughout the volume of the film. In the case of electrochemical power systems, conversion processes in active oxide films, such as the $NiOOH/Ni(OH)_2$ system, have an important bearing on the charge–discharge efficiency. They determine the dynamics of charging and discharging processes and therefore the power density of the system. The same is true for electrochemically active conducting polymer films, such as polyaniline or polypyrrole. Similar factors of charge cycle efficiency and cycle dynamics are important in the case of electrochromic films. The common fundamental characteristics in all of these electrochemical systems is the dynamics of the overall film conversion process and its dependence on the combined rates of electron and ion transport through the volume of the film.

Whereas the film conversion processes discussed above are characterized by their reversibility, as reflected by the large number of oxidation/reduction cycles typically possible, some lasting morphological changes can also be induced by electrochemical cycling. Ellipsometric and, particularly, spectroellipsometric measurements can reveal such morphological changes induced by electrochemical cycling. We discuss in this section first, reversible film conversion processes as studied by ellipsometry, and next, ellipsometric observations of structural effects in cycled conducting polymer films.

A. Ellipsometric Studies of Film Conversion

The following general cases of combined electron–ion propagation modes through the volume of an electrochemically active film are expected:

Mode A When the rate of the electrochemical perturbation (e.g., rate of change of applied potential) is relatively slow and the effective electronic and ionic conductivities of the film are both relatively high, propagation of the conversion process will be uniform within the volume of the film. The simplest way to quantify the requirement for this case to hold would be $\tau > \rho d_{\text{film}} C_{\text{film}}$, where τ is the typical duration of the perturbation, ρ the dominant resistivity in the film (ionic or electronic), and C_{film} the capacitance of the film per unit cross-sectional area associated with the film charging process.

Mode B When $\tau < \rho_{\text{ion}} d_{\text{film}} C_{\text{film}}$ and ρ_{elec} is negligible compared with ρ_{ion}, the film conversion process is expected to propagate in the form of a front moving from the film–electrolyte interface toward the substrate–film interface.

Mode C When $\tau < \rho_{\text{elec}} d_{\text{film}} C_{\text{film}}$, and ρ_{ion} is negligible compared with ρ_{elec}, the film conversion process is expected to propagate in the form of a front moving from the substrate–film interface toward the film–electrolyte interface.

Clearly, identification of the mode of propagation of the film conversion process can shed important light on the component process which limits the rate of film conversion. The contribution of ellipsometry to the study of the propagation mode in film conversion processes was pioneered by DeSmet and coworkers [38]. They demonstrated that different, readily distinguishable ψ vs. Δ plots will be obtained during the electrochemical conversion of a film, depending on which of the three different propagation modes described above applies.

Simulation of the *ellipsometric conversion curve* (i.e., the ψ vs. Δ plot expected during an electrochemical conversion process taking place according to one of the three modes described above) can be performed based on the film thickness and the optical properties of its fully reduced and fully oxidized forms. These required properties can be derived ellipsometrically during growth of the film. DeSmet, Ord, and co-workers used this ellipsometric diagnostics extensively to study electrochemical film conversion mechanisms in anodic oxide films on metal electrodes. For some recent applications of this approach to the study of electrochemical conversion processes in anodic oxide films on vanadium and on tungsten, see Ref. 39. Earlier ellipsometric work by DeSmet and Ord on conversion processes in oxide films is described in Ref. 1.

Adaptation of the ellipsometric study of DeSmet and Ord to films of conducting polymers which undergo similar electrochemical conversion processes was first described by Gottesfeld et al. [40]. A simulated mode A ellipsometric conversion curve (at $\lambda = 550$ nm), given in Ref. 40 for the conversion of PANI films in HCl solutions, was based on a gradual (1000 step) shift between the complex refractive index of the fully reduced form of the film ($n_{film} = 1.48 - 0.013i$, measured at 0 V vs. Ag/AgCl) and the fully oxidized form of the film ($n_{film} = 1.30 - 0.060i$ measured at 0.50 V), taking place uniformly throughout the volume of the film and assuming constant thickness during the conversion. Film thickness and optical properties of the fully reduced (0 V) and fully oxidized (0.5 V) forms of PANI were evaluated from ellipsometric data collected in the preceding PANI film growth experiment. Simulations of modes B and C of film conversion were based on a dual-layer film model, with a sharp boundary between the layers. One hundred gradual steps were used in which the thickness of the oxidized layer increases at the expense of the reduced layer of the film, or vice versa. Ellipsometric conversion curves recorded during the anodic conversion at 50 mV/s of two polyaniline films, 115 and 150 nm thick, clearly indicated a case of uniform propagation (mode A). These results are presented in Figure 21. In subsequent ellipsometric work on the electrochemical growth and conversion of films of PANI on Pt electrodes, Robertson et al. [16] showed that PANI films grown galvanostatically on Pt substrates from perchloric acid solutions were associated with a different film conversion pattern. The ellipsometric conversion curves reported by Robertson et al. are shown in Figure 22. Robertson et al. have not attempted a complete fit of their measured ellipsometric conversion curves according to modes A to C described above, and noticed only that the measured curve did not conform to the uniform conversion mode reported by Gottesfeld et al. [40]. They suggested that the origin of the difference in the conversion mode is the higher density of the films grown and studied by them. It is interesting to notice that the curvature of the U-shaped conversion curves in Figure 22 [16] conforms qualitatively to a process of anodic conversion starting from the metal–film substrate and a process of cathodic

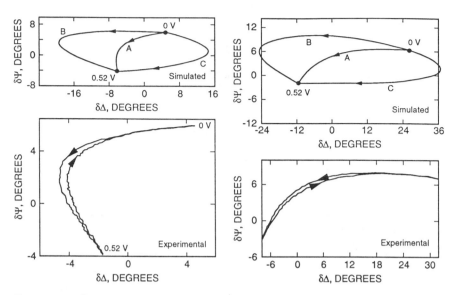

FIGURE 21 Simulated (top) and experimental (bottom) ellipsometric conversion curves for the electrochemical conversion of a polyaniline film on Pt in 2 *M* HCl, recorded at 550 nm and 60° incidence during a 50-mV/s scan between 0 and 0.52 V vs. a silver wire electrode. The left-side panels correspond to a film thickness of 115 nm, those on the right side correspond to a film thickness of 150 nm. The simulated curves A, B, and C correspond to the respective conversion modes described in this section. In both cases the experimental curves clearly correspond to mode A. (Reprinted by permission from the Electrochemical Society.)

conversion starting from the film–solution interface. Although the finer details may be more complicated, as suggested by Robertson et al., such opposite directionality of the anodic and cathodic film conversion processes may be quite readily understood. A PANI film may exhibit high electronic resistivity when fully reduced (Robertson et al. reported that high cathodic overpotentials were required for complete galvanostatic reduction), and therefore the anodic conversion would propagate from the metal–film interface. On the other hand, when fully doped, a dense film will exhibit significant electronic conductivity but could be associated with lower ionic conductivity caused by a significant barrier to ionic insertion. Therefore, a front propagating from the film–electrolyte interface may well be the mode of cathodic conversion in a denser film of PANI. When the directionality of propagation is reversed in the cathodic vs. the anodic conversion process, the two conversion curves would overlap to first approximation, because the intermediate optical states will be similar during the two opposite processes. This is observed qualitatively in Figure 22.

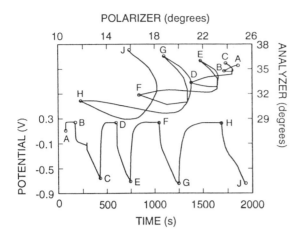

POLARIZER (degrees)

TIME (s)

FIGURE 22 Variations of the potential (below) and the ellipsometric conversion curves (above) during galvanostatic growth and conversion of a film of PANI grown in sulfuric acid solution (see Ref. 16). The very different pattern of the conversion curve here versus that shown for PANI in Figure 21 has been interpreted as a result of the more compact PANI film obtained in sulfuric acid. (Reprinted by permission from the Electrochemical Society.)

In two recent ellipsometric investigations of film conversion processes in conducting polymer films, Lee et al. [41,42] have demonstrated applicability of this type of ellipsometric study to other conducting polymer systems. They also broadened the spectrum of propagation modes considered in electrochemical conversion processes by adding the following case.

Mode D   Directional propagation of film conversion occurs as in modes B or C above, but with graded film oxidation and film reduction along the thickness dimension as determined by diffusion from the leading interface into the volume of the film. Schemes for modes A, B, and D, as given in Ref. 41, are shown in Figure 23. The last authors have shown that the direction of both oxidative and reductive conversions in films of polypyrrole (PPy) in 0.2 M aqueous KCl solutions [41] is from the film–electrolyte interface toward the metal–film interface. Alternating anodic and cathodic potential steps of 1-s duration enabled to record the time sequence of the conversion process by collecting the ellipsometric readings with a time resolution of 37 ms. While readily noticing the directionality of propagation by the similarity of the experimental ellipsometric conversion curves to simulated curves for mode B, the authors further advanced to show that they could improve the fit between model and experiment by assuming mode D. For the multilayer modeling of mode D, 20 stages indexed by i and representing sequential steps of film conversion were employed to repre-

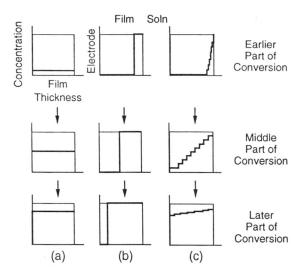

FIGURE 23 Schematic presentation of the three film conversion models: (a) homogeneous film–conversion (mode A); (b) propagation from the solution–film interface with sharp boundary (mode B); (c) graded propagation from film–solution
interface (mode D) (see Ref. 41). (Reprinted by permission from the Electrochemical Society.)

sent the oxidation, and 20 similar steps to represent the reduction. A Nernst
diffusion layer from the solution–film interface toward the electrode–film interface was assumed, where the thickness of the layer at different stages of the
conversion was taken to be $i/10$ of the total film thickness ($i < 10$). This growing
diffusion layer was itself divided into eight discrete sublayers, indexed by j, to
make a stepped concentration profile resemble a linear diffusion profile. The
refractive index at each stage i in layer j was taken to be proportional to the
fractional concentration of the oxidized and reduced form, $f_{o,ij}$ and $f_{r,ij}$, respectively, and to the refractive indices, n_O and n_R. The complete set of equations
employed is given in Ref. 41. This procedure enabled simulation of the experimental ψ vs. Δ plots based on known solutions for ellipsometry of stratified
planar isotropic structures sandwiched between two semiinfinite phases (see Ref.
2). Simulated ellipsometric conversion curves and experimental results for the
electrochemical conversion of PPy films are given in Figure 24. The results
demonstrate clearly that the introduction of the diffusion layer for both the oxidized and reduced species within the converted film improves the fit to the
ellipsometric conversion curve measured. The authors went on to show that, by
consideration of the times corresponding to the sequential steps of propagation,

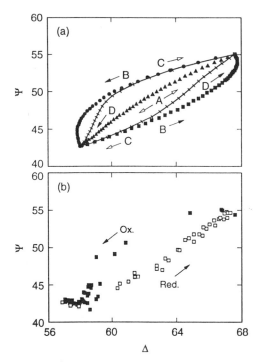

FIGURE 24 (a) Simulated curves for the ellipsometric conversion of a film of poly-pyrrole on a Pt electrode (in aqueous 0.2 *M* KCl solution). λ = 632.8 nm, angle of incidence 67°. Optical properties in the fully oxidized and fully reduced states are, respectively, n_{film} = 1.60 − 0.40i, d_{film} = 94.0 and n_{film} = 1.72 − 0.27i, d_{film} = 91.5 nm. Simulated curves A, B, C, and D correspond to the respective conversion modes (A, triangles; B and C, squares and circles; and D, ⑂⑂⑂). (b) Experimental ellipsometric conversion curves for this PPy film subjected to multipotential steps of 1 s width between −0.25 and +0.6 V vs. SCE. The experimental results conform best to mode D, with propagation originating from the film–solution interface. (Reprinted by permission from the Electrochemical Society.)

the effective diffusion coefficient for the oxidized and reduced forms of PPy could be evaluated. The effective diffusion coefficient evaluated in this way for the reduced form was 1.7 10^{-10} cm²/s, which agreed well with previous reports based on other types of experiments. The conversion directionality from the solution–film interface inward, as demonstrated by Lee et al. ellipsometrically for PPy [41], would be common for conducting polymer films of sufficient doping levels, having typically effective electronic conductivities which are significantly higher than their effective ionic conductivities. A reversed directionality (i.e., propagation from the film–substrate interface) is expected only when the electronic conductivity of the film drops significantly.

A case of site propagation from the metal–film interface has been demonstrated ellipsometrically by Lee et al. for the conversion process in thionine films grown on glassy carbon electrodes [42]. These results are shown in Figure 25. The multilayer directional propagation mode (mode D) again gave the best fit to experiment in the potential step experiments conducted. Lee et al. also showed, however, that the form of the ellipsometric conversion curve conforms better to mode A (i.e., uniform propagation) when slower potential scanning experiments were employed. This finding is a direct demonstration of the dependence of the mode of spatial propagation on the time scale of the experiment, τ.

The thionine film described in Ref. 42 is in a highly reduced (undoped) state under the highest cathodic potentials applied, as reflected by the very low imaginary component of n_{film} measured for the cathodic form. Therefore, it could be expected that the very limited electronic conductivity would force spatial propagation during the anodic process to originate from the electrode–film interface. However, less expectedly, the same direction of propagation also took place during the reduction of the oxidized form of the thionine film. This would suggest that the oxidized form of thionine is also associated with very low electronic conductivity, and the much higher value of k_{film} (0.24) measured for the oxidized state apparently originates from localized electronic states that do not contribute to electronic conductivity. Another possibility, however, is that the directionality for reduction of the oxidized form is dictated by an impedance localized at the film–substrate interface caused by poor physical contact between film and substrate. Effects of poor film adhesion have been suggested before as a possible source of such localized interfacial impedance [43]. (In fact, Lee et al. point out that studies of thionine films on other substrates have concluded on the reverse mode of propagation, from the film–solution interface inward.) When considering such possible effects of poor physical contact, it becomes clear that knowledge of the electronic conductivity in the polymer (or oxide) phase at a given state of oxidation cannot serve as absolute basis for predicting the direction of site propagation. The ellipsometric technique is an effective means to probe this directionality directly and therefore serves as an important source of information on such electrochemical processes.

B. Ellipsometric Observations of Structural Effects Induced by Film Multicycling

Some reversible structural variations caused by the cycling of conducting polymer films between oxidized and reduced forms have been suggested before based on the scan-rate independent hysteresis observed in cyclic voltammograms

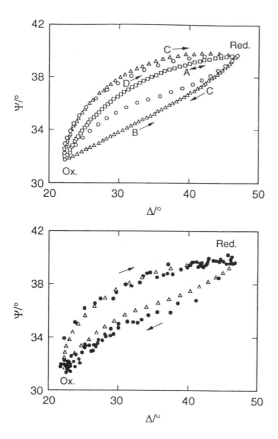

FIGURE 25 (a) Simulated ellipsometric conversion curves for a film of thionine on a GC electrode subjected to oxidation–reduction cycles between 0.8 and −0.25 V vs. SCE. Film properties at the two extreme potentials are at 0.8 V, n_{film} = 1.70 − 0.24i, d_{film} = 81.0 nm, and at −0.25 V, n_{film} = 1.65 − 0.04i, d_{film} = 71.0 nm. Simulated curves A, B, C, and D correspond to the respective conversion modes (A, squares; B and C, triangles; D, circles). (b) Experimental results for a multi-potential step experiment with 50 s pulse width applied to the thionine film designated by circles, and model calculations for mode D designated by triangles, showing good fit to mode D of conversion, with propagation originating from the substrate–film interface (see Ref. 24). (Reprinted by permission from Elsevier.)

(CVs) of conducting polymer films [44]. In particular, variations in the form of the CV have been observed between the first cycle and subsequent cycles applied to a freshly grown film [44], and ascribed to some undefined variations in film structure induced by initiation of potential cycling. Ellipsometry has the capability to examine cycle-induced changes in film properties. One example for such an observation has been described by Beckstead et al. [45] who investigated the growth and cycling of a film of Prussian blue at a wavelength of 633 nm. Beckstead et al. described a change in the optical properties of a film of Prussian blue under the same applied potential, induced by a single potential cycle to the cathodic limit and back. Following a single cycle from 0 V vs. Hg/Hg_2SO_4 to -0.8 V and back to 0 V, the change in optical properties measured at 0 V was from $\hat{n}_{film} = 1.48 - 0.26i$ to $\hat{n}_{film} = 1.43 - 0.34i$. A slight drop in film thickness was also detected together with this apparent change in $\hat{n}_{film}$ following a potential cycle of this type. This report was probably first to describe an ellipsometric observation of a phenomenon of interest in practical applications of such films (e.g., electrochromic systems or electrochemical capacitor applications). Such applications require prolonged potential multicycling, bringing up questions on the limitations of structural stability under such probing conditions. The two trends shown in Ref. 45 were (a) an increase in k_{film} (i.e., increase in film absorptivity) and (b) a slight decrease in film thickness caused by the potential multicycling routine. Such effects are expected, in principle, when the film undergoes what may be described as an electrochemically induced densification. A similar effect has recently been described by Sabatani et al. for films of polyaniline (PANI) cycled in acid solutions [17]. The authors of the last paper have demonstrated from single-wavelength measurements and, particularly, from spectroscopic ellipsometry that potential multicycling brings about an ''electrochemical annealing'' effect in PANI films. The spectrum of an as-grown film of PANI, recorded spectroellipsometrically under an applied anodic potential, was compared to the spectrum of the same film at the same anodic potential following single application of a cathodic bias for 2 min. As shown in Figure 26, following such a single potential cycle the spectroellipsometer detects, under the same applied anodic potential (same doping level), a significant shift of the major near IR absorption peak to lower photon energies. The shift of this spectral feature to lower photon energies corresponds to improved long-range order in the PANI film. This conclusion is based on the higher degree of charge carrier (polaron) delocalization expected as a result of such higher order, which would lead to a shift of the near-IR feature as observed. The interpretation given by Sabatani et al. for the restructuring phenomenon following single subjection of PANI to cathodic (undoping) potentials [17] was that elimination of electronic and ionic charge from the film (undoping) facilitates interparticle contact and subsequent interparticle bonding. This leads to a denser structure in the PANI film, increasing the optical and, presumably, also the dc conductivity. In this

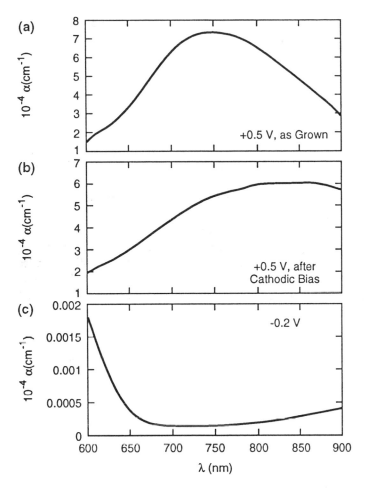

FIGURE 26 Spectra of the absorption coefficient of a PANI film evaluated from spectroellipsometric measurements. (a) Film grown on a gold electrode substrate in 1 *M* HClO₄ measured immediately after growth under bias of +0.5 V vs. SCE; (b) same film, measured under the same bias, after short subjection to a cathodic potential of −0.2 V; (c) spectrum of the cathodic form of the film (measured with the spectroellipsometer at −0.2 V), given for comparison. The significant shift of the spectrum at 0.5 V following a short temporary application of a cathodic bias, indicates an electrochemical annealing process in the film (see Ref. 17). (Reprinted by permission from the Royal Society of Chemistry.)

case the spectroellipsometric measurement did not detect significant changes in film thickness following electrochemical annealing, suggesting a restructuring effect dominated by enhanced interparticle bonding without appreciable inter-particle distance variations.

V. ELLIPSOMETRIC STUDIES OF ORGANIC MONOLAYERS AND MULTILAYERS

The preparation of monolayers on electrode surfaces by adsorption from solution has become an important tool in the control of interfacial properties in electro-chemical systems. In particular, monolayers of controlled dimensions on elec-trode surfaces have provided excellent model systems for studies of the depen-dence of electron transfer rates on electron tunneling range [46]. Potential applications in electrochemical sensors have also been described [47]. Other than in electrochemistry, such surface films have potential applications in many areas, including catalysis, corrosion protection, lubrication, adhesion, nonlinear optical devices, and artificial biomembranes [48]. The need for the detailed character-ization of such monomolecular layers is common to all of these areas of study. We discuss in this section applications of ellipsometry to the study of Langmuir–Blodgett films and self-assembled monolayers in a range of solid–liquid, solid–air, and liquid–gas interfaces. Results obtained for some nonelectrochemical sys-tems are included here because of their relevance to the nature and use of such monolayers at electrode–solution interfaces.

Ellipsometry has been used as a standard technique to determine the thick-ness of monolayers and multilayers of various self-assembled or Langmuir–Blodgett (LB) films [49]. In most of these studies, only the variation of the phase parameter, Δ, as a result of monolayer formation is measured ex situ, and the film thickness is derived assuming a transparent film with a (real) refractive index of 1.5. This approach is most easily justified for adsorption on oxide surfaces, where the film–substrate electronic interaction is minimal in many cases and does not affect the ellipsometric reading (see Ref. 1, sec. Vb). How-ever, even in such cases the accuracy of thickness determination according to Δ measurements is limited. Gun and co-workers reported ellipsometric studies on a multilayer of LB films with different lengths of fatty acids formed on an oxide surface [50]. Single-wavelength (546.1-nm) ellipsometric measurements were performed. The dependence of Δ on the estimated thickness of stacked LB films is given in Figure 27. It shows a linear dependence of $\delta\Delta$ on estimated film thicknesses up to 100 Å (three to five monolayers), where the estimate of film thickness is based on molecular dimensions. It is evident that, per given estimated film thickness, the change in Δ for the shorter-chain fatty acids is relatively higher than for the longer-chain fatty acids. This was explained by the relatively stronger contribution of the highly polarizable carboxylate head-

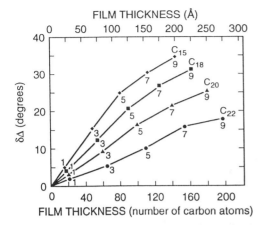

FIGURE 27 Measured change of Δ for stacked LB monolayers as function of total film thickness estimated from molecular dimensions. The number of LB monolayers stacked is shown in the figure (see Ref. 50). (Reprinted by permission from Academic Press.)

group to the change in refractive index in the case of the shorter-chain molecules. This example shows that even when the assumption on negligible electronic interaction with the substrate probably holds, as in a case of LB films of this type on an oxide surface, the refractive index has to be well known to enable accurate film thickness determination by such δΔ measurements. Critical studies on thickness determination for monolayers on oxide surfaces using single-wavelength (632.8-nm) ellipsometry were performed by Wasserman and co-workers [51], who compared ellipsometric results with results of low-angle x-ray (1.5 to 1.7 Å) reflectivity measurements. Self-assembled monolayers of alkylsiloxane were prepared by adsorbing alkyltrichlorsilanes [$Cl_3Si(CH_2)_nR$] from solution onto silicon–silicon dioxide substrates. The comparison of the thicknesses of alkylsiloxane monolayers, as measured by ellipsometry and by x-ray reflectometry is given in Figure 28. It shows systematically a slightly higher (by 0.14 nm) thickness determined by ellipsometry. The reason for this discrepancy was explained by the inclusion of the Si atom of the alkylsiloxane group in the ellipsometrically determined thickness, whereas x-ray reflectometry presumably only measures the thickness of the hydrocarbon part of the alkylsiloxane group and excludes the Si atom because of the similarity of its electron density to that of the silicon oxide substrate.

Chemisorption on metal surfaces generates a more complex optical problem because of perturbation of the optical properties of the ''skin'' of the metal substrate and the addition of optical transitions associated with the substrate–

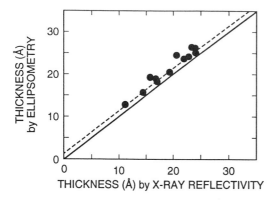

adsorbate complex (see ref. 1, sec. Vb). Consequently, demonstration of the validity of the simplified assumptions that $\delta\Delta$ is proportional to film thickness and that the film is transparent with $n = 1.5$ has to be performed carefully. Porter and co-workers used single-wavelength (632.8-nm) null-detection ellipsometry to determine the thickness of alkyl thiols [$HS(CH_2)_nCH_3$] with $n = 1, 3, 5, 7, 9, 11, 13, 15, 19,$ and 21, self-assembled on Au [52]. The gold substrate was prepared by the evaporation on the native oxide of a crystalline silicon surface (i.e., ensuring a high-quality smooth substrate) and the alkyl thiol monolayers were formed by immersing the gold substrate in 1 mM solutions. The thicknesses determined from a three-phase optical model (air/organic film/Au substrate), assuming a real refractive index of 1.45 for the film and using the measured complex refractive index for the bare gold substrate, are shown in Figure 29. The measured thicknesses show two apparent regions of dependence on the number of carbon atoms (n) in the alkyl chain. Below $n = 9$ there is a weak dependence of the ellipsometric thickness on n. Above $n = 11$, the ellipsometric thicknesses increase linearly with n but tend to deviate from those calculated for a fully extended chain normal to the surface. These apparent discrepancies were accounted for by two factors. First, the chemisorption process (Au + RSH — AuSR + H$^+$, R = alkyl) induces changes in the complex refractive index of the surface of the gold substrate, and therefore invalidates the single-film model. Second, as the chain length increases, the packing density of the monolayer may become higher, resulting in a refractive index higher than 1.45 [52]. A similar

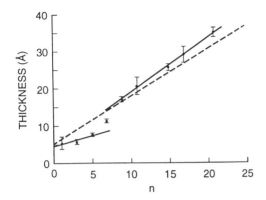

FIGURE 29 Film thicknesses for *n*-alkyl thiols adsorbed at gold. The ellipsometry determined thickness (from Δ measurements alone) is given by the solid line. The dashed line gives the estimated thickness for a fully extended chain normal to the surface. (Reprinted with permission from the American Chemical Society.)

study performed by Bain and co-workers [53] used single-wavelength ellipsometry (632.8 nm) to determine the thicknesses of several self-assembled monolayers, such as *n*-alkylthiols, ω-mercaptocarboxylic acids, and dialkyl disulfides. Ellipsometric thicknesses of monolayers of *n*-alkylthiols prepared under conditions similar to those used in [52] are shown in Figure 30. In general, the ellipsometric thickness was below that predicted for a fully extended, all-trans configuration with normal orientation. However, the linear increase in the ellipsometric thickness with increase in chain length seemed to be qualitatively substantiated, leading to a suggestion on a possible tilt of the molecular orientation to 30° from the normal.

Recently, Rubinstein and co-workers prepared selective ion-binding self-assembled monolayers of thiobis(ethylacetoacetate) which were characterized by electrochemistry, contact-angle measurement, and ellipsometry [47]. In particular, they observed unexpectedly high ellipsometric effects upon metal ion binding to the immobilized monolayer complexant. By measuring not just the phase parameter Δ but also the amplitude parameter ψ, the latter group demonstrated clearly that some of the optical effects observed as a result of processes within adsorbed monolayers cannot be accounted for by simplistic models based on a nonabsorbing single film. This work also demonstrated that a minimal requirement to confirm the validity of such simple assumptions in single-wavelength ellipsometry would be to take measurements of δψ in addition to δΔ measurements. According to the magnitude and sign of δψ, the assumption of transparency or semitransparency in the adsorbed layer can be probed (see Ref. 1, sec. IVb).

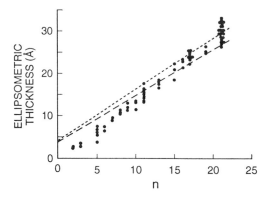

FIGURE 30 Ellipsometric derivation of the thickness of monolayers of *n*-alkane-thiols on gold. The short-dashed line corresponds to the thickness expected theoretically for a close-packed monolayer oriented normal to the surface, whereas the long-dashed line corresponds to such a monolayer oriented 30° from the normal. Note the similarity of the deviations for *n* < 10 here and in Figure 29. The model of a single transparent film employed is less satisfactory the thinner the layer (Ref. 53). (Reprinted with permission from the American Chemical Society.)

Kim and Vedam studied the orientation of pyridine molecules adsorbed from pure water on a Ag surface [18]. Using Kretschmann's configuration [54], they performed spectroscopic ellipsometry measurements of the surface plasmon polariton (SPP) associated with the Ag substrate surface, following changes in the SPP with increased solution concentration of pyridine. (The Kretschmann configuration [54] enhances significantly the sensitivity of the ellipsometric measurement but can be used only with a limited number of metal substrate surfaces.) To interpret the changes in the optical properties of the interface due to pyridine adsorption, a five-phase optical model was used, which consisted of: glass/Ag film/Ag and pyridine mixture/pyridine/H_2O. The optical properties of the pyridine layer were assumed identical to those of neat bulk pyridine. The need to include the mixed Ag/pyridine layer to account for the measured ellipsometric effects demonstrated the significance in this case, as in other cases of chemisorption on metal substrates, of electronic perturbations and/or effects of a metal roughness layer on the ellipsometric effect measured. The thickness of the pyridine layer evaluated from this five-layer model exhibited an abrupt increase as the concentration exceeded 10^{-4} M, indicating a configurational transition in the adsorbed pyridine layer (Figure 31). At pyridine concentrations smaller than 10^{-4} M, the pyridine ring apparently favors an orientation parallel to the substrate, whereas at higher concentrations the pyridine ring switches to normal orientation.

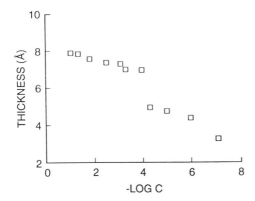

FIGURE 31 Thickness of a pyridine monolayer on silver derived from spectroellipsometric measurements in the Kretchmann configuration. A mixed Ag/pyridine layer between the substrate and the monolayer of pyridine was required to fit the results. An abrupt rise in the thickness of the monolayer as the solution concentration exceeds 10^{-4} M is ascribed to molecular reorientation from parallel to normal to the surface (see Ref. 18).

The ellipsometric technique has also been used to study monolayers and multilayers of adsorbed biomolecules on electrode surfaces. Szucs and co-workers reported ellipsometric and electrochemical studies on monolayers of glucose oxidase [55] and cytochrome c adsorbed on gold electrode surfaces [56]. Cytochrome c was adsorbed on a bare gold electrode and on a surface modified with 4,4'-dipyridyl disulfide. A few selected wavelengths (546.1, 632.8, 790 nm) were used for the ellipsometric measurements to prevent photodecomposition. The evaluated thickness of the adsorbed cytochrome c layer on the bare gold electrode was 1.8 to 2.2 nm, with a refractive index of 1.48 to 1.50. This thickness is about half the expected native molecular thickness of cytochrome c. The interpretation for the smaller thickness derived was that the cytochrome c adsorbed on a bare gold electrode is unfolded. In contrast, the thickness of the cytochrome c monolayer adsorbed on the gold surface modified with 4,4'-dipyridyl disulfide was found ellipsometrically as 3.5 to 4.0 nm. The refractive index of 1.41 to 1.42 was closer to values expected for the native cytochrome c. This seems to indicate [56] that cytochrome c is adsorbed on the modified gold surface without significant structural changes. Consequences regarding electron transfer to the adsorbed cytochrome c molecule were discussed [56].

Ellipsometry has also been used to study the interfacial structure of liquid–air systems: for example, Langmuir films at the air–water interface. We describe here some results of such studies which seem to have relevance for the study of monomolecular films of this type in electrochemical interfaces. Kawaguchi and co-workers [57] reported an ellipsometric study of polymeric monolayers such as poly(ethylene oxide) (PEO), poly(tetrahydrofuran) (PTHF), poly(vinyl acetate) (PVAC), poly(methyl methacrylate) (PMMA), and poly(γ-methyl-L-glutamate) (PMLG) at the air–water interface. The thicknesses of polymeric monolayers at the air–water interface showed a strong dependence on the hydrophilicity of the polymer and on surface pressure. Kawaguchi and co-workers attempted in this work to calculate the amount of polymer transferred onto solid substrates from the air–water interface in a LB film preparation routine. Rasing and co-workers [58] used ellipsometry to study two-dimensional phase transitions of pentadecanoic acid at the air–water interface. The structure of a monolayer of pentadecanoic acid at the air–water interface was studied for the gas, liquid, and condensed states of the film. Single-wavelength (632.8 nm) ellipsometry with an angle of incidence of 75° was employed and an accuracy in $\delta\Delta$ of 10^{-4} rad was achieved. The authors demonstrated in this last work that $\delta\Delta$ measurements reflect well the different regions of the isotherm for an organic monolayer on water, providing estimates of domain sizes. Kim and co-workers [59] studied using polarization modulation ellipsometry the structure of monolayers of fatty acids of different chain lengths at the air–water interface using different cationic species in the solution phase. $\delta\Delta$ was found to be linearly proportional to the chain length as seen in Figure 32. In addition, a higher ionic polarizability of the cationic species resulted in higher changes in Δ for a given chain length, suggesting inclusion of the cation in the surface monolayer. The

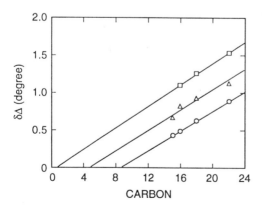

FIGURE 32 Measured $\delta\Delta$ vs. carbon number for fatty acid monolayers adsorbed on aqueous 1 m*M* PbCl$_2$ (squares), 1 m*M* CdCl$_2$ (triangles), and 10 m*M* HCl (circles) (see Ref. 59). (Reprinted with permission from the American Chemical Society.)

issue of optical anisotropy in such layers is also treated in the latter work. Paudler and co-workers [60] treated rigorously a uniaxial, ultrathin film using multiple-angle ellipsometry. The anisotropic optical nature, expected in principle for a well-oriented monolayer, is discussed analytically and numerically, providing a relationship between the measured ellipsometric parameters and the angle of incidence, refractive indices in the lateral and perpendicular directions, and film thickness. From such calculations it was demonstrated that multiple-angle ellipsometric data could not provide a unique film thickness in the case of anisotropic films. The refractive indices of behenic acid monolayers at the air–water interface were thus obtained by multiple-angle ellipsometry measurements but assuming film thickness as determined by x-ray measurements. The values derived, $n_x = 1.47$ and $n_z = 1.54$, showed that the optical anisotropy of fatty acid monolayers is not very large. The anisotropic nature of the fatty acid (behenic acid) was studied using infrared spectroscopic polarization modulation ellipsometry by Benferhat and co-workers [61]. The sensitivity of IR ellipsometric spectra to anisotropy in the organic monolayer is much better. The presence of CH_2 stretching modes in the spectrum showed that the layers are not well oriented perpendicularly to the substrate up to a thickness of three monolayers. Beyond that thickness the molecular orientation is easily distinguished from the ellipsometric spectra. This study [61] has demonstrated the importance of the extension of ellipsometric studies to the infrared region for revealing the structure of organic films several monolayers thick. The special instrumental aspects of IR ellipsometry have not been discussed in this chapter and the reader is referred to Ref. 62 for recent demonstrations of its application.

The main conclusions from the ellipsometric studies of adsorbed monolayers described above can be summarized as follows:

1. Single-wavelength ellipsometry can provide only approximate thicknesses for adsorbed monolayers according to $\delta\Delta$ measurements. It is particularly effective in this simple mode for cases of physisorption on nonmetallic substrates.
2. Significant chemisorptive interactions (e.g., in the case of self-assembled monolayers of alkyl thiols on gold electrode surfaces) could cause strong deviations from the simple model usually employed of a nonadsorbing layer with a real refractive index close to $n = 1.5$. Such effects of electronic film–substrate interactions are most significant in the ellipsometry of ultrathin monomolecular layers, while diminishing in relative significance beyond a film thickness of 10 nm (e.g., for stacked LB layers or for polymeric monolayers).
3. Optical anisotropy in organic monolayers is most easily revealed and analyzed with IR ellipsometry.

VI. THEORY AND SIMULATION TECHNIQUES FOR ELLIPSOMETRY

A. Equations of Ellipsometry

We start out considering a planar interface between two semi-infinite media, the ambient and the substrate (Fig. 33). Let z denote the direction perpendicular to the interface and assume that a linearly polarized plane wave is incident on the boundary between the two media. If the electric field vector of the wave lies on the plane of incidence, as in Figure 33, we say that the wave has p polarization; if the electric field vector is perpendicular to the plane of incidence, we say that the wave has s polarization. The fundamental equations describing the reflection and refraction of the light over the interface can be found in most books on electromagnetic theory or optics. Of these, the equations most fundamental to ellipsometry are the Fresnel relations for p and s polarization, given by

$$\frac{E_{rp}}{E_{ip}} = r_p = \frac{\hat{n}_1 \cos \phi_0 - n_0 \cos \phi_1}{\hat{n}_1 \cos \phi_0 + n_0 \cos \phi_1} \tag{14}$$

and

$$\frac{E_{rs}}{E_{is}} = r_s = \frac{n_0 \cos \phi_0 - \hat{n}_1 \cos \phi_1}{n_0 \cos \phi_0 + \hat{n}_1 \cos \phi_1} \tag{15}$$

The Fresnel coefficients r_p and r_s represent the ratios of the reflected to the incident complex amplitudes of the electric field vectors. In these equations n_0 and $\hat{n}_1$ are the indices of refraction of the ambient and the substrate, respec-

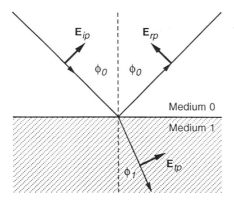

FIGURE 33 Schematic of the basic ellipsometric experiment for reflection from an ambient–substrate interface. The ambient is denoted with subscript 0 and the substrate with subscript 1.

tively, and ϕ_0 and ϕ_1 are the angle of incidence and the angle of refraction. The complex index of refraction, $\hat{n}$ (where the caret is used to denote that the quantity is complex), is normally written in terms of its real, n, and imaginary, k, parts as $\hat{n} = n - ik$. (An equivalent description employs the complex dielectric constant, $\epsilon = \epsilon' - i\epsilon''$, where $\epsilon = \hat{n}^2$.)

These two equations are at the heart of all ellipsometric measurements. Indeed, reflection ellipsometry is a technique based on the measurement of the quotient ρ of the Fresnel coefficients r_p and r_s. In ellipsometry, it is customary to write this quotient in the form

$$\rho = \frac{r_p}{r_s} = \tan \psi \, \exp(i\Delta) \tag{16}$$

where $\tan \psi$ represents the amplitude change in ρ and Δ the phase change. In many experiments it is also possible to measure the reflectance, which is defined by the quantities $R_p = |r_p|^2$ and $R_s = |r_s|^2$, corresponding to those fractions of the total intensity reflected with polarizations p and s, respectively.

The simplest model for an ambient–film–substrate system consists of a single, homogeneous film sandwiched between the ambient and the substrate. A somewhat more complicated model is shown in Figure 34, where a stratified structure consisting of multiple films is sandwiched between the ambient and the substrate. In general, we assume that the optical properties of each layer have spatial homogeneity (i.e., are constant within the extent of the corresponding layer). The ellipsometric equations describing this model are discussed in the appendix to this chapter.

B. Basic Problem of Thin-Film Ellipsometry

In one of the basic experiments in reflection ellipsometry (e.g., rotating element mode) one normally shines a light beam of known polarization state on a film–substrate system and measures the quantities $\tan \psi$ and $\cos \Delta$ of equation (16), reflecting changes in the polarization state of the reflected light. Such changes are produced by the combined effect of the film and the substrate and therefore have information about their optical properties, as indicated by equations (14) and (15) and their counterparts for the film–substrate system. If one has previously done an experiment with the bare substrate, leading to a knowledge of its optical constants, it should be possible to back out of the experimental data, with appropriate modifications of equations (14) to (16) as discussed in the appendix, the optical constants, and the thickness of the film. Thus the fundamental questions to be asked are: (a) what are the optical properties of the film? and (b) what is its thickness?

These questions place ellipsometry into a general class of problems known in physics as inverse scattering problems. The main idea is as follows. We have

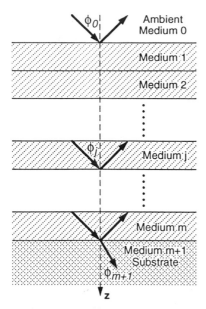

Stratified Isotropic Structure

FIGURE 34 Schematic of the basic ellipsometric experiment for reflection from an ambient–multilayer–substrate system. The ambient is denoted with subscript 0 and the substrate with subscript $m + 1$. The intervening m layers are denoted with subscripts 1 to m.

a scattering system whose properties we want to determine; a known signal is made to impinge on the system and the scattered signal is measured. From both of these known signals one would like to extract as much information as possible about the nature of the system.

Because the properties of the system are, to a large degree, unknown, the analysis of inverse scattering data generally has the following structure:

1. A physical model of the system is developed; for example, one assumes that the film consists of a homogeneous, single-phase, perfectly flat layer.
2. The physical model is then cast into a set of mathematical equations that describe the interaction between the incident wave and the model.
3. A set of parameters is assumed for the model, such as the value of the real and imaginary parts of the index of refraction and the thickness of the film.

4. Given the incident wave, the mathematical equations, together with the assumed values of the parameters, are used to predict a scattered wave.
5. The results of the model for the scattered wave are compared with the corresponding quantities from the experimentally observed scattered wave.
6. A new set of parameters is constructed and the process is repeated until a "best fit" is obtained.

Of course, the determination of when a best fit has been obtained is intimately related to the methods employed to predict the new set of parameters. Obviously, the process is, in the great majority of cases, iterative and its convergence and stability depend on the algorithms employed in the solution.

In thin-film reflection ellipsometry we invariably resort to physical models that consist of layered, planar structures. In addition, the majority of the models also assume that the individual layers are linear and isotropic with respect to the scattering of light. The ellipsometric equations describing such a system are discussed in the appendix.

C. Analysis of Thin-Film Growth Experiments at Fixed Wavelength

We consider first a common problem in the characterization of thin films by ellipsometric means, the in situ measurement of the thickness and index of refraction of a growing film. The physical model we will assume is that at any stage of the growth the film can be approximated by a single, uniform, isotropic layer whose properties can be described mathematically by the equations described in the appendix. (In this case, using the standard convention, we relabel the media through which the light has to travel so that the ambient is still denoted by the subscript 0, but the film and the substrate are now denoted with the subscripts 1 and 2, respectively.)

To separate the properties of the film from those of the substrate, we first determine the index of refraction of the substrate, $\hat{n}_2$, by measuring the ellipsometric angles for the bare substrate and using equations (14) to (16). We next take a series of M ellipsometric measurement of the growth of the film at different (unknown) thicknesses $d_1, d_2, \ldots, d_M$. Let $\{\psi_1, \psi_2, \ldots, \psi_M, \Delta_1, \Delta_2, \ldots, \Delta_M\}$ be the experimental ellipsometric angles for the M measurements. We want to determine the index of refraction of the film and the M thicknesses d_i. Our model will then be that of a uniform layer whose index of refraction, $\hat{n}_1$, is the same for all of the growth stages. This model, which may or may not be very realistic depending on each case, is the simplest model that often leads to a meaningful interpretation of the experiments. We have a total of $M + 2$ unknowns, namely, the M values of the thickness and the real and imaginary parts of the index of refraction of the film. On the other hand, we have a total of $2M$ knowns, given by the measured ellipsometric angles.

The solution of the problem is often carried out by combining the knowns and unknowns into a figure-of-merit function which measures the agreement between the data and the model. The figure-of-merit or cost function is normally defined so that it is positive or zero. When it is zero the model leads to values that are in perfect agreement with the experimentally measured data. For the present type of problem one commonly uses as a cost function the maximum likelihood function [63]

$$\chi^2 (n_1, k_1, d_1, d_2, \ldots, d_M) = \sum_{i=1}^{M} \left[\frac{(\psi_i - \psi_i^{calc})^2}{(\sigma_i^\psi)^2} + \frac{(\Delta_i - \Delta_i^{calc})^2}{(\sigma_i^\Delta)^2} \right] \qquad (17)$$

where the superscript calc denotes values calculated from the model. The quantities σ_i^ψ and σ_i^Δ represent weight factors that can be related to the experimental errors. Since many of the standard ellipsometry experiments actually measure tan ψ and cos Δ, it is also common to use the expression (see, e.g., Ref. 64)

$$\chi^2 = \sum [(\tan \psi_i - \tan \psi_i^{calc})^2 + (\cos \Delta_i - \cos \Delta_i^{calc})^2] \qquad (18)$$

which is equivalent to equation (17) when one considers that in most computers the numerical accuracy of the trigonometric functions is several significant figures better than the experimentally measured values of tan ψ and cos Δ. It should be pointed out that some authors [64] use the symbol σ to denote the function χ^2.

The aim of the procedure is to minimize the function χ^2, in the sense of least-squares fitting [65], by varying the unknown parameters. Since the functional dependence of the ellipsometric angles on the unknown parameters (e.g., index of refraction and the thickness of the film) is nonlinear, the minimum of χ^2 must be calculated by an appropriate nonlinear algorithm. One of the most popular techniques for the minimization of χ^2 is the Levenberg–Marquardt method. The reader can find an excellent discussion of its implementation, together with ready-to-use subroutines, in Ref. 65. Care must be taken to give appropriate weights to those parts of the experimental data corresponding to large uncertainty regions. This is particularly important when the measured ellipsometric angle Δ is near 180°. In this case, cos Δ, which in rotating element ellipsometry is the experimentally measured quantity, is such that very small changes in its value correspond to relatively large changes in the angle, leading to larger uncertainties than for regions where cos Δ varies more rapidly.

When additional experimental quantities are measured, for example the reflectance of the film, it is straightforward to modify equations (17) and (18) to include the additional data. Thus, when the reflectances R_p and R_s are also measured, one would employ the following modification to equation (17):

$$\chi^2 = \sum \left[\frac{(\psi_i - \psi_i^{calc})^2}{(\sigma_i^{\psi})^2} + \frac{(\Delta_i - \Delta_i^{calc})^2}{(\sigma_i^{\Delta})^2} \right.$$
$$\left. + \frac{(R_{pi} - R_{pi}^{calc})^2}{(\sigma_i^{Rp})^2} + \frac{(R_{si} - R_{si}^{calc})^2}{(\sigma_i^{Rs})^2} \right] \quad (19)$$

Once one has determined the optimum set of parameters from the previous procedure, representing the best fit of the model to the experimental data, one can ask about the level of confidence that can be associated with those parameters. To assign confidence limits to the optimum model parameters, it is necessary to go beyond least-squares fitting. This can be done using Monte Carlo simulations to obtain a probability distribution for the parameters [65] from which one can estimate the errors or the validity of the model.

Suppose now that the confidence limits on the parameter estimates indicate poor agreement between the model we have chosen and the experimental data. In our example, this would mean that the simple model that assumes that the film grows uniformly, with constant values for the real and imaginary parts of the index of refraction, is not appropriate for the experiment we carried out. We must then change or extend the model. Let us consider a couple of possibilities.

In one case we can assume that at a given stage of the growth, the film's index of refraction, within any given layer, is a function of the distance perpendicular to the interface, so that, effectively, there is a gradient in the values of n and k that can be described in terms of the following expressions:

$$n(z) = n_0 + Az \quad \text{and} \quad k(z) = k_0 + Bz \quad (20)$$

Here A and B are constants. One can then use these equations in the mathematical formulation of the model to obtain expressions for ψ^{calc} and Δ^{calc} that can be incorporated into equation (15). Unfortunately, the expressions in equation (20) lead to a very complicated formalism for the ellipsometric angles to make them impractical for implementation in computer programs. As a result, we often resort to a simpler scheme—but just as accurate—based on stratified structures. Thus, in this model, we subdivide the film into a series of uniform layers of different indices of refraction. Instead of letting each layer have a varying index of refraction, we require that they have a constant value for n and k but these values increase or decrease from one layer to the next (Fig. 35). Under these circumstances, we can use the formalism developed in the appendix to obtain values for ψ^{calc} and Δ^{calc}.

An important consideration for these types of models is the accounting of the number of knowns versus unknowns. Thus, if, as before, we carry out M

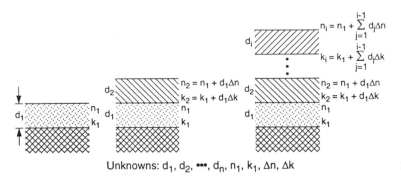

Unknowns: $d_1, d_2, \cdots, d_n, n_1, k_1, \Delta n, \Delta k$

FIGURE 35 Linear model for the optical constants of an inhomogeneous multi-layer system. The values of n and k increase linearly from layer to layer.

measurements and for each growth stage we add a new layer with three un-knowns (d_i, n_i, and k_i), the total number of unknowns will be $3M$, whereas we have only $2M$ known quantities. This presents a mathematical problem that can be solved by an appropriate modification of the model[*]—for example, assuming that n and k vary from layer to layer according to a fixed functional form, say linear in the distance from the interface (Fig. 35), so that the unknowns are the M values of the thicknesses of the layers and the slopes and intercepts of the linear relations for n and k.

D. Effective Medium Theory [66–68]

An alternative for the ellipsometric characterization of film growth is afforded by effective medium theory. Suppose, for example, that we suspect that the film grows by nucleation at the liquid–film interface. We would then expect some significant portions of the film, particularly for very thin films, to be constituted by a mixture of two or more phases: say, a solid, porous phase immersed in the solution (Fig. 36), with different optical constants. Effective medium theory allows us to calculate the optical properties of the composite, assuming that we know those of the constituent phases.

In effective medium theory we consider a system consisting of inclusions of a phase whose dielectric constant is ϵ_1, immersed in a phase of dielectric constant ϵ_0. Assume that the corresponding volume fractions of the two phases are η_0 and η_1, with $\eta_0 + \eta_1 = 1$. One can then show [66,67] that the dielectric

---

[*]As we will see below, this problem can also be tackled by means of the maximum entropy method.

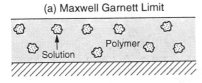

(a) Maxwell Garnett Limit

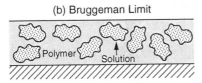

(b) Bruggeman Limit

FIGURE 36 Schematic of films that can be described with effective medium theory: (a) Maxwell Garnett limit; the volume fraction of one medium is larger than that of the other medium; (b) Bruggeman limit; the volume fractions of both media are comparable.

constant of the composite medium, assuming spherical inclusions, is given by

$$
\epsilon_{eff} = \epsilon_0 \left[\frac{1 + 2\eta_1 \left(\dfrac{\epsilon_1 - \epsilon_0}{\epsilon_1 + 2\epsilon_0} \right)}{1 - \eta_1 \left(\dfrac{\epsilon_1 - \epsilon_0}{\epsilon_1 + 2\epsilon_0} \right)} \right] \tag{21}
$$

This equation is most applicable in the Maxwell Garnett limit of effective medium theory [67]. In this case, $\eta_1 \ll 1$ and ϵ_0 corresponds to the dielectric constant of the matrix, in which a number of small particles of dielectric constant ϵ_1 are embedded (Fig. 36a).

Another common case is given by the Bruggeman limit [68], in which the two phases have comparable volume fractions (Fig. 36b). The equation for the effective dielectric constant is obtained from the assumption that a particle of dielectric constant ϵ_i, where $i = 1, 2$, is surrounded by a medium of dielectric constant ϵ_{eff}, leading to

$$
\eta_0 \frac{\epsilon_0 - \epsilon_{eff}}{\epsilon_0 + 2\epsilon_{eff}} + \eta_1 \frac{\epsilon_1 - \epsilon_{eff}}{\epsilon_1 + 2\epsilon_{eff}} = 0 \tag{22}
$$

Webman et al. [66] have given the following algorithm for the numerical solution of equation (22):

1. Calculate $x = \epsilon_1/\epsilon_0$.
2. Calculate $a = \frac{1}{4}[(3\eta_0 - 1)(1 - x) + x]$.
3. Calculate $f = a \pm (a^2 + \frac{1}{2}x)^{1/2}$.

4.   Calculate $\epsilon_{\text{eff}} = \epsilon_0 f.$

This procedure is valid for $0.05 < |x| < 20$.

Equations (21) and (22) can be incorporated into the procedure used to minimize equation (17). For example, suppose that we use a two-phase model in which we assume that ϵ_0 and ϵ_1 are constant but where the volume fraction η_1 changes as the film grows. Then one would let η_1 vary as one of the unknown parameters in equation (17). In the Bruggeman limit, we would use equation (22) to calculate ϵ_{eff} for the film, and from it we would get the index of refraction of the film, which, in turn, would be incorporated into the ellipsometric equations that determine ψ and Δ. An alternative to this procedure is to solve the ellipsometric equations directly for η_1 and the thickness of the film using standard techniques, such as Newton's method [65]. This can be done in this case, assuming that we know ϵ_0 and ϵ_1, because the number of unknowns, η_1 and the thickness, is equal to the number of knowns, ψ and Δ. Equation (9) has been used with this procedure to determine the volume fraction, thickness, and optical constants of electrochemically grown conducting polymers [12a].

E.   Spectroellipsometry

1.   Method of Arwin and Aspnes

As described elsewhere in this chapter, in spectroscopic ellipsometry ψ and Δ are determined as functions of the wavelength (see Section II). In this case, for a given film thickness, the experiments produce a set of ellipsometric angles $\psi(\lambda_i)$ and $\Delta(\lambda_i)$, where λ_i is the wavelength.

A number of procedures have been advanced for the interpretation of these experiments in terms of the optical spectrum of the film and its thickness. In one approach [24] one uses a model with a uniform, homogeneous film whose thickness is first guessed. Then for each λ_i one solves from $\psi(\lambda_i)$ and $\Delta(\lambda_i)$ for the real and imaginary parts of the index of refraction, $n(\lambda_i)$ and $k(\lambda_i)$, or, equivalently, the dielectric function, $\epsilon'(\lambda_i)$ and $\epsilon''(\lambda_i)$. If the guess for the thickness is wrong, artifacts appear in the optical spectra. These can arise from prominent features in the substrate spectrum. Such spurious features, which are apparent because they show up in the form of discontinuities or sharp cusps, are not expected to appear for a small range of guessed thicknesses around the correct value.

The general formulation of Arwin and Aspnes' procedure [24] is quite simple. They assume a model consisting of a substrate, a film, and the ambient, with dielectric functions ϵ_s, ϵ, and ϵ_a. The ratio of the Fresnel coefficients, equation (16), is expressed as

$$\rho_{\text{meas}} = \rho(\epsilon_s, \epsilon, \epsilon_a, d, \lambda, \phi_0) \tag{23}$$

where d is the thickness, λ the wavelength of the light, ϕ_0 the angle of incidence, and ρ_{meas} the quantity measured by the instrument. The unknowns are d and ϵ. Let $<d>$ be a guess for the thickness near the true value d. Then one can use Newton's method to invert equation (23) to obtain an estimate for the dielectric function, $<\epsilon>$. Equation (23) is then expanded to first order about the exact values to argue that if $<\epsilon>$ is to correspond to the correct solution, it cannot show system-related artifacts, such as optical structure originating from the spectrum of the substrate. The spurious interference of the substrate spectrum with the net spectrum of the film–substrate system is determined from the second derivatives of $<\epsilon>$ as a function of the energy of the incident light. This procedure works best in those cases in which the substrate has strong features that do not overlap with any features in the film spectrum.

Figure 37 shows an example of this procedure applied to spectroscopic ellipsometry of galvanostatically grown polyaniline on a gold substrate (Y.-T. Kim, unpublished results). The plot shows the second derivative of the calculated dielectric function of the film as a function of photon energy. The different curves correspond to different values of the thickness of the film. Because the spectrum of the gold substrate has a strong feature around 2.5 eV and presumably the film does not, according to the Arwin–Aspnes criterion one must choose that thickness corresponding to a featureless spectrum for the second derivative of the film dielectric function. This corresponds to the curve labeled by the

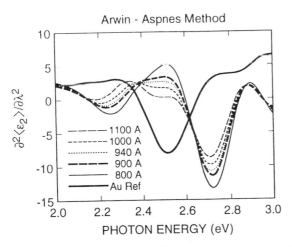

FIGURE 37 Plot of the second derivative of ϵ_2 as a function of photon energy for a PANI film on a gold substrate. The different curves correspond to different assumed thicknesses of the film.

thickness of 940 Å in Figure 37. The corresponding calculated spectrum is shown in Figure 38.

It is important to note that this approach requires the assumption that the spectrum of the film is featureless around the region where the spectrum of the substrate exhibits at least one strong feature. This may not necessarily be the case for all films and all substrates. In addition, because relatively large changes in the thickness of the film can be accommodated by relatively small changes in the dielectric function of the film (see below), it is not unambiguously clear which thickness to choose from those that show featureless spectra.

On the other hand, there are examples in which the Arwin–Aspnes technique leads to results in excellent agreement with other, independent methods. One example is given by the results of Kim et al. [7b], where the dielectric function of polypyrrole (on gold) was obtained by the Arwin–Aspnes method. The results were confirmed by a Lorentz oscillator model. In this work the Lorentz oscillators were used to parameterize the dielectric function of polypyrrole and deduce the electronic transition energies. Thus the general form of the dielectric function was written as

$$\epsilon = \epsilon_\infty + \sum_n \frac{f_n^2}{E_n^2 - E^2 - iE\Gamma_n}$$

where ϵ is the complex dielectric function, ϵ_∞ the infinite photon energy dielectric function, and E_n the nth resonance energy of the electronic system. The oscillator strength f_n is a measure of the relative probability of an electronic

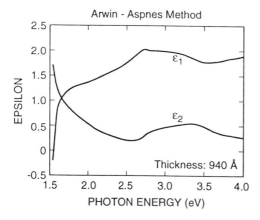

FIGURE 38 Optical spectrum of PANI on gold obtained using the Arwin–Aspnes method. The experimental data are the same as those of Figure 37, with a film thickness of 940 Å.

transition associated with the nth oscillator resonance, and Γ_n is the damping coefficient for the nth oscillator resonance. This analysis, whose results are shown in Figure 12, provided a thickness of 475 ± 20 Å, in good agreement with the value of 490 Å that was obtained with the Arwin–Aspnes method. In this case the dielectric function of the film is featureless around the region where the dielectric function of the gold substrate exhibits at least one strong feature. Because this technique cannot always distinguish clearly between the substrate and film features, as in the case of polyaniline above, it is recommended to use another independent approach, such as the Lorentz oscillator methods or the maximum entropy technique described below, to confirm the thickness and dielectric function of the film.

2. Maximum Entropy Method

Recently we have proposed that the analysis of spectroellipsometric data can be carried out by means of the maximum entropy method. This formalism, introduced by Jaynes [69] and Tribus [70], is normally employed in the determination of unknown Bayesian probabilities. To illustrate the principles of the method, let us consider a system with a set of N possible experimental outcomes with probabilities p_i. We will show below how these probabilities can be related to the measured ellipsometric angles ψ and Δ. The probabilities are assumed to be normalized, so that $\Sigma_i \, p_i = 1$. Following the formalism of information theory [71] one can define the entropy of such a set of probabilities by

$$S = -\sum_i p_i \log p_i \tag{24}$$

where the log function denotes the natural logarithm.

The maximum entropy formalism [72,73] uses this definition of the entropy as a tool to determine the probability distribution p_i. Jaynes suggested that the probabilities be chosen by finding the maximum value of the entropy, subject to any constraints that the system must satisfy. The main requirement of the approach is that the probability distribution be positive definite and additive.

In the case of spectroscopic ellipsometry we proceed as follows. For definiteness, we consider an experiment in which one measures the ellipsometric angles ψ and Δ as a function of the wavelength or the photon energy for a single film on a substrate of known optical constants. As a model for this system we assume that the film does not change thickness during the experiment and that its optical constants and thickness are related to the ellipsometric angles through the standard ellipsometric equations for a single film sandwiched between ambient and substrate media as described in the appendix. If the measurements consist of N different pairs of ellipsometric angles, the unknowns are $n_1, \ldots, n_N$, $k_1, \ldots, k_N$, and d, namely, the complex index of refraction for each wavelength and the thickness of the film. Although the number of unknowns is one more than the number of knowns, the maximum entropy formalism allows us to solve the problem unambiguously. To do this we assign a discrete set of probabilities as follows.

Noting that the real and imaginary parts of the index of refraction, as well as the thickness of the film, are always positive definite, we set the probabilites to be defined by

$$
p_i = \begin{cases} \dfrac{n_i}{A}, & 0 \le i \le N \\[2mm] \dfrac{k_{i-N}}{A}, & N + 1 \le i \le 2N \\[2mm] \dfrac{d}{A}, & i = 2N + 1 \end{cases}
\tag{25}
$$

The normalization constant A is given by

$$
A = \sum_i (n_i + k_i) + d
\tag{26}
$$

The formal problem involves the computation of the maximum entropy distribution p_j ($j = 1, \ldots, 2N + 1$) from the experimental set of ellipsometric angles $\psi(\lambda_i)$ and $\Delta(\lambda_i)$ ($i = 1, \ldots, N$). The experimental angles are related to the unknowns through a model of the form

$$
\psi^{calc} = F_1(p) \qquad \Delta^{calc} = F_2(p)
\tag{27}
$$

where the functions F_1 and F_2 are obtained from the standard equations of ellipsometry derived in the appendix. The variables p in equation (27) stand for all the appropriate variables entering into the description of the model, such as the indices of refraction and the film thickness. To be able to use the data, we construct a constraint function χ^2 given by equation (17), which measures the deviations of the model from the experimental data. The computational problem is then to find the distribution p_j which maximizes the entropy S [equation (24)] subject to a constraint of the form

$$
\chi^2 \le \chi^2_{max}
\tag{28}
$$

where χ^2_{max} is chosen to reflect the experimental accuracy (e.g., χ^2_{max} is commonly set to the number of experimentally measured values).

Although a number of algorithms have been proposed for the solution of this problem [12a], because the equations leading to the functions F_1 and F_2 are highly nonlinear, we have chosen to implement a nonlinear programming algorithm developed at the Systems Optimization Laboratory of Stanford University [74]. This algorithm is well suited for our problem because it does not appear to suffer the numerical instabilities that other algorithms exhibit.

Figure 39 reports the results of using the maximum entropy method on the single-film analysis of the galvanostatically grown polyaniline described in Figures 37 and 38. The plot shows the resulting dielectric function as a function of photon energy. However, this spectrum, which is quite similar to that of Figure 38, leads to a thickness of 517 nm. This value is consistent with an independent fit in which we used effective medium theory and the maximum likelihood approach [cf. equation (17)] to determine the spectrum and the thickness of the film. The resulting thickness for this independent fit is 518 nm and the corresponding spectrum is shown in Figure 40.

Although the single-film model described above leads to excellent fits to the experimental data, we decided to proceed further with the analysis of the polyaniline–gold system by applying the maximum entropy method to a two-film model. In this case we assumed that besides the ambient and the mixed gold substrate, the actual film consisted of two different layers, a pure-polymer layer in contact with the ambient and a gold–polymer layer in contact with the gold substrate. This model was designed to investigate the effect of roughness at the gold–polymer interface on the spectroellipsometric spectrum. Its motivation rests on the possibility that since the polymer was grown electrochemically, the growth process may have filled a roughened gold surface with polymer, effectively creating a layer composed of an unknown mixture of gold and polymer.

The maximum entropy approach proceeds in the same manner as described above for the single-film case except that the functions F_1 and F_2 are obtained

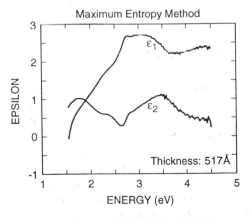

FIGURE 39 Optical spectrum of PANI on gold obtained using the maximum entropy method. The experimental data are the same as those of Figure 37. The thickness of 517 Å was obtained as part of the fitting procedure.

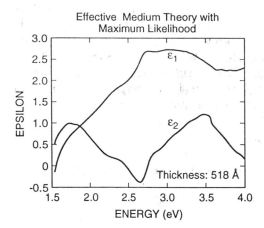

FIGURE 40 Optical spectrum of PANI on gold obtained using the maximum likelihood method combined with effective medium theory. The experimental data are the same as those of Figure 37. The thickness of 518 Å was obtained by minimizing the maximum likelihood function [equation (18)] as a function of film thickness.

from the two-film equations similar to those of the appendix. However, one of the characteristics of the maximum entropy method is that it leads to that probability distribution, p_i, that maximizes the entropy consistent with any other information supplied in the form of explicit constraints. In our case the only additional information we have supplied is that the ellipsometric angles calculated by the model be consistent with the experimental data through equation (28). One could, of course, introduce additional assumptions to the two-film model that would require, for example, that one or both of the films have a predetermined minimum thickness. Since we do not have other additional information that supports this or other similar assumptions, we decided not to restrict any of the parameters that describe the two-film model.

The results of the maximum entropy method for the two-film model were somewhat surprising in that the solution we obtained corresponded to the single-film solution described above. If we restricted both films to have a nonzero thickness we found that the solution returned by the maximum entropy method consisted of two films whose optical properties were identical and whose total thickness (sum of the two individual thicknesses) was equal to the result we had obtained with the single-film model. In addition, the quality of the fit in both cases (comparing the values of Δ and ψ calculated from the model to those obtained from the experiment) was almost identical to each other and to that of the single-film model.

An important conclusion from this study is that in the absence of additional experimental information, such as the reflectivity of the film or an independent measurement of its thickness, any model consisting of more than one layer does not predict ellipsometric angles that are in better agreement with experiment than those predicted by a single-film model. This means that the single-film model is just as consistent with the experimentally measured ellipsometric angles as a model with more layers and that the spectroscopic ellipsometry experiments that produce values of Δ and ψ only at a single angle of incidence cannot differentiate between models with different numbers of layers. On the other hand, our simulations indicate that one can distinguish morphological features of the films (e.g., the difference between one and more layers) with spectroscopic ellipsometry when multiple angles of incidence are used or with other, additional experimental information, such as reflectivity measurements.

APPENDIX

Reflection and Transmission by Stratified, Isotropic Structures [2]

Let us consider a stratified structure consisting of a series of parallel layers, denoted by the numbers, $1, 2, \ldots, j, \ldots, m$ (Fig. 34). The semi-infinite ambient medium is denoted by the number 0; the substrate, denoted by the number $m + 1$, is also semi-infinite. We will also assume that all the layers obey isotropic linear optics and that the complex index of refraction of the jth layer is $\hat{n}_j$ and its thickness is d_j; the indices of refraction of the ambient and the substrate are denoted by n_0 and $\hat{n}_{m+1}$, respectively.

A monochromatic plane wave incident on the first layer generates a wave reflected into the ambient and, after passing through all the intervening layers, a wave transmitted into the substrate. The electric field within each layer is made up of two waves: one moving toward the next layer in the positive direction (downward in Fig. 34), which we will denote by $\mathbf{E}^+$, and the other moving toward the previous layer, in the negative direction, denoted by $\mathbf{E}^-$. All the wave vectors lie on the same plane—the plane of incidence. Moreover, in any one of the layers the angles between the wave vectors of the positive and negative waves and the normal to the interface are equal (i.e., the angle of incidence equals the angle of reflection).

When the incident wave is linearly polarized, with its electric vector vibrating in the plane of incidence (p polarization) or perpendicular to it (s polarization), all the subsequent plane waves excited by it in any of the layers will have the same polarization. Since a wave with an arbitrary polarization can be written as a linear combination of *s* and *p* waves, we will concentrate on these two types of linearly polarized waves.

Let $E^+(z)$ and $E^-(z)$ denote the complex amplitudes of the waves moving along the positive and negative directions at an arbitrary plane at a distance z from the interface between the top layer and the ambient. The total electric field at the plane z can be described by a two-component vector of the form

$$\mathbf{E}(z) = \begin{bmatrix} E^+(z) \\ E^-(z) \end{bmatrix} \tag{A1}$$

If we now consider the electric field on two different planes z' and z'', parallel to the layers, because of the linear behavior of the media, $\mathbf{E}(z')$ and $\mathbf{E}(z'')$ must be related by a mapping of the form

$$\mathbf{E}(z') = \mathbf{S}\mathbf{E}(z'') \tag{A2}$$

where $\mathbf{S}$ denotes a 2×2 matrix which characterizes that part of the stratified structure confined between the two planes at z' and z''.

Let us now consider the form of the matrix $\mathbf{S}$ obtained when the two planes z' and z'' are immediately before and immediately after the interface between two adjacent layers. Thus we let $z' = z_j^{-0}$ and $z'' = z_j^{+0}$, where the superscript -0 denotes immediately before and $+0$ immediately after the interface between the $(j-1)$th and the jth layers. Instead of the symbol $\mathbf{S}$ for the matrix we use $\mathbf{I}_{j-1,j}$, denoting the 2×2 transmission matrix characteristic of the interface. Thus we have

$$\mathbf{E}(z_j^{-0}) = \mathbf{I}_{j-1,j}\mathbf{E}(z_j^{+0}) \tag{A3}$$

On the other hand, if we pick z' and z'' to be just inside the planes forming the boundaries to the jth layer, we will have $z' = z_j^{+0}$ and $z'' = z_j^{+0} + d_j$, and

$$\mathbf{E}(z_j^{+0}) = \mathbf{L}_j\mathbf{E}(z_j^{+0} + d_j) \tag{A4}$$

where $\mathbf{L}_j$ is a propagation matrix characteristic of the jth layer. The equation connecting the wave impinging upon the first layer and the wave transmitted into the substrate is

$$\mathbf{E}(z_1^{-0}) = \mathbf{S}\mathbf{E}(z_{m+1}^{+0}) \tag{A5}$$

where the matrix $\mathbf{S}$ is known as the scattering matrix. It is straightforward to show [2] that it has the form

$$\mathbf{S} = \mathbf{I}_{01}\mathbf{L}_1\mathbf{I}_{12} \cdots \mathbf{I}_{j-1,j}\mathbf{L}_j \cdots \mathbf{I}_{m-1,m}\mathbf{L}_m\mathbf{I}_{m,m+1} \tag{A6}$$

The matrices $\mathbf{I}_{j-1,j}$ and $\mathbf{L}_j$ are given in terms of the Fresnel coefficients by

$$\mathbf{I}_{j-1,j} = \frac{1}{t_{j-1,j}} \begin{bmatrix} 1 & r_{j-1,j} \\ r_{j-1,j} & 1 \end{bmatrix} \qquad \mathbf{L}_j = \begin{bmatrix} \exp(i\beta_j) & 0 \\ 0 & \exp(-i\beta_j) \end{bmatrix} \tag{A7}$$

where $t_{j-1,j}$ is the Fresnel coefficient for transmission, given by

$$t_{p,j-1,j} = \frac{2\hat{n}_{j-1} \cos \phi_{j-1}}{\hat{n}_j \cos \phi_{j-1} + \hat{n}_{j-1} \cos \phi_j} \tag{A8}$$

$$t_{s,j-1,j} = \frac{2\hat{n}_{j-1} \cos \phi_{j-1}}{\hat{n}_{j-1} \cos \phi_{j-1} + \hat{n}_j \cos \phi_j}$$

for p and s polarizations, respectively. The symbol $r_{j-1,j}$ stands for the Fresnel coefficient for reflection, given in equations (14) and (15) for $j = 1$. The quantity β_j denotes the change in the phase of the wave, along the z direction, between the beginning and the end of the jth layer; it is given by

$$\beta_j = \frac{2\pi}{\lambda} d_j \hat{n}_j \cos \phi_j \tag{A9}$$

As an example, consider the case of a single uniform layer. From equations (A6) and (A7) we have

$$
\begin{aligned}
\mathbf{S} = \mathbf{I}_{01}\mathbf{L}_1\mathbf{I}_{12} &= \left\{ \frac{1}{t_{01}} \begin{bmatrix} 1 & r_{01} \\ r_{01} & 1 \end{bmatrix} \right\} \begin{bmatrix} \exp(i\beta_1) & 0 \\ 0 & \exp(-i\beta_1) \end{bmatrix} \\
&\quad \left\{ \frac{1}{t_{12}} \begin{bmatrix} 1 & r_{12} \\ r_{12} & 1 \end{bmatrix} \right\} \\
&= \frac{1}{t_{01}t_{12}} \begin{bmatrix} \exp(i\beta_1) + r_{01}r_{12}\exp(-i\beta_1) & r_{12}\exp(i\beta_1) + r_{01}\exp(-i\beta_1) \\ r_{01}\exp(i\beta_1) + r_{12}\exp(-i\beta_1) & r_{01}r_{12}\exp(i\beta_1) + \exp(-i\beta_1) \end{bmatrix} \\
&= \begin{bmatrix} S_{11} & S_{12} \\ S_{21} & S_{22} \end{bmatrix}
\end{aligned}
\tag{A10}
$$

Consider a wave impinging upon the interface 01, being reflected back into the ambient and transmitted all the way into the substrate, passing through the interface 12. We then have

$$\begin{bmatrix} E_0^+ \\ E_0^- \end{bmatrix} = \begin{bmatrix} S_{11} & S_{12} \\ S_{21} & S_{22} \end{bmatrix} \begin{bmatrix} E_2^+ \\ 0 \end{bmatrix} \tag{A11}$$

where the vector on the right has a component in the positive direction only because inside the substrate there are no reflected waves. To associate these equations with the ellipsometric experiment, we must calculate the reflection coefficients for p and s polarizations and take their ratios, as in Eq. (16). Thus, for either polarization we have [cf. equations (A10) and (A11)]

$$r = \frac{E_0^-}{E_0^+} = \frac{S_{21}}{S_{11}} = \frac{r_{01} + r_{12}\exp(-2i\beta_1)}{1 + r_{01}r_{12}\exp(-2i\beta_1)} \tag{A12}$$

where the coefficients r_{01} are given by equations (14) and (15), with similar expressions for r_{12}. Equations (16) and (A12) allow us to calculate the ellipsometric angles ψ and Δ for this model and to compare them with the measured values.

REFERENCES

1. S. Gottesfeld, in *Electroanalytical Chemistry* (A. J. Bard, ed.), Vol. 15, Marcel Dekker, New York, 1989, p. 143.
2. R. M. A. Azzam and N. M. Bashara, *Ellipsometry and Polarized Light*, North-Holland, New York, 1977.
3. R. A. Collins and Y.-T. Kim, *Anal. Chem.*, *62*: 887A (1990).
4. R. Muller, in *Techniques for Characterization of Electrodes and Electrochemical Processes* (R. Varma and J. R. Selman, eds.), Wiley, New York, 1991, p. 31.
5. R. J. Archer, *Manual on Ellipsometry*, Gaertner Company, Chicago, 1968.
6. (a) R. H. Muller, *Surface Sci.*, *56*: 19 (1976). (b) P. S. Hauge, *Surface Sci.*, *96*: 108 (1980).
7. (a) Y.-T. Kim, R. W. Collins, and K. Vedam, *Surface Sci.*, *223*: 341 (1990). (b) Y.-T. Kim, R. W. Collins, K. Vedam, and D. L. Allara, *J. Electrochem. Soc.*, *138*: 3266 (1991).
8. (a) R. F. Cohn, J. W. Wagner, and J. Kruger, *J. Electrochem. Soc.*, *135*: 1033 (1988). (b) R. F. Cohn, J. W. Wagner, and J. Kruger, *Appl. Opt.*, *27*: 4664 (1988). (c) R. F. Cohn and J. W. Wagner, *Appl. Opt.*, *28*: 3187 (1988). (d) R. F. Cohn, *Appl. Opt.*, *29*: 304 (1990).
9. D. E. Aspnes and A. A. Studna, *Appl. Opt.*, *14*: 220 (1975).
10. (a) K. Sugimoto and S. Matsuda, *J. Electrochem Soc.*, *130*: 2313 (1983). (b) K. Sugimoto, S. Matsuda, Y. Ogiwara, and K. Kitamura, *J. Electrochem. Soc.*, *132*: 1791 (1985).
11. C. C. Streinz, J. W. Wagner, J. Kruger, and P. J. Moran, *J. Electrochem. Soc.*, *139*: 711 (1992).
12. (a) A. Redondo, E. A. Ticianelli, and S. Gottesfeld, *Synth. Metals*, *29*: E265 (1989). (b) J. Rishpon, A. Redondo, C. Derouin, and S. Gottesfeld, *J. Electroanal. Chem.*, *294*: 73 (1990). (c) I. Rubinstein, J. Rishpon, E. Sabatani, A. Redondo, and S. Gottesfeld, *J. Am. Chem. Soc.*, *112*: 6135 (1990).
13. (a) S. N. Jasperson, and S. E. Schnatterly, *Rev. Sci. Instrum.*, *40*: 761 (1969). (b) S. N. Jasperson, D. K. Burge, and R. C. O'Handley, *Surface Sci.*, *37*: 548 (1973). (c) V. M. Bermudez and V. H. Ritz, *Appl. Opt.*, *17*: 542 (1978).
14. G. E. Jellison and D. H. Lowndes, *Appl. Opt.*, *24*: 2948 (1985).
15. C. M. Carlin, L. J. Kepley, and A. J. Bard, *J. Electrochem. Soc.*, *132*: 353 (1985).
16. G. R. J. Robertson, J. L. Ord, D. J. DeSemet, and M. A. Hopper, *J. Electrochem. Soc.*, *136*: 3380 (1989).
17. E. Sabatani, A. Redondo, J. Rishpon, A. Rudge, I. Rubinstein, and S. Gottesfeld, *J. Chem. Soc. Faraday Trans.*, *89*(2): 287 (1993).
18. Y.-T. Kim, Ph.D. thesis, Pennsylvania State University, University Park, Pa., 1991.
19. A. Redondo, E. Ticianelli, and S. Gottesfeld, *Mol. Cryst. Liq. Cryst.*, *160*: 185 (1988).
20. W.-K. Paik and J. O'M. Bockris, *Surface Sci.*, *28*: 61 (1971).

21. A. Hamnett and A. R. Hillman, *J. Electrochem. Soc.*, *135*: 2517 (1988).
22. A. R. Hillman, D. A. Taylor, A. Hamnett, and S. J. Higgins, *J. Electroanal. Chem.*, *266*: 423 (1989).
23. (a) S. J. Higgins and A. Hamnett, *Electrochim. Acta*, *36*: 2123 (1991). (b) P. A. Christensen and A. Hamnett, *Electrochim. Acta*, *36*: 1263 (1991).
24. H. Arwin and D. E. Aspnes, *Thin Solid Films*, *113*: 101 (1984). H. Arwin, D. E. Aspnes, R. Bjorklund, I. Lundstrom, *Synthetic Metals*, *6*: 309 (1983).
25. C. Tian, G. Jin, F. Chao, M. Costa, and J. P. Roger, *Thin Solid Films*, in press.
26. S. J. Roser, R. M. Richardson, M. J. Swann, and A. R. Hillman, *J. Chem. Soc. Faraday Trans.*, *87*(17): 2863 (1991). See also R. M. Robertson et al., *Faraday Discussions*, *94*:XXX (1992).
27. J. Rishpon and S. Gottesfeld, *Biosensors Bioelectron.*, *6*: 143 (1991).
28. Y.-T. Chin and B. D. Cahan, *J. Electrochem. Soc.*, *139*: 2432 (1992).
29. S. Gottesfeld, G. Maia, J. B. Floriano, G. Tremiliosi-Filho, E. A. Ticianelli, and E. R. Gonzalez, *J. Electrochem. Soc.*, *138*: 3219 (1991).
30. S. Gottesfeld, M. Yaniv, D. Laser, and S. Srinivasan, *J. Phys. Paris, Colloq.*, *C5*: 145 (1977).
31. C. Barbero and R. Kotz, *J. Electrochem. Soc.*, *140*: 1 (1993).
32. L. J. Kepley and A. J. Bard, *Anal. Chem.*, *60*: 1459 (1988).
33. A. M. Brodsky and M. I. Urbach, *Prog. Surface Sci.*, *33*: 991 (1990).
34. D. Oelkrug, M. Fritz, and H. Stauch, *J. Electrochem. Soc.*, *139*: 2419 (1992).
35. W. S. Kruijt, M. Sluyters-Rehbach, and J. H. Sluyters, *J. Electroanal. Chem.*, *285*: 117 (1990).
36. A. M. Foontokov, V. E. Kazarinov, M. I. Urbach, and A. Tadjeddine, *Surface Sci.*, *239*: 59 (1990).
37. S. P. Armes, M. Aldissi, M. Hawley, J. G. Beery, and S. Gottesfeld, *Langmuir*, *7*: 1447 (1991).
38. (a) J. C. Clayton and D. J. DeSmet, *J. Electrochem. Soc.*, *123*: 174 (1976). (b) D. J. DeSmet, *Electrochim. Acta*, *21*: 1137 (1976).
39. (a) J. L. Ord, S. D. Bishop, and D. J. DeSmet, *J. Electrochem. Soc.*, *138*: 208 (1991). (b) J. L. Ord and D. J. DeSmet, *J. Electrochem. Soc.*, *139*: 728 (1992).
40. S. Gottesfeld, A. Redondo, and S. W. Feldberg, *J. Electrochem. Soc.*, *134*: 271 (1987).
41. C. Lee, J. Kwak, and A. J. Bard, *J. Electrochem. Soc.*, *136*: 3720 (1989).
42. C. Lee, J. Kwak, L. J. Kepley, and A. J. Bard, *J. Electroanal. Chem.*, *282*: 239 (1990).
43. S. Gottesfeld, A. Redondo, I. Rubinstein, and S. W. Feldberg, *J. Electroanal. Chem.*, *265*: 15 (1989).
44. S. W. Feldberg and I. Rubinstein, *J. Electroanal. Chem.*, *240*: 1 (1988).
45. D. J. Beckstead, D. J. DeSmet, and J. O. Ord, *J. Electrochem. Soc.*, *136*: 1927 (1989).
46. (a) C. E. D. Chidsey, *Science*, *251*, 919 (1991). (b) E. Katz, N. Itzhak, and I. Willner, *Langmuir*, *9*: 1392 (1993).
47. S. Steinberg, Y. Tor, E. Sabatani, and I. Rubinstein, *J. Am. Chem. Soc.*, *113*: 5176 (1991).

48. J. D. Swalen, D. L. Allara, J. D. Andrade, E. A. Chandross, S. Garoff, J. Israelachvili, T. J. McCarthy, R. Murray, R. F. Pease, J. F. Rabolt, K. J. Wynne, and H. Yu, *Langmuir*, *3*: 932 (1987).

49. A. Ullman, *An Introduction of Ultrathin Organic Films; From Langmuir–Blodgett to Self-Assembly*, Academic Press, New York, 1991.

50. J. Gun, R. Isovoco, and J. Sagiv, *J. Colloid Interface Sci.*, *101*: 201 (1984).

51. S. R. Wasserman, G. M. Whitesides, I. M. Tidsewell, B. M. Ocko, P. S. Pershan, and J. D. Axe, *J. Am. Chem. Soc.*, *111*: 5852 (1989).

52. M. D. Porter, T. B. Bright, D. L. Allara, and C. E. D. Chidesy, *J. Am. Chem. Soc.*, *109*: 3559 (1987).

53. C. D. Bain, E. B. Troughton, Y.-T. Tao, J. Evall, G. M. Whitesides, and R. G. Nuzzo, *J. Am. Chem. Soc.*, *111*: 321 (1989).

54. E. Kretschmann, *Z. Phys.*, *241*: 313 (1971).

55. A. Szucs, G. D. Hitchens, and J. O'M. Bockris, *J. Electrochem. Soc.*, *136*: 3748 (1988).

56. A. Szucs, G. D. Hitchens, and J. O'M. Bockris, *Electrochim. Acta*, *37*: 403 (1992).

57. (a) M. Kawaguchi, M. Tohyama, Y. Mutoh, and A. Takahashi, *Langmuir*, *4*: 407 (1988). (b) M. Kawaguchi, M. Tohyama, and A. Takahashi, *Langmuir*, *4*: 411 (1988).

58. T. Rasing, H. Hsiung, Y. R. Shen, and M. W. Kim, *Phys. Rev. A*, *137*: 2732 (1988).

59. M. W. Kim, B. B. Sauer, H. Yu, M. Yazdanian, and G. Zografi, *Langmuir*, *6*: 236 (1990).

60. M. Paudler, J. Ruths, and H. Riegler, *Langmuir*, *8*: 184 (1992).

61. R. Benferhat, B. Drevellon, and P. Robin, *Thin Solid Films*, *156*: 295 (1988).

62. (a) J. F. Rabolt, F. C. Burns, N. E. Schlotter, and J. D. Swalen, *J. Chem. Phys.*, *78*: 946 (1983). (b) R. T. Graf, J. L. Koenig, and H. Ishida, *Anal. Chem.*, *58*: 64 (1986). (c) F. Ferrieu, *Rev. Sci. Instrum.*, *60*, 3212 (1989). (d) V. A. Yakovlev, M. Li, and E. A. Irene, *J. Opt. Soc. Am. A*, *10*: 509 (1993).

63. H. Cramér, *Mathematical Methods of Statistics*, Princeton University Press, Princeton, N.J., 1971, p. 424.

64. P. J. McMarr and J. R. Blanco, *Appl. Opt.*, *27*: 4265 (1988).

65. W. H. Press, S. A. Teukolsky, W. T. Vetterling, and B. P. Flannery, *Numerical Recipes in Fortran*, 2nd ed. Cambridge University Press, Cambridge, 1992, pp. 650, 675, 684.

66. (a) I. Webman, J. Jortner, and M. H. Cohn, *Phys. Rev. B.*, *15*: 5712 (1970). (b) D. Stroud, *Phys. Rev. B*, *12*: 3368 (1975). (c) R. Landauer, *J. Appl. Phys.*, *23*: 779 (1952).

67. J. C. Maxwell Garnett, *Phil. Trans. Roy. Soc. (London)*, *203*: 385 (1904); *205*: 237 (1906).

68. D. A. G. Bruggeman, *Ann. Phys. Ser. 5*, *24*: 636 (1935); *25*: 644 (1936).

69. E. T. Jaynes, *Phys. Rev.*, *106*: 620 (1957).

70. M. Tribus, *J. Appl. Mech.*, *28*: 1 (1961).

71. C. E. Shannon, *Bell Syst. Tech. J.*, *27*: 379, 623 (1948).

72. E. T. Jaynes, *IEEE Trans. Syst. Sci. Cybern.*, *SSC-4*: 227 (1968).

73. (a) S. F. Gull, and G. J. Daniel, *Nature*, *272*: 686 (1978). (b) R. B. Evans, in *The Maximum Entropy Formalism* (R. D. Levine and M. Tribus, eds.), MIT Press, Cam-

bridge, Mass., 1979. (c) J. Skilling and S. F. Gull, in *Maximum Entropy and Bayesian Methods in Inverse Problems* (C. R. Smith and W. T. Grady, eds.), D. Reidel, Dordrecht, The Netherlands, 1985. (d) J. Skilling, in *Maximum Entropy and Bayesian Methods in Applied Statistics* (J. H. Justice, ed.), Cambridge University Press, Cambridge, 1986. (e) N. A. Farrow and F. P. Ottensmeyer, in *Maximum Entropy and Bayesian Methods* (J. Skilling, ed., Kluwer, Dordrecht, The Netherlands, 1989. (f) A. B. Templeman and L. Xingsi, in *Maximum Entropy and Bayesian Methods* (J. Skilling, ed.), Kluwer, Dordrecht, The Netherlands, 1989.

74. P. E. Gill, W. Murray, M. A. Saunders, and M. H. Wright, *Systems Optimization Laboratory, Technical Report 86-2*, Stanford University, Stanford, Calif., 1986.

10

Analysis of Surface Layers on Well-Defined Electrode Surfaces

ARTHUR T. HUBBARD
University of Cincinnati, Cincinnati, Ohio

EUGENE Y. CAO
Medical College of Ohio, Toledo, Ohio

DONALD A. STERN
Chevron Chemical Company, Kingwood, Texas

I. INTRODUCTION

Understanding the nature of the interface formed between an electrode and the surrounding solution at an atomic and molecular level is the motivation for the rapidly growing field of surface electrochemistry [1–6]. Information regarding the interfacial structure and composition, such as elemental composition, molecular orientation, and mode of surface bonding, and how the two properties relate to the interfacial reactivity, is essential to understand fundamental electrochemical phenomena.

Modern ultrahigh-vacuum (UHV) technology and surface spectroscopy techniques developed in the last two decades have made surface electrochemical studies possible. The integration of UHV surface spectroscopic techniques with electrochemical methods of analysis provides a unique way of revealing many aspects of interfacial behavior at an atomic and molecular level. Atomically clean and structurally ordered single-crystal surfaces can be prepared in a UHV environment. Several techniques can be used to characterize the substrate and adsorbed species: The long–range crystallographic structural order of the electrode surface and adsorbed layer can be determined by low–energy electron diffraction (LEED) [7]; the elemental composition can be measured by Auger electron spectoscopy (AES) [8-11] and by x-ray photoelectron spectroscopy (XPS or ESCA) [12,13]; the vibrational spectrum of the adsorbate can be characterized by high-resolution electron energy loss spectroscopy (HREELS) [14] and surface Fourier transform infrared spectroscopy (FTIR) [15]. Scanning tunneling microscopy (STM) [16–18] and atomic force microscopy (AFM) [19] have also become widely used tools to determine and characterize the real-space surface structure and morphology. Angular distribution Auger microscopy (ADAM) [20,21] is emerging as a promising technique for probing atomic structure of the surface. In view of the complexity of interfacial processes, no surface technique will be able to supply, by itself, a complete description of electrochemical systems. A multitechnique approach has been perceived as being most appropriate for the fundamental studies in electrochemical surface science. Combining complementary surface-sensitive techniques with electrochemical methods such as cyclic voltammetry (CV) and chronocoulometry provide the means to investigate the electrode–electrolyte interfacial structure, composition, and reactivity at an atomic and molecular level.

Recent atomic, ionic, and molecular adsorption studies have revealed that many chemisorbed layers formed from aqueous solutions are stable under UHV conditions [5,6]. The structure and composition as well as the chemical and electrochemical reactivities of the adsorbed layers are found to be influenced by the electrode potential, solution pH, adsorbate concentration, crystallographic structure, and composition of the electrode surface, temperature, and in the case of molecular adsorption, molecular structure of the adsorbate.

Results of several studies, including ionic adsorption from aqueous solutions, electrodeposition, and molecular adsorption at atomically clean and structurally ordered single-crystal electrode surfaces, are presented in this chapter. The advantages of using the multitechnique approach to characterizing these systems to better understand the chemical and electrochemical processes occurring at the electrode–electrolyte interface will be illustrated.

II. EXPERIMENTAL ASPECTS

Oriented single crystals provide surfaces that can be brought to a well-defined state by appropriate cleaning procedures. A well-defined surface is one that exhibits an atomically ordered surface structure as evidenced by LEED and verified to be free of detectable contaminants by Auger spectroscopy. Ultrahigh vacuum pressures (less than 10^{-8} torr) are required for surface cleanliness; the number of residual gas molecules in vacuum must be low enough to maintain a surface free of contaminants for the duration of an immersion experiment. Low pressures are also required to provide an adequate mean free path for the electron spectroscopies used to characterize the crystal.

When a well-defined state has been achieved, the crystal is transferred to the solution cell for atmospheric pressure immersion experiments. Adsorption from aqueous solutions at controlled potential and pH may be followed by electrochemical characterization using voltammetric techniques to explore the electrochemical reactivity of the adsorbed layer. Following emersion of the crystal from the solution containing the subject adsorbate and evacuation, UHV surface characterization of the adsorbed layer proceeds. Elemental composition and packing density determinations are carried out by Auger spectroscopy [11], while the long-range order of the adsorbed layer is investigated by LEED, and the vibrational spectrum of the adsorbed layer is observed by HREELS, which can help identify the adsorbate molecular structure, conformation, and intra- and intermolecular interactions, and determine the mode of attachment to the surface. These complementary techniques are chosen because they possess a high degree of surface sensitivity and do not perturb significantly the structure and composition of the adsorbed layer. This method of surface characterization is appropriate for adsorbate systems that are strongly chemisorbed to the substrate and are therefore stable in solution and vacuum.

A schematic diagram of the surface electrochemistry apparatus used in the adsorption studies in shown in Figure 1. The apparatus consists of two inter-

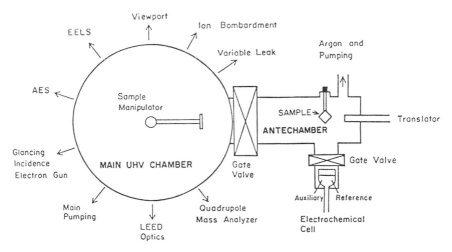

FIGURE 1 Schematic diagram of surface electrochemistry apparatus.

connected stainless steel vacuum chambers. The main ultrahigh vacuum chamber contains the surface-sensitive instrumentation (LEED, AES, and HREELS), as well as the ion-bombardment cage used to clean the sample and a quadrupole mass analyzer to monitor the residual gas in the system. When isolated from the main chamber by means of a gate valve, the secondary chamber (antechamber) can be brought up to ambient pressure with Ar gas in order to carry out the immersion experiments using electrochemical techniques. The adsorbed layer is formed by immersing the sample in an aqueous electrolyte containing the subject compound at a specific electrode potential. After the adsorbate solution is drained away, the sample is rinsed with dilute electrolyte at the same potential to remove excess solute. Following emersion from the dilute electrolyte, the antechamber is evacuated and the sample with the adsorbed layer is transferred back into the main UHV chamber for surface characterization.

III. IONIC LAYERS

A. Halides

Halogens adsorbed from ionic solutions at Pt(111) and Ag(111) surfaces produce stable chemisorbed layers. In general, the adsorbed species are present as neutral atoms rather than as anions, and therefore the adsorption can be described as a redox process:

$$X \rightarrow X \text{ (adsorbed)} + e^- \qquad (X = Br, I, Cl, F, \ldots) \qquad (1)$$

The adsorption process and the structure and chemical properties of the adsorbed layer are strongly dependent on the electrode potential, solution pH, and concentration [22–26]. The structure and composition of Pt(111) and Ag(111) surfaces following immersion into aqueous solutions containing Br^-, I^-, Cl^-, and F^- at controlled potential and pH are described in this section.

The adsorption behavior of Br onto Pt(111) is strongly influenced by the solution pH and electrode potential [22]. Adsorption of Br from aqueous KBr and $CaBr_2$ solutions (pH 4 and 6) occurs primarily in the potential range from -0.1 to 0.5 V (vs. Ag/AgCl); the symmetry of adsorbed Br is Pt(111) (3×3)– Br, and the packing density is $\frac{4}{9}$. The packing density of adsorbed species as a function of electrode potential at pH 6 is shown in Figure 2. Cation packing density displays a minimum where Br packing density shows a maximum. Reductive desorption of Br as Br^- anions occurs at potentials more negative than -0.1 V, as evidenced by a sharp and reversible adsorption–desorption peak near -0.1 V in the voltammetric scan, which corresponds to the structural transition from Pt(111) (3×3)–Br to Pt(111) (4×4)–Br. At pH 4 only the (3×3) structure is observed between 0 and 0.6 V, and at pH 8 no ordered Br overlayer can be observed because cation and OH^- adsorption become predominant. The

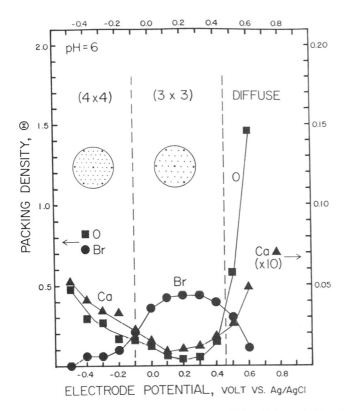

FIGURE 2 Adsorption profile of Br at Pt(111) from 0.05 mM aqueous bromide solutions at pH 6 (CaBr$_2$): packing density and structure as a function of potential. (From Ref. 22.)

adsorption is stronger in acidic solutions than in alkaline solutions, as shown in Figure 3. Competition with OH adsorption at higher pH could account for the behavior observed. Surface oxidation occurs in two stages. In the first stage, the surface oxidation is reversible and the structural order of Pt(111) is retained. In the second stage of the oxidation, the surface becomes irreversibly disordered. Cation and oxygen adsorption become significant at relatively positive potentials and high pH. As a result of the presence of strongly hydrated cations such as Ca^{2+}, the Pt surface is hydrophilic in Br$^-$ solutions.

Iodine adsorption on Pt(111) occurs in aqueous KI solution throughout the double-layer potential range [23]. Within different potential ranges, the adsorbed layer displays distinct adlattice structures and packing densities corresponding to different adsorption states. As shown in Figure 4, at relatively neg-

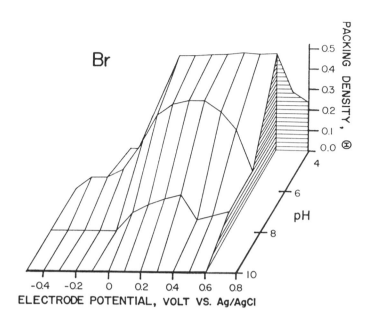

FIGURE 3 Isometric projection of adsorption profile of Br adsorbed at Pt(111) from CaBr₂ solutions. (From Ref. 22.)

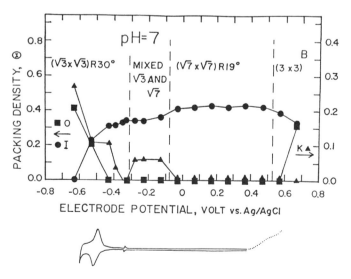

FIGURE 4 Adsorption profile of I at Pt(111) from 0.1 mM KI solutions at pH 7: packing density and surface structure as a function of potential. Corresponding CV scan is shown below profile. $A = 0.77$ cm². (From Ref. 23.)

ative potentials, the adlattice is Pt(111)($\sqrt{3} \times \sqrt{3}$)R30°–I with a packing density of $\theta_I = \frac{1}{3}$. At potentials in midrange, the structure is Pt(111) ($\sqrt{7} \times \sqrt{7}$)R19.1°–I, $\theta_I = \frac{3}{7}$. A Pt(111)(3 × 3)–I adlattice, with $\theta_I = \frac{4}{9}$, is observed at relatively positive potentials. Corresponding LEED patterns are shown in Figure 5. The structural transition from ($\sqrt{3} \times \sqrt{3}$)R30°–I to ($\sqrt{7} \times \sqrt{7}$)R19.1°–I is characterized by a sharp and reversible voltammetric peak near −0.3 V, attributed to the reductive desorption–oxidative adsorption of iodine. A voltammetric feature is not observed corresponding to the transition from ($\sqrt{7} \times \sqrt{7}$)R19.1°–I to (3 × 3)–I because of the relatively small change in the I packing density ($\frac{3}{7}$ to $\frac{4}{9}$ in coverage of iodine). The $\sqrt{3}$ structure is formed at relatively negative potentials through a redox process described in Eq. (1). In this structure (Fig. 6A) iodine atoms probably occupy equivalent threefold sites on the Pt(111) lattice, while for the $\sqrt{7}$ structure (Fig. 6B) the iodine atoms occupy two types of lattice sites: threefold and onefold. Recent experiments involving ADAM and STM also indicate the existence of the one-

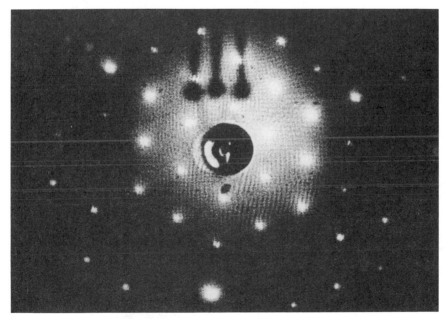

(A)

FIGURE 5 LEED patterns after immersion of Pt(111) into 0.1 mM HI solutions: (A) Pt(111)(3 × 3)–I, 83 eV, electrode potential is 0.4 V; (B) Pt(111)($\sqrt{7} \times \sqrt{7}$)R19.1°–I, 73 eV, electrode potential is −0.2 V; (C) Pt(111)($\sqrt{3} \times \sqrt{3}$)R30°–I, 67 eV, electrode potential is −0.34 V. (From Ref. 23.)

(B)

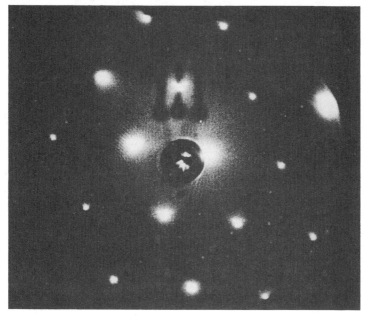

(C)

FIGURE 5 Continued

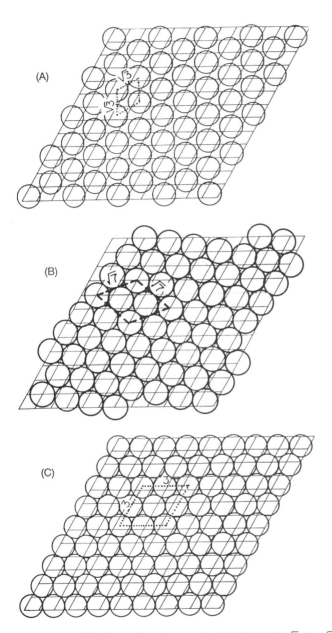

Figure 6 Surface structures of (A) Pt(111)($\sqrt{3} \times \sqrt{3}$)R30°–I, $\theta_i = \frac{1}{3}$, (B) Pt(111)($\sqrt{7} \times \sqrt{7}$)R19.1°–I, $\theta_i = \frac{3}{7}$, and (C) Pt(111)(3 × 3)–I, $\theta_i = \frac{4}{9}$. (From Ref. 23.)

fold site attachment of the $\sqrt{7}$ iodine atom on the Pt(111) surface [27–30]. The solution pH affects the potential range of the adsorption process as expected but has little affect on the packing density and structure of the adsorbed layer. The adsorption of iodine is evidently strong enough to suppress adsorption of OH and other oxygen species at Pt(111).

A remarkable characteristic of iodine adsorption of Pt(111) is that the Pt(111)($\sqrt{7} \times \sqrt{7}$)R19.1°–I and the Pt(111)(3 $\times$ 3)–I adlattice are strongly hydrophobic, while the Pt(111)($\sqrt{3} \times \sqrt{3}$)R30°–I structure is hydrophilic. Evidently, the close-packed (3 $\times$ 3) and ($\sqrt{7} \times \sqrt{7}$) iodine layer prevent the Pt surface from interacting with water molecules, while the ($\sqrt{3} \times \sqrt{3}$) layer does not. The surface is hydrophilic at negative extremes of potential where the iodine layer is desorbed, and at positive extremes of potential where oxidation removes the iodine layer [23].

Chlorine adsorption at Pt(111) from aqueous solution is much weaker than that of Br or I. A study of Pt(111) in Cl$^-$ solutions containing Ca^{2+} showed that the packing density and structure of the adsorbed layer are dependent on Cl$^-$ concentration and electrode potential [24]. Only under very restricted conditions

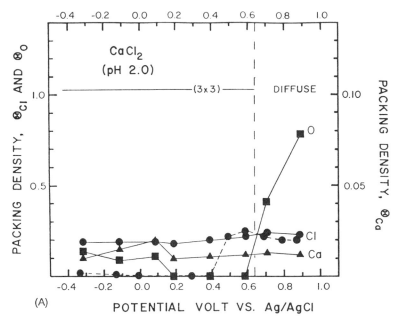

FIGURE 7 Adsorption profiles of Cl at Pt(111) from 0.05 mM CaCl$_2$ (dashed curve) or 10 mM HCl (solid curve) at (A) pH 2 and (B) pH 6; packing density as a function of electrode potential. (From Ref. 24.)

is a stable and ordered Cl layer formed at the surface: at concentrations exceeding 10 mM (pH 2) an ordered Pt(111)(3 × 3)–Cl ($\theta_{Cl} = \frac{1}{4}$) structure is formed in the potential range -0.4 to 0.65 V (Fig. 7A). A less stable layer forms at more negative potentials than -0.4 V and is susceptible to beam damage from LEED and AES. At lower Cl$^-$ concentration or higher pH, Cl adsorption becomes less predominant, and retention of Ca^{2+} by the surface is potential dependent, as shown in Figure 7B [24].

Fluorine does not strongly chemisorb at Pt from aqueous solutions [24]. When immersed into aqueous F$^-$ solutions, the surface properties of the Pt electrode are dominated primarily by hydrogen and OH adsorption.

Strong adsorption of Cl, Br, and I at Ag(111) surfaces occurs from aqueous KCl, KBr, and KI solutions, respectively [25,26]. Adsorption takes place through

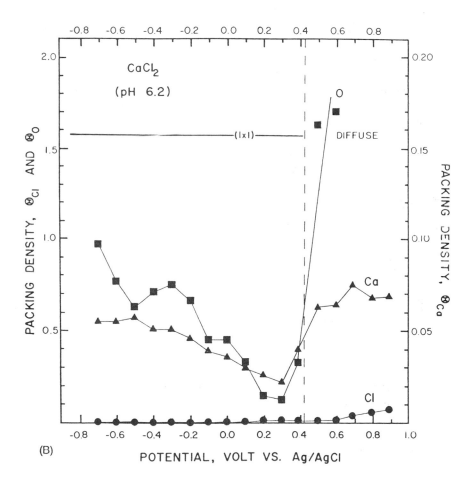

(B)

a redox process as described in equation (1). Oxidative adsorption (or reductive desorption) of these halogens at the Ag(111) surface gives rise to a very broad voltammetric feature spanning much of the accessible potential range. Detailed descriptions of these experiments can be found in Ref. 25.

It is worth mentioning that halogen adsorption at Pt(111) from aqueous solution differs substantially from halogen adsorption from the gas–phase [31]. The relatively reactive halogens, F_2 and Cl_2, strongly adsorb on Pt(111) from the gas phase, as indicated by high sticking coefficients [32], high thermal desorption temperatures [32–34], and large increases in the work function [33]. Br and I, on the other hand, are less strongly adsorbed because the Br and I layers exhibited lower thermal desorption temperatures and smaller work function changes [35,36]. In the case of adsorption from solution, Br and I tend to be more interactive with the surface, and adsorption takes place over a wide range of electrode potentials, while F chemisorption did not occur and an ordered Cl layer formed only under very limited conditions. A possible explanation for the differences between gas- and liquid-phase adsorption can be based on differences in electronic structure. The electronic structures of the reactive halogens such as F and Cl favor very limited and selective adsorption states. In solution, these reactive halogens can easily form stable states within the solution rather than interacting with the electrode. By contrast, the d electrons of Br and I may couple with the d electrons of the Pt atoms on the surface, which may be energetically quite favorable. The d-electron coupling may produce more than one binding state, and this is probably the reason why adsorbed Br and I have different adsorption sites on the surface at different electrochemical potentials.

B. Cyanide and Thiocyanate

Ordered adsorbed cyanide layers are formed from aqueous KCN solutions at Pt(111) [37]. LEED and AES indicate that the packing density and overlayer structure are potential dependent. At potentials more positive than -0.45 V, the structure is Pt(111)$(2\sqrt{3} \times 2\sqrt{3})$R30°–CN with $\theta_{CN} = \frac{7}{12}$, while at potentials more negative than -0.45 V, the structure is Pt(111) $(\sqrt{13} \times \sqrt{13})$R14°–CN with $\theta_{CN} = \frac{7}{13}$. The packing density suggests that the CN moiety is vertically oriented and attaches to the surface through the carbon atom [37–39]. The $2\sqrt{3}$ CN structure is independent of solution concentration from 10^{-4} M to 10^{-1} M. The acid–base behavior of ordered CN layers has been described in Ref. 38.

A recent study by LEED, AES, and HREELS reveals that thiocyanate forms ordered structures on Pt(111) from aqueous KSCN solutions with the vertically oriented SCN moiety bonded through either the sulfur or nitrogen atoms [40]. A (2 × 2) LEED pattern is observed (Fig. 8A) and the packing density of the layer, determined by AES, is $\theta_{SCN} = \frac{1}{2}$; therefore, the adsorbed structure is a combination of Pt(111)(1 × 2)–SCN and Pt(111)(2 × 1)–SCN. The carbon-to-sulfur Auger signal ratio is essentially 1:1, indicative of vertically bonded SCN. Adsorbed SCN from pH 3 solutions is found to be protonated as indicated by the presence of N—H (Pt—SCNH) and C—H (Pt—NCHS) stretching and bending bands in the HREELS spectra (Fig. 9). Although the packing density is not strongly potential dependent, the HREELS spectra depend on the electrode potential during adsorption. At positive potentials (near 0.6 V), where electrooxidation of unadsorbed SCN⁻ ion begins, a prominent new HREELS peak emerges near 1145 cm^{-1}, assignable to S=C=N stretching. In addition, the sulfur content of the adsorbed layer is reduced by half, compared to that at negative potentials, suggesting that electrooxidation of SCN⁻ ions leads to formation of an adsorbed layer containing cyanide species (Pt—C≡NH and Pt—N≡CH) and a similar proportion of a sulfur-containing species (Pt—N=C=S). Heating the (1 × 2)-SCN layer to 700°C desorbs all the C and N and produces a sharp (2 × 2) LEED pattern with quarter coverage S ($\theta_S = 0.25$) corresponding to Pt(111)(2 × 2)–S (Fig. 8B).

Contrary to the adsorption behavior of SCN at Pt(111), protonation of adsorbed SCN at Ag(111) does not occur even at pH 3 [40]. The HREELS spectrum of SCN adsorbed from acidic solutions at Ag(111) (Fig. 10) reveals a striking potential dependence as well as the absence of N—H stretching. A prominent C—S stretching band at 772 cm^{-1} is observed after adsorption at −0.3 V, whereas a prominent C≡N stretching band is observed following adsorption at +0.14 V. An ordered Ag(111)(2 × 3√3, rectangular)–SCN structure with $\theta_{SCN} = \frac{1}{4}$ is observed from −0.42 to 0.1 V, while a diffuse 1 × 1 LEED pattern is observed at potentials more negative of −0.42 V. The LEED pattern is completely diffuse for adsorption at +0.14 V, which is also the potential region where the formation and re-reduction of a series of soluble thiocyanate complexes occurs [41]:

$$Ag + xSCN^- \rightarrow Ag(SCN)_x^{1-x} + e^- \tag{2}$$

where x = 0, 1, 2, ... The LEED and HREELS observations indicate the formation of a distinct AgSCN solid phase.

IV. ELECTRODEPOSITED LAYERS

Electrodeposition and electrocrystallization have been the subject of numerous interesting studies [42]. In the process of electrodeposition onto an atomically clean single-crystal surface, the initial deposition usually experiences an energy advantage (an "underpotential") relative to bulk deposition, particularly the underpotential deposition (UPD), often occurs in more than one stage for most combinations of deposited metal and dissimilar substrate. Electrodeposition is a complex chemical process which is controlled by many variables. The atomic

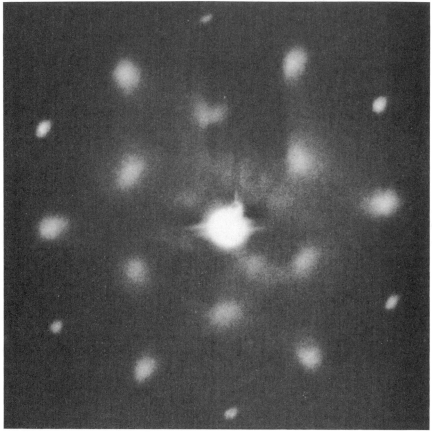

(A)

FIGURE 8 LEED patterns of (A) adsorbed SCN⁻ at Pt(111), 87.1 eV, corresponding to Pt(111)(1 × 2)–SCN and (B) following heating of (1 × 2)–SCN layer to about 700°C for 3 min, 66.4 eV, forming a Pt(111)(2 × 2)–S pattern. Experimental conditions: adsorption from 0.2 mM KSCN solution in 2 mM HF (pH 3) at −0.2 V. (From Ref. 40.)

structure of the underpotential deposited layer at each stage of deposition is strongly affected by the structure of the substrate, nature of the substrate and electrode potential. In this section, several studies of metal electrodeposition onto well-ordered single-crystal electrode surfaces will be presented to demonstrate how these conditions affect the chemical and physical properties of the deposited layer, and also to show the existence of specific surface reactions in

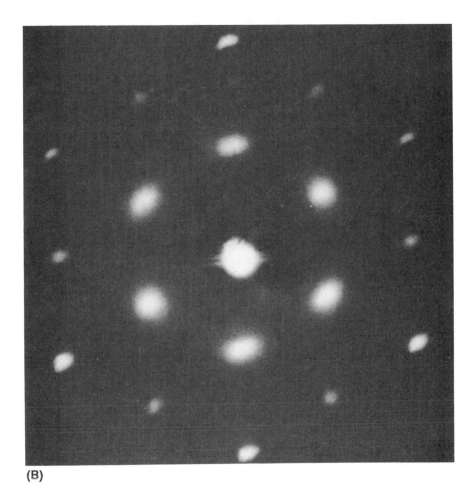

(B)

these electrodeposition processes that are very difficult to observe in conventional in situ experiments.

A. Electrodeposition at Pt

1. Ag on Pt(111)($\sqrt{7} \times \sqrt{7}$)R19.1°–Iodine

The electrodeposition of Ag onto Pt(111) is a good example to illustrate that the structure and composition of the deposited layer are strongly dependent on the electrochemical potential [43,44]. Prior to deposition, the surface was pretreated with high-purity iodine vapor to form a stable Pt(111) ($\sqrt{7} \times \sqrt{7}$)R19.1°–I adlattice, which serves to passivate the Pt surface from attack by electrolyte and residual gases that would complicate the deposition

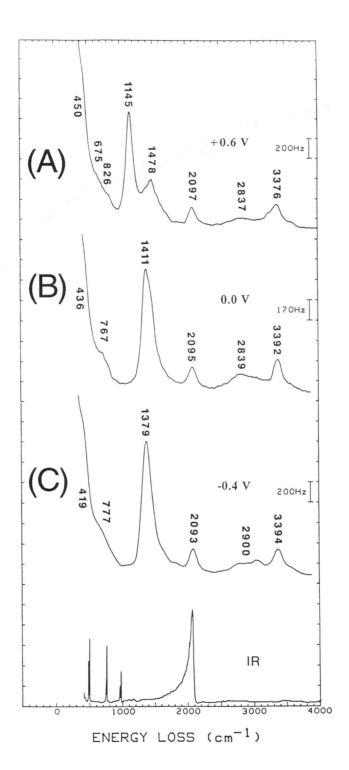

process [43]. As illustrated in Figure 11, electrodeposition of Ag occurs in four distinct ranges of electrode potential: there are three UDP processes in which deposition occurs at potentials more positive than the equilibrium potential of the Ag^+/Ag half-cell, followed by bulk Ag electrodeposition. The iodine layer remains attached to the surface during multiple cycles of electrodeposition–dissolution of Ag layers, as indicated by the iodine Auger signal, which does not decrease as the amount of deposited Ag increases (in contrast to the signals due to the Pt substrate) [43]. The I/Ag Auger signal ratio increases sharply with decreasing angle of incidence of the primary electron beam [44]. This suggests that the Ag atoms are able to pass through the iodine layer and attach to the Pt substrate. LEED patterns observed in the first UPD peak contained the beams of the Pt(111)$(\sqrt{7} \times \sqrt{7})$R19.1°–I pattern as well as the beams of the Pt(111) (3×3)–Ag, I pattern. At this stage of electrodeposition, the Ag monolayer forms in patches, leaving parts of the $\sqrt{7}$-iodine layer temporarily unaffected, until finally the (3×3)–Ag deposit covers the entire surface. The LEED patterns corresponding to the second and third UPD processes suggest structures in which the Ag and I layers form coincidence lattices on the Pt(111) surfaces; multiple domains are present in which the Ag layer has an oblique unit mesh and the iodine atoms are in registry with the Ag [45].

Combining the LEED, Auger, and voltammetric measurements with digital simulation of the LEED patterns reveal the specific structure associated with each stage of the UPD process. For example, the second UPD process forms the epitaxial layer of oblique close-packed Ag having a layer of iodine atoms adsorbed on top of the layer, as shown in Figure 12.

The coverage of the adsorbed Ag layer after the first UPD peak is about half a monolayer ($\theta_{Ag} \approx 0.44$). The second UPD peak produces a coverage of $\theta_{Ag} \approx 0.82$ and the LEED patterns show split spots indicative of multiple domains having oblique structures. As Ag deposition proceeds into the third peak, the Ag coverage approaches a monolayer. After the third peak, the coverage is about 1.5 to 1.7, and a second Ag layer is formed.

2. Ag on Pt(100)–I Adlattice

Electrodeposition of metallic layers is remarkably sensitive to the structure and composition of the adsorbed layer at the substrate surface. Electrodeposition of

---

FIGURE 9 EELS spectra of adsorbed SCN^- at Pt(111) at 0.6 V (A), 0.0 V (B), and -0.4 V (C). Shown for comparison is the IR spectrum of solid KSCN in Nujol [59] (bottom curve). The Nujol peaks have been subtracted from the spectra. Experimental conditions: Adsorption from 0.1 mM KSCN in 10 mM KF/HF (pH 3) and rinsed with 2 mM HF (pH 3) solution. The EELS incidence and detection angles were 62° from the surface normal; beam energy 4 eV; beam current, 200 pA; resolution, 80 to 120 cm^{-1} (FWHM). (From Ref. 40.)

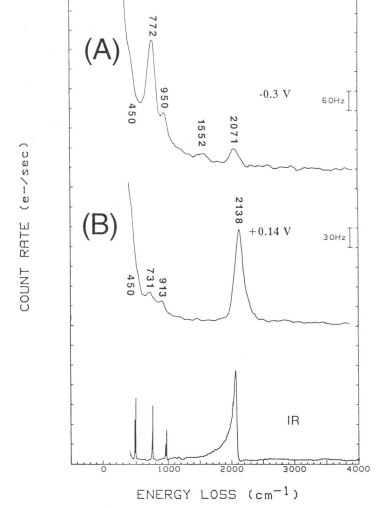

FIGURE 10 EELS spectra of SCN⁻ adsorbed at Ag(111) at −0.3 V (A) and 0.14 V (B). IR spectrum of solid KSCN in Nujol [59] is shown in the bottom curve, with the Nujol peaks subtracted from the spectrum. Experimental conditions: spectrum (B) was obtained by immersing the sample at −0.06 V and scanning to 0.14 V. Other experimental conditions as given in Figure 9. (From Ref. 40.)

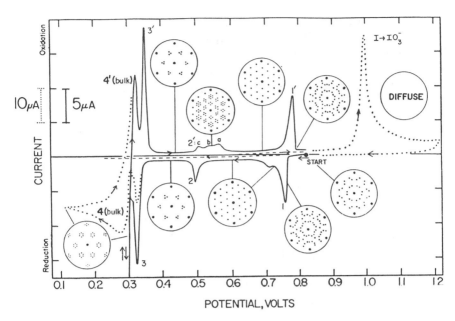

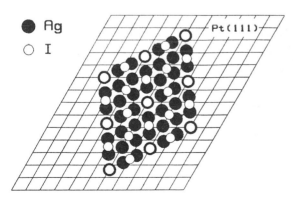

FIGURE 12 Structural model of the AgI bilayer formed after the second UDP process of Ag deposition at Pt(111)–I. The structure illustrates the 3 × 3 symmetry of the iodine lattice with respect to the Ag monolayer and the virtual incommensuracy of the Ag monolayer with respect to the Pt(111) substrate. (From Ref. 45.)

Ag onto an iodine-pretreated Pt(100) surface shows that the nature of an adlattice of iodine has a strong influence on the structure and composition of the electrodeposited Ag layer [46]. The iodine pretreated Pt(100) surface produces three distinct adlattice structures, as shown in Figure 13: (a) Pt(100) $[c(\sqrt{2} \times 2\sqrt{2})]R45°-I$, $\theta_I \approx 0.5$, (b) an incommensurate layer, $\theta_I \approx 0.52$, and (c) Pt(100)$[c(\sqrt{2} \times 5\sqrt{2})]R45°-I$, $\theta_I \approx 0.6$. Silver deposition onto the $c(\sqrt{2} \times 2\sqrt{2})R45°$ structure is characterized by the two narrow UPD peaks, as shown in Figure 14. The LEED patterns obtained at various stages of Ag electrodeposition reveal that the deposition is an orderly process. The deposited layer is stable and well ordered. However, electrodeposition of Ag onto the incommensurate iodine layer no longer produces a well-ordered epitaxial layer. Although the influence of packing density and structure of adlattice are difficult to separate, the difference in the structure of the iodine adlattice is probably responsible for the different epitaxial layers, because the I packing density difference was extremely small (0.02 nmol/cm²). Similarly, Ag deposition onto Pt(100) $[c(\sqrt{2} \times 5\sqrt{2})]R45°-I$ adlattice forms unstable and less ordered polycrystalline rather than monoatomic layers. The I packing density difference is probably the predominant factor in this case. Comparison of electrodeposition of Ag onto Pt(111) and Pt(100) with iodine adlattices indicates that the substrate structure imposes a strong influence on the nature of deposited overlayers. An interesting observation in these experiments is the fact that the iodine atoms remain attached to the surface during deposition or dissolution of Ag, which

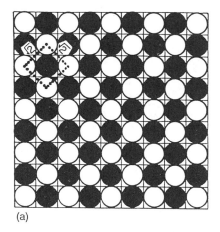

(a)

FIGURE 13 Surface structural models of Ag deposited on Pt(100)–I: (a) Pt(100)$(\sqrt{2} \times \sqrt{2})R45°$–Ag,I ($\theta_{Ag} = \theta_I = 0.50$); (b) Pt(100)$(10\sqrt{2} \times 10\sqrt{2})$R45°–Ag,I ($\theta_{Ag} = 0.42$, $\theta_I = 0.85$); (c) Pt(100)$(\sqrt{34} \times \sqrt{34})$R31°–Ag,I ($\theta_{Ag} = 0.47$, $\theta_I = 1.06$). (From Ref. 46.)

suggests that Ag is able to proceed through the iodine adlattice to bond directly to the Pt surface.

3. Electrodeposition of Other Metals

Electrodeposition of Cu onto Pt(111) pretreated with I_2 vapor takes place in two narrow, closely spaced UPD potential ranges 0.2 V more positive than bulk deposition [47]. Underpotential deposition of Cu on Pt(111)($\sqrt{7} \times \sqrt{7}$)R19.1°–I

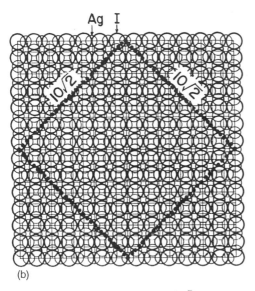

(b)

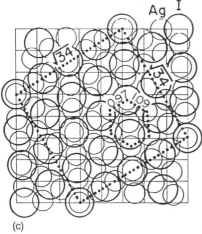

(c)

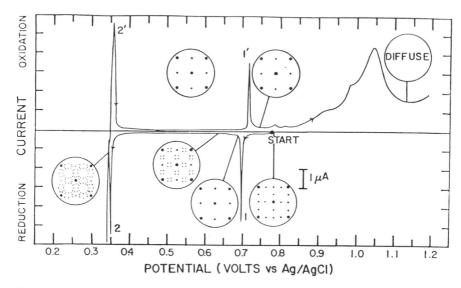

FIGURE 14 Cyclic voltammogram and LEED patterns for deposition of Ag at the Pt(100)[c($\sqrt{2}$ × 2$\sqrt{2}$)]R45°–I surface. Experimental conditions: 1 mM Ag$^+$ in 1 M HClO$_4$. Scan rate, 2 mV/s. A = 1.12 cm^2. (From Ref. 46.)

forms two ordered structures. The initial UPD process forms essentially the same (3 × 3) superlattice as for Ag, Pt(111)(3 × 3)–Cu,I, for which $\theta_{Cu} = \theta_I = \frac{4}{9}$ (Fig. 15A). However, the second UPD process yields Pt(111)(10 × 10)–Cu,I, with $\theta_{Cu} = \frac{8}{9}$ and $\theta_I = \frac{4}{9}$ (Fig. 15B). The most striking difference between Cu and Ag deposition at Pt(111) is that electrodeposition of bulk Cu does not grow layer by layer, but instead, forms separate granules of Cu. While the LEED pattern indicates Pt(111)(10 × 10)–Cu,I, the Pt substrate Auger signal remains unchanged, even after deposition of the equivalent of hundreds of monolayers. The substrate Auger signal would have decreased if the deposition had occurred in the form of monolayers.

Studies of electrodeposition of Pb onto Pt(111) from Cl$^-$ and Br$^-$ aqueous solutions illustrate how the nature of the supporting electrolyte anion affects the electrodeposition process [48,49]. The voltammetry of electrodeposition of Pb onto Pt(111) from solutions containing PbCl$_2$ and HCl is shown in Figure 16. The deposition is represented by 10 UPD processes, followed by bulk deposition of Pb. The LEED patterns obtained at various stages during the potential scan demonstrate that the electrodeposited layer is ordered and undergoes a series of structural transitions as Pb is deposited. Prior to the onset of Pb deposition, adsorbed Cl atoms from the Cl$^-$ electrolyte are present and form a Pt(111)

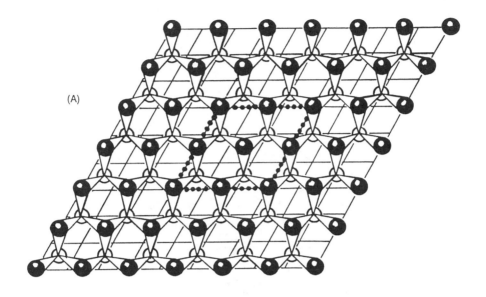

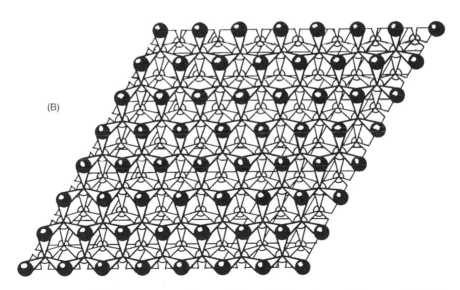

FIGURE 15 Surface structural models of Cu deposited on Pt(111) $(\sqrt{7} \times \sqrt{7})$R19.1°–I. (A) Pt(111)(3 × 3)–I,Cu ($\theta_{Cu} = \theta_I = \frac{4}{9}$); Cu(111) layer forming a (3 × 3) coincidence lattice with Pt(111); solid circles represent I atoms and open circles for Cu atoms: (B) Pt(111)(10 × 10)–I,Cu ($\theta_{Cu} = \frac{8}{9}$, $\theta_I = \frac{4}{9}$). (From Ref. 47.)

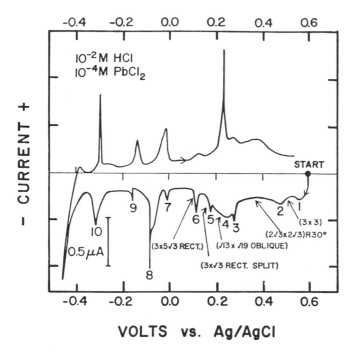

FIGURE 16 Cyclic voltammogram of Pt(111) in aqueous solution of 0.1 m*M* PbCl$_2$, in 10 m*M* HCl. Scan rate 0.5 mV/s. *A* = 0.77 cm^2. (From Ref. 48.)

(3 × 3)–Cl adlattice [48]. Cl occupies the topmost layer of the surface at all Pb coverages. Spontaneous deposition of Pb onto the Pt(111)(3 × 3)–Cl surface in Cl$^-$ media occurs to produce a series of complex superlattices. At open-circuit potential, the adsorbed layer forms a Pt(111)(2$\sqrt{3}$ × 2$\sqrt{3}$)R30° structure; however, the instability of the structure has thus far prevented complete identification. Similarly, electrodeposition of Pb onto Pt(111) from Br$^-$ media produces several ordered UPD superlattices [49]. However, the Pb deposits are stable only below $\theta_{Pb} = \frac{1}{2}$ in Br$^-$ media, evidently due to the strong adsorption of Br at Pt, which destabilizes the Pb deposits. A series of hexagonal and oblique structures are also observed in Br$^-$ media. The deposition of Pb from Br$^-$/I$^-$ mixture is analogous to the electrodeposition of Ag onto the Pt(111)($\sqrt{7}$ × $\sqrt{7}$)R19.1°–I surface, as discussed previously. Deposits of Pb form a rectangular structure Pt(111)(4 × 3)–Pb,I, for which the ideal packing densities are $\theta_{Pb} = \frac{1}{4}$ and $\theta_I = \frac{3}{4}$, and a hexagonal superlattice, Pt(111)($\sqrt{13}$ × $\sqrt{13}$)R14°–Pb,I, for which $\theta_{Pb} = \frac{7}{13}$ and $\theta_I = \frac{3}{13}$.

B. Electrodeposition at Other Substrates

The process of electrodeposition of Bi, Pb, Tl, or Cu at the Ag(111) surface involves noticeably smaller underpotentials than for depositions of Ag, Pb, or Cu at Pt(111) [50]. However, only adsorbed Bi layers manifest sufficient stability at open-circuit potential to permit the determination of the structure of the electrodeposited layer. The UPD Bi layer from acetate buffer–electrolyte is hexagonally close-packed in a Ag(111)($2\sqrt{3} \times 2\sqrt{3}$)R30°–Bi monolayer with a coverage of $\theta_{Bi} = \frac{7}{12}$, in which the ($2\sqrt{3} \times 2\sqrt{3}$)R30° unit cells are present in strip-shaped domains averaging 8.5 silver unit meshes in width and virtually unlimited length. Electrodeposition of Cu onto Ag(111) is a gradual process with no UPD peaks. The Cu deposit is stable at open-circuit potential and consists of a polycrystalline monolayer. Deposition of Pb at Ag(111) from acetate, perchlorate, chloride, bromide, or iodide electrolytes, followed by characterization of UHV, yields ordered layers of PbO. The structure of the layers varies with the nature of the electrolyte anion [50]. Evidently, oxidation of the electrodeposited Pb layer occurs during open-circuit contact with liquid electrolyte just prior to evacuation.

In a recent study of Ru(001) immersed into aqueous electrolytes containing Ag^+ ion and methane, Ag is deposited onto well-ordered Ru(001) and oxidized Ru surfaces [51]. Cyclic voltammetry of both clean Ru(001) and perchlorate-oxidized Ru electrodes in electrolyte containing Ag^+ indicates that Ag deposition occurs in a broad potential range from +0.2 to −0.5 V. The overlayer of Ag forms a (1 × 1) structure.

V. MOLECULAR LAYERS

Insight into the nature of molecular layers formed from solution at single-crystal electrode surfaces has been gained by combining surface analytical techniques with electrochemical methods of analysis [1,2]. The use of multiple techniques enables the investigation of several surface properties of the adsorbed species, such as molecular packing density, long-range structural order, molecular structure, orientation and mode of surface attachment, and the chemical and electrochemical reactivity of the surface layer. Types of organic compounds studied thus far at structurally ordered and atomically clean electrode surfaces include alcohols, aldehydes, alkenes, alkynes, amines, amino acids, aromatics, carboxylic acids, cyanides, thiocyanides, mercaptans, and five- and six-membered heteroaromatics. The remainder of this chapter is devoted to a discussion of recent experimental results of several types of organic compounds adsorbed at well-defined surfaces that illustrate how the adsorbate structure, orientation, and mode of attachment correlate with the observed surface chemistry and electrochemical reactivity.

Many molecules chemisorb at electrode surfaces with sufficient strength to permit the transfer of the adsorbed layer from one solution to another, a phenomenon demonstrated by using thin-layer electrochemical methods [52]. The stability of several classes of molecules in vacuum has also been demonstrated. One such example is 2,5-dihydroxy-4-methyl benzyl mercaptan (DMBM) adsorbed from aqueous solution at Pt(111) [53], which is attached to the surface through the sulfur atom in a vertical orientation with the reversibly electroactive hydroquinone moiety pendant. The cyclic voltammogram of an adsorbed layer of DMBM, formed by immersing Pt(111) in a dilute solution of the molecule followed by rinsing of the electrode with pure electrolyte, is shown in Figure 17A (solid curve). The reversible oxidation–reduction peaks clearly suggest that the adsorbed molecule does not rinse away and the hydroquinone moiety is pendant. Virtually identical voltammetric behavior (Fig. 17A, dotted

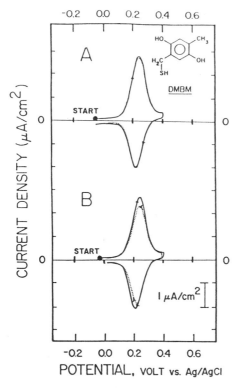

FIGURE 17 Cyclic voltammetry of adsorbed DMBM at Pt(111). (A) Solid curve, immersion into 0.7 m*M* DMBM followed by rinsing with 10 m*M* trifluoroacetic acid (TFA); dotted curve: as above followed by 1 h under vacuum prior to voltammetry. (B) Solid curve, first voltammetric scan; dotted curve, second scan. Scan rate, 5 mV/s. $A = 0.77$ cm^2. (From Ref. 53.)

curve) is observed when the DMBM layer adsorbed from solution and rinsed in pure electrolyte is followed by a 1-h exposure to UHV and subsequently reimmersed in pure electrolyte. The similarity in the two voltammetric curves strongly suggests that the adsorbed DMBM layer is not significantly perturbed by vacuum. This stability in vacuum has also been demonstrated with many other chemisorbed molecules and thus permits the study of the qualitative and quantitative properties of molecular layers using surface spectroscopies, including AES, LEED, HREELS, and IR.

A. Influence of Electrode Potential

The influence of electrode potential on the surface chemistry of several classes of organic compounds at Pt and Ag electrode surfaces have recently been investigated [54–58]. Electron energy loss spectroscopy has played a pivotal role in determining the orientation, mode of surface attachment, and conformation by revealing the vibrational bands of the adsorbed species. It is an especially powerful technique for compounds having a functional group whose vibrational bands change in frequency and/or intensity with changes in the electrode potential applied during adsorption or pH of the adsorbate or rinsing solution. The HREELS spectra of nicotinic acid (NA) adsorbed from acidic solution at Pt(111) shows such a potential dependence, as illustrated in Figures 18A and B. The vibrational bands characteristic of a free carboxylic acid group are clearly observed in the HREELS spectrum of NA adsorbed at a negative electrode potential (Fig. 18A): a prominent O—H stretching band at 3566 cm^{-1} and a C=O stretching band at 1748 cm^{-1}. The HREELS spectrum closely resembles the mid-IR spectrum of NA vapor [59] (Fig. 18A, lower curve), which indicates that the pyridine ring structure is not significantly perturbed upon adsorption. The O—H stretching band at 3567 cm^{-1} is strongly attenuated when NA is adsorbed from acidic solutions at positive potentials, as shown in Figure 18B, although the remainder of the spectrum closely resembles the HREELS spectrum of adsorbed NA at negative potentials, including the C=O stretching band at 1733 cm^{-1}. These observations suggest that the carboxylic acid moiety coordinates to the surface at positive potentials through the carboxylate oxygen with loss of the acid proton.

Treatment of the adsorbed pendant carboxylic acid NA layer with base results in significant retention of K$^+$ ions (as indicated by Auger spectroscopy) as well as major changes in the HREELS spectrum (Fig. 18C). The O—H stretching band vanishes when the NA layer is rinsed with base and reappears when the surface is rinsed with acid, as expected for a free carboxylic acid O—H group. The HREELS spectrum following the base rinse also exhibits two prominent peaks at 1612 and 1379 cm^{-1}, partially attributable to asymmetric and symmetric carboxylate OCO stretching modes, which are also observed in the potassium nicotinate IR spectrum (Fig. 18C, bottom curve).

To summarize the potential-dependent behavior of adsorbed NA, the following structural models are proposed:

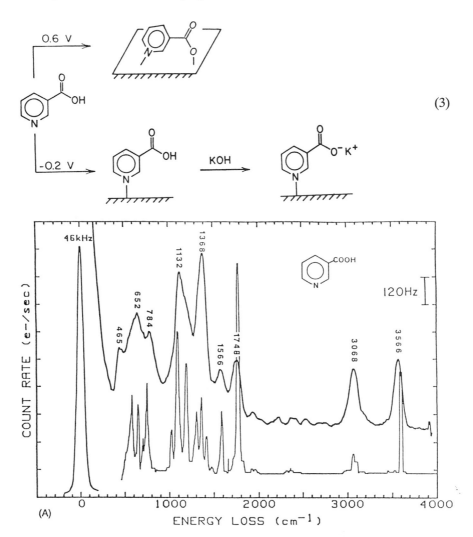

(3)

(A)

FIGURE 18 Vibrational spectra of NA: (A) EELS spectrum of NA adsorbed at Pt(111), −0.2 V, pH 3; (B) EELS spectrum of NA at Pt(111), 0.6 V, pH 3; (C) EELS spectrum of NA at Pt(111), −0.3 V, rinsed at pH 10. Lower curve in (A) and (B) is the mid-IR spectrum of NA vapor [59]. Lower curve in (C) is the mid-IR spectrum of solid KNA. Experimental conditions: adsorption from 1 m*M* NA in 10 m*M* KF, pH 3 [(A) and (B)] or pH 7 (C), followed by rinsing with 2 m*M* HF, pH 3 [(A) and (B)] or 0.1 m*M* KOH, pH 10 (C); EELS conditions as in Figure 9. (From Ref. 54.)

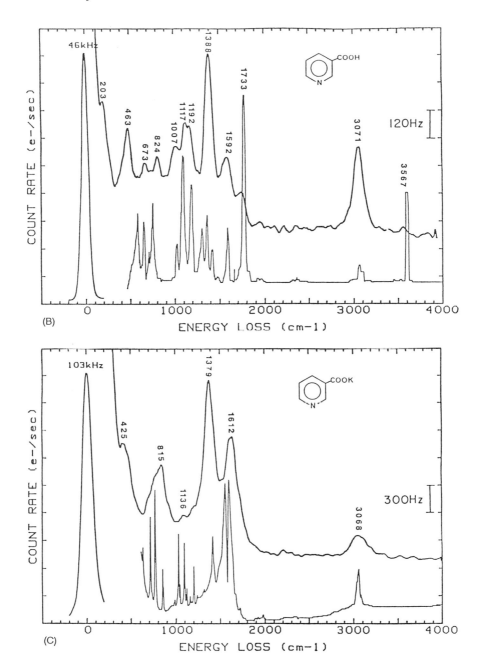

The intensity of the O—H stretching band varies smoothly from maximum to minimum as the electrode potential applied during adsorption is varied from negative to positive, as shown in Figure 19. As can also be seen from the figure, the NA packing density is virtually constant over the range of potentials where the O—H stretch intensity undergoes its transition; therefore, the attenuation of the O—H vibration is not due to desorption of NA. The normalized O—H stretch intensity (O—H/C—H signal ratio) is essentially constant from −0.5 to −0.1 V. The decrease in packing density at potentials more negative than −0.2 V could be due to interference from hydrogen adsorption at the Pt(111) surface. Similarly, the acute drop in the packing density for adsorption potentials greater than −0.5 V could be due to surface oxidation. The observation that the packing density is virtually constant throughout the transition indicates that very slight changes in molecular orientation occur.

Studies of NA and other pyridinecarboxylic acids at Pt(111) have revealed that the primary mode of attachment is through the pyridine nitrogen atom and that coordination of the carboxylic acid moiety to the surface is dependent on

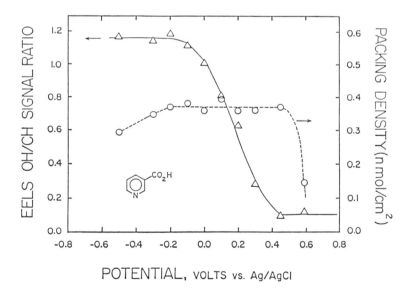

POTENTIAL, VOLTS vs. Ag/AgCl

FIGURE 19 Potential dependence of molecular packing density (O) and O—H/ C—H signal ratio (Δ) for NA adsorption at Pt(111). The packing density was determined from Auger data, while the O—H/C—H ratio was obtained from the EELS spectra. Experimental conditions: adsorption from 1 mM NA in 10 mM KF (adjusted to pH 7), followed by rinsing with 2 mM HF (pH 3); EELS conditions as in Figure 9; Auger conditions: electron beam at normal incidence, 100 nA, 2000 eV. (From Ref. 54.)

the location of the acid group on the pyridine ring and, as is the case with NA, on the applied electrode potential [54]. Carboxylic acid moieties in the para position of pyridine cannot readily coordinate with the surface because of the relatively large distance between the acid moiety and the Pt surface, and a free carboxylic acid moiety is observed at all potentials studied. By comparison, carboxylic acid moieties in the ortho position are coordinated extensively to the surface at all potentials. These results show that the surface chemistry is greatly influenced by the structure of the molecular adsorbate.

Electrode potential not only influences the mode of surface coordination of organic adsorbates, as described above, but also affects the conformation of the adsorbate. The HREELS spectrum of a carboxylic acid derivative of the multinitrogen heteroaromatic 2-pyrazinecarboxylic acid (PZCA) adsorbed at various electrode potentials at Pt(111) is shown in Figure 20, along with the IR spectrum of PZCA vapor [60] (bottom curve) [55]. Two distinct O—H stretching bands are observed in the HREELS spectrum at −0.2 V (Fig. 20A): a 3567-cm^{-1} band assignable to the free O—H stretching mode, and a 3408^{-1} band assignable to intramolecular hydrogen bonding (O—H····N) stretch between the carboxylic acid moiety and the nitrogen atom farthest from the surface in the adsorbed state, as depicted in Eq. (4):

$$(4)$$

$$(5)$$

$$(6)$$

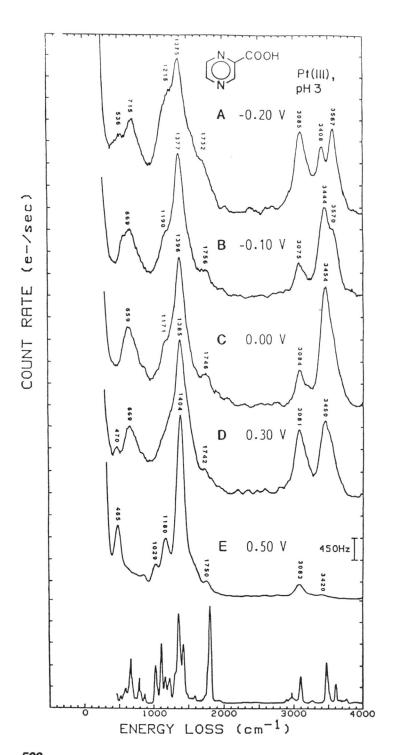

Increasing the potential from -0.2 to 0.0 V results in a corresponding increase in the intramolecular hydrogen-bonded O—H conformation and a decrease in the free acid form, as indicated by the increase in the hydrogen-bonded O—H stretch HREELS band and a decrease in the free O—H stretching band [Fig. 20C and Eq. (5)]. The influence of the positive potential probably shifts electron density from the pyrazine ring to the surface, thereby localizing the electron density at the pendant ring nitrogen atom, making it a stronger proton acceptor and thus favoring stronger intramolecular hydrogen bonding. The HREELS O—H stretching band disappears when the electrode potential is increased to 0.5 V (Fig. 20E), and is attributed to the surface coordination of the carboxylic moiety as illustrated in Eq. (6). This change in surface attachment with increasing potentials is similar to that observed for NA. The results discussed above show that HREELS is quite useful in determining adsorbate conformation.

B. Influence of Adsorbate Molecular Structure

Molecular structure can have a profound impact on the mode of surface attachment, orientation, chemical and electrochemical reactivity, and molecular packing density of the adsorbed species. Comparison of two structurally similar molecules, nicotinic acid (3-pyridinecarboxylic acid) and benzoic acid (BA), adsorbed at Pt(111) under similar experimental conditions, reveals major differences in their surface chemistry. As discussed in Section V.A, NA adsorbs at Pt(111) through the ring nitrogen atom, with the ring nearly perpendicular to the surface and with the carboxylic acid group pendant under conditions of acidic solution and negative electrode potential. The molecular packing density determined by Auger spectroscopy for NA under those conditions is 0.36 nmol/cm^2, while the molecular packing density for BA under similar experimental conditions is 0.21 nmol/cm^2. This profound difference in packing density can be attributed to a difference in adsorbate molecular orientation: BA is postulated to adsorb in a horizontal orientation with the aromatic ring parallel to the Pt surface, with the carboxylate moiety also coordinated to the surface [54], while NA is perpendicular to the surface, as discussed above.

Further evidence supporting the proposed surface orientations comes from the HREELs results shown in Figures 18A (NA) and 21 (BA), with the corresponding vapor-phase IR spectrum shown below each HREELS spectrum. As

---

FIGURE 20 EELS spectra of PZCA at Pt(111) at -0.20 V (A), -0.10 V (B), 0.0 V (C), 0.30 V (D), 0.5 V (E). The lowest curve is the gas-phase IR spectrum of PZCA [60]. Experimental conditions: adsorption from 1 mM PZCA solution (10 mM KF/HF, pH 3) followed by rinsing with 2 mM HF, pH 3. Other conditions as in Figure 9. (From Ref. 55.)

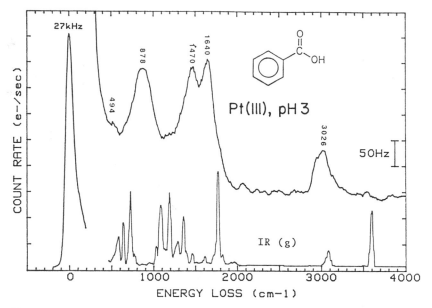

FIGURE 21 EELS spectrum of BA adsorbed at Pt(111). Bottom curve is the gas-phase IR spectrum of BA [61]. Experimental conditions: adsorption from 1 m*M* BA in 10 m*M* KF/HF (pH 3), followed by rinsing with 2 m*M* HF (pH 3) at −0.1 V. Other conditions as in Figure 9. (From Ref. 54.)

discussed earlier, the HREELS spectrum of adsorbed NA closely resembles the NA vapor-phase IR spectrum, suggesting that the NA molecular structure is not perturbed upon adsorption. By comparison, the HREELS spectrum of adsorbed Ba does not resemble the BA vapor-phase IR spectrum [61]: the characteristic carboxylic acid O—H stretching band at about 3600 cm^{-1} and the C=O stretching band at about 1760 cm^{-1} are barely visible. In addition, there is very little correspondence between the remaining frequencies in the HREELS and IR spectra, except for the C—H stretching band at 3026 cm^{-1}, which indicates perturbations of the molecular structure due to interactions between the adsorbate and the Pt surface.

Another indication that adsorbed BA does not have a pendant carboxylic acid moiety comes from surface basification experiments, wherein the adsorbed layer formed from an acidic solution and at a negative electrode potential is rinsed with a basic solution (10^{-4} *M* KOH). Adsorbed BA does not retain K$^+$ ions under these conditions, while adsorbed NA does.

Further evidence for the difference in surface orientation of the two compounds is demonstrated by the cyclic voltammetry of adsorbed BA (Fig. 22A)

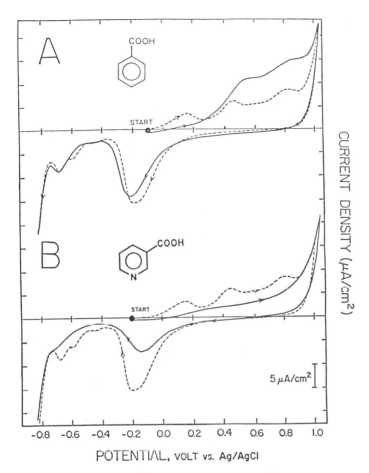

FIGURE 22 Cyclic voltammograms of adsorbed BA (solid curve, A) and NA (solid curve, B) layers at Pt(111), and clean Pt(111) (dashed curves) in 10 mM KF electrolyte (pH 7). Experimental conditions: adsorption from 1 mM aqueous solution (10 mM KF/HF) at -0.1 V for BA and -0.2 V for NA, followed by rinsing with pure electrolyte at the same potential. Scan rate, 5 mV/s. $A = 0.77$ cm^2. (From Ref. 54.)

and NA (Fig. 22B), which show substantial differences in electrooxidation reactivity of the two adsorbates. [The dashed curve in both figures is the cyclic voltammogram of clean Pt(111) in pure electrolyte]. The broad peaks between 0.4 and 1.0 V in Figure 22A (solid curve) are due to the electrochemical oxidation of adsorbed BA. The number of electrons required to oxidatively desorb a BA molecule, n_{ox}, was determined using the equation

$$n_{ox} = \frac{Q_{ox} - Q_b}{FA\Gamma} \tag{7}$$

where Q_{ox} is the total charge due to irreversible oxidation of adsorbed BA and the electrode surface, Q_b the charge for the oxidation of the bare electrode surface, F the Faraday constant, A the geometric area of the electrode, and Γ the molecular packing density. Integration of the anodic current due to the oxidation of the BA layer yields an n_{ox} value of 28, which is in agreement with the expected n_{ox} value of 29 for complete oxidation of BA to CO_2:

$$+ \ 12 \ H_2O \longrightarrow 7 \ CO_2 + 29 \ H^+ + 29 \ e^- \tag{8}$$

This is further evidence that BA adsorbs in a horizontal orientation at Pt(111).

Adsorbed NA passivates the Pt surface toward oxidation and with respect to the adsorption of H atoms (reduction), as shown by the decreases in anodic as well as cathodic current (Fig. 22B, solid curve), relative to what would be observed in the absence of adsorbed NA (Fig. 22B, dashed curve). This inertness toward electrochemical oxidation is further evidence that NA adsorbs at Pt(111) through the ring nitrogen atom with the ring in a nearly vertical orientation.

C. Influence of the Nature of the Substrate

Results from a comparative study of the adsorption behavior of aromatic thiols and alkyl mercaptans at Pt(111) and Ag(111) [58] have shown that both the thiols and mercaptans are attached to each substrate primarily through the sulfur atom with the remainder of the molecule pendant; they exhibit similar packing densities; and they retain their framework molecular structures on the surface, as evidenced by the similarity of the HREELS spectra of each adsorbate at both substrates. The major difference in adsorption behavior of thiols and mercaptans at Pt(111) and Ag(111) is the greater tendency to form monolayers having long-range order at Ag(111) than at Pt(111).

This is illustrated by the HREELS spectra shown in Figure 23 for adsorbed thiophenol (TP) at Ag(111) and Pt(111), as well as the IR spectrum of liquid TP (bottom curve) for comparison. The similarity of the vibrational band frequencies and intensities for both spectra indicate that TP adsorbs at both surfaces in a similar form. In addition, the similarity of the adsorbed spectra with the

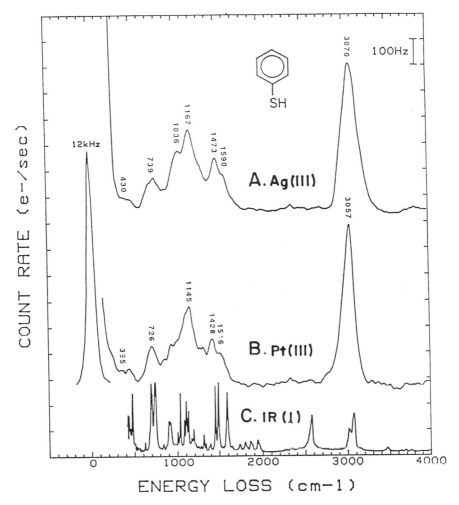

FIGURE 23 Vibrational spectra of TP. A and B are the EELS spectra of TP at Ag(111) (pH 3, −0.20 V) and Pt(111) (pH 2, 0.0 V), respectively, from a saturated (about 0.1 m*M*) aqueous solution of TP. Aqueous solutions contained 2 m*M* HF (pH 3) or 10 m*M* TFA (pH 2). Other conditions as in Figure 9. (From Ref. 58.)

liquid-phase IR spectrum suggests that the framework molecular structure of adsorbed TP is similar to that of liquid TP, except for the absence of the S—H stretch of adsorbed TP (seen at 2567 cm^{-1} in the IR spectrum). These results suggest that dissociation of sulfhydryl hydrogen occurs upon adsorption and that attachment to both surfaces occurs primarily through the sulfur atom with the remainder of the molecule pendant:

$$\text{(structure)} \quad \xrightarrow[\text{Pt (III)}]{\text{Ag(III), or}} \quad \text{(structure)} \quad + \; H^+ \; + \; e^- \qquad (9)$$

  SH

                                    S

                                    ///////

Differences in the long-range order of TP at Ag(111) and Pt(111) are observed: a sharp $(\sqrt{7} \times \sqrt{31}, 88°)R40.9°$ LEED pattern is formed for adsorbed TP at Ag(111), shown in Figure 24A, whereas a diffuse LEED pattern is observed for TP at Pt(111). The surface structure model for adsorbed TP adsorbed at Ag(111) is shown in Figure 24B. Other mercaptans, such as benzyl mercaptan and 2-mercaptoethanesulfonic acid sodium salt, also form adsorbed layers exhibiting long-range order at Ag(111) but not at Pt(111). A possible explanation for the observed differences in long-range order for TP and related compounds at Ag(111) and Pt(111) is that the differences in the surface free energy among the adsorption sites (i.e., one-, two-, or threefold sites) are significantly larger at Ag(111) than at Pt(111) [62].

A comparative study of pyridines adsorbed from solution at Pt(111) [54] and Ag(111) [58] electrode surfaces have shown that pyridine chemisorbs at Pt(111) but not at Ag(111). The results of pyridine adsorbed at Pt(111) can be summarized as follows:

1. Pyridine adsorbs in a nearly vertical orientation through the nitrogen atom with a tilt angle of about 71° between the plane of the ring and the Pt surface.

2. The molecular packing density is 0.46 nmol/cm^2 and is virtually independent of solution concentration.

3. It forms an ordered, incommensurate surface structure, Pt(111) $(3.324 \times 4.738, 77.1°)R34°$–PYR.

4. It passivates the Pt(111) surface toward surface oxidation (OH adsorption) and reduction (H adsorption).

5. It exhibits the same packing density and molecular orientation independent of the solution pH from pH 0 to 10 and also independent of adsorption electrode potential within the range between H$_2$ evolution and Pt oxidation.

By contrast, no carbon or nitrogen Auger signal was detected following a pyridine immersion experiment, in which Ag(111) was immersed into a 2 mM pyridine solution and rinsed with pure electrolyte. The possibility that the adsorbed pyridine layer is stable in solution but not in vacuum has been ruled out based on the results of an adsorption study of a reversibly electroactive pyridine derivative, (3-pyridyl)hydroquinone (3PHQ) [58]. (3-Pyridyl)hydroquinone is

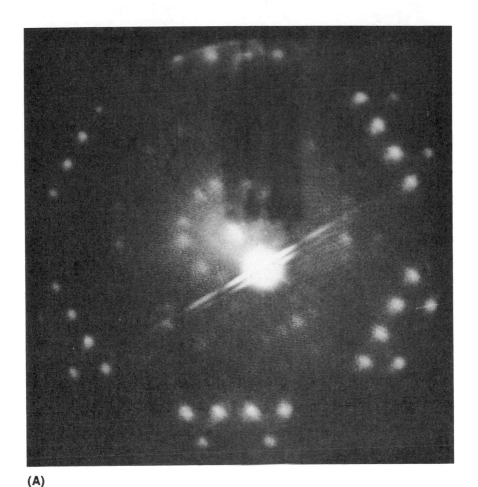

(A)

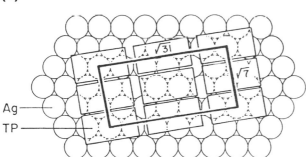

(B)

FIGURE 24 LEED pattern (A) and surface structure (B) of TP adsorbed at Ag(111). The pattern is $(\sqrt{7} \times \sqrt{31}, 88°)$R40.9°. The TP packing density in this surface structure is 0.544 nmol/cm², corresponding to 4 TP molecules and 17 Ag atoms per unit cell. Experimental conditions: LEED beam energy 23.4 eV; adsorption from TP saturated solution (2 m*M* HF, pH 3) at −0.10 V. (From Ref. 58.)

strongly chemisorbed at Pt(111) and attaches to the surface through the nitrogen atom with the hydroquinone moiety pendant [54,63]. In addition, the pendant hydroquinone moiety of the adsorbed molecule is reversibly electroactive, as shown in Figure 25. By comparison, immersion of Ag(111) into 0.5 mM 3PHQ solution at pH 4 followed by rinsing with pure electrolyte and a subsequent cyclic voltammetric scan produces no detectable current, due to the redox of 3PHQ.

VI. CONCLUDING REMARKS

The results presented in this chapter illustrate how the multitechnique approach provides collaborative evidence to characterize complex electrode–adsorbate systems. This methodology can be applied to many other solid–liquid interfacial systems to study fundamental areas of surface science, including surface reaction kinetics and mechanisms, surface acid–base chemistry, surface reaction product distributions, electrical double-layer phenomena, and surface redox processes.

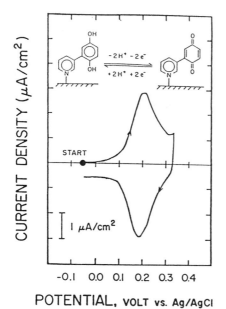

FIGURE 25 Cyclic voltammogram of adsorbed 3PHQ at Pt(111) in 10 mM KF/HF (pH 4) electrolyte. Experimental conditions: adsorption from a 0.5 mM 3PHQ solution (10 mM KF/HF, pH 4) at −0.04 V, followed by rinsing with 1 mM HF. Scan rate, 5 mV/s. A = 0.77 cm^2. (From Ref. 54.)

Progress in these areas can lead to further understanding of electrocatalytic phenomena, surface passivation, mechanisms of corrosion, and other areas involving surface electrochemistry.

ACKNOWLEDGMENTS

Acknowledgment is made to the National Science Foundation, the Air Force Office of Scientific Research, and the Gas Research Institute for financial support for this work. The assistance of Arthur Case, Frank Douglas, Richard Shaw, and Vickie Townsend is gratefully acknowledged.

REFERENCES

1. A. T. Hubbard, Electrochemistry at well-characterized surfaces, *Chem. Rev.*, *88*: 633 (1988).
2. A. T. Hubbard, Surface electrochemistry (Kendall Award Address, American Chemical Society, 1989), *Langmuir*, *6*: 97 (1990).
3. D. M. Kolb, UHV techniques in the study of electrode surfaces, *Z. Phys. Chem.*, *154*: 179 (1987).
4. P. N. Ross and F. T. Wagner, The application of surface physics techniques to the study of electrochemical systems, in *Advances in Electrochemistry and Electrochemical Engineering* (H. Gerischer and C. W. Tobias, eds.), Wiley, New York, 1985, Vol. 13, p. 69.
5. A. T. Hubbard, Electrochemistry of well-defined surfaces, *Acc. Chem. Res.*, *13*: 177 (1980.
6. A. T. Hubbard, J. L. Stickney, M. P. Soriaga, V. K. F. Chia, S. D. Rosasco, B. C. Schardt, T. Solomun, D. Song, J. H. White, and A. Wieckowski, Electrochemical processes at well-defined surfaces, *J. Electroanal. Chem.*, *168*: 43 (1984).
7. L. J. Clarke, *Surface Crystallography: An Introduction to Low Energy Electron Diffraction*, Wiley, New York, 1985.
8. J. C. Tracy, in *Electron Emission Spectroscopy* (W. Dekeyser et al., eds.), D. Reidel, Dordrecht, The Netherlands, 1983, p. 295.
9. P. W. Palmberg, Quantitative auger electron spectroscopy using elemental sensitivity factors, *J. Vacuum Sci. Technol.*, *13*: 214 (1976).
10. J. A. Schoeffel and A. T. Hubbard, Quantitative elemental analysis of substituted hydrocarbon monolayers on platinum by auger electron spectroscopy with electrochemical calibration, *Anal. Chem.*, *49*: 2330 (1977).
11. N. Batina, D. G. Frank, J. Y. Gui, B. E. Kahn, C.-H. Lin, F. Lu, J. W. McCargar, G. N. Salaita, D. A. Stern, D. C. Zapien, and A. T. Hubbard, Oriented adsorption at well-defined electrode surfaces studied by Auger, LEED and EELS spectroscopy, Proc. 4th International Fischer Symp., *Electrochim. Acta*, *34*: 1031 (1989).
12. N. Winograd and S. W. Gaarestroom, in *Physical Methods in Modern Chemical Analysis* (T. Kuwana, ed.), Vol. 2, 1978, p. 115.

13. P. K. Ghosh, *Introduction to Photoelectron Spectroscopy*, Wiley-Interscience, New York, 1983.
14. H. Ibach and D. L. Mills, *Electron Energy Loss Spectroscopy and Surface Vibrations*, Academic Press, New York, 1982.
15. B. E. Hayden, in *Vibrational Spectroscopy of Molecules on Surfaces* (J. T. Yates and T. E. Madey, eds.), Plenum Press, New York, 1987, p. 267.
16. G. Bining, H. Rohrer, C. Gerber, and E. Weibel, (7×7) Reconstruction on Si(111) resolved in real space, *Phys. Rev. Lett.*, *50*: 120 (1983).
17. R. C. Engstrom and C. M. Pharr, Scanning electrochemical microscopy, *Anal. Chem.*, *61*: 1099A (1989).
18. M. P. Green, K. J. Hanson, D. A. Scherson, X. Xin, M. Richter, P. N. Ross, Jr., R. Can, and I. Lindan, In situ scanning tunneling microscopy studies of the underpotential deposition of lead on Au(111), *J. Phys. Chem.*, *93*: 2181 (1989).
19. D. Rugar and P. Hansma, Atomic force microscopy, *Phys. Today*, *43*(10): 23 (1990).
20. D. G. Frank, N. Batina, T. Golden, F. Lu, and A. T. Hubbard, Imaging surface atomic structure by means of auger electrons, *Science*, *247*: 182 (1990).
21. A. T. Hubbard, D. G. Frank, O. M. R. Chyan, and T. Golden, Imaging surface atomic structure by means of auger electrons, *J. Vacuum Sci. Technol.*, *B8*: 1329 (1990).
22. G. N. Salaita, D. A. Stern, F. Lu, H. Baltruschat, B. S. Schardt, J. L. Stickney, M. P. Soriaga, D. G. Frank, and A. T. Hubbard, Structure and composition of a platinum (111) surface as a function of pH and electrode potential in aqueous bromide solutions, *Langmuir*, *2*: 828 (1986).
23. F. Lu, G. N. Salaita, H. Baltruschat, and A. T. Hubbard, Adlattice structure and hydrophobicity of Pt(111) in aqueous potassium iodide solutions: influence of pH and electrode potential, *J. Electroanal. Chem.*, *222*: 305 (1987).
24. D. A. Stern, H. Baltruschat, G. N. Salaita, M. Martinez, J. L. Stickney, D. Song, S. K. Lewis, D. G. Frank, and A. T. Hubbard, Characterization of single-crystal electrode surfaces as a function of potential and pH by auger spectroscopy and LEED: Pt(111) in aqueous $CaCl_2$ and HCl solutions, *J. Electroanal. Chem.*, *217*: 101 (1987).
25. G. N. Salaita, F. Lu, L. Laguren-Davidson, and A. T. Hubbard, Structure and composition of Ag(111) surfaces as a function of electrode potential in aqueous halide solutions, *J. Electroanal. Chem.*, *229*: 1 (1987).
26. A. T. Hubbard, Electrochemistry at well-characterized surfaces, *Comprehensive Chemical Kinetics* (C. H. Bamford, D. F. H. Tipper, and R. G. Compton, eds.), Vol. 28, Elsevier, Amsterdam, 1988, p. 1.
27. D. G. Frank, O. M. R. Chyan, T. Golden, and A. T. Hubbard, Probing three distinct iodine monolayer structures at Pt(111) by means of angular distribution auger microscopy (ADAM): results agree with scanning tunneling microscopy, *J. Phys. Chem.*, *97*: 3829 (1993).
28. B. C. Schardt, S. L. Yau, and F. Rinaldi, Atomic resolution imaging of adsorbates on metal surfaces in air: iodine adsorption on Pt(111), *Science*, *243*: 1050 (1989).
29. S. L. Lau, C. M. Vitus, and B. C. Schardt, In situ scanning tunneling microscopy of adsorbates on electrode surfaces: image of $(\sqrt{3} \times \sqrt{3})R30°$-iodine adlattice on platinum (111), *J. Am. Chem. Soc.*, *112*: 3677 (1990).

30. S. C. Chang, S. L. Yau, B. C. Schardt, and M. J. Weaver, Comparisons between scanning tunneling microscopy and outer-sphere electron-transfer rates at Pt(111) surfaces coated with ordered iodine layers, *J. Phys. Chem.*, *95*: 4787 (1991).

31. M. Grunze and P. A. Dowben, A review of halocarbon and halogen adsorption with particular reference to iron surfaces, *Appl. Surface Sci.*, *10*: 209 (1982), and references therein.

32. E. Bechtold, Adsorption of fluorine on Pt(111), *Appl. Surface Sci.*, *7*: 231 (1981).

33. H. H. Farrell, The coadsorption of I and Cl on Pt(111), *Surface Sci.*, *100*: 613 (1980).

34. W. Erley, Chlorine Adsorption on the (111) faces of Pd and Pt, *Surface Sci.*, *94*: 281 (1980).

35. E. Bertel, K. Schwaha, and F. P. Netzer, The adsorption of bromine on Pt(111): observation of an irreversible order–disorder transition, *Surface Sci.*, *83*: 439 (1979).

36. T. E. Felter and A. T. Hubbard, LEED and electrochemistry of iodine on Pt(100) and Pt(111), *J. Electroanal. Chem.*, *100*: 473 (1979).

37. J. L. Stickney, S. D. Rosasco, G. N. Salaita, and A. T. Hubbard, Ordered ionic layers formed on Pt(111) from aqueous solutions, *Langmuir*, *1*: 66 (1985).

38. D. G. Frank, J. Y. Katekaru, S. D. Rosasco, G. N. Salaita, B. C. Schardt, M. P. Soriaga, D. A. Stern, J. L. Stickney, and A. T. Hubbard, pH and potential dependence of the electrical double layer at well-defined electrode surfaces: Cs$^+$ and Ca^{2+} at Pt(111)(2$\sqrt{3}$ × 2$\sqrt{3}$)R30°–CN$^-$, *Langmuir*, *1*: 587 (1985).

39. O. A. Reutov, I. P. Beletskaya, and A. L. Kurts, in *Ambient Anions* (J. P. Micheal, ed.), Consultants Bureau, New York, 1983, p. 205.

40. E. Y. Cao, P. Gao, J. Y. Gui, F. Lu, D. A. Stern, and A. T. Hubbard, Adsorption and electrochemistry of SCN$^-$: comparative studies of Ag(111) and Pt(111) electrodes by means of AES, CV, HREELS and LEED, *J. Electroanal. Chem.*, *339*: 311 (1992).

41. J. K. Foley, S. Pons, and J. J. Smith, Fourier transform infrared spectroelectrochemical studies of anodic processes in thiocyanate solutions, *Langmuir*, *1*: 697 (1985).

42. M. Fleischmann and H. R. Thrisk, in *Advances in Electrochemistry and Electrochemical Engineering* (P. Delahay, ed.), Interscience, New York, 1963.

43. A. T. Hubbard, J. L. Stickney, S. D. Rosasco, M. P. Soriaga, and D. Song, Electrodeposition on a well-defined surface: silver on Pt(111)($\sqrt{7}$ × $\sqrt{7}$), *J. Electroanal. Chem.*, *150*: 165 (1983).

44. J. L. Stickney, S. D. Rosasco, D. Song, and A. T. Hubbard, Superlattices formed by electrodeposition of silver on iodine-pretreated Pt(111): studies by LEED, auger spectroscopy and electrochemistry, *Surface Sci.*, *130*: 326 (1983).

45. D. G. Frank, O. M. R. Chyan, T. Golden, and A. T. Hubbard, Auger emission angular distributions from a silver monolayer in the presence and absence of an iodine overlayer at Pt(111): evidence for the predominance of inhomogeneous inelastic scattering of Auger electgrons by atoms, *J. Phys. Chem.*, *98*: 1895 (1994).

46. J. L. Stickney, S. D. Rosasco, B. C. Schardt, and A. T. Hubbard, Electrodeposition of silver on Pt(100) surfaces containing iodine adlattices: studies by low-energy

electron diffraction, auger spectroscopy, and thermal desorption, *J. Phys. Chem.*, *88*: 251 (1984).

47. J. L. Stickney, S. D. Rosasco, and A. T. Hubbard, Electrodeposition of copper on platinum (111) surfaces pretreated with iodine: studies by LEED, auger spectroscopy and electrochemistry, *J. Electrochem. Soc.*, *131*: 260 (1984).

48. B. C. Schardt, J. L. Stickney, D. A. Stern, A. Wieckowski, D. C. Zapien, and A. T. Hubbard, Electrodeposition of Pb onto Pt(111) in aqueous chloride solutions: studies by LEED and auger spectroscopy, *Surface Sci.*, *175*: 520 (1986).

49. B. C. Schardt, J. L. Stickney, D. A. Stern, A. Wieckowski, D. C. Zapien, and A. T. Hubbard, Electrodeposition of Pb onto Pt(111) in aqueous bromide solutions: studies by LEED and auger spectroscopy, *Langmuir*, *3*: 239 (1987).

50. L. Laguren-Davidson, F. Lu, G. N. Salaita, and A. T. Hubbard, Electrodeposition of Tl, Pb, Bi and Cu on Ag(111) from aqueous solutions: studies by auger spectroscopy and LEED, *Langmuir*, *4*: 224 (1988).

51. E. Y. Cao, D. A. Stern, J. Y. Gui, and A. T. Hubbard, Studies of Ru(001) Electrodes in aqueous electrolytes containing silver ions and methane: LEED, HREELS, auger spectroscopy and electrochemistry, *J. Electroanal. Chem.*, *354*: 71 (1993).

52. A. T. Hubbard, Electrochemistry in thin layers of solution. *CRC Crit. Rev. Anal. Chem.*, *3*: 201 (1973).

53. D. A. Stern, E. Wellner, G. N. Salaita, L. Laguren-Davidson, F. Lu, N. Batina, D. G. Frank, D. C. Zapien, N. Walton, and A. T. Hubbard, Adsorbed thiophenol and related compounds studied at Pt(111) electrodes by EELS, auger spectroscopy and cyclic voltammetry, *J. Am. Chem. Soc.*, *110*: 4885 (1988).

54. D. A. Stern, L. Laguren-Davidson, D. G. Frank, J. Y. Gui, C.-H. Lin, F. Lu, G. N. Salaita, N. Walton, D. C. Zapien, and A. T. Hubbard, Potential-dependent surface chemistry of 3-pyridine carboxylic acid (niacin) and related compounds at Pt(111) electrodes, *J. Am. Chem. Soc.*, *111*: 877 (1989).

55. S. A. Chaffins, J. Y. Gui, C.-H. Lin, F. Lu, G. N. Salaita, D. A. Stern, and A. T. Hubbard, Multi-nitrogen heteroaromatics studied at Pt(111) surfaces by EELS, auger spectroscopy and electrochemistry: pyrazine, pyrimidine, pyridazine, 1,3,5-triazine and carboxylic acid derivatives, *Langmuir*, *6*: 1273 (1990).

56. J. Y. Gui, D. Stern, C.-H. Lin, P. Gao, and A. T. Hubbard, Potential-dependent surface chemistry of hydroxypyridines adsorbed at Pt(111) electrodes studied by EELS, LEED, AES, and electrochemistry, *Langmuir*, (honoring Arthur Adamson), *7*: 3183 (1991).

57. J. Y. Gui, F. Lu, D. A. Stern, and A. T. Hubbard, Surface chemistry of mercapto-pyridines at Ag(111) electrodes studied by EELS, LEED, auger spectroscopy and electrochemistry, *J. Electroanal. Chem.*, *292*: 245 (1990).

58. J. Y. Gui, D. A. Stern, D. G. Frank, F. Lu, D. C. Zapien, and A. T. Hubbard, Adsorption and surface structural chemistry of thiophenol, benzyl mercaptan and alkyl mercaptans: comparative studies at Ag(111) and Pt(111) electrodes by means of auger, EELS, LEED and electrochemistry, *Langmuir*, *7*: 955 (1991).

59. C. J. Pouchert, *The Aldrich Library of FTIR Spectra*, Aldrich, Milwaukee, Wis., 1985.

60. *Sadtler Standard Infrared Vapor Spectra*, Sadtler Research Laboratories, Philadelphia, 1986.

61. R. A. Nyquist, *The Interpretation of Vapor-Phase Spectra*, Vol. 2, Sadtler Research Laboratories, Philadelphia, 1984.
62. A. T. Hubbard and J. Y. Gui, Molecular electrochemistry at well-defined surfaces, *J. Chim. Phys.*, *88*: 1547 (1991).
63. D. C. Zapien, J. Y. Gui, D. A. Stern, and A. T. Hubbard, Surface electrochemistry and molecular orientation: studies of pyridyl hydroquinone adsorbed at Pt(111) by cyclic voltammetry, auger electron spectroscopy and electron energy loss spectroscopy, *J. Electroanal. Chem.* (honoring Roger Parsons), *330*: 469 (1992).

11

Electrodeposition of II–VI Semiconductors

GARY HODES
Weizmann Institute of Science
Rehovot, Israel

I. INTRODUCTION AND SCOPE

Electroplating or electrodeposition of semiconductors (abbreviated here as ED semiconductors) has been the subject of several general reviews in the past decade [1–3] that cover the field up to and including 1986. There are also a number of recent reviews dealing with specific topics: mechanistic aspects of II–VI semiconductors [4], CdTe and CuInSe$_2$ used as solar cells [5], as well as a short general review that includes some semiconductors among other materials [6].

The original purpose of the present review was to bring the interested reader up to date in the general field of semiconductor electrodeposition. In the process of writing, however, it became clear that this would require more than a single chapter, and it was decided to limit the review to II–VI semiconductors, which was the group investigated most intensively.

To avoid needless repetition of material given in previous reviews, certain obvious topics, covered in earlier reviews, will not be discussed here. These include general principles of electroplating and alloy plating, as well as uses for the ED films. It is sufficient to mention here that much of the impetus for studying ED semiconductors comes from the search for large-area semiconductor films for use as solar [photovoltaic (PV) and photoelectrochemical (PEC)] cells.

The work discussed here will cover the subject matter from 1987 through 1992. Earlier work will be used noncomprehensively as background material or where it is considered useful for other purposes. The term *electrodeposition* will be covered in a broad sense, and the review includes methods such as anodization (which is strictly electrochemical conversion rather than ED) and ED of precursor metal and chalcogen films followed by annealing. Electrophoresis has not been included, as no work on electrophoretic deposition of II–VI materials appears to have been carried out in the time frame covered by this review. Reference 114 gives an earlier example of this method for depositing CdS and CdSe films.

The material is grouped according to techniques of electrodeposition. The choice of substrate onto which the semiconductor film is deposited is also discussed. As an overview of the activity in the field over the past 6 years, Table 1 gives a fairly comprehensive listing of II–VI semiconductors which have been electrodeposited during that period, with references. Table 1 should serve a valuable function in directing the reader to the relevant literature, and as such, constitutes in itself an important part of the value of this review.

II. SUBSTRATE MATERIAL

There are several factors that dictate the choice of substrate material, depending on the intended purpose of the semiconductor film. Since this purpose is predominantly for solar cells, the most obvious factor is that the substrate forms an ohmic contact with the semiconductor. (An exception to this is when the semiconductor is deposited onto another semiconductor film, such as ED CdTe onto CdS films, for CdTe/CdS PV cells.) Ohmic contacts to n-type semiconductors normally present few problems, and Ti or transparent conducting glass based on SnO_2 or $In_2O_3:SnO_2$ films generally make adequate ohmic contacts with n-type semiconductors. Films of p-type semiconductors are more problematic, which is unfortunate, since most of the solid-state PV polycrystalline thin-film studies are for cells based on p-CuInSe$_2$ or p-CdTe. Substrates for p-CdTe have been predominantly CdS-coated conducting glass. The back ohmic contact to the ED CdTe has been made with evaporated Au or Ni [33,115] and also with evaporated Cu-doped p-ZnTe [35]. The use of CdHgTe, either as a buffer layer between the CdTe and substrate, or instead of CdTe, has been shown to facilitate ohmic contact formation because of the lower resistivity of the former [33,115]. In most cases the CdTe has been etched to form a Te-rich film at the contacting surface, which improved ohmic contact [116]. Conducting glass has often been used as a substrate since it allows optical transmission measurements

TABLE 1 Electroplated II–VI Semiconductors from 1987 to 1992 by Reference Number[a]

	1987	1988	1989	1990	1991	1992
CdTe	7–13	31–35	45–58	64–67	79–86	98–101
CdSe	14–22	36–39	(46), (55)	68, 69	(79), 87–92	102, 103
CdS	23–26	(36), 40–42	59	(69), 70–72	(79), 93	104–106
ZnS		(42)	60	73		
ZnSe	27,28		(46)			
ZnTe		43	(46), 61			107
Alloys	29 Cd(Se,Te)	44 (Cd, Zn)Se	62 Cd(Se,Te)	74 Cd(Se,Te)	94 Cd(Se,Te)	108 Cd(S,Se)
	30 Cd(Se,Te)	(33) (Cd,Hg)Te	(61) (Cd,Zn)Te	75 Cd(S, Se)	95 Cd(Se,Te)	109 (Cd,Zn)S
		(42)(Cd,Zn)S	63 (Cd,Hg)Te	76 (Cd,Zn)S	96 (Cd,Zn)Se	110(Cd,Zn)Se
				77 (Cd,Hg)Te (incl. HgTe)	97 (Cd,Zn)S (also Cd-Bi-S)	111(Cd,Zn)Se
				78 (Cd,Hg)Te (incl. HgTe)	(79)(Cd,Hg)Te	112(Cd,Hg)Te
						113(Cd,Hg)Se

[a]Numbers in parentheses refer to references that have already appeared in the table for a different semiconductor. A number of references deal with more than one semiconductor.

of the films to be made. Cowache et al. described a cathodic pretreatment of SnO_2–glass to form ca. 20 nm of elemental Sn on the SnO_2. When immersed in an aqueous solution containing TeO_2, this Sn layer converts to a Sn–Te compound, which improved the homogeneity of the subsequently ED n-CdTe film and resulted in good stoichiometry and adhesion of the CdTe as long as the Sn layer was not substantially thicker than 20 nm [49].

For PEC cells, where n-type semiconductors have been used predominantly, Ti is the most commonly used substrate. Ti was used as a substrate for the original preparation of ED CdSe photoanodes [117]. The reason for this was the chemical, and no less important, electrochemical inertness of Ti. Electrochemical inertness means that Ti is a very poor electrocatalyst for most electrochemical reactions. Another way of expressing this is that relatively large potentials in both directions can be imposed on a Ti electrode with little current flow. To illustrate the importance of this property, consider a CdSe film on (a) a Ti substrate and (b) a Pt substrate in a PEC. At the point of maximum power, the CdSe substrate potential will be at ca. -0.5 V with respect to the redox potential in solution. Any exposed substrate (e.g., pinholes in the CdSe film or uncovered parts of the substrate) can, in principle, give a cathodic reduction current, which will reduce the anodic photocurrent at that potential. The greater the electrocatalytic activity of the substrate, the greater the cathodic current, resulting in a lower net photoanodic current. Pt, which is, in general, a good electrocatalyst, will thus give a smaller photocurrent as the photovoltage increases (this is apart from normal considerations of fill factor), while Ti, a poor electrocatalyst, will not cause any noticeable reduction in photocurrent (assuming typical photocurrents in the mA cm^{-2} range) until a much more negative photovoltage has been reached. The electrochemical shorting of electrolyte to substrate parallels shorting of the second semiconductor layer (or metal layer in a Schottky cell) through pinholes to the substrate in a solid-state cell. The fact that pinholes do not necessarily cause shorting in a PEC if the substrate exhibits poor electrocatalytic activity is one reason for the reduced demands on quality of semiconductor films used in a PEC compared with a solid-state cell. For the same reason, Cr (as Cr-plated steel) was also used successfully for ED n-Cd chalcogenides [118], although Ti has continued to be the popular choice as substrate material (Cr is not so widely available as bulk sheet metal).

Good adhesion of the ED film to the substrate is another important property. There is often more art and experience rather than science in this aspect. Adherent films are usually obtained on Ti substrates. This may be explained, at least in part, by chemical reaction at the interface between semiconductor film and oxidized Ti surface; such a reaction was found only for adherent films of CdSe, while no such reaction was seen for nonadherent films [119].

Finally, although the properties of the films themselves are, in most cases, considered to be independent of the nature of the substrate (assuming a polycrystalline substrate), at least one work on ED CdTe has noted a major difference in the morphology of films deposited on Ti and on conducting glass [120], although no explanation was given for the difference. In both cases, the films were found to be amorphous by x-ray diffraction (XRD). (The term *amorphous*, using XRD, also includes very small crystallites or domains < ca. 2 nm.)

III. ELECTRODEPOSITION OF II–VI SEMICONDUCTORS BY REDUCTION OF XO_2 COMPOUNDS

This seems to be the earliest and most studied method for the ED of compound semiconductors. Gobrecht et al. [121] reported the deposition of CdSe films (using a Se anode) and Ag_2Se (using H_2SeO_3) and suggested that the mechanism proceeded by co-deposition of the elements. Pacauskas et al. [122] used an electrolyte containing H_2SeO_3 and $CdSO_4$ in dilute H_2SO_4 and cathodes of Cu or Pt. Hodes et al. [117] described photoelectrodes of CdSe deposited from a SeO_2-containing electrolyte onto Ti substrates (Ti was used for reasons described earlier). The as-deposited films were annealed to improve their photoactivity, which was very poor in the as-deposited form.

This method has been used, with great success, as evidenced by subsequent use of these films for photovoltaic cells, to deposit CdTe from solutions containing TeO_2 as reported by Panicker et al. [116] and Kroger [123]. They studied the properties of the deposited films as well as the mechanism of ED. Since unlike CdSe or CdS, CdTe can be obtained readily in either conductivity type, they found the conductivity type to be dependent on the deposition potential (Cd or Te rich) as well as on added impurities (In gave *n*-type while Cu gave *p*-type). The crystal size, estimated from XRD, was reported to be 50 to 100 nm and independent of deposition temperature; the XRD results could be interpreted in terms of an apparent increase in crystal size with increase in deposition temperature—from ca. 4 nm at 35°C to ca. 9 nm at 90°C. Our own (unpublished) XRD measurements on CdTe deposited by this method show crystal sizes to be no smaller than ca. 20 nm, even for deposition at 35°C.

They proposed a mechanism for the deposition based on initial deposition of Te followed by underpotential deposition (UPD) of Cd on the Te to form CdTe. This deposition occurs at a potential less cathodic than that needed to deposit Cd because of the gain in free energy due to compound formation between Cd and Te. This general mechanism has been verified by other groups [4,46,124–126] and is widely accepted to be correct, with some modifications and additional insights.

Mori and Rajeshwar [45] studied the nucleation of Te and CdTe on glassy carbon and found that the nucleation and growth of both materials are not fundamentally different (i.e., Cd^{2+} does not radically change the nucleation and growth pattern of Te other than to give CdTe, rather than Te, as the final product). They also postulated a two-dimensional growth mechanism with instantaneous nucleation. At Au electrodes, they found that the six-electron reduction of Te to H_2Te could also occur (this reaction did not occur on glassy carbon) with follow-up reactions—either chemical or electrochemical—to give Te [127].

Sella et al. [126] considered adsorption of $HTeO_2^+$ (the active Te species in solution) at the cathode to be a controlling step, not just diffusion of the $HTeO_2^+$ species, which had been believed to be the only rate-controlling step. They suggested that Cd^{2+} (or alkali metal cations) can displace adsorbed $HTeO_2^+$, explaining the shift of the Te deposition potential in the presence of Cd^{2+}. Dennison and Webster, however, found that this shift was caused by the sulfate anion rather by the cation, and suggested that it was due mainly to ion pairing between the sulfate and $HTeO_2^+$ ions, with a possible contribution from a pH change due to the added sulfate [101]. Deslouis et al. [57] carried out electrohydrodynamical impedance experiments and postulated that a slow surface process is involved in CdTe formation. They found this slow surface step to be consistent with adsorption of a Cd-containing species prior to its incorporation into the crystal network. Whatever the exact nature of this slow surface step, its consequence is that the surface concentration of $HTeO_2^+$ will be nonzero (unlike the case for pure Te deposition). The result is a decrease in the CdTe growth rate [57,126], which can explain why relatively smooth CdTe films were formed, contrasted with dendritic Te films (in the absence of Cd^{2+}, which acts as a leveling agent) characteristic of pure mass transport control [57]. Note, however, the possibility of a chemical reaction between $HTeO_2^+$ and deposited Cd to give Te [116]. This would occur preferentially at higher TeO_2 concentrations and possibly lower deposition temperatures (due to the slower solid-state reaction between Cd and Te).

The Cd^{2+} discharge has been treated both as a separate reaction with subsequent solid state reaction to give CdTe [56]:

$$Cd^{2+} + 2e^- \rightarrow Cd^0 \qquad Cd^0 + Te^0 \rightarrow CdTe \tag{1}$$

or as a one-step reaction with Te [46,126]:

$$Cd^{2+} + Te^0 + 2e^- \rightarrow CdTe \tag{2}$$

The UPD of Cd on Te implies that bonding between the two elements occurs upon discharge (although some further rearrangement of the bonding may occur subsequently), which lends support to reaction (2); reaction (1), assuming the formation of free Cd^0, would not be expected to lead to UPD of Cd apart from an initial tiny quantity as allowed by the Nernst equation (initial very low Cd/Cd^{2+} ratio), and which is therefore not correctly UPD, although it may appear to be so experimentally.

The fact that this technique for electroplating CdTe is usually carried out at relatively high temperatures probably reflects the kinetic limitations on the solid-state reaction between Cd and Te at low temperatures. In fact, Dennison and Webster found that at room temperature, the reaction between Cd and Te is incomplete [101].

To sum up, the overall mechanism for CdTe deposition entails deposition of Te with subsequent UPD of Cd on the deposited Te to form CdTe. Since the Te "substrate" is being continually formed anew, the UPD of Cd also occurs continually wherever Te has been deposited. The high concentration of Cd and low concentration of TeO_2 ensures that it is much more likely for Cd, rather than more Te, to be deposited on Te, therefore maintaining good stoichiometry of the films. The chemical bonding between alternate "layers" that is inherent to a UPD mechanism indicates a good crystallinity should be obtained. In fact, Stafsudd found that CdTe deposited onto the (111) and (iii) faces of single-crystal CdTe is epitaxial with the substrate (O. Stafsudd, private communication).

The mechanism of CdSe deposition seems from the literature to be different from that of CdTe, although Loizos et al. [87] have suggested that the mechanisms are similar. Mishra and Rajeshwar have compared the deposition mechanisms for CdSe and CdTe using cyclic photovoltammetry [4,46] and conclude, in contrast to earlier suggested mechanisms [128,129], that the predominant reaction is a direct six-electron reduction:

$$Cd^{2+} + H_2SeO_3 + 4H^+ \rightarrow CdSe + 3H_2O \tag{3}$$

The difference in morphology between CdSe and CdTe deposited from these solutions lends indirect support for a different mechanism in the case of the two reactions. CdTe films, which form by a controlled atom-by-atom deposition, are relatively smooth, while the CdSe films are often porous and cauliflower-like. However, as noted by Tomkiewicz et al. [129], the porous CdSe layer is deposited on an initial thin compact layer. They explained this by a reduction of Cd^{2+} to Cd followed by chemical reaction between Cd^0 and H_2SeO_3 to give CdSe, which forms the compact layer, and a chemical reaction between electrogenerated selenide and Cd^{2+} to form the porous layer. Rajeshwar has pointed out [4] that the formation of CdSe by the initial deposition of Cd^0 would be ruled out since the Cd^{2+}-induced reduction wave in the cyclic voltammogram of Cd^{2+}–SeO_2 solutions occurs at a more positive potential than that of Cd^{2+} or SeO_2 alone. (In contrast, note that for CdTe, addition of Cd^{2+} shifts the TeO_2 wave more negative, again indicating a possible difference in deposition mechanism.) However, the work of Tomkiewicz et al. does suggest that two different mechanisms may be applicable to the CdSe deposition corresponding to the initial (compact layer) and bulk (porous layer) structures. It is tempting to assume that the mechanism of the compact layer formation is similar to that of CdTe (although the shift of the TeO_2 and SeO_2 reduction waves in opposite directions on addition of Cd^{2+} suggests against this), whereas the porous layer is due to the reaction between selenide and Cd^{2+} [128,129], which would be expected to lead to such a morphology.

There is a strong similarity between the I–V characteristics for CdTe deposition (fig. 5, Ref. 116) and those for CdSe when the SeO_2 concentration is low (fig. 1, Ref. 87). In particular, for both cases the plateau region, where optimum deposition occurs, becomes narrower as the $Se(Te)O_2$ concentration increases. The relatively smooth morphology and appreciable PEC effects of the as-deposited CdSe films from solutions of low SeO_2 concentration [87] suggest a change in the predominant mechanism for CdSe deposition depending on SeO_2 concentration and again lend support for the same basic mechanism in the Se and Te systems when the SeO_2 and TeO_2 concentrations are similar.

It should be noted that CdSe deposition is usually carried out from baths of lower pH (<1) than CdTe (ca. 2). It has been reported that CdSe does not form from Cd^{2+}–SeO_2 solutions at pH > 1 (for high SeO_2 concentrations under stirring), but only gray, hexagonal Se [56], although, in another study, red amorphous Se mixed with CdSe has been found at pH 2.2, under what appear to be similar conditions [87].

To conclude the mechanistic studies of CdSe deposition, it must be kept in mind that the initially deposited layers may be substrate dependent, as has been shown by the work of Skyllas-Kazacos and Miller [128], and this should be considered when comparing otherwise similar experiments carried out on different substrates. Most mechanistic studies have been carried out on the initial deposited layer on various substrates, and the results of these studies need not necessarily be applicable to the bulk of the deposited films. What is still needed are more mechanistic studies carried out on predeposited CdSe. In view of the importance of the concentration of SeO_2 (stirring increases the effective concentration of SeO_2 at the cathode), another important point to consider when comparing results is whether or not stirring was employed.

Pandey et al. [91,92] have compared a singly deposited film of CdSe with one deposited in three separate platings with an anneal and etch step between each plating. They report that the three-step method results in less porous films with a larger crystal size and with better PEC behavior. They explain the improvement by (a) removal of excess Se due to the annealing and (b) deposition of CdSe in the pores of the previously deposited film, thereby reducing the porosity. They note that the etch step—where an anodic bias is imposed on the film before each redeposition—is essential to obtain good PEC behavior.

The XO_2 bath can also be used for the deposition of CdS using an aqueous solution of SO_2 (H_2SO_3), although this appears to have been mentioned only once and briefly in the literature, where cubic CdS films were obtained using an acidic Na_2SO_3-based electrolyte [130].

Relatively little work has been done on the corresponding Zn–chalcogenides. Sing and Rai [27,28] have deposited ZnSe from electrolytes containing SeO_2 and a zinc salt. The mechanism of deposition of ZnSe and ZnTe has been studied by Mishra and Rajeshwar using cyclic photovoltammetry [4,46], who find that the mechanism is similar to that for the corresponding Cd–chalcogenide. Mondal et al. [107] have deposited ZnTe onto CdTe–CdS–ITO–glass as a contact for PV cells. They used a higher pH than usual (3 to 4) and the Te was added as Na_2TeO_3 (TeO_2 in NaOH). From XRD measurements they obtained strongly textured {111} films with crystal size—estimated from peak widths— of at least several tens of nanometers. At pH values below 3, free Te was deposited. The Te concentration was maintained very low (<1 mM) to prevent preferential deposition of Te even at a higher pH. Cu doping of the ZnTe was carried out by codeposition of Cu^{2+} ions complexed with triethanolamine. Other than XRD, no other characterization of the ZnTe film itself was described; PV cell parameters of the overall cell were given.

Of particular interest is the finding from the first two groups above that the ED ZnSe is *p*-type, in contrast to ZnSe formed by other methods, which is usually *n*-type. The large-bandgap II–VI semiconductors (of which ZnSe is an example) normally exhibit only one conductivity type because of self-compensation. This fundamental difference between ED ZnSe and other forms of the semiconductor is of interest, not only for reasons of basic science, but also because the existence of a large-bandgap *p*-type semiconductor for use as a window material in heterojunction photovoltaic cells would allow the use of *n*-type absorber materials (up to now, these are usually *p*-type) with the advantages of better control over doping and, in particular, better and more easily formed ohmic contacts using *n*- rather than *p*-type absorbers.

There have been some recent adaptations of this basic electrodeposition method. Gregory et al. [80,86] have described a method they term *electrochemical atomic layer epitaxy* (ECALE) for CdTe. This method is based on the UPD of *both* elements of a binary semiconductor. Since reductive UPD of Te from TeO_2 would occur only at potentials where Cd would redissolve from the film, oxidative UPD of telluride ion, formed by reduction of TeO_2 at sufficiently negative potentials, was employed. Each elemental layer was deposited from a solution containing only that element. Thus Cd was UPD from a Cd^{2+} solution (depositing up to a maximum of a monolayer of Cd) followed by rinsing and then UPD Te from a TeO_2-containing solution. This method is reminiscent of the (nonelectrochemical) method of successive ion-layer adsorption described by Nicolau et al. [131], with additional potential control possible due to the UPD mechanism. In both cases, three dips in different solutions are required per (approximately) monolayer of semiconductor formed (the intermediate rinse step is considered as a dip step). This makes these techniques very slow for anything other than very thin films, but has the advantage in principle of imparting very fine control on the molecular level.

Another variation of this plating bath—sequential monolayer electrodeposition—has been described by Kressin et al. [88]. They reasoned that the excess Se, which is usually deposited with CdSe from SeO_2 baths, could be eliminated by depositing a submonolayer with a large excess of Cd and then anodically redissolving the excess Cd. The deposition was carried out by continually scanning the potential and by using sufficiently low concentrations of SeO_2 (0.3 mM) that submonolayer amounts of Se were formed per cycle. The resulting CdSe was found to be stoichiometric with no Se excess.

These results should be compared with those of Loizos et al. [87], who also obtained stoichiometric CdSe if the SeO_2 concentration was low enough (< ca. 1 mM), with the advantage in the former case that potential control is less critical. For higher SeO_2 concentrations (3 mM), the cyclic sweep method gave more stoichiometric films than those formed at constant potential.

An interesting difference between the films of Kressin et al. and most other films deposited from SeO_2 solutions is apparent from XRD measurements: the former are a mixture of hexagonal and cubic CdSe, while as-deposited CdSe films from this electrolyte are usually found to be only cubic. This difference cannot be explained by a substrate effect (the substrate used was Ti, which has commonly been used in many other studies), and it would be of interest to study it further. The crystal size estimated from peak broadening—assuming broadening due only to crystal size—is ca. 30 nm for the cubic crystals and considerably larger for the hexagonal ones. Also, the XRD results show a texturing of the as-deposited films exposing the {002} plane, which is lost after annealing.

Superlattices of CdSe–ZnSe have been electrodeposited using two different strategies. In one, a high concentration of Zn^{2+}, together with a low concentration of Cd^{2+}, was employed together with potential modulation in order to overcome the preferential deposition of CdSe over ZnSe [111]. While alloys of (Cd,Zn)Se could be deposited by this method, only partial modulation was obtained, with both Cd and Zn present in all parts of the film, but to varying extents. A more successful method employed a flow technique whereby first ZnSe was deposited from a bath containing Zn^{2+} and SeO_2 (no Cd^{2+}), the substrate was rinsed with a solution of H_2SO_4, then CdSe was deposited from a Zn-free bath [110]. XPS showed that the composition changed from almost totally ZnSe to almost totally CdSe over a period of ca. 10 nm.

There has been considerable recent activity in deposition of alloys of II–VI semiconductors, as is evident from Table I (''alloy'' is used here instead of the more accurate term ''solid solution'' for simplicity). The XO_2 method has been used predominantly for depositing either (Hg,Cd)Te (MCT—mercury cadmium telluride) for use in CdS/Cd(Hg)Te heterojunction cells, or Cd(Se,Te) for PEC cells.

The use of MCT to reduce the contact resistance to p-CdTe cells was described in Section II [33,115]. Mori et al. [77] studied the mechanism of MCT (as well as HgTe) deposition from TeO_2 baths. They conclude that for HgTe, the mechanism involves discharge of $HTeO_2^+$ species at elemental Hg and, at sufficiently cathodic potentials where H_2Te can form, chemical reaction between electrogenerated H_2Te and Hg^{2+} ions. They observed anodic stripping of Te and suggested that this implies slow reaction between Hg and Te—slower than for the CdTe case. Using a combination of optical transmission and Auger elemental analysis, they found the ternary MCT films to be nonhomogeneous, with CdTe, HgTe, and MCT all present, although without any free elemental Cd, Hg, or Te.

Neumann-Spallart el al. [63] studied ED films of MCT on Ti and SnO_2–glass substrates with emphasis on characterization of the films. They showed a relation between the Hg/Cd ratio in the films and Hg concentration in the electrolyte. They also provided information on the morphology of the films (relatively Hg-rich films are less smooth than those with low Hg concentrations), preferential texturing as a function of Hg content and resistivity (ca. $10^5 \Omega \cdot cm$ without any apparent strong dependence on Hg content in the range of compositions studied by them).

Loizos et al. [56] studied the mechanism of Cd(Se,Te) deposition and obtained alloys (at pH 2.2; note the earlier discussion concerning lack of CdSe deposition at this pH) only at low SeO_2 concentrations (<3 mM). Under these conditions of low SeO_2 concentrations, they proposed that the deposition mechanism is similar to that for CdTe (i.e., underpotential deposition of Cd on predeposited Te or Se to give the compound). This should be compared with previous work which showed that Cd(Se,Te) alloys could be obtained at higher SeO_2 concentrations (20 mM), although these experiments were carried out at lower pH and without stirring [130].

Gonzales-Velasco and Rodriguez [75] deposited Cd(S,Se) from an acid bath containing both selenite and thiosulfate as sources of Se and S. As explained in Section IV it is likely that the use of thiosulfate in acid conditions is equivalent to using sulfite (i.e., SO_2 in aqueous solution). That paper concentrated on PEC characterization of the films of different composition.

Pandey et al. [44] have reported on CdSe films treated in a solution of $ZnSO_4$ to give (Cd,Zn)Se. AES was used to show the presence of Zn in the films, but no XRD characterization was given to support the presence of a solid solution. It is not unlikely that the measured Zn was present as occluded $ZnSO_4$ from the solution.

Very little has been reported on the XO_2-based bath in alkaline solutions. Hodes et al. [130] described Cd(Se,Te) photoanodes plated from an alkaline bath containing SeO_2, TeO_2, and cyanide. The main difference between these films and corresponding ones plated from an acid bath (in both cases, after annealing) was a several-times-smaller grain size (ca. 1 μm) for the alkaline-plated films, which could explain the lower PEC conversion efficiencies obtained with those films compared with the acid ones. The alkaline-plated films were more compact (again after annealing—no comparison was made of the as-deposited films) than the acid-plated ones. Sella et al. [126] noted that the CdTe plated from alkaline TeO_2 baths was powdery. One difference in the two baths above was the use of a surfactant in the former, which might act as a leveling agent, similar to the effect of Cd^{2+} in acidic TeO_2 solutions [126]. It appears that the success of the acidic bath has resulted in a lack of interest in the alternative alkaline bath, and it would certainly be useful to have more studies on this system.

We conclude with some data on defect levels in these films. Basol and Stafsudd [132] described a relatively deep trap level (0.56 eV below E_c) measured on Schottky contacts to as-deposited n-type CdTe. So et al. [10] used admittance spectroscopy and photocapacitance measurements on CdTe Schottky devices (which had been annealed at 250 or 300 °C). They found two donor levels (E_a = 0.12 eV and 0.19 to 0.26 eV) and possibly another one or two more shallow donor levels. Pandey et al. [21] used space-charge-limited current behavior in Schottky contacts to CdSe (annealed in air at 340°C) to determine trap depth and density in the films. They measured traps 0.17 to 0.20 eV below E_c with a density of ca. 7×10^{15} cm^{-3} and attributed them to native defects or defect-impurity complexes.

Gutiérrez and Salvador [20] used photocurrent spectroscopy of CdSe photoanodes to study the effective doping density (N_D) and diffusion length (L_P) in ED CdSe films as a function both of annealing in a small concentration of O_2 and various etching treatments. They presented a model of neutralization of Se vacancies (V_{Se}) at grain boundaries which lowers N_D. Without O_2 in the annealing atmosphere, N_D tends to increase with increase in temperature due to increase in V_{Se}, giving a value of N_D that is too high for optimal photovoltaic purposes. This is in accord with the extensive literature on the effect of annealing polycrystalline CdTe and $CuInSe_2$-based photovoltaic cells in air to decrease V_{Se} and even change the conductivity type [133].

IV. ELECTRODEPOSITION FROM AQUEOUS ELECTROLYTES OF ZERO-VALENT CHALCOGEN COMPOUNDS

Almost all the reported work in this section involves deposition from thiosulfate ($S_2O_3^{2-}$) or selenosulfate ($SeSO_3^{2-}$) solutions. Since there is no corresponding Te compound, this method cannot be used for depositing tellurides.

McCann and Skyllas-Kazacos [134] and Power et al. [135] described the deposition of CdS from $Na_2S_2O_3$-containing solutions. The former used alkaline solutions in which the Cd^{2+} was complexed with ammonia–EDTA. Within a limited range of potentials and ammonia concentrations, CdS was deposited. The mechanism appeared to involve formation of sulfide ions as an intermediate. PEC activity of the films was described. Power et al. used a noncomplexed Cd^{2+} solution (actually, Cd^{2+} forms a weak complex with thiosulfate). At the natural pH of the $CdSO_4/Na_2S_2O_3$ solution (6.7), only Cd metal was deposited. At lower values of pH (2.8), where thiosulfate decomposed according to the reaction

$$S_2O_3^{2-} \longrightarrow S \text{ (colloidal)} + SO_3^{2-} \tag{4}$$

films of CdS were obtained. They carried out the deposition under conditions of continual potential cycling, where excess Cd could be redissolved anodically at low enough concentrations of Cd^{2+}, no excess Cd was formed, but the reaction to form CdS was a cathodic one, and at low concentrations of Cd^{2+}, cycling was presumably not essential.

Fatas et al. [25] and Morris et al. [105] studied the deposition of CdS using pulse electroplating. The former used an ammonium salt to complex the Cd^{2+} (and also glycerol to give more uniform films), while the latter worked in the absence of a complexant. Both used acidic solutions where the decomposition of thiosulfate occurred. The conditions under which CdS formed were very different in the two cases. In the former, the onset of cathodic current was ca. 0.5 V negative of the onset measured by the latter, probably due to complexation by the ammonium salt. Cd was deposited during the cathodic pulse and CdS formed during the anodic pulse in the former case, while CdS—with some Cd^0— formed during the cathodic pulse and the Cd^0 was anodically redissolved during the anodic pulse in the latter case (similar to the results reported in Ref. 135). The mechanism suggested by Fatas et al. was reduction of Cd^{2+} to Cd^0 during the cathodic pulse and anodization of the Cd metal by sulfide, formed during the cathodic cycle by reduction of colloidal S, during the anodic cycle. Morris et al. assumed, on the basis of the results of Power et al. [135], that the reaction proceeded by reduction of Cd^{2+} and colloidal S. From consideration of the disproportionation reaction of thiosulfate into sulfur and sulfite, it seems that reduction of sulfite is more probable than that of S in the colloidal state. If this is indeed the case, this method is essentially the same as the XO_2-based method, since sulfite is the anion of sulfurous acid–SO_2 in water.

The properties of the CdS films described in Ref. 105 and those prepared in a similar way by Jayachandran et al. but with dc plating [59] are for the most part similar and differ considerably from those described in Ref. 25. The former give compact films of hexagonal CdS (the XRD results of Ref. 59 could be explained by either a partially textured hexagonal structure or a less-textured cubic one, while Ref. 105 shows a nontextured hexagonal structure) of crystal size ca. 60 nm. Reference 25 describes much more porous films (as might be anticipated since the reaction forming CdS is really anodization) of cubic non-texture CdS with a crystal size estimated from the XRD results of the same order as given in the other two papers (with the possibility of much small crystallites mixed with the larger ones, as the shape of the XRD peaks suggests). Optical spectra of the films (on conducting glass) gave values of E_g that were all higher—to a greater or lesser extent—than the normal value for CdS [2.48, 2.53 (2.42), and 2.63 (2.43) eV for Refs. 59, 105, and 25, respectively—the parenthetical figures represent the values after annealing]. Such increased values of E_g, and their reduction to normal values after annealing, are characteristic of quantum size effects due to very small crystal size (discussed in more detail in Section V). However, this explanation seems unlikely here, at least for the films described in Refs. 105 and 59, since the measured crystal size is an order of magnitude larger than would be required for such an effect to occur. Reference 59 also describes films deposited on Ti and Al (all the other films were on conducting glass—either SnO_2 or ITO). There were some differences in morphology for the different substrates as well as different XRD patterns (the metals gave weak, nontextured hexagonal patterns but, as with the SnO_2–substrate, moderately sharp peaks). Some electrical parameters such as N_D, mobility, resistivity and trap density were also reported.

Skyllas-Kazacos and Miller first used a selenosulfite-based electrolyte to deposit CdSe [136]. Their reasoning was based on the expectation that excess Se, which so often was found in films of CdSe deposited from SeO_2 solutions, would not be likely to occur in films deposited from a zero-valent Se compound, because there would be no high-valency Se species present that could oxidize electrogenerated selenide ion. In addition, any free Se that may form would redissolve in the excess sulfite (selenosulfite, $SeSO_3^{2-}$, is prepared by dissolving Se in excess sulfite). $SeSO_3^{2-}$ is reduced to selenide ion:

$$SeSO_3^{2-} + 2e^- \longrightarrow Se^{2-} + SO_3^{2-} \tag{5}$$

which can then react with Cd^{2+}. Using an EDTA–ammonia complex, the Cd-reduction potential was kept more negative than that needed to reduce the selenosulfate. In the as-deposited state, the films were moderately active as photoanodes and obviously better than as-deposited films from the SeO_2 bath. In addition, the films could be deposited under much less critical conditions of concentrations and mass transfer than with the SeO_2 bath.

Cocivera et al. [137] described a variation of this method using nitrilotriacetic acid (NTA) as a complexant instead of EDTA–ammonia, and at a lower pH. They obtained smooth, stoichiometric films of CdSe in contrast to the porous, nonstoichiometric films often obtained from the SeO_2 baths. In a follow-up study of the deposition mechanism [138], they at first believed that the mechanism involved either reduction of a Cd-selenosulfate complex directly to CdSe or reduction of Cd^{2+} to Cd^0 followed by chemical reaction of the latter with selenosulfate. In a more recent mechanistic study [38], they concluded that the reaction occurred by reduction of selenosulfate to selenide ion followed by chemical reaction with Cd^{2+} to form CdSe. Although selenosulfate reduction in the absence of Cd occurred at more negative potentials than CdSe deposition, the mechanism above was explained by a positive shift of the selenosulfate reduction due to the follow-up chemical reaction. The selenosulfate apparently had an inhibiting effect on the reduction of Cd ions to Cd metal, explained by adsorption of a selenium species on the electrode (i.e., poisoning of the electrode for Cd deposition).

XRD analyses of the films described by Szabo and Cocivera [138] indicated that they were either amorphous or comprised of very small crystallites. Annealing at 480°C resulted in a hexagonal XRD pattern with preferential basal-plane texturing. SEM studies showed that relatively smooth films were obtained which were subject to cracking after annealing. The as-deposited films gave <1% conversion efficiency in a PEC that increased to ca. 5 to 7% after annealing. Fantini et al. [37] and Wynands and Cocivera [89] also found very broad XRD peaks characteristic of amorphous or very small crystalline material as did Cerdeira et al. [39]. The latter calculated a crystallite size of 5.6 nm based on XRD peak broadening due to small crystal size. They compared ED CdSe (from a selenosulfate bath) with chemically deposited films (CD CdSe) deposited from essentially the same bath. It had been shown that the CD CdSe films exhibit quantum-size effects (QSEs) due to the very small size of the crystallites in the films (<10 nm), while the ED CdSe films did not [139]. Cerdeira et al. concluded that the lack of QSEs in the ED CdSe was due to better electrical connection between the crystallites, since the resistivity of the ED films was much less than that of the CD films. We have also found very broad XRD peaks for the as-deposited CdSe but attributed this to strain in the crystals [139]. The reason for this is that TEM studies of the films show the crystallites in them to be in the range of tens to hundreds of nanometers (which would be too large to exhibit QSEs). In addition, the XRD peak widths are strongly dependent on the angle measured (i.e., on the particular reflection), which suggests that strain, rather than crystal size, is the cause of the peak broadening, although this could also be due to a crystal shape that is not symmetric in three dimensions. A further property of the as-deposited CdSe films, which could be explained by strain, among other possibilities, is shown by the optical transmission spectra. A more gradual transmission (absorption) onset is found than expected for a direct-bandgap semiconductor [37,139], which sharpens to a normal onset after annealing [37].

Decker et al. [18] studied the morphology of growing CdSe films deposited from both the SeO_2 and selenosulfate baths onto Si coated with a thin Ti film. The former bath nucleated by smooth, hemispherical, isolated islands of a few hundred nanometers which coalesced into irregular, globular structures of several micrometers, giving films with a porous structure. The latter, in contrast, produced smaller (ca. 100 nm) isolated nucleii with occasional large (a few micrometers) aggregates of crystals (typically, ca. 0.5 μm) which coalesced into relatively smooth, nonporous films as they grew thicker. This difference in growth modes suggests that in the case of the SeO_2-based solution, CdSe prefers to grow on CdSe than on the substrate, while this is not obviously the case for the selenosulfate bath.

Decker et al. [14,16,18] have exploited the relatively smooth nature of the selenosulfate-deposited CdSe to measure optical reflectance of the films as they grow. From such experiments they were able to calculate the film thickness in situ, faradaic efficiency of the deposition, and refractive index as a function of wavelength of the light.

Besides the earlier annealing studies mentioned above [138], Szabo and Cocivera [19] studied the effects of various annealing ambients on N_D and L_P measured by photocurrent spectroscopy. As is so often the case for the II–VI films, the optimum ambient (highest L_P value) was found to be air, and they suggested several models to explain this, the most likely being filling of V_{Se} by oxygen, leading to a reduction in N_D and a wider space-charge layer. Cerdeira et al. [39] also studied the effect of annealing on their CdSe films. They found a gradual sharpening of the photoluminescence (PL) and XRD peaks with increasing annealing temperature as well as a change of crystal structure from cubic to predominantly hexagonal between 300 and 450°C.

Wynands and Cocivera [89] studied adhesion of ED CdSe by depositing the films onto ITO/glass which had been pretreated with Se, Cd, or In either by ED of a thin layer of the element onto the ITO or by thermal treatment in the elemental vapor. While the films on the untreated ITO contained cracks and flaked off after annealing, the films on the treated ITO (irrespective of which treatment) were crack-free and adherent. Heating the ITO in Se vapor was shown to cause partial conversion of the ITO into In and Sn selenides. Roughening of the surface occurred, which in itself would cause better adhesion. They also found that the treated substrate changed the doping characteristics of the CdSe [102]. Using Hall effect and resistivity measurements, they studied the films after annealing at 500°C in N_2 and measured values of N_D from 10^9 to 10^{19} cm^{-3}, resistivities from 1 to 10^8 Ω·cm, and mobilities in the range 1 to 50 cm^2/V·s. By increasing the amount of Cd or In electrodeposited on the ITO, the value of N_D increased and the resistivity decreased, due to doping by diffusion of these donors.

There have been several reports on deposition of ZnS from both alkaline [60] and acid [42,73] baths. Lokhande et al. [60] used an alkaline bath with the Zn complexed by EDTA. Their voltammetric data show that ZnS is formed at potentials considerably less cathodic than required for Zn deposition. This suggests that the deposition occurs either by UPD of Zn on predeposited S or by a pathway involving both Zn^{2+} and thiosulfate concurrently. At the potentials employed, it is unlikely that sulfide ion is involved. The XRD of these films indicated that no elemental Zn or S was co-deposited and that the grain size was fine (no XRD pattern was shown, so it is unclear just what the crystal size was). Lokhande et al. also deposited ZnS films from an acid bath [42]. In this case the free elements were co-deposited with ZnS. Also, the measured value of E_g (from transmission spectra) of the acid-plated ZnS was lower (3.55 eV) than that from the alkaline bath (3.70 eV, which is closer to the normal literature values). The films were smooth (on a 500× magnification scale) and adhered well to the different substrates used (Ti, steel, SnO_2–glass). Using a similar acid bath, Sanders [73] found that the films on SnO_2–glass were (200) textured. As he pointed out, this is somewhat unusual since if there is texturing in II–VI films, it is usually with the {111}—or equivalent {0001} for the hexagonal modification—planes exposed. From the considerable number of other studies on SnO_2–glass substrates, there seems to be no reason to expect an orienting effect from the substrate. The XRD pattern also indicated a crystal size of at least some tens of nanometers for these films.

Ternary alloys—mainly (Cd,Zn)S—have been deposited by this technique. Lokhande et al. [42] used acidic solutions to deposit CdZn sulfides. XRD showed mixtures of different phases—CdS, ZnS, Cd, Zn, and S—without any clear evidence for solid solution formation. Optical spectra were interpreted to give values of E_g that varied from 2.4 eV (CdS) to 3.55 eV (ZnS), depending on the composition. Jayachandran et al. [76] reported solid solutions of (Cd,Zn)S of hexagonal structure and with grain size of the alloys smaller than that of pure CdS.

Morris and Vanderveen [109] were unable to reproduce the results of Ref. 42. By using low concentrations of Cd^{2+} and high concentrations of Zn^{2+}, they obtained films of (Cd,Zn)S whose compositions were sensitive to plating parameters. The films were composed of hexagonal crystals with a grain size—measured from XRD—of ca. 160 nm, which increased to ca. 200 nm after annealing. This grain size was apparently insensitive to Zn content up the maximum Zn content studied ($Cd_{0.53}Zn_{0.47}S$). The films were adherent, clear, and moderately smooth—they appeared somewhat porous from the SEM data. The resistivity of the films increased with Zn content. Of interest is the fact that annealing reduced the resistivity of CdS by about three orders of magnitude but did not affect the Zn-containing films. E_g measured from optical spectra were consistently about 0.1 eV higher for the as-deposited films compared to the annealed ones for all compositions studied. Some PV cell characterizations were carried out for the films deposited on p-CdTe.

Darkowski and Grabowski [96] studied the deposition of (Cd,Zn)Se from selenosulfate solutions. They suggested a mechanism whereby selenosulfate is reduced to sulfide, which then reacts chemically with Cd^{2+} and Zn^{2+}. Complexation of the Cd with EDTA and Zn with ammonia was used—if only ammonia itself was used, nonstoichiometric films resulted. The films were in most cases

Se rich with increased lattice spacing for the alloys compared with pure CdSe, contrary to what would be expected. It was suggested that this may be due to Se excess or Zn interstitials. Some PEC properties of the films were reported.

A different method involving zero-valent Se and Te was reported by Skyllas-Kazacos [140]. The chalcogen was present as $SeCN^-$ or (presumably) $TeCN^-$. No free Se or Te could be deposited in the potential range studied, and the mechanism was presumed to proceed via reduction of the Se(Te)CN to Se(Te) ion. Mixed CdSe–CdTe films were deposited, but XRD showed these films to be mixtures of the two binaries and not a ternary. The XRD peaks were broad, corresponding to a crystal size of less than 10 nm. Some PEC results using these films were given.

V. DEPOSITION FROM NONAQUEOUS SOLVENTS CONTAINING ELEMENTAL CHALCOGEN

This method was first reported by Baranski and Fawcett [141]. It is based on cathodic deposition from organic solvents containing dissolved metal ions and elemental S or Se. Films of CdS, CdSe, HgS, PbS, Tl_2S, Bi_2S_3, Cu_2S, NiS, and CoS were reported. At that time the deposition was believed to proceed by deposition of a layer of metal followed by chemical reaction with dissolved chalcogen.

In subsequent studies of the mechanism of the deposition of CdS [142–144], it was suggested that the deposition proceeded by two mechanisms: the first on bare metal surfaces and a different one on CdS-covered surfaces. The deposition on bare metal (Au,Pt,Hg) was seen as a prepeak corresponding to approximately a monolayer. This was interpreted either as reduction of a metal sulfide—formed by chemisorption of sulfur on the substrate—together with Cd^{2+} to give CdS [142,143] or underpotential deposition of Cd on the bare metal [144]. Since, as was shown in Ref. 142 and as would be expected from chemical intuition, sulfur is chemisorbed on the surface of these metals, the UPD of Cd would be on a film of chemisorbed sulfur, resulting in a monolayer of CdS that is bonded to the substrate (there is probably some rearrangement of the substrate–sulfur bonding). A possibly important difference in the experimental conditions of the study of Baranski et al. [142] and that of Roe et al. [144] is that the former did not observe a peak due to reduction of metal sulfide in DMSO if the sulfur concentration was low (ca. 1 mM), while the latter worked (and observed a peak) under conditions where the sulfur concentration was less than 1 mM. Although this may appear to be a discrepancy between the two studies, there are other differences in the conditions which may be responsible— in particular, the different anions used in the two studies [$Cd(CH_3SO_3)_2$ in Ref. 142 and $CdCl_2$ in Ref. 144]. As described below, the anion has an important role in determining the nature of the CdS films. The deposition of ZnS and PbS onto bare metal surfaces was also found to proceed by the same mechanism as that of CdS [142].

According to the studies above, the mechanism of CdS deposition on substrates that have already been covered with CdS (i.e., the vast bulk of the film) switches to reduction of sulfur to sulfide, followed by chemical reaction between sulfide and cadmium ions. This mechanism seems to be the case for deposition from a diethylene glycol (DEG) solution containing $Cd(ClO_4)_2$, where Tafel plots for sulfur reduction are independent of Cd^{2+} concentration, even if this is zero, over a range of current densities [143]. It is less obvious—although still possible—in the other cases studied. Thus Baranski et al. [142] note that they do not observe reduction of sulfur on CdS-covered electrodes from DMSO solutions containing $Cd(CH_3SO_3)_2$; for the mechanism above to be valid in this case, the sulfur reduction would have to be catalyzed by Cd^{2+}. The results of Roe et al. [144] show that the onset of CdS deposition on CdS occurs at about the same potential as does sulfur reduction for very small currents (which is in support of the mechanism above) but seems to follow more closely the voltammetric characteristics of Cd deposition at higher currents, suggesting a mechanism whereby deposited Cd^0 reacts with sulfur (either from solution or at the surface), although this does not necessarily disprove the sulfur-reduction mechanism. The observation by Baranski and Fawcett [143] that in the presence of moderate concentrations of Cl^- ion, Cd^0 is deposited along with CdS (at least from DEG solutions), supports a mechanism involving deposition of Cd metal as a first step under these conditions.

The fact that Cd^0 was not co-deposited with CdS if the Cl^- (or Br^-) concentration was either very low or high was explained by assuming that adsorbed sulfur blocked the adsorption of Cd^{2+} [143]. At low $[Cl^-]$, complete coverage of the surface with sulfur was assumed. At intermediate values of $[Cl^-]$, Cl^- could be co-adsorbed at the surface, allowing bridging adsorption of Cd^{2+}. At high $[Cl^-]$, the Cd^{2+} reduction potential was shifted sufficiently negative due to complex formation so that no free Cd could be formed. In this and other papers by this group [145,146], other anion effects were reported. Of particular note was the fact that films deposited from Cl^- solutions were either highly (from DMSO solutions on Pt) or moderately (from DEG solutions on Mo) textured, with only the {002} peak seen in the XRD pattern of the former and a mixture of the {100} and {002} peaks in the latter. No preferential texturing was seen in the films deposited from perchlorate solutions. In all cases, the hexagonal structure was obtained (this was not obvious from the results of the DMSO–Pt experiments [145], where only one peak was seen, but can be inferred from the other results). Grain size could not be reliably measured from the XRD spectra because of the apparently large stresses in the Cl^--bath films, apparently due to inclusion of solvent into these films. However, XRD peak widths of Cl^--bath-deposited films decreased on annealing at 200°C, while those from perchlorate baths did not. The grain size of the these films is discussed further below. Additives such as water and thiourea were found to act as leveling agents, giving smoother (but often less adherent) films [144]. Tl doping was found to decrease the resistivity of the films [145]. Some electrical properties of the films, including photoconductivity, were measured [146]. Of particular interest were the very long decay times found for the photoconductivity. This was ascribed to a high density of trapping centers.

Mondon [147] studied the reduction of S_8 in DMSO and reaction of the electrogenerated polysulfide species with Cd^{2+} by optical absorption spectroscopy. He concluded that the two-step reduction of S_8 resulted in S_8^{2-} at the first step and S_4^{2-} at the second step. These ions can dismutate and react further to give other polysulfide species. Since the different polysulfide species absorb in different regions of the spectrum, their disappearance on addition of Cd^{2+} could be monitored spectroscopically. Of particular note was the observation that a certain concentration of Cd^{2+} would discolor a solution containing twice that concentration of polysulfide ion, which was interpreted as formation of a soluble cadmium polysulfide complex of general composition $[Cd(S_x)_2]^{2-}$, where x could be 4, 6, or 8. The stability of these complexes increased as x became smaller. This result suggests that the mechanism of CdS formation may be further complicated by intermediate formation of these complexes, although such complexes might form only under conditions of relatively high local polysulfide concentrations (i.e., at high current densities and low free Cd^{2+} concentrations).

Balakrishnan and Rastogi studied the deposition of CdS from DMSO–$CdCl_2$–S solutions as a function of substrate, Cd^{2+} concentration, and plating current density [40,72]. In contrast to the results of Ref. 143, they reported that the stoichiometry of the CdS does not vary appreciably over a range of different deposition conditions. Based on this observation, they concluded that the mechanism did not proceed by deposition of Cd metal followed by reaction with sulfide ions. They reasoned that if the foregoing mechanism were valid, free Cd would be found in the CdS films since they found the sulfurization of free Cd to be much slower than the deposition of Cd based on a separate experiment (which was not described). There is also the question of whether deposited Cd would react with sulfur or with sulfide ion. The authors present optical spectroscopic data of the electrodeposition solutions after various periods of deposition which indicate the presence of various polysulfide ions in the solution. This is unexpected since free (poly)sulfide ions in solution should react rapidly with the relatively large concentrations of Cd^{2+}. It may be that the concentrations of both (poly)sulfide and free Cd^{2+} are low (the latter due to complexation by DMSO and Cl^-), preventing bulk precipitation of CdS in the solution. It should be noted that the chemical reaction between in situ–formed Cd and sulfide (or dissolved sulfur) will probably be considerably faster than the same reaction with a relatively thick Cd film, so in the absence of details of the foregoing experiment on sulfurization of free Cd, chemical reaction between electrodeposited Cd and dissolved sulfur (or—less likely—sulfide ions) should not be ruled out.

The study cited above also showed a strong dependence of the crystallographic texturing of the CdS films on the experimental parameters. An increase in either Cd^{2+} concentration or plating current density resulted in films that were more preferentially textured (on SnO_2–glass or steel substrates; on Cd–Cr–glass, the films were {101} textured even at lower currents or Cd^{2+} concentrations). The effect of current density appears to be very strong. Thus a film deposited on steel at 2 mA/cm^2 showed a mixture of {101}, {110}, and {200} reflections, while when the current was increased to 3 mA/cm^2, the film was predominantly {002} textured. These results were explained in a general way by changes in the adsorption of the various ionic and atomic species on the growing film under different growth conditions. The crystal size estimated from the XRD peak widths appear to be on the order of 20 nm (that peak broadening was due to crystal size and not strain was clear from the small dependence of peak width of the {101} family of planes on angle in most of the results). The authors infer a grain size of 0.2 to 0.5 μm from SEM studies. It seems that as so often occurs in ED films, the grains are composed of many individual crystallites.

The CdS films described above were found to be coherent, even for very thin films (< 30 nm), adherent, and crack-free (for large current densities and Cd^{2+} concentrations; cracked films were obtained at low currents and low Cd^{2+} concentrations). This is contrary to the findings of Baranski et al. [145] that noncracked films were obtained from the DMSO–CdCl$_2$ bath only if the plating current was reduced during deposition. Another apparent discrepancy between the two papers was the finding that the initial growth rate of the CdS films was exponential under conditions of low-to-medium Cd/S ratios in one case [72] and linear in the other under what seemed to be similar conditions of Cd and S concentrations [145]. In view of the apparent large sensitivity of different properties of the films prepared by this plating technique to experimental conditions (with the notable exception of composition), care must be taken in comparing different experimental results.

Fatas et al. [26] have compared CdS deposition from DMSO and propylene carbonate (PC) electrolytes. Since the Cd and S concentrations in the two baths were probably different, this makes a comparison between the two solvents difficult due to the sensitivity of the CdS to composition described above. They found that the deposition potential in the PC solution was much lower than that in DMSO and also was stable during deposition, unlike the DMSO bath, which exhibited an increase in potential with time. These observations were connected with the greater solvation of the species in DMSO, which requires more energy for the desolvation step. The films deposited from PC did not crack as did those from DMSO. Consideration of the SEM data showing a smooth but cracked structure in the latter case compared with a more dendritic structure in the former case can explain this difference; cracking is not required in a dendritic structure as it is in a coherent one to relieve strain, since there already are many phase boundaries in the structure. Crystal size was measured from XRD peak broadening. The size did not appear to be very dependent on either substrate (steel or SnO_2–glass) or on the solvent (despite the very different macroscopic morphologies as seen by SEM). The size varied between ca. 11 and 25 nm, with an increase in deposition temperature and a decrease in time of deposition giving larger crystals.

Rastogi and Balakrishnan have reported the deposition of CdTe from DMSO solutions containing $CdCl_2$ and elemental Te [51]. The low solubility of Te in this solvent (<1 mM even at elevated temperatures) necessitates the use of low current densities; 1 mA/cm^2, a commonly used current density for CdS and even CdSe, gave films of CdTe which were powdery and contained free Cd. The crystallinity of the CdTe films was very dependent on plating current within the range 0.1 to 1.0 mA/cm^2. Up to ca. 0.15 mA/cm^2, nonoriented, hexagonal CdTe crystals were deposited. Between 0.15 and 0.25 mA/cm^2, oriented, hexagonal crystals were obtained, which from SEM pictures appear to be about 0.5 μm in size. As the current increased, the films became more polycrystalline and mixtures of both hexagonal and cubic CdTe (the latter being the more commonly obtained modification) were formed. As the authors pointed out, the orientation of the films should not be due to a substrate effect (the substrates were SnO_2–glass, which are polycrystalline). They reasoned that the use of an organic solvent with a low dielectric constant (compared to water) would promote ion–ion rather than ion–solvent interaction, which implies stronger interaction between atoms of a growing nucleus than would be the case in an aqueous solution. (This may be modified by the complexing action between DMSO and many ions, including Cd^{2+}.) These growing nucleii then acted as growth sites for further deposition which occurred by charge transfer to Cd^{2+} ions followed by surface diffusion of (presumably) Cd^0 to a CdTe nucleus at low currents and direct discharge of the ions at the nucleii at higher currents. At lower currents, there were fewer and larger nucleii with better defined faces. This is likely to be due to the thermodynamic instability of very small nuclei, which means that they can dissolve if the rate of deposition is not sufficient to increase their size beyond a critical size in a short time.

Balakrishnan and Rastogi also described the deposition of phosphorus-doped CdTe by adding elemental P to their deposition solution [85]. The P/Te ratio in the films was essentially the same as that in the solutions over a fairly wide range of concentrations. The P concentration in the films appeared, from AES results, to be much higher than normal doping concentrations. Also, AES showed that considerable amounts of S were present in the films, which was attributed to inclusion of solvent (at the high vacuum at which AES measurements are carried out, it could be argued that any free solvent would be removed; it is possible that the DMSO is strongly adsorbed in the film and would not be readily removed). Electrical characterization of the films showed that P-free films were *n*-type, with resistivity = 30 $\Omega\cdot$cm, which initially increases with P content up to ca. 10^4 $\Omega\cdot$cm as the *n*-type films are compensated and eventually type-converted to *p*-type. Further increase in P content results in a gradual decrease in resistivity, giving a minimum value of ca. 10 $\Omega\cdot$cm with corresponding acceptor concentration of 8×10^{16} cm^{-3}. They also found the trap density to be an order of magnitude less than that found for CdTe deposited from aqueous baths. A side effect of the P doping was an increase in crystal size up to ca. 0.8 μm attributed to the effect of P on the number of nucleation sites.

Golan et al. [103] studied the nucleation of CdSe, deposited from solutions of DMSO–Cd(ClO$_4$)$_2$–Se onto gold substrates using TEM with electron diffraction (ED). The gold was in the form of evaporated thin films on glass with crystal sizes of tens to hundreds of nm nanometers. By using a selected-area aperture in the ED technique, single gold crystallites could be investigated. Thus the deposition was studied in effect on {111} single-crystal gold surfaces. The first monolayer of nanocrystals (equivalent to a film of ca. 5 nm thickness for complete coverage, and not to be confused with a monolayer of material coverage, which would be a fraction of 1 nm) was found to be epitaxially oriented with the basal plane of the hexagonal CdSe parallel to the gold {111} surface and with uniform azimuthal orientation. The CdSe nanocrystals were 5.1 nm in size with a standard deviation of 1.5 nm. Continuation of deposition led to a second layer of crystals that were no longer oriented. The close lattice match between the CdSe and gold (a mismatch of 0.6% between the relevant CdSe and gold lattice spacings in a 2:3 ratio) was suggested as a possible reason for the epitaxy.

While the size of the CdSe nanocrystals was independent of the deposition parameters, the distribution of the crystals on the gold varied considerably and reproducibly with change in deposition conditions. For a constant charge density passed (therefore, the same amount of CdSe assumed deposited), low plating current densities or high temperatures led to large (ca. 50 nm) aggregates of nanocrystals. These aggregates became smaller as the current was increased or temperature decreased until eventually only isolated crystals were formed (without any appreciable change in the total amount of CdSe or size of crystals, except for short pulses of relatively high current, when the plating current efficiency dropped). This behavior was probably due to the increased chance for surface diffusion of the crystals either at higher temperatures or lower currents (therefore, longer times). It should be noted that the films were stable at room temperature even over the period of a year.

Baranski et al. reported some properties of mixed Cd(S,Se) films deposited from DMSO–CdCl$_2$ solutions [148]. As Se content increased (by increasing the Se/S ratio in the bath), the films became smoother and less cracked. They apparently also became amorphous, as seen by the loss of the XRD pattern as the Se content increased. Interestingly, while the XRD peak intensity decreased for small amounts of Se in the film, the peak width did not appear to increase appreciably as would be expected if the effect was a gradual decrease in crystal size. Some PEC properties of the films were given. Cd(S,Se) films were also deposited by Loufty and Ng [149] using elemental sulfur as usual, but using SeO_2 as the Se source. They described PEC characteristics of the films.

Hodes et al. [36,69] found that films of CdS and CdSe prepared by this technique were composed of aggregated nanocrystals, typically between 4 and 15 nm in diameter depending on deposition conditions, and exhibited quantum-size effects (QSEs). QSEs are observed in semiconductors when the dimension of the semiconductor is comparable to, or smaller than, the bulk exciton diameter. This size depends on the semiconductor, but for CdSe and CdS it begins at ca. 10 nm (somewhat smaller for CdS and larger for CdSe). It is manifested by an increase in the optical bandgap, due to increase in the electron and hole energy levels. The increase in E_g was ca. 0.2 eV for CdSe and CdS deposited from chloride (or other halide-containing solutions) and ca. 0.1 eV for films deposited from perchlorate baths (in the absence of halide). The crystal size was ca. 5 nm (halide solutions) and ca. 10 nm (perchlorate solutions), although other variables, in particular temperature and film thickness, affected crystal size. Annealing the films caused crystal growth and a decrease in E_g until the normal bulk value was reached. These spectral shifts had been reported quite frequently, either directly or indirectly, by other workers using this plating technique for Cd chalcogenides as well as for other metal chalcogenides, but without a convincing explanation for their origin.

These films were photoactive in a PEC configuration but not in a solid-state PV configuration. This was explained by penetration of the somewhat porous film by electrolyte, allowing photogenerated charge to be collected efficiently at the surface of individual crystallites [36,106]. Separation of charge occurred not by a built-in space-charge layer as in other PV cells, but by different electrochemical kinetics for electron and hole transfer into the electrolyte. Films normally behaved as n-type semiconductors (with respect to the direction of photocurrent flow), but could also be made to behave as p-type. This is a consequence of kinetically determined current flow. A single dopant in a 5-nm semiconductor crystal would lead to a heavily degenerate material (assuming that the concept of doping in bulk semiconductors can be applied to these nanocrystals, which is a highly suspect assumption). It is clear that some common semiconductor concepts need to be reconsidered in dealing with such films.

VI. DEPOSITION FROM NONAQUEOUS SOLVENTS CONTAINING COMPOUND CHALCOGENS

This technique was first described by Darkowski and Cocivera [150]. They used tri-n-butylphosphine telluride (PT), in which Te is present as the zero-valent state, as the Te source. Cd was present as the perchlorate and the solvent was propylene carbonate (PC). The deposition was carried out at 100°C on Ti substrates. Cyclic voltammetry of this system showed that PT was reduced at more positive potentials than Cd^{2+} itself, and that the CdTe deposition potential was shifted slightly positive of the PT reduction potential. This was interpreted as due to interaction between PT and Cd^{2+}, such interaction to form complexes having been described previously for Cd^{2+} and the corresponding phosphine selenide. At relatively positive potentials (where Cd^{2+} by itself was not reduced), Te-poor films were obtained while, at more negative potentials (where Cd^{2+} was reduced), the films were reasonably stoichiometric. This was shown in a later study to be due to the relatively low BPT/Cd ratios used initially; high ratios resulted in little compositional dependence on potential [12].

The as-deposited films were *p*-type, measured by the PEC response in the deposition solution. They were smooth with a tendency to crack as the films became thicker. XRD did not show a well-defined crystal structure for the as-deposited films; annealing at 400°C resulted in a cubic structure with a crystal size of ca. 50 nm estimated from the XRD peak widths.

Illumination of the film during deposition had a major effect on the deposition [12]. The plating current, which dropped during deposition (at a constant potential) in the dark, was much more stable with time under illumination. The drop in plating current in the dark was attributed to the *p*-type conductivity of the growing film (cathodic bias is in the reverse saturation direction of a *p*-type semiconductor–electrolyte junction) and the increasing resistance of the film. Illumination will allow photocurrent to pass in this potential region and there may also be a reduction in the resistance due to photoconductivity. The large variations in film composition described above for films deposited in the dark were not found under illumination. The composition was dependent on the Cd^{2+} concentration in the electrolyte under illumination, but not in the dark. This latter finding could be due to the higher average plating currents used under illumination (the plating was normally carried out under constant potential), which would mean that the conditions are closer to diffusion limitation of the Cd^{2+}, but not the PT, which was present at much higher concentrations. Plating efficiencies >95% were found.

It was suggested that the main mechanism of deposition was reduction of a Cd–BPT complex directly to CdTe. Variations of the stoichiometry were attributed to reduction of BPT to give telluride ions (these could react further in a number of ways to give excess Te) or formation of Cd metal by cadmium ion reduction. If free Cd^0 is indeed formed, the possibility of chemical reaction between the free metal and BPT should also be considered. Solar cells were fabricated using these films deposited on CdS–ITO–glass [7,9]. Efficiencies greater than 5% were obtained after an air anneal of the junction at 400°C.

Various doping treatments were investigated for these CdTe films [66,67,81]. Dopants used were Cu, Ag, Te, Cd, and In. They were introduced into the films by different techniques. In particular, co-deposition during plating or electromigration from an aqueous solution of the desired dopant were the two main techniques employed. The resistivities as well as activation energies of the dopant levels were measured for annealed samples. For the undoped samples, the resistivity was on the order of 10^6 $\Omega\cdot$cm, which decreased by 10 to 50 times on illumination. The resistivity decreased depending on dopant and method of doping. The electromigration technique was able to reduce the resistivity many orders of magnitude (to as low as 10^{-2} $\Omega\cdot$cm for Cu doping), depending on the amount of dopant introduced (i.e., electromigration current for a fixed time). It was reasoned that the electrical properties of the doped films were controlled by grain boundary states. This is expected since the electromigration (or even simple electrodeposition of the dopant at the grain surfaces) will first affect the surface properties. Interestingly, while the photosensitivity normally decreased with decreasing dark resistivity, for films where Cd was diffused from the vapor phase, the photosensitivity increased even though the dark conductivity dropped. In this case, the films were type-converted to n-type.

Sanders and Cocivera reported deposition of CdSe using the corresponding selenide compound (tri-n-butylphosphine selenide, PS) [15]. They found that using PC as a solvent gave poor films (mostly nonstoichiometric and only thin films obtained). These problems were overcome by using DEG as a solvent. The films were deposited at 160°C. The density of the films was low (ca. 60% of the bulk CdSe density). This would suggest a porous structure, although SEM micrographs show the films to be fairly smooth and coherent. Also, a small degree of cracking was seen, which became more severe after annealing. As noted earlier, cracking occurs in smooth, coherent films and is not expected in rough films, which a large porosity implies. Cyclic voltammetry showed that the CdSe deposition occurred slightly positive of the Cd deposition and with a larger peak current, while only a small PS reduction current was observed at about the same potential. Based on this, the mechanism suggested to be most probable was reduction of some type of Cd–PS cómplex (as noted previously, such complexes are known to exist). Another possibility given was reduction of Cd^{2+} to Cd^0 followed by chemical reaction between the Cd and PS (it had been noted that a Cd rod immersed in the solution became coated with a red-brown film, which may be presumed to be either CdSe or elemental Se). The current–voltage characteristics were similar for both bare Ti and CdSe-coated Ti, and the results are thus valid for both the initial nucleation steps and for the film growth. PEC characterization of the annealed films was described and moderate conversion efficiencies were obtained.

Preusser and Cocivera extended this method to CdS by using an antimony–sulfur analog of the phosphine chalcogenides, triphenylstibine sulfide

(Ph$_3$SbS, TSS) [23]. Films were deposited onto a number of different substrates from an electrolyte containing cadmium and lithium perchlorates and TSS in PC at room temperatures ranging from room temperature to 90°C and current densities between 50 and 100 μA/cm^2. An ac component was superimposed on the dc current. The effect of this component was unclear except that it appeared to improve the PEC response of the CdS films.

The mechanism of the deposition was studied both on bare Ti [23] and on CdS-covered Ti [41]. In the former case, deposition of CdS occurred more positive than the TSS reduction potential and slightly more positive than the Cd reduction potential. As for the case of CdSe above, a mechanism involving reduction of a Cd–TSS complex was suggested. Because of the small difference in both potentials and peak currents between CdS and Cd^0 formation, the formation of Cd, followed by reaction with TSS to give CdS, is also possible. The latter mechanism was suggested for deposition on CdS [41] on the basis of a more thorough study. The chemical reaction between Cd metal and the deposition solution was shown to occur. It was suggested that Cd^{2+} reduction proceeded via two separate electron transfer steps. The film composition depended on the Cd/TSS ratio in the electrolyte. A high Cd/TSS ratio gave Cd-rich films (supporting the participation of Cd^0 in the mechanism), while a low ratio gave films that were S-rich, the amount of excess S increasing slowly with decrease in the ratio. Since TSS itself was not found to be reduced under these conditions, the formation of excess S implies that the reaction between Cd and TSS forms not just CdS but also elemental S. The addition of excess TSS was found to decrease the peak current, which could be explained either by a Cd–TSS complex which was less active to reduction than Cd^{2+}, or as seems more likely, by inhibition of the reduction due to adsorption of TSS at the electrode.

Various methods were used to characterize the CdS films [23]. No Sb was found to be incorporated into the films from elemental analyses. The plating current efficiency was >88% based on a two-electron transfer. From XRD, a crystallite size of 16 nm was estimated, although it is not clear if this is for as-deposited samples or after annealing at 200°C. The latter may be more likely since in a later paper [71], a crystal size of 12 nm is given, and annealing at 200°C is expected to have an appreciable but not major effect on the crystal size. TEM showed the films to be coherent (with a small amount of cracking that increased after annealing), although not very smooth. Optical spectra gave a value for E_g of 2.63 eV, which increased to 2.67 eV after annealing at 200°C. In another study by this group [66], annealing at 400°C reduced E_g to 2.35 eV. The high value for the as-deposited films is indicative of the quantum-size effect described earlier, although a crystal size of 12 to 16 nm would not give such an effect (a reduction in size of about three times would be required) and the increase in E_g after annealing at 200°C is also difficult to understand on this basis. A moderate PEC response of the films was demonstrated.

A study of the electrical characteristics of these films was made [71]. The resistivity of the films was on the order of 10^7 Ω-cm, measured both directly and by analysis of the current–voltage behavior of the growing film, assuming that the increase in voltage (for constant current) during deposition was due to the resistivity of the growing film. Activation energies were calculated from the temperature dependence of resistivity. They were found to be somewhat dependent on various solution treatments, which was explained by adsorption of the solution species on the grain surfaces throughout the presumably porous film. Mott–Schottky plots were linear with considerable frequency dispersion. A value for N_D of ca. 2×10^{21} cm^{-3} was measured from these plots, which as pointed out by the authors, was unlikely. This value could be explained, to some extent, by surface area effects (the capacitances were very large). It may be that such an analysis is not valid for films made up of such small crystallites, where band bending in the crystals may be less important than the potential barriers between the individual crystallites. Values of N_D, measured from the dependence of photocurrent quantum efficiency on potential, were ca. 10^{20} cm^{-3} and did not change after annealing in H_2 at 200°C. Although appreciably lower than N_D measured from the Mott–Schottky plots, these values still appear to be unrealistically high. Diffusion lengths after this annealing treatment were found to be 10 nm or less. Since annealing at 200°C should cause only a small growth in crystal size, it is probable that the diffusion length is limited by the small crystallite size.

Colyer and Cocivera [112] deposited films of $Cd_xHg_{1-x}Te$ (where $1 > x > 0.65$) using this method. The plating bath contained $Cd(ClO_4)_2$, HgI_2, ethylene diamine (to complex the metal ions and slow down the formation of Te apparently catalyzed by the free ions), PT, and $LiClO_4$ (as supporting electrolyte) in PC. They obtained smooth, adherent films on ITO–glass provided that the plating current density was not greater than 0.05 mA/cm^2 (higher values gave powdery films). The composition of the films depended on the various plating parameters; higher Hg/Cd ratios, lower temperatures, and more negative plating potentials all led to films with a higher concentration of Hg. While the first is obvious, the temperature and potential (or current density) dependence is contrary to what would be expected if the deposition was diffusion controlled by low Hg concentrations. The very low current densities employed also suggests that diffusion control is not operative in this case. The plating current efficiencies were high: >90%. Annealing the films at temperatures greater than ca. 300°C led to Hg loss (at 400°C, all the Hg was removed from the films). XRD of an as-deposited and annealed (200 or 300°C) film was shown ($x = 0.9$) and crystal sizes of 7.5 and 22 nm were calculated for the 200 and 300°C annealed films, respectively. An approximate crystal size of 5 nm could also be estimated for the as-deposited film. Optical absorption spectra were measured and a variation of E_g from 1.03-1.32 eV was observed (for the 300°C annealed films) over a composition range of $x = 0.75$ to 0.93, respectively. For $x = 0.72$, values for E_g of 1.22 eV (as deposited) and 1.15 eV (300°C annealed with x increasing to 0.75 due to loss of Hg) were found. The as-deposited absorption spectrum could also be interpreted to give a value of $E_g = $ ca. 1.53 eV, since interference effects could account for the apparent absorption at 1.22 eV. This value would be reasonable for a film of ca. 5-nm crystallites, assuming that the increase was due to a QSE.

A different technique, using $TeCl_4$ in an organic solvent (ethylene glycol), first suggested by Engelken and Van Doren [125], was described by Gore and Pandey [32]. In this case, the Te is in the +4 valency state, unlike the zero-valent Te used above. In this respect it resembles more the aqueous TeO_2 bath. The deposition was carried out from a solution of $CdCl_2$ and $TeCl_4$ in ethylene glycol at 160°C onto Ni substrates. Near-stoichiometric or Cd-rich films were obtained, depending on the deposition conditions. SEM indicated a grain size of 0.2 to 2 μm with a granular morphology. As expected for such a morphology, no cracking was observed. Both hot-probes and PEC measurements showed that the films were n-type as deposited. Relatively good PEC characteristics were obtained from the as-deposited films. Addition of KI to the deposition solution was found to improve the adhesion of the films to the substrate [55]. It was suggested that adsorption of iodide on the electrode affects the interaction between Cd and Te ions and the substrate. XRD results [54] showed that below a deposition temperature of 130°C, no CdTe was formed, and above this temperature, the crystal size increased with increasing temperature. A rough estimation of the crystallite size from the XRD data gives values of 7 and 20 nm for deposition temperatures of 160 and 175°C respectively. This is a large size difference for such a small temperature change. Strong {111}, {220}, and {311} reflections were observed. Cd, Te, and TeO_2, which were the main constituents of the deposit at a deposition temperature of 130°C, were not found in films deposited at 160°C and higher.

The mechanism of this deposition technique was studied [99]. The appearance first of a $TeCl_4$ reduction wave followed at more negative potentials by a UPD of Cd suggests a mechanism similar to that occurring for the aqueous TeO_2 system (i.e., UPD of Cd on continually forming Te to give CdTe). As for the aqueous system, the films become increasingly Cd-rich as the deposition potential is made more negative, as Cd is deposited at (or more negative than) its equilibrium potential. SEM studies showed the films to be made up of aggregates (a few hundred nanometers in size) of crystallites a few tens of nanometers in size. Films grown at high current densities were smoother than those grown at lower currents, and in this case, as for most other smooth, coherent films, were found to contain cracks, in contrast to the more granular films deposited at low currents.

VII. ANODIZATION

This technique, where a metal is anodized in a solution containing anions of the desired chalcogen, is well known for oxides, such as in anodized aluminum, where the Al metal is covered with a thin oxide film. Miller and Heller [151] used this method to form photoanodes of CdS by anodizing Cd in a sulfide-containing solution. Peter [152] studied this reaction in detail. He found three distinct regions in the I–V characteristics as the potential of the Cd electrode was swept in the anodic direction. The first, a fairly sharp, structured peak, was associated with formation of up to two monolayers of CdS. A plateau region, where the current was more or less constant with increasing anodic potential, followed the monolayer peak. The electric field across the growing CdS film in this region was estimated to be $>10^6$ V/cm and a high field growth whereby Cd^{2+} ions diffused in the CdS was postulated. From C–V measurements, the surface roughness of the film at this stage was estimated at ca. 1.2 to 1.3 and a value calculated for N_D of 1.6×10^{25} m^{-3}. The third region, which began after the CdS thickness reached ca. 5 nm, exhibited a sharp increase in current (the transpassive region). The increase in current corresponded to a sharp drop in the field across the film due to formation of a porous film, and growth in this region was considered to be controlled by a diffusion process. Once this porous structure formed, further growth was believed to occur by migration of Cd^{2+} ions across the barrier layer formed in the plateau region. Since the transpassive region began at a well-defined potential, it was reasoned that initiation of pores was not a chemical process but was dependent on the potential or field at the interface. The mechanism of this growth was explained by tunneling of electrons from electrolyte species and/or from the valence band or surface states in the CdS to the Cd metal (either directly or via the conduction band of the CdS).

The work of Yeh et al. [153] supported many of the conclusions of Peter. They studied the growth of CdS films under both constant-current and constant-potential conditions and concluded that the kinetics of growth were similar for both cases. Up to a film thickness of 5 nm (the plateau region described by Peter), the kinetics could be explained by a high-field-assisted formation of CdS. They suggested that there would be no appreciable space-charge layer in this thickness to affect the kinetics, an arguable assumption if the high value of N_D measured by Peter was correct.

Birss and Lee [154] studied the initial monolayer region (in an electrolyte of pH 14; the studies above used a lower pH of ca. 9). They found that the initial deposition occurred at the thermodynamic potential, implying that neither a UPD nor a nucleation overpotential mechanism was applicable. They found, as did Peter, that the "monolayer peak" was structured and calculated that a half-monolayer was first formed, followed by formation of a full monolayer and finally a second full monolayer. They suggested that once the first half-monolayer formed, lateral repulsion between adjacent surface ions induced a place exchange process (between underlying Cd and surface S). The same occurs for the second half-monolayer. Finally, a full monolayer forms since the larger fields at the more positive potentials could overcome the lateral repulsion.

Krebs et al. [93] carried out some physical characterizations of anodized CdS. By electron diffraction, they found the structure to be hexagonal. The diffuse nature of the diffraction patterns, together with SEM observations by Peter [152] that the crystal size was less than 50 nm, indicates a small crystal size. Auger analyses showed the films to be essentially pure CdS; there was no evidence of oxides. Miller et al. [155] had indicated that their films contained some hydroxide. However, the latter carried out the anodization at a pH of ca. 14, compared to a pH of 9 used by Krebs et al., which would increase the chance of hydroxide incorporation in the films. In considering the mechanism of growth of the various stages of the film, Krebs et al. believed the porous film to be formed by anodic dissolution of Cd to Cd^{2+} and subsequent precipitation of CdS by reaction of the Cd^{2+} and sulfide ions.

In another study, Krebs and Heusler studied the PEC properties of the compact CdS film (i.e., before onset of porous growth) [104]. They found that no photopotential was generated for film thicknesses of less than 1.35 to 1.7 nm (note that Miller et al. [155] also found hardly any photoeffect for very thin films). They explained this by an accumulation layer at the Cd/CdS interface. Presumably, this negative band bending prevented positive band bending at the electrolyte/CdS side of the very thin film. The high doping densities measured during the initial growth stage [104,152] could also explain the lack of photopotential since such highly doped material would not be expected to give much photoeffect. Electron-hole recombination at the CdS-Cd interface offers yet another explanation. For thicker films, where photoeffects were obtained, ring-disk measurements showed that sulfide oxidation consumed only 5% of the photocurrent; most was used in dissolution of the Cd to Cd^{2+} ions (which underwent subsequent precipitation to give CdS). From the spectral dependence of the photocurrent, they calculated a value for E_g of 2.40 to 2.45 eV, but which was characteristic of an indirect transition (or an amorphous material). They also made $C–V$ measurements and found that the value of N_D dropped from ca. 10^{19} cm^{-3} at a film thickness of 3.15 nm to 2.5×10^{18} cm^{-3} at 4.5 nm. They noted that the Mott–Schottky plots were linear over only a small potential region (ca. 50 mV). These very high values of N_D (found also by Peter [152]), if correct, suggest that Cd (either in the atomic form or, more likely, as an ion) diffuses readily into the compact layer as described previously, but that sulfide ion (probably HS$^-$ in the case of the relatively low-pH solutions used in most of these studies) does not readily diffuse into the film from the other side.

Ham et al. studied the formation of CdTe [47] and CdSe [90] by anodizing Cd in telluride and selenide solutions and compared the corresponding anodizations for formation of CdS, CdSe, and CdTe [90]. In all three cases, the initial peak in the voltammogram was at the same potential and could be correlated with formation of $Cd(OH)_2$, which subsequently reacted with chalcogenide ions in solution to give CdX. Cyclic photovoltammetry showed a photoeffect (n-type for both CdSe and CdTe) already at the monolayer formation stage (note the contrast between these findings and the apparent lack of photoresponse for very thin CdS films discussed above). While the Cd–Se system exhibited I–V behavior which was similar to that for CdS (the potential at which transpassive behavior—steep increase in current—began was less positive than that for CdS, i.e., the plateau region was less extensive), the onset of the transpassive region in the Cd–Te system was not well defined and two partially resolved peaks, which were ascribed to telluride and tellurium oxidation, followed the "monolayer" peak. The energy levels of CdTe with respect to the telluride/tellurium potential (and thus the transpassive onset potential) given in fig. 12, Ref. 90, do not reflect the values of open-circuit voltage obtainable with this system in general (>0.7 V [156]). Thus it is likely that as for CdS and CdSe, the transpassive potential for CdTe is fairly close to the valence-band edge. However, the presence of the oxidation peaks of tellurium species on CdTe clearly contrasts the absence of the equivalent peaks for the S and Se systems. This was explained by tunneling of electrons through a very thin CdTe film (see below) to the Cd substrate—in effect causing metallic behavior of the Cd/CdTe electrode. The formation of the porous films beyond the transpassive potential for CdS and CdSe was explained by the Fermi level of the semiconductor crossing into the valence band at potentials more positive than the transpassive potential (which, as mentioned above, is close to the valence band). This induces degeneracy in the semiconducting film, with the result that further increase in potential is dropped mainly across the Helmholtz layer and causes increasing Cd dissolutions to Cd^{2+}, which reacts with chalcogenide to give the porous film.

Further characterization of the above CdSe and CdTe films was reported. Photocurrent spectroscopy allowed estimation of E_g, and values of ca. 1.5 and 1.7 eV were found for CdTe and CdSe, respectively. PL of CdSe gave a signal centered at 1.51 eV, indicating the presence of recombination sites a little lower than the conduction band level. AES analyses showed that significant amounts of oxygen were present in the films; this was ascribed to the presence of $Cd(OH)_2$ or CdO. From a thermodynamic point of view, $Cd(OH)_2$ is not expected to be present in the films, due to its much greater solubility product compared with the Cd–chalcogenides [any $Cd(OH)_2$ present in a chalcogenide–ion solution will be converted to the chalcogenide]. There may, however, be kinetic reasons for its existence. SEM studies of the CdS and CdSe films showed them to be smooth up to the transpassive region followed by a nodular growth. In the case of CdTe, a needlelike growth was obtained. Also, XPS + sputter etch data for the CdTe films showed them to be very thin, leading to the conclusion that anodic film growth was much slower for CdTe than for CdS or CdSe.

VIII. ELECTRODEPOSITION FROM MOLTEN SALTS

This method has been employed to a limited extent for II–VI semiconductors. It possesses the general advantage over the other methods that because of the high temperatures usually employed (400 to 500°C has been typical), the deposits exhibit good crystallinity. However, the same high temperatures make the technique less convenient to work with.

Yamamoto and Yamaguchi [157] used the technique to deposit ZnSe films on single-crystal Si and Ge substrates. On {111} Ge, epitaxial {111} ZnSe films ca. 5 μm thick were deposited. Polycrystalline films were deposited at current densities higher than ca. 3 mA/cm^2 and also on {111} Si at all current densities.

Markov and Ilieva [158] deposited CdTe on Cu from a solution of CdCl$_2$ and TeO$_2$ in a KCl–LiCl eutectic. The Cu substrate reacted with both Te and deposited Cd, and mixtures of Cu$_2$Te, Cu$_2$Cd, and CdTe were obtained with a needlelike morphology. In a similar study, Zakhar'yash et al. [58] found elemental Te also formed by chemical reaction between electrogenerated telluride ions and TeO$_2$ (similar to the formation of excess Se in CdSe films deposited from aqueous SeO$_2$ baths).

Minoura et al. [159] deposited CdSe from the same eutectic as above containing CdCl$_2$ and Na$_2$SeO$_3$ at 450°C. XRD spectra exhibited sharp peaks, although the crystal size was somewhat smaller for films deposited at more negative potentials. CdO was found in films deposited at these negative potentials. Films grown at relatively positive potentials were textured predominantly {0002}, but this preferential texturing was lost at the more negative potentials. It is clear that there is a fundamental difference between the films deposited in the two different potential regimes, and this will be discussed below. The current efficiency was found to be 180%, which translates into a three-electron transfer (allowing for some loss in current efficiency) instead of the six electron transfer expected assuming a purely electrochemical mechanism. No explanation was given for this phenomenon. SEM studies showed the films to be made up of hexagonal needles, typically 5 to 10 μm in length and from <1 to 5 μm in cross section. The needles were often hollow. The films deposited at negative potentials were composed of smaller crystals at the film surface, as indicated by XRD. A more coherent morphology was obtained using pulsed electrolysis. Good photocurrents were obtained from the films in a PEC.

Minoura et al. [22] studied the mechanism of the CdSe deposition above. They found that two different mechanisms were operative. At lower potentials, Na$_2$SeO$_3$ (or SeO$_2$) was reduced to metallic Se. The UPD of Cd then occurred on this Se deposit. At more negative potentials, Se^0 was reduced to selenide, which reacted chemically with Cd^{2+} ions in the electrolyte to form CdSe. Also, free Cd could be deposited which reacted with any oxygen present to form CdO, explaining the presence of this phase in films deposited at negative potentials. At even more negative potentials, however, the amount of CdO in the films decreased somewhat; this was explained by the dominant selenide ion formation at more negative potentials. The change in crystallographic texture and crystal size at more negative potentials can both be associated with this mechanistic change.

Minoura et al. [70] made a similar study of CdS deposition, using Na_2SO_3 as the sulfur source. CdS deposition occurred chemically in the electrolyte due to disproportionation of sulfite ions to sulfide (note that this may give a clue to the high current efficiencies obtained in the CdSe bath, although no bulk precipitation was reported in that case). CdS was only (irreproducibly) deposited over a very narrow potential range. The mechanism suggested was Cd deposition followed by reaction of the Cd with an unidentified sulfur source. This mechanism could explain the narrow potential range for deposition. At more positive potentials, Cd did not deposit (there was a positive shift of ca. 0.1 V in the Cd deposition potential induced by the sulfite). At more negative potentials, mainly Cd was deposited in the liquid state and did not adhere to the substrate. The positive shift of the Cd deposition potential suggests UPD deposition of Cd on sulfur (this was not believed to occur due to the absence of sulfur deposition from Cd-free electrolytes; however, just a monolayer, or even less, of sulfur is required to allow UPD of Cd with subsequent continuous deposition.

The films were found by XRD to be composed of hexagonal crystals with no preferential growth direction. They were seen by SEM to be of a similar hollow needle structure as the CdSe films described above. By adding elemental Se to the electrolyte, a brownish deposit was obtained which was identified by XRD as a solid solution of Cd(S,Se).

IX. DEPOSITION BY A MULTISTAGE PROCESS

This section covers films deposited by separate deposition of the components of the final film, followed by annealing at elevated temperatures. Two different techniques have been used for semiconductor deposition by this general method; (a) deposition of a metal, alloy, or stacked layers of metals followed by reactive annealing in a chalcogenide-containing atmosphere, and (b) deposition of stacked layers of metal(s) and elemental chalcogen followed by annealing in an inert atmosphere to react the metal(s) and chalcogen. The former has been much more thoroughly explored for the ternary chalcopyrite semiconductors than for II–VI materials. The latter has been reported for II–VI semiconductors in a few papers.

Shih and Qiu [160] described CdTe films formed by depositing first Te (from a TeO_2 bath) followed by Cd (from a $CdSO_4$ bath) on Mo substrates. The

Te–Cd film of ca. 3 μm thickness was then heated in nitrogen. Above 370°C, single-phase CdTe was formed.

Basol and Kapur [43] formed ZnTe films by depositing first Te, then Zn (total thickness 0.83 μm), onto either ITO–glass or Mo-coated glass. The Te was deposited from TeO_2 in H_2SO_4 and the Zn from $ZnSO_4$, $ZnCl_2$, and boric acid, both at room temperature and at constant current. It was noted that if Zn were plated before Te, the Zn dissolved in the acidic Te electrolyte. The films were annealed for 90 mn at 450°C in flowing forming gas (N_2–H_2). XRD showed the resulting brick-red films to be polycrystalline ZnTe with no major elemental phases present. SEM showed the films to be rough but coherent, with an apparent grain size of a few hundred nanometers. The deposition of 5 nm of Cu on top of the stacked layers before annealing resulted in larger (1 to 2 μm), well-defined grains. This was attributed to the well-known ability of Cu to act as a recrystallization agent in II–VI compounds. Transmission spectra showed a sharp absorption edge with a calculated E_g value of ca. 2.1 eV (compared to the normal value for ZnTe of 2.24 eV). However, there was a considerable amount of scattering, with a value for transmission in the subband gap region of 5% given.

Basol et al. [61] also described films of (Cd,Zn)Te deposited by this technique on conducting glass. They stressed the need to obtain smooth, continuous films of the separate layers to obtain a uniform stoichiometry. For all the layers, unspecified additives were employed to obtain such films. They also noted that Zn plated better on Cd than on Te—hence the order of deposition: Te, Cd, and Zn. The films were heated for times varying from 30 mn to 2 h and at temperatures from 350 to 580°C. XRD showed the formation of cubic (Zn,Cd)Te at all concentrations from pure ZnTe to CdTe. The composition was controlled by the amounts of Cd and Zn in the stacked layers. The resistivity varied from 3 × 10^3 Ω·cm for ZnTe to 10^5 Ω·cm for $Cd_{0.6}Zn_{0.4}$Te. For Cu-doped films, the corresponding values were 0.6 and 7 × 10^3 Ω·cm. Optical transmission spectra showed the films to be less specularly reflecting (therefore rougher) than corresponding evaporated films. However, transmission in the subbandgap region was ca. 60 to 70% and the bandgaps calculated from the spectra agreed with literature values (both properties contrasting with those found above for ZnTe [43] and suggesting optimization of the preparation conditions). PV cells were made from these films ($Cd_{0.9}Zn_{0.1}$Te/CdS) with a conversion efficiency of ca. 3.7%.

REFERENCES

1. G. F. Fulop and R. M. Taylor, *Annu. Rev. Mater. Sci.*, *15*: 197 (1985).
2. C. D. Lokhande and S. H. Pawar, *Phys. Status Solidi (a)*, *111*: 17 (1989).

3.   R. C. DeMattei and R. S. Feigelson, in *Electrochemistry of Semiconductors and Electronics* (J. McHardy and F. Ludwig, eds.), Noyes Publications, Park Ridge, N.J., 1992, pp. 1–52.

4.   K. Rajeshwar, *Adv. Mater.*, *4*: 23 (1992).

5.   B. M. Basol, *Doga. Turk. J. Phys.*, *16*: 107 (1992).

6.   P. C. Searson, *Solar Energy Mater. Solar Cells*, *27*: 377 (1992).

7.   A. Darkowski and M. Cocivera, *J. Electrochem. Soc.*, *134*: 226 (1987).

8.   M. W. Verbrugga and C. W. Tobias, *J. Electrochem. Soc.*, *134*: 3104 (1987).

9.   J. von Windheim, A. Darkowski, and M. Cocivera, *Can. J. Phys.*, *65*: 1053 (1987).

10. S. M. So, W. Hwang, and C. H. Liu, *J. Appl. Phys.*, *61*: 2234 (1987).

11. G. Maurin and D. Pottier, *J. Mater. Sci. Lett.*, *6*: 817 (1987).

12. J. von Windheim and M. Cocivera, *J. Electrochem. Soc.*, *134*: 440 (1987).

13. R. D. Engelken, *J. Electrochem. Soc.*, *134*: 832 (1987).

14. F. Decker, N. G. Ferreira, and M. Fracastoro-Decker, *J. Electrochem. Soc.*, *134*: 1499 (1987).

15. B. W. Sanders and M. Cocivera, *J. Electrochem. Soc.*, *134*: 1075 (1987).

16. M. Fracastoro-Decker, J. L. S. Ferreira, N. V. Gomes, and F. Decker, *Thin Solid Films*, *147*: 291 (1987).

17. R. K. Pandey, A. J. N. Rooz, and S. K. Kulkarni, *Thin Solid Films*, *150*: 51 (1987).

18. F. Decker, J. R. Moro, J. L. S. Ferreira, and M. Vanzi, *Ber. Bunsenges. Phys. Chem.*, *91*: 408 (1987).

19. J. P. Szabo and M. Cocivera, *J. Appl. Phys.*, *61*: 4820 (1987).

20. M. T. Guttiérez and P. Salvador, *Solar Energy Mater.*, *15*: 99 (1987).

21. R. K. Pandey, R. B. Gore, and A. J. N. Rooz, *J. Phys. D.*, *20*: 1059 (1987).

22. H. Minoura, T. Negoro, M. Kitakata, and Y. Ueno, *Thin Solid Films*, *147*: 65 (1987).

23. S. Preusser and M. Cocivera, *Solar Energy Mater.*, *15*: 175 (1987).

24. S. M. Aliwi and A. M. Aushana, *J. Solar Energy Res.*, *5*: 55 (1987).

25. E. Fatas, P. Herrasti, F. Arjona, and E. G. Camarero, *J. Electrochem. Soc.*, *134*: 2799 (1987).

26. E. Fatas, P. Herrasti, F. Arjona, E. Garcia Camarero, and J. A. Medina, *Electrochim. Acta.*, *32*: 139 (1987).

27. K. Singh and J. P. Rai, *J. Mater. Sci.*, *22*: 132 (1987).

28. K. Singh and J. P. Rai, *Phys. Status Solidi (a)*, *99*: 257 (1987).

29. M. T. Guttiérez and J. Ortega, *Proc. 7th ECPV Solar Energy Conf.*, 1987, p. 1214.

30. K. Uosaki, N. Karube, T. Kadowaki, S. Sato, and H. Kita, *J. Chem. Soc. Jpn.*, *11*: 2006 (1987).

31. P. Sircar, *Appl. Phys. Lett.*, *53*: 1184 (1988).

32. R. B. Gore and R. K. Pandey, *Thin Solid Films*, *164*: 255 (1988).

33. B. M. Basol, *Solar Cells*, *23*: 69 (1988).

34. V. Ramanathan, L. Russell, C. H. Liu, and P. V. Meyers, *Proc. 20th IEEE PV Spec. Conf.*, Vol. 2, 1988, p. 1417.

35. P. V. Meyers, *Solar Cells*, *24*: 35 (1988).

36. G. Hodes and A. Albu-Yaron, *Proc. Electrochem. Soc.*, *88-14*: 298 (1988).

37. M. C. A. Fantini, J. R. Moro, and F. Decker, *Solar Energy Mater.*, *17*: 247 (1988).

38. J. P. Szabo and M. Cocivera, *Can. J. Chem.*, *66*: 1065 (1988).
39. F. Cerdeira, I. Torriani, P. Motisuke, V. Lemos, and F. Decker, *Appl. Phys. A*, *46*: 107 (1988).
40. K. S. Balakrishnan and A. C. Rastogi, *Thin Solid Films*, *163*: 279 (1988).
41. S. Preusser and M. Cocivera, *J. Electroanal. Chem.*, *252*: 139 (1988).
42. C. D. Lokhande, V. S. Yermune, and S. H. Pawar, *Mater. Chem. Phys.*, *20*: 285 (1988).
43. B. M. Basol and V. K. Kapur, *Thin Solid Films*, *165*: 237 (1988).
44. R. K. Pandey, A. J. N. Rooz, S. R. Kumar, and S. K. Kulkarni, *Semicond. Sci. Technol.*, *3*: 729 (1988).
45. E. Mori and K. Rajeshwar, *J. Electroanal. Chem.*, *258*: 415 (1989).
46. K. K. Mishra and K. Rajeshwar, *J. Electroanal. Chem.*, *273*: 169 (1989).
47. D. Ham, K. K. Mishra, A. Weiss, and K. Rajeshwar, *Chem. Mater.*, *1*: 619 (1989).
48. J. Touskova, D. Kindl, and J. Tousek, *Sol. Energy Mater.*, *18*: 377 (1989).
49. P. Cowache, D. Lincot, and J. Vedel, *J. Electrochem. Soc.*, *136*: 1646 (1989).
50. K. S. Balakrishnan and A. C. Rastogi, *Proc. 9th ECPV Solar Energy Conf.*, 1989, p. 500.
51. A. C. Rastogi and K. S. Balakrishnan, *J. Electrochem. Soc.*, *136*: 1502 (1989).
52. L. E. Lyons, G. C. Morris, and R. K. Tandon, *Solar Energy Mater.*, *18*: 315 (1989).
53. I. Radhakrishna, K. R. Murali, K. Nagaraja Rao, and V. K. Venkatesan, *Proc. SPIE Int. Soc. Opt. Eng.*, *1149*: 134 (1989).
54. R. B. Gore, R. K. Pandey, and S. K. Kulkarni, *J. Appl. Phys.*, *65*: 2693 (1989).
55. R. B. Gore and R. K. Pandey, *Solar Energy Mater.*, *18*: 159 (1989).
56. Z. Loizos, N. Spyrellis, G. Maurin, and D. Pottier, *J. Electroanal. Chem.*, *269*: 399 (1989).
57. C. Deslouis, G. Maurin, N. Pebere, and B. Tribollet, *Electrochim. Acta*, *34*: 1229 (1989).
58. S. M. Zakhar'yash, V. P. Khan, and V. I. Amosov, *Rasplavy*, 29 (1989).
59. M. Jayachandran, M. J. Chokalingham, and V. K. Venkatesan, *J. Mater. Sci. Lett.*, *8*: 563 (1989).
60. C. D. Lokhande, M. S. Jadhav, and S. H. Pawar, *J. Electrochem. Soc.*, *136*: 2756 (1989).
61. B. M. Basol, V. K. Kapur, and M. L. Ferris, *J. Appl. Phys.*, *66*: 1816 (1989).
62. M. T. Gutiérrez and J. Ortega, *Solar Energy Mater.*, *19*, 383 (1989).
63. M. Neumann-Spallart, G. Tamizhmani, A. Boutry-Forveille, and C. Levy-Clement, *Thin Solid Films*, *169*: 315 (1989).
64. J. Touskova, D. Kindl, J. Kovanda, and J. Tousek, *Electrotech. Obz.*, *79*: 462 (1990).
65. G. S. Sanyal, A. Mondal, K. C. Mondal, B. Ghosh, H. Saha, and M. K. Mukherjee, *Solar Energy Mater.*, *20*: 395 (1990).
66. J. A. von Windheim and M. Cocivera, *J. Phys. D.*, *23*: 581 (1990).
67. J. A. von Windheim, I. Renaud, and M. Cocivera, *J. Appl. Phys.*, *67*: 4167 (1990).
68. K. R. Murali, I. Radhakrishna, K. Nagaraja Rao, and V. K. Venkatesan, *J. Mater. Sci.*, *25*: 3521 (1990).
69. G. Hodes, T. Engelhard, A. Albu-Yaron, and A. Pettford-Long, *Mater. Res. Soc. Symp. Proc.*, *164*: 81 (1990).

70. H. Minoura, Y. Takeichi, H. Furuta, T. Sugiura, and Y. Ueno, *J. Mater. Sci.*, *25*: 472 (1990).
71. S. Preusser and M. Cocivera, *Solar Energy Mater.*, *20*: 1 (1990).
72. K. S. Balakrishnan and A. C. Rastogi, *Solar Energy Mater.*, *20*: 417 (1990).
73. B. W. Sanders, *J. Cryst. Growth*, *100*: 405 (1990).
74. M. T. Guttiérez and J. Ortega, *Solar Energy Mater.*, *20*: 387 (1990).
75. J. Gonzalez-Velasco and I. Rodriguez, *Solar Energy Mater.*, *20*: 167 (1990).
76. M. Jayachandran, V. K. Venkatesan, T. Mahalingam, and V. Vinni, *Proc. SPIE*, *1284*: 260 (1990).
77. E. Mori, K. K. Mishra, and K. Rajeshwar, *J. Electrochem. Soc.*, *137*: 1100 (1990).
78. S. N. Sahu and C. Sanchez, *Solid State Commun.*, *73*: 597 (1990).
79. J. A. von Windheim, H. Wynands, and M. Cocivera, *J. Electrochem. Soc.*, *138*: 3435 (1991).
80. B. W. Gregory, D. W. Suggs, and J. L. Stickney, *J. Electrochem. Soc.*, *138*: 1279 (1991).
81. J. A. von Windheim and M. Cocivera, *J. Electrochem. Soc.*, *138*: 250 (1991).
82. A. K. Turner et al., *Solar Energy Mater.*, *23*: 388 (1991).
83. A. K. Turner, J. M. Woodcock, M. E. Ozsan, and J. G. Summers, *Proc. 10th ECPV Solar Energy Conf.*, 1991, p. 791.
84. S. Bonilla and E. A. Dalchiele, *Thin Solid Films*, *204*: 397 (1991).
85. K. S. Balakrishnan and A. C. Rastogi, *Solar Energy Mater.*, *23*: 61 (1991).
86. B. W. Gregory and J. L. Stickney, *J. Electroanal. Chem.*, *300*: 543 (1991).
87. Z. Loizos, N. Spyrellis, and G. Maurin, *Thin Solid Films*, *204*: 139 (1991).
88. A. M. Kressin, V. V. Doan, J. D. Klein, and M. J. Sailor, *Chem. Mater.*, *3*: 1015 (1991).
89. H. Wynands and M. Cocivera, *Chem. Mater.*, *3*: 143 (1991).
90. D. Ham, K. K. Mishra, and K. Rajeshwar, *J. Electrochem. Soc.*, *138*: 100 (1991).
91. R. K. Pandey, S. R. Kumar, A. J. N. Rooz, and S. Chandra, *J. Mater. Sci.*, *26*: 3617 (1991).
92. R. K. Pandey, S. R. Kumar, A. J. N. Rooz, and S. Chandra, *Thin Solid Films*, *200*: 1 (1991).
93. M. Krebs, M. I. Sosa, and K. E. Heusler, *J. Electroanal. Chem.*, *301*: 101 (1991).
94. M. T. Gutiérrez, *Solar Energy Mater.*, *21*: 283 (1991).
95. S. Moorthy Babu, T. Rajalakshmi, R. Dhanasekaran, and J. P. Ramasamy, *J. Cryst. Growth*, *110*: 423 (1991).
96. A. Darkowski and A. Grabowski, *Solar Energy Mater.*, *23*: 75 (1991).
97. C. D. Lokhande, V. S. Yermune, and S. H. Pawar, *J. Electrochem. Soc.*, *138*: 624 (1991).
98. J. A. von Windheim and M. Cocivera, *J. Phys. Chem. Solids*, *53*: 31 (1992).
99. R. K. Pandey, G. Razzini, and L. P. Bicelli, *Solar Energy Mater., Solar Cells*, *26*: 285 (1992).
100. J. H. Armstrong, B. R. Lanning, and M. S. Misra, *AIP Conf. Proc. 268* (R. Noufi, ed.), 1992, p. 177.
101. S. Dennison and S. Webster, *J. Electroanal. Chem.*, *333*: 287 (1992).
102. H. Wynands and M. Cocivera, *J. Electrochem. Soc.*, *139*: 2052 (1992).

103. Y. Golan, L. Margulis, I. Rubinstein, and G. Hodes, *Langmuir, 8*: 749 (1992).
104. M. Krebs and K. E. Heusler, *Electrochim. Acta, 37*: 1371 (1992).
105. G. C. Morris and R. Vanderveen, *Solar Energy Mater. Solar Cells, 27*: 305 (1992).
106. G. Hodes, I. D. J. Howell, and L. M. Peter, *J. Electrochem. Soc., 139*: 3136 (1992).
107. A. Mondal, B. E. McCandless, and R. W. Birkmire, *Solar Energy Mater. Solar Cells, 26*: 181 (1992).
108. I. Rodriguez and J. Gonzalez-Velasco, *J. Mater. Sci., 27*: 747 (1992).
109. G. C. Morris and R. Vanderveen, *Solar Energy Mater. Solar Cells, 26*: 217 (1992).
110. C. Wei and K. Rajeshwar, *J. Electrochem. Soc., 139*: L40 (1992).
111. V. Krishnan, D. Ham, K. K. Mishra, and K. Rajeshwar, *J. Electrochem. Soc., 139*: 23 (1992).
112. C. L. Colyer and M. Cocivera, *J. Electrochem. Soc., 139*: 406 (1992).
113. S. Weng and M. Cocivera, *Chem. Mater., 4*: 615 (1992).
114. Y. Ueno, H. Minoura, T. Nishikawa, and M. Tsuiki, *J. Electrochem. Soc., 130*: 43 (1983).
115. B. M. Basol and E. S. Tseng, *Appl. Phys. Lett., 48*: 946 (1986).
116. M. P. R. Panicker, M. Knaster, and F. A. Kroger, *J. Electrochem. Soc., 125*: 566 (1978).
117. G. Hodes, J. Manassen, and D. Cahen, *Nature, 261*: 403 (1976).
118. G. Hodes, J. Manassen, and D. Cahen, Israel patent 58,440, Mar. 1, 1983.
119. A. Albu-Yaron, D. Cahen, and G. Hodes, *Thin Solid Films, 112*: 349 (1984).
120. R. N. Bhattacharya and K. Rajeshwar, *J. Electrochem. Soc., 131*: 2032 (1984).
121. Von H. Gobrecht, H. D. Liess, and A. Tausend, *Ber. Bunsenges. Phys. Chem., 67*: 930 (1963).
122. E. Pacauskas, J. Janickis, and A. Saudargaite, *Lietuvos TSR Mokslu Akad. Darbai Ser. B*, 75 (1969).
123. F. A. Kroger, *J. Electrochem. Soc., 125*: 2028 (1978).
124. R. D. Engelken and T. P. Van Doren, *J. Electrochem. Soc., 132*: 2904 (1985).
125. R. D. Engelken and T. P. Van Doren, *J. Electrochem Soc., 132*: 2910 (1985).
126. C. Sella, P. Boncorps, and J. Vedel, *J. Electrochem. Soc., 133*: 2043 (1986).
127. E. Mori, C. K. Baker, J. R. Reynolds, and K. Rajeshwar, *J. Electroanal. Chem., 252*: 441 (1988).
128. M. Skyllas-Kazacos and B. Miller, *J. Electrochem. Soc., 127*: 869 (1980).
129. M. Tomkiewicz, I. Ling, and W. S. Parsons, *J. Electrochem. Soc., 129*: 2016 (1982).
130. G. Hodes, J. Manassen, S. Neagu, D. Cahen, and Y. Mirovsky, *Thin Solid Films, 90*: 433 (1982).
131. Y. F. Nicolau, M. Dupuy, and M. Brunel, *J. Electrochem. Soc., 137*: 2915 (1990).
132. B. M. Basol and O. M. Stafsudd, *Solid State Electron., 24*: 121 (1981).
133. D. Cahen and R. Noufi, *Appl. Phys. Lett., 54*: 558 (1989).
134. J. F. McCann and M. Skyllas-Kazacos, *J. Electroanal. Chem., 119*: 409 (1981).
135. G. P. Power, D. R. Peggs, and A. J. Parker, *Electrochim. Acta, 26*: 681 (1981).
136. M. Skyllas-Kazacos and B. Miller, *J. Electrochem. Soc., 127*: 2378 (1980).
137. M. Cocivera, A. Darkowski, and B. Love, *J. Electrochem. Soc., 131*: 2514 (1984).
138. J. P. Szabo and M. Cocivera, *J. Electrochem. Soc., 133*: 1247 (1986).

139. G. Hodes, A. Albu-Yaron, F. Decker, and P. Motisuke, *Phys. Rev. B*, *36*: 4215 (1987).
140. M. Skyllas-Kazacos, *J. Electroanal. Chem.*, *148*: 233 (1983).
141. A. S. Baranski and W. R. Fawcett, *J. Electrochem. Soc.*, *127*: 766 (1980).
142. A. S. Baranski, W. R. Fawcett, and A. C. McDonald, *J. Electroanal. Chem.*, *160*: 271 (1984).
143. A. S. Baranski and W. R. Fawcett, *J. Electrochem. Soc.*, *131*: 2509 (1984).
144. D. K. Roe, L. Wenzhao, and H. Gerischer, *J. Electroanal. Chem.*, *136*: 323 (1982).
145. A. S. Baranski, W. R. Fawcett, A. C. McDonald, R. M. de Nobriga, and J. R. MacDonald, *J. Electrochem. Soc.*, *128*: 963 (1981).
146. A. S. Baranski, M. S. Bennett, and W. R. Fawcett, *J. Appl. Phys.*, *54*: 6390 (1983).
147. F. Mondon, *J. Electrochem. Soc.*, *132*: 319 (1985).
148. A. S. Baranski, W. R. Fawcett, K. Gatner, A. C. McDonald, J. R. MacDonald, and M. Selen, *J. Electrochem. Soc.*, *130*: 579 (1983).
149. R. O. Loufty and D. S. Ng, *Solar Energy Mater.*, *11*: 319 (1984).
150. A. Darkowski and M. Cocivera, *J. Electrochem. Soc.*, *132*: 2768 (1985).
151. B. Miller and A. Heller, *Nature*, *262*: 680 (1976).
152. L. M. Peter, *Electrochim. Acta*, *23*: 165 (1978).
153. L. S. Yeh, P. G. Hudson, and A. Damjanovic, *J. Appl. Electrochem.*, *12*: 153 (1982).
154. V. I. Birss and L. E. Kee, *J. Electrochem. Soc.*, *133*: 2097 (1986).
155. B. Miller, S. Menezes, and A. Heller, *J. Electroanal. Chem.*, *94*: 85 (1978).
156. A. B. Ellis, S. W. Kaiser, J. M. Bolts, and M. S. Wrighton, *J. Am. Chem. Soc.*, *99*: 2839 (1977).
157. A. Yamamoto and M. Yamaguchi, *Jpn. J. Appl. Phys.*, *14*: 561 (1975).
158. I. Markov and M. Ilieva, *Thin Solid Films*, *74*: 109 (1980).
159. H. Minoura, T. Negoro, M. Kitakata, and Y. Ueno, *Solar Energy Mater.*, *12*: 335 (1985).
160. I. Shih and C. X. Qui, *Mater. Lett.*, *3*: 446 (1985).

12

Electronically Conducting Soluble Polymers

A. F. Diaz
IBM Almaden Research Center, San Jose, California

My T. Nguyen
Polychrome Corporation, Carlstadt, New Jersey

Mario Leclerc
University of Montreal, Montreal, Quebec, Canada

I. INTRODUCTION

The study of conductive polymers for technological applications continues to receive much attention. The availability films of polypyrrole, polythiophene, and polyaniline via the electrosynthetic route and the interesting electroactive properties characteristic of the π-conjugation in the polymer backbone has drawn a large number of laboratories into this field of research. In particular, the demonstration that films of these materials on electrodes could be switched electrochemically between the insulating and conductive state with a corresponding change in the electrical, optical, and chemical properties attract the attention of investigators from different disciplines. Much of this work has been summarized in numerous review articles. Many of these articles focus on the parent polymers (no substituents), such as polyaniline, polyacetylene, the five-membered heteroaromatic (pyrrole, thiophene, indole, etc.) and the phenylene polymers and provide a general description of the preparation methods, the polymer structure and morphology, and the chemical, physical, and electrochemical properties [1–10]. There are many other classes of polymers with conductive properties besides these and they are reviewed in some of the articles [2,6]. In particular, the reviews by Naarmann et al. [6,8] provide a good summary of the novel chemical routes to many of these polymers, and a fairly complete coverage of the different polymers with active centers. Some of the articles emphasize the active properties of these materials and their potential application to current technologies [3,5,7,8]. These articles summarize many feasibility studies that demonstrate the potential applications of these materials. Finally, the topic of processible or soluble polymers was summarized briefly [2].

In this chapter we have summarized the recent progress made with the derivatized polythiophenes, polypyrroles, polyanilines, and polyphenylene vinylenes and we emphasis those polymers that are soluble. In this summary we include the preparation and characterization methods, the solution and melt-processing methods, and the applications that have been considered. This is not intended to be a complete review of the area of conductive polymers, nor to include all the classes of polymers that can be found in the literature with electroactive properties.

II. SYNTHESIS AND CHEMICAL CHARACTERIZATION

The interest in the development of conjugated polymers with improved properties continues to surge because their optical, electrical, and electrochemical properties show great promise for use in several technology applications. However, before these materials can affect a technology effectively, they must be processible, available in commercial quantities and reasonably well characterized. In addition, the polymers must have good mechanical properties and much better performance reliability than is currently demonstrated. In the first part of this chapter, recent developments in the synthesis of soluble conjugated aromatic polymers are reviewed. These are the polymers that can be solution processed to produce films and coatings.

A. Polythiophenes

Polythiophene is prepared directly from thiophene by several electrochemical and chemical oxidation methods [11,12]. The polymerization reaction is described as shown in Figure 1 [12–14]. Polythiophene in the neutral state is insulating and can be oxidized (''doped'') to the conducting form with a conductivity exceeding 100 S/cm^2. However, the polymer is virtually insoluble in every solvent and is difficult to process. As is often the case with aromatic polymers, the chain stiffness and strong interchain interactions render these ma-

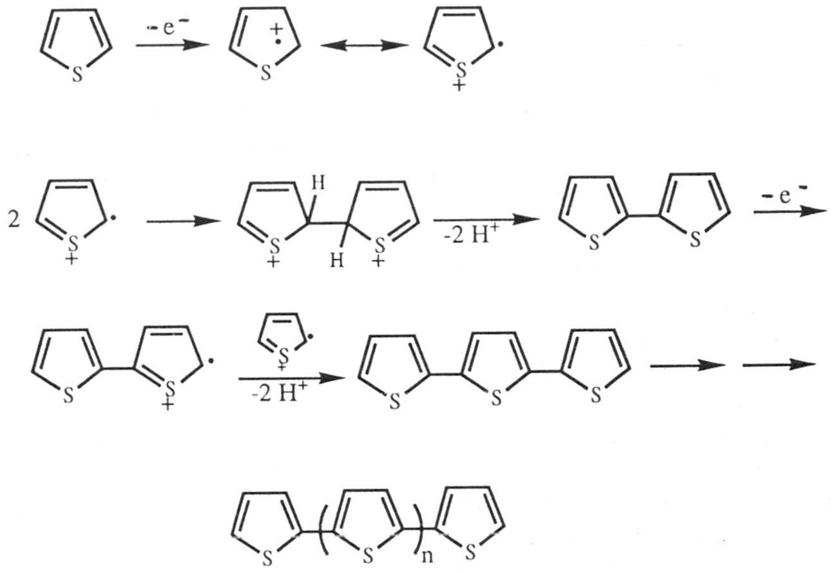

FIGURE 1 Electrochemical or chemical oxidation of thiophene.

terials insoluble. These materials also cannot be melt processed because they are thermally unstable at the melt temperatures.

With substituents on the polymer the interchain interactions in rigid rod polymers [15] are known to be reduced, and this fact has been used successfully to solubilize polythiophenes. Indeed, the poly(3-alkylthiophenes) (with > 3 carbon atoms in the alkyl group) were the first conductive polymers to be solubilized, and they are now considered among the most promising polymers for technical applications. Poly(3-alkylthiophenes), with a number-average molecular weight of ca. 5000, were first synthesized via a Grignard reaction (Fig. 2) [16]. Higher-molecular-weight polymers are prepared by the electrooxidation of 3-alkylthiophenes [17–19] or by chemical oxidation using FeCl$_3$ [20]. The latter method is a good way to prepare large quantities of poly(3-alkylthiophenes) in

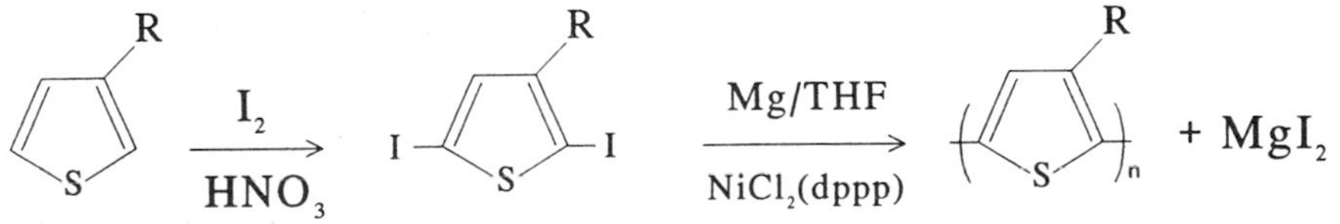

FIGURE 2 Polymerization of 3-alkylthiophenes via Grignard couplings.

good yield (ca. 70%). The polymerization reactions generate the neutral polymers, which are soluble in common organic solvents such as chloroform, tetrahydrofuran, and toluene. The corresponding oxidized forms are still poorly soluble in these solvents. In line with the polymerization scheme shown in Figure 1, the structural analysis of these novel materials reveal that the thiophene rings are primarily α-α' coupled like the parent polymer. However, careful nuclear magnetic resonance (NMR) analyses have shown that the poly(3-alkylthiophenes) have structural units other than those expected from simple coupling of adjacent thiophene moieties in 2,5'-positions [21–23]. In principle, four different regiochemical structures (triads) may occur in these polymers, depending on the relative reactivity of the 2- and 5-positions (Fig. 3). The various coupling modes—head to tail, head to head, and tail to tail—produce different characteristic peaks in the ^{1}H and ^{13}C NMR spectra. As shown in the ^{13}C spectra in Figure 4, a perfectly α-α'-coupled poly(3-hexylthiophene) can exhibit up to 16 peaks for the four aromatic carbon atoms of the thiophene ring. The head-to-tail and tail-to-tail couplings allow a coplanar conformation of the backbone, while the head-to-head coupling forces the backbone to twist to relieve the steric interactions between the repeat units. The polymerization reactions normally produce poly(3-alkylthiophenes) which are 70 to 90% head-to-tail coupled and which have some crystallinity. Many x-ray diffraction measurements have been performed on these materials, and they reveal a coherent ordering from the anti-coplanar conformation of the main polymer chain [24–27]. On the other hand, poly(3,3'-dihexyl-2,2'-bithiophene) is primarily head-to-head coupled and nonplanar. The polymer has a shorter conjugation length and the corresponding change in the physical properties [22]. Poly(3-alkylthiophenes) with almost complete head-to-tail couplings were recently prepared [28,29]. Finally, soluble and highly conjugated polythiophenes can also be prepared from monosubstituted bithiophenes [30] or terthiophenes [31,32].

Substituents other than alkyl groups have been used to modify the physical properties of the polymers while retaining their solubility for processing applications. The solubility of polythiophenes can be extended to water with the incorporation of sulfonate side groups [33,34]. The oxidation potentials of soluble polythiophenes with alkoxy substituents are lower, and consequently, these polymers are more stable in the oxidized (conducting) state [35–40]. The alkoxy substituents impose less steric crowding than the alkyl substituents; thus poly[(3-alkoxy)-4-methylthiophenes] [38], poly(3,3'-dialkoxy-2-2'-bithiophenes) [39], and poly(alkylenedioxythiophenes) [40] are highly conjugated. However, with two long-chain alkoxy substituents on each repeat unit, the conjugation length of the polymer is again reduced and the physical properties change accordingly [38]. Other soluble polythiophenes have been prepared with fluoroalkyl [41], aryl [42–44], ether [45–47], ethylmercapto [48], or ester [49] substituents. It is

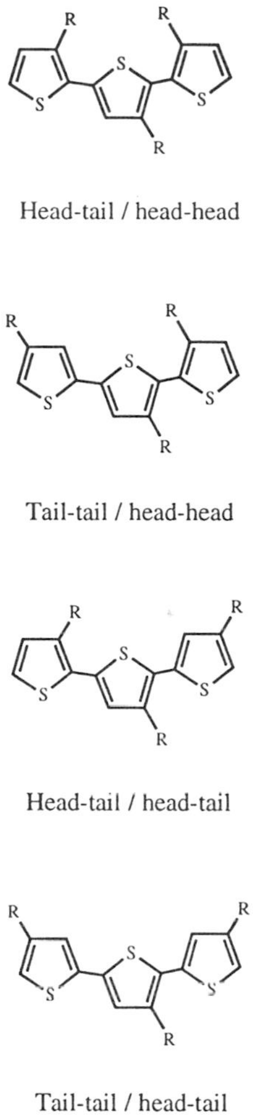

Head-tail / head-head

Tail-tail / head-head

Head-tail / head-tail

Tail-tail / head-tail

FIGURE 3 Different regiochemical structures in poly(3-alkylthiophenes).

also possible to attach redox groups such as bipyriridyl [50], benzoquinone [51], or viologen [52], and chiral groups [53,54] onto polythiophene.

B. Polypyrroles

Like the polythiophenes, the polypyrroles and the substituted polypyrroles are prepared by electrochemical or chemical oxidation following the reaction scheme shown in Figure 1 [55,56]. Polypyrrole is also difficult to process and

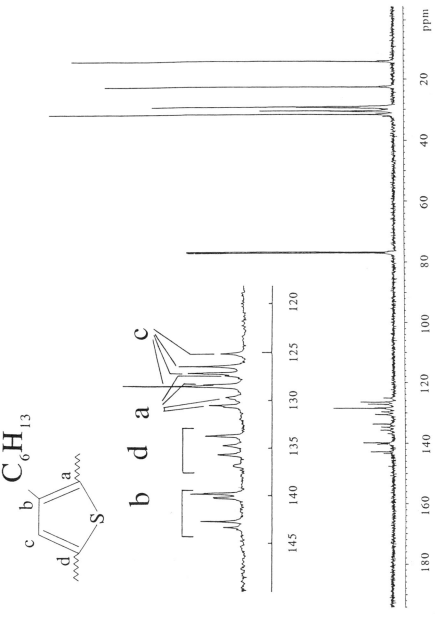

Figure 4 ^{13}C NMR spectrum of poly(3-hexylthiophene).

characterize because it is infusible and insoluble in all solvents. N-substituted polypyrroles were the first examples of substituted conducting polymers to be prepared and they were prepared electrochemically. These polymers do not have ring–ring coplanarity, due to the steric interactions of the substituent. Therefore, they have reduced conjugation lengths and the corresponding change in the physical properties [57,58]. Nevertheless, substitution on the pyrrole nitrogen remains a convenient synthetic route to derivatized polypyrroles and has been used to introduce redox [59,60] and chiral substituents [61,62]. A novel synthesis based on Pd-catalyzed coupling reactions has recently allowed the preparation of well-defined substituted and unsubstituted oligopyrroles [63–65] whose structure, electrochemistry [66], and conjugation length [67,68] could be studied.

Following the success in making processible poly(3-alkylthiophenes), 3-substituted polypyrroles were also synthesized by the electrochemical or chemical oxidation of the corresponding monomers. The neutral ketopyrrole polymer [69] and poly(3-alkylpyrroles) [70,71] are soluble in common organic solvents. Substitution at the 3-position does not really affect the ring–ring coplanarity; thus the polymers remain highly conjugated. Although no detailed structural analyses have yet been reported for these polymers, they probably have an anti-coplanar structure by analogy with the 3-substituted polythiophenes. Other functional groups that have been introduced in the 3-position of polypyrrole are alkylthio [72], carboxylic [73], chiral amino acid [74], aryl [75], and alkylsulfonated groups [76]. With the latter groups, the polymer is water soluble. Finally, a recent publication reports the preparation of soluble N-substituted ring-fused polypyrroles where the substitution is on the "spacer" ring, thus minimizing the steric hindrance to coplanarity [77].

C. Polyanilines

Polyaniline is certainly the oldest conjugated polymer and was first reported in 1862 by Letheby [78]. In 1910, polyaniline was described as an octamer existing in four different states [79] and was later analyzed for its electrical properties by the group of Jozefowicz [80]. Polyaniline has attracted much interest because it is reasonably stable toward water and oxygen and has attractive electrical and optical properties [81–86]. Polyaniline is represented by the following general formula: $—[(B—NH—B—NH)_y(B—N=Q=N)_{1-y}]_n$, where B denotes a benzenoid reduced unit and Q, a quinoid oxidized unit. This is structure A in Figure 5. The polyaniline segments exist in three discrete oxidation levels, corresponding to the case where y equals 1 (leucoemeraldine), y equals 0.5 (emeraldine), or y equals 0 (pernigraniline) [87]. All the other average oxidation states are a mixture of any two of these oxidation states. The emeraldine base form (A) is particularly interesting since it may undergo reversible protonic acid doping, as shown in Figure 5. This produces reversible changes in the optical and

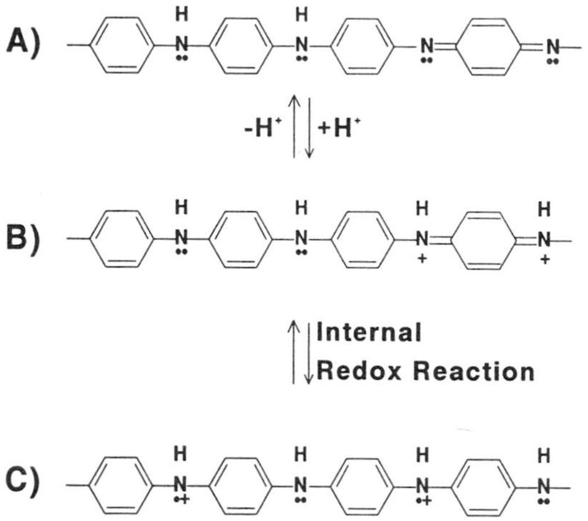

FIGURE 5 Protonic acid doping in emeraldine.

electrical properties [88,89]. Structure C represents the emeraldine salt form. It is conducting and is produced directly from the electrochemical or chemical oxidation of aniline in aqueous acidic conditions. The emeraldine base (A) is insulating and is obtained by deprotonation of the emeraldine salt in aqueous basic solution [90]. The structures shown in Figure 5 were confirmed by comparisons with well-defined model compounds [91]. Various synthetic procedures were recently reported which improve the physical properties of the polyaniline [92,93].

The fully reduced polyaniline (leucoemeraldine) (Fig. 6A) [87] and the fully oxidized polyaniline (pernigraniline) (Fig. 6D) [94] are both insulating and were recently chemically synthesized from the emeraldine state (Fig. 6B). The electrooxidation of aniline in acidic solutions produces the polymer, which is in the various oxidation states shown in Figure 6. Even the unstable intermediate state (Fig. 6C) that represents the fully oxidized polyaniline (pernigraniline) in a protonated form was recently stabilized and confirmed to be present in highly acidic conditions [95,96].

The emeraldine base form is soluble in N-methylpyrrolidinone [97] and in concentrated protonic acids such as H_2SO_4 [98]. To improve the solubility of polyaniline, especially in organic solvents, derivatives of polyaniline containing different alkyl [99–103], alkoxy [104–106], and aryl substituents [107,108] have been synthesized and characterized. In general, they are reasonably soluble in organic solvents such as chloroform and tetrahydrofuran. They exhibit acceptable electrical conductivity but have relatively low molecular weights compared to polyaniline. Finally, several water-soluble derivatives of polyaniline have been

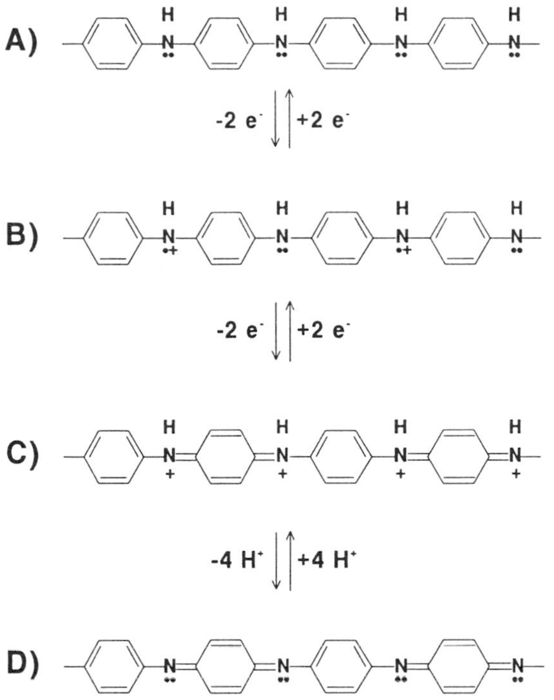

A)

-2 e$^-$ $\downarrow\uparrow$ $+2$ e$^-$

B)

-2 e$^-$ $\downarrow\uparrow$ $+2$ e$^-$

C)

-4 H$^+$ $\downarrow\uparrow$ $+4$ H$^+$

D)

FIGURE 6 Electrochemical reactions in polyaniline.

reported. The sulfonation of polyaniline produces a water-soluble polymer that is also self-doped [109–112]. The polymerization of diphenylaminesulfonic acid [113,114] and the copolymerization of diphenylaminesulfonic acid and aniline produces soluble polymers where in the latter, the solubility and conductivity of the polymer can be adjusted in a regular manner by varying the monomer ratio in the copolymer [101,112–114]. The derivatization of polyaniline is not limited to the phenyl ring but can also be accomplished on the nitrogen atom. Indeed, N-substituted polyanilines with pendent alkylsulfonate groups have been prepared. These materials are soluble in water and have low conductivities compared to polyaniline [115–117]. The proper balance between solubility and electrical conductivity may be in the copolymers of N-alkylated and unsubstituted anilines [113,114].

D. Polyphenylenes and Related Polymers

Taking the lead from the progress with the other conductive polymers, soluble neutral alkyl-substituted polyparaphenylenes were recently prepared by a new palladium-catalyzed polymerization method [118,119]. Moreover, the bandgap

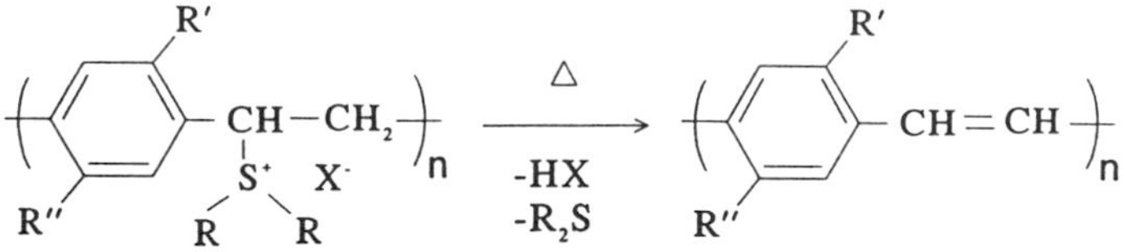

FIGURE 7 Synthesis of poly(phenylene vinylenes) via the pyrolysis of sulfonium salt precursors.

and oxidation potential of the conjugated polyphenylenes can be adjusted to lower levels by incorporating the vinylene group in the polymer chain [120]. Indeed, as shown in Figure 7, highly conjugated poly(phenylene vinylenes) were prepared via the pyrolysis of the soluble precursor with the desired alkyl group in the 2- and 5-positions [121,122]. Poly(2,5-dimethoxyphenylene vinylene) was prepared in this way [123], and with long, flexible side chains, the polymers are soluble [124,125]. Soluble derivatives of the poly(furanylene vinylene) [126] and poly(thienylene vinylene) [127–129] polymers were also prepared in this way. Finally, processible poly[1,4-bis(2-thienyl)phenylenes)] with properties intermediate between those of the polyphenylenes and polythiophenes were recently synthesized [130].

III. SOLUTION AND MELT PROCESSING

As mentioned above, the substituted polymers are processible and can now be considered for use in practical devices. The soluble polymers can be spin-coated or cast from solutions to produce coatings or free-standing films. The high-molecular-weight polymers form films, coatings, and even fibers with good mechanical properties. Since most of the substituted polymers are soluble only when in the neutral and insulating form, they must be processed in this form and then oxidized if the conductive form is desired. The emeraldine form of polyaniline is processible in acidic solutions and produces films and fibers in the conductive form [131]. Moreover, with the right protonic acid (e.g., dodecylbenzenesulfonic acid) it is possible to obtain the emeraldine salt, which is conductive and soluble in common nonpolar or weakly polar organic solvents [132]. The process used affects the morphology of the resulting material. For instance, the as-synthesized emeraldine salt (ES-I) has a crystal structure which is different from the one for emeraldine salt (ES-II), which is prepared by the protonation of the solution-processed emeraldine base [133,134].

Polyaniline [131,135,136], poly(phenylene vinylenes) [137,138], poly-(thienylene vinylene) and poly(3-alkylthiophenes) [139–141] were recently processed in the form of stretch-oriented films or fibers. Draw ratios up to 20 were obtained with poly(thienylene vinylene) [142,143]. These oriented materials exhibit anisotropic mechanical, optical, and electrical properties. Variations in the melt temperature of the poly(3-alkylthiophenes) were observed with the length of the side chain [139], and this provides some latitude for selecting the process temperature. Finally, both polyaniline [136] and poly(3-alkylthiophenes) [144] can be stabilized in solution as gels, and more interestingly, spherical particles of conducting polymers have been prepared by dispersion polymerization [145–147]. These particles are 100 to 500 nm in diameter, form stable suspensions, and have conductivities close to the values for the electropolymerized materials.

A. Blends and Composites

For many applications it is necessary to combine the conjugated polymer with a common thermoplastic to obtain a material that has good mechanical properties, is low in cost, and has the desired optical, electrical, or electrochemical properties. Whether the resulting mixtures are composites (segregated mixtures) or blends (dispersions at a molecular level) will depend on the polymer–polymer compatibility in the solid state [148]. Since in most studies the complete characterization of the mixtures and the compatibility of the components are not reported, our comments on these mixtures containing the conjugated polymers do not differentiate between blends and composites.

Poly(3-alkylthiophenes) and their copolymers have been solution and melt blended with various common polymers, such as polystyrene [148,149], polyethylene [149], and poly(ethylene vinyl acetate) [149–151]. Subsequent oxidation of the blend produces materials with electrical conductivities as high as 0.1 S/cm and with only 5 to 10% (v/v) conductive polymer in the blend [143,146,150]. Polyaniline has been blended with several synthetic polymers to produce stable, conductive materials, including some fibers with poly(*p*-phenylene terephthalamide) [152,153].[†] Polyaniline has also been solution processed with polymers such as polyethylene, poly(methyl methacrylate), nylons, and poly(vinyl chloride) [132,154,155].

B. Langmuir–Blodgett Films

The Langmuir–Blodgett technique is a very powerful method for preparing ultrathin layers of amphiphilic molecules with well-defined molecular organization. This technique has been used to manipulate various conjugated molecules and polymers to design novel polymeric architectures with control at the molecular level. For example, it was shown that the polymerization of a mixed monolayer

---

[†]An interesting discussion about soluble vs. dispersed conducting polymers is presented in Ref. 153.

of pyrrole and 3-hexadecylpyrrole gives a copolymer with a structural organization substantially different from bulk polymerized films [156]. Similar results were also obtained with octadecyl 4-methylpyrrole-3-carboxylate and 4-methyl-1-octadecylpyrrole-3-carboxylic acid [158]. Alternatively, films have been produced using preformed polymers. Thin films of poly(2-octadecoxyaniline) [157,158] and poly(3,4-dibutoxythiophene) were prepared by this technique [159]. The films of the former exhibit anisotropic conductivity [158]. The monolayers of the pure conjugated polymers are often not sufficiently stable to be transferred onto substrates; thus mixed monolayers containing poly(3-alkylthiophenes) and amphiphilic molecules, such as stearic acid [160,161], arachidic acid [162], and poly(isobutyl methacrylate) [163], were prepared instead. Finally, monolayers of polyion complexes consisting of a mixture of stearylamine and poly-(thiophene-3-acetic acid) or sulfonated polyaniline have also been reported [163].

IV. APPLICATIONS

A. Light-Emitting Diodes

Light-emitting diode (LED) prototypes that use a conductive polymer have received much attention after the first report by Burroughes and co-workers in 1990 [164]. The basic structure of the LED is shown in Figure 8. It consists of a positive hole-injecting electrode with a high work function, ϕ, such as In-SnO$_2$ (ITO) or an electroactive polymer, a negative electron-injecting electrode with a low ϕ such as Al, In, Mg, or Ca, and the light-emitting conductive polymer film sandwiched between these two electrodes [165–168]. These films are fabricated one layer at a time and then encapsulated in transparent glass or

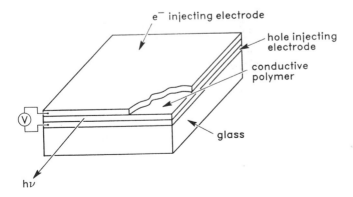

FIGURE 8 Schematic sketch of polymer-based electroluminescent device.

flexible polymeric materials. The conductive polymer films are solution cast [166,169,170], vacuum deposited [171], or generated directly by the polymerization reaction [164,167,168,172–174].

In this layered structure the injected holes and electrons migrate across the polymer layer, combine to form excitons, which then decay with photon emission as shown in Figure 9 [164,175,176]. The LED prototype consisting of ITO/poly(p-phenylene vinylene) (PPV)/Al emitted a green-yellow light (2.5 eV) at 10 V forward voltage with a quantum efficiency μ (photons/electron-injected) near 0.05% [164]. The LED prototype constructed by Heeger and co-workers [166] using ITO/poly[2-methoxy-5-(2'-ethylhexoyl)-1,4-phenylenevinylene]

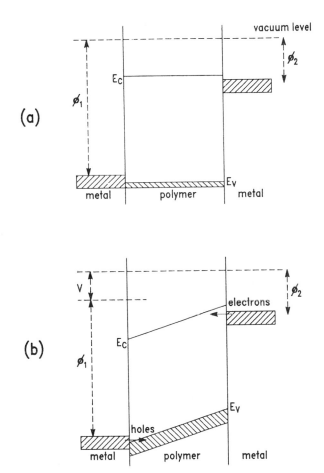

FIGURE 9 Band scheme for a polymer LED under (a) flat-band conditions, no external bias, only ϕ_1–ϕ_2, and (b) with forward bias V. (From Ref. 165.)

(MEH-PPV)/In emits a yellow-orange light (2.1 eV) at 9 V forward voltage with an intensity that decreases when the temperature is lowered or the forward voltage is increased. The μ is ca. 0.05%. The photoluminescence spectrum emitted from this device is essentially the same as the electroluminescence spectrum of the polymer with the emission energy just below the bandgap (Fig. 10). Furthermore, $\ln(I/V^2)$ is linear with $1/V$, which is consistent with the tunneling theory. This suggests that the electroluminescence originates from the recombination of the opposite charge polarons (radical cations) generated by the injection of electrons and holes, and that the electron injection from the rectifying metal contact into the gap states of the positive polaron majority carriers is a dominating process [160]. It was shown further that μ can be enhanced by using lower-ϕ materials at the electron-injecting electrode. This regulates the hole and electron injection rate so that they combine in the bulk of the emissive polymer layer. For example, using Ca (ϕ = 3.0 eV) in place of In (ϕ = 4.2 eV) as the negative electron-ejecting electrode in a ITO/MEH-PPV/Ca prototype increased μ to ca. 1% and light emission begins at 4 V forward bias [166]. The use of Ca is not practical, however, because of its sensitivity to ambient conditions.

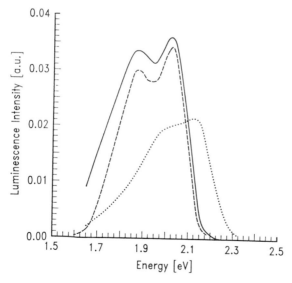

FIGURE 10 Electroluminescence intensity vs. photon energy at 300 K (dotted) and 90 K (solid) for indium/MEH-PPV diode under a forward bias of 13 V dc with 3 V ac superposed at 681 Hz; photoluminescence spectrum (dashed) at 90 K shown for comparison. (From Ref. 166.)

Alternatively, μ can be enhanced by placing an electron-transport layer between the emissive conductive polymer layer and the negative electrode as shown in Figure 11 [176]. This approach was demonstrated by Friend and co-workers using [2-(4-biphenylyl)-5-(4-*tert*-butylphenyl)-1,3,4,-oxadiazole] in poly(methyl methacrylate) (PBO/PMMA) as the transport layer [176]. Finally, μ can also be enhanced by adjusting the conjugation length along the polymer backbone as was demonstrated with poly(3-alkylthiophene) [171] or by incorporating structural units with different π–π* energies into the polymer. With the shorter conjugation lengths, the mobility of the excitons is reduced. New processible and stable conductive polymers are also being used as the electron-injecting electrode. Films containing polyaniline [169] and poly(carbonate-co-styrylamine) [177] have replaced In-SnO$_2$ successfully as the hole-injecting electrode. A list of the light-emitting diodes that have been fabricated with conductive polymers and their electroluminescent characteristics are summarized in Table 1.

The use of conductive polymers in light-emitting diodes offers several important advantages over inorganic and molecular organic materials. The polymers can be processed into flexible films with good mechanical properties and large surface areas. Their spectral response can be adjusted and they are less expensive to produce [178]. On the other hand, their long-term stability needs improvement. With the availability of synthetic routes to derivatized polymers with modified physical properties, it is not unreasonable to expect that the needed improvements can be accomplished in the near future. Low-cost all-plastic LEDs with different sizes, shapes, and colors will become a reality in the days to come.

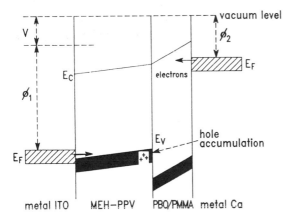

FIGURE 11 Band scheme for LED with a (PBO/PMMA) electron injection layer. (From Ref. 176.)

TABLE 1 Light-Emitting Diodes of Various Conductive Polymers

Polymer[a]	− Electrode	+ Electrode	Colors[b]	Q.Y. (%)	Ref.
PPV	ITO	Al	G/Y	0.05	164
PPP	ITO	Al	Blue	0.05	167
MEH-PPV	ITO	In	Y/O	0.05	166
MEH-PPV	PANI	Ca	Y/O	1.0	166
PPV-PDMPPV	ITO	Al	G/Y to O/R	0.01–0.3	173
PTH	ITO	MgIn	O/R	0.3	170

[a]PPV, poly(*p*-phenylene vinylene); PPP, poly-*p*-phenylene; MEH-PPV, poly[2-methoxy-5-(2′-ethylhexoxy)-1,4-phenylene vinylene]; PPV-PDMPPV, poly(*p*-phenylene vinylene-co-dimethylphenylene divinylene); PTH, poly(3-alkylthiophene).
[b]G, green; Y, yellow; B, blue; O, orange; R, red.

B. Electrochromic Display Devices

Conductive polymers are electrochromic and undergo color change when they are switched between the neutral and oxidized states. They are particularly suited for use in electrochromatic display devices (ECDs) where the color displayed is controlled by the applied voltage. The basic structure of ECD device is a layered structure containing three different films, a transparent conductive layer (e.g., In-SnO$_2$ or SnO$_2$) on a glass or plastic which is also an electrode, the electrochromic layer supported on the second electrode, and an ion-conducting electrolyte which separates these two electrochromic layers (electrodes) physically. This design is shown in Figure 12. Many materials display electrochromism, including metal oxides (WO$_3$, MoO$_3$, IrO$_2$), organic, and intercalated compounds (viologens, Prussian blue, phthalocyanines) [179]. These materials can be devised into cathodic and anodic electrochromic materials, depending on the coloration mechanism. However, the commercialization of ECDs using these

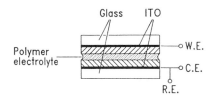

☑ Anodic electrochromic film

☒ Cathodic electrochromic film

FIGURE 12 Structure of an electrochromic smart window device.

materials is still limited by high cost of fabrication (vacuum evaporation, sputtering) and high costs of certain materials such as IrO_2. In the case of ECDs using viologen or Prussian blue, the stability of these materials to repeated cycles is a limitation. Polyaniline, polythiophene, polypyrrole, and their derivatives have been tested in prototype ECDs and shown to be quite stable to the switching process in the absence of impurities in the electrolyte. These polymers can function well as the complementary electrode in the electrochromic display device [180–184]. These polymers can be prepared as thin films by less expensive procedures such as solution casting or electrodeposition. An electrochromic display device consisting of WO_3/poly(2-acrylamido-2-methyl-1-propanesulfonic acid)/poly(N-benzylaniline) (PBAN) was fabricated by Nguyen and co-workers [185]. This device shows reversible color changes from transparent clear yellow to deep blue when the applied potential is changed from -0.1 V to 1.0 V. The colorization results from the insertion of H^+ into WO_3 (transparent) to form $(WO_3)H^+$ (dark blue) and the oxidation of poly(N-benzylaniline) (clear yellow) to the oxidized states (blue) with the incorporation of the anions.

This device could be cycled more than 10^5 times between the bleached and colored states without significant degradation. But the long-term stability is eventually limited by the slow dissolution of the WO_3 and the transparent conductive layer by the acid electrolyte. Electrochromic display devices using polyaniline [186] poly(diphenylamine) [187], polypyrrole [181], and polythiophene [182] were also fabricated with different counterelectrodes and electrolytes. Although conductive polymers offer some advantages over other materials in term of coloration, processibility, and lower material costs, adhesion to the support electrode and long-term stability remain as serious limitations.

C. Rechargeable Batteries

Rechargeable batteries are among the first commercial products to incorporate conductive polymers [188,189]. The basic structure of a rechargeable battery is again a layered structure much like the LED and ECD devices. The more successful batteries have been constructed using polypyrrole, polythiophene, polyaniline, or their derivatives as cathodes, and Li metal or Li–Al alloy as the anodes. The electrolyte separator can be an organic solution of a lithium salt, such as propylene carbonate–$LiBF_4$, or a solid polymer electrolyte, such as poly(ethylene oxide)–or poly(methoxyethoxyphosphazene)–lithium salt [188,189]. These commercial batteries have been fabricated in different shapes and sizes, such as a button, cyclinder, or credit card. Bridgestone commercialized a button-type, 3-V battery with a polyaniline cathode and a Li–Al alloy anode. Similarly, BASF brought to the market a button-type, 3-V rechargeable battery using polypyrrole. These batteries compare well with the Ni–Cd battery in terms of long life, low power, and reliability and are used as backup batteries for static random access memories, cellular phones, hand-held calculators, and wristwatches [188]. The development of commercial high-current-density rechargeable batteries using conductive polymers is still a challenge for scientists and engineers.

D. Biochemical and Chemical Sensors

Conductive polymers have been used successfully in the development of enzyme biosensors for the detection of glucose, penicillin, and other biologically important substances in biomedical analysis. Polypyrrole and polyaniline films are particularly appropriate for this application because they are environmentally stable and enzymes can be immobilized on them by electrodeposition [190–195]. In general, enzyme biosensors measure a current or resistance change in the conductive polymer film in response to the products of the enzymatic reaction. A biosensor prototype for the detection of penicillin was fabricated using polypyrrole and penicillinase membrane by Nishizawa and co-workers [196]. As shown in Figure 13, the sensor design has a thin film of polypyrrole coated on a hydrophobically pretreated array electrode. This film was coated with a cross-linked penicillinase membrane that was generated from a mixture of penicillanse, bovine serum albumin, and glutaraldehyde. The coated electrode array was then placed in a flow-through cell supported with an O-ring for measurements. This biosensor functions as a pH-sensitive device, where the resistance of the polypyrrole film responds to the solution pH. During analysis, penicillinase catalyzes the hydrolysis of penicillin to penicilloic acid and acidifies the

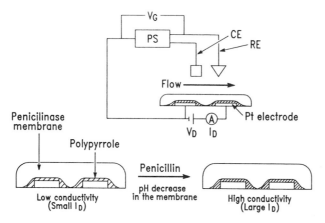

FIGURE 13 Principle of the penicillin sensor and the configuration of electrochemical apparatus for in situ conductivity measurement. PS, potentiostat; CE, Pt counterelectrode; RE, reference electrode (SCE); V_G, gate voltage; V_D and I_D, drain voltage and current. (From Ref. 196.)

polypyrrole film. As shown in Figure 14, the current response of the polypyrrole film is proportional to the concentration of penicillin in the solution. This biosensor is sufficiently stable that it can be used to analyze several samples without significant degradation.

In a similar configuration, a glucose biosensor prototype was fabricated by Contractor and co-workers [197] using polyaniline with surface-immobilized glucose oxidase. The operation of this sensor is also based on the change in the resistance of the polyaniline film as the solution pH changes from the enzymatic oxidation of glucose to gluconic acid. The resistance response of the device increases linearly with the glucose concentration up to 10 mM. This sensor shows an initial loss in activity before reaching a stable level, then remains stable within 10% for several measurements. The initial loss of activity is attributed to the leaching out or deactivation of the loosely bound enzyme. Recently, a glucose biosensor using polypyrrole with covalently bonded glucose oxidase enzyme was reported by Lowe and co-workers [190]. The stability of this biosensor is nearly 200-fold greater than the sensor that physically entraps enzyme. Electrodes coated with polyaniline or polypyrrole for use in biosensor applications have been commercialized by EG & G PARC.

Conductive polymers have also been used for development of gas sensors and this topic has been reviewed by Zotti [195]. In general, the operation of conductive polymer gas sensors is based on the decrease in the film resistance

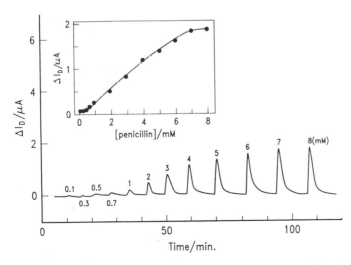

FIGURE 14 I_D responses of the penicillin sensor upon injection of various concentrations of penicillin solutions. Experimental conditions are the same as those in Figure 12. (From Ref. 196.)

when the polymer is oxidized. For example, in the sensor constructed by Hanawa and co-workers [198] neutral polypyrrole is used to detect oxidant gases such as SO_2, NO_2, and I_2. Gas sensors using polypyrrole in the oxidized form were used to detect reducing gases such as NH_3, NO_2, and H_2S [199–201]. Other electrode sensors which are based on conductive polymers have been applied to the detection of alcohol and humidity [202,203].

E. Microlithography

Conductive polymers, particularly polyaniline, polythiophene, and their derivatives, were found suitable for microlithiographic applications. Angelopoulos and co-workers [204–207] reported that polyaniline can be made into a high-resolution negative conducting resist with 0.25-μm lines with a conductivity of 0.1 S/cm. The microlithographic process of polyaniline involved spin-coating the *N*-methylpyrrolidinone solution containing undoped polyaniline and triphenylsulfonium hexafluoroantimonate, then exposing the films to either UV radiation through a 240-nm narrowband filter or to a 25-keV electron beam through a photomask. In the exposed regions of the film the onium salt photodegrades, producing acid that dopes polyaniline and renders it insoluble. In the unexposed regions the polymer remains soluble and is removed by washing with *N*-methylpyrrolidone. In a more recent report, the same group describe the use of a polyaniline–polyacid blend which is water soluble and can be used as a surface discharge layer for electron-beam lithography [207]. A similar procedure was applied to polythiophene and its derivatives [208]. Recently, Abdou and co-workers [209] also reported that patterns of micron resolution on the doped poly(alkylthiophene) films could be generated directly by laser. In the exposed regions the polymer becomes insoluble to organic solvents, possibly due to photo-induced cross-linking. Similar results were also found with the doped poly(aniline-co-*N*-allylaniline) and poly(aniline-co-*N*-benzylaniline) films [210]. The presence of photo-cross-linkable functional segments in these copolymers allows the films to be patterned using UV-visible light to produce structures with sharp edges.

F. Other Applications

Conductive polymers have also been considered for use in many other applications that were not covered here. This includes a variety of unrelated unique applications such as electromagnetic (EMF) shields, conductive adhesives, electrostatic dissipators, antistatic films, paints, and fibers. From the initial feasibility studies, conductive polymers perform well in many of these applications. As may be expected, they have also been considered for use in electronic devices such as Schottky diodes, metal–insulator–semiconductor diodes and metal–insulator–semiconductor field–effect transistors [211–213]. However, the motivation for using conductive polymers in commercial devices is low because the carrier mobility in these polymers is much lower than in inorganic semiconductors [214].

V. SUMMARY

It has become evident that conductive polymers with intermediate conductivities will be chosen most easily for microelectronic and other technical applications. Most applications require the conducting polymers to possess good mechanical properties, processing latitude, and performance reliability. This combination of properties is not attained by the parent polymers which have the highest conductivities. These properties can only be achieved by derivatizing the polymers to enable processing and to enhance compatibility for formulating blends and composites with other polymers which exhibit the necessary mechanical properties. The focus has shifted from that of 10 to 15 years ago, when most of the experimental work was devoted to understanding the transport mechanism in these polymers, toward the synthesis of polymers with the highest conductivities possible. Since the involvement of synthetic chemists, good progress has been made in producing derivatized conductive polymers with modified properties. However, the successful development of applications and devices requires enhanced stability and long-term performance reliability of these polymers.

ACKNOWLEDGMENTS

M. L. thanks S. Holdcroft for kindly providing a sample of poly(3-hexylthiophene) for NMR analysis.

REFERENCES

1. E. M. Genies, A. Boyle, M. Lapkowski, and C. Tsintavis, *Synth. Metals.*, *36*: 139 (1990).
2. J. R. Reynolds, C. K. Baker, C. A. Jolly, P. A. Poropatic, and J. P. Jose, in *Conductive Polymers and Plastics* (J. M. Margolis, ed.), Chapman & Hall, London, 1989, Chap. 1.
3. M. Goosey, in *Specialty Polymers* (R. W. Dyson, ed.), Blackie, Glasgow, 1987.
4. G. P. Evans, *Adv. Electrochem. Sci. Eng.*, *1*: 1 (1990).
5. A. F. Diaz, J. F. Rubinson, and H. B. Mark, Jr., *Adv. Polym. Sci.*, *84*: 113 (1988).
6. H. Naarmann and N. Theophilou, in *Electroresponsive Molecular and Polymeric Systems* (T. A. Skotheim, ed.), Marcel Dekker, New York, 1988, Chap. 1.
7. M. Gauthier, M. Armand, and D. Muller, in *Electroresponsive Molecular Polymeric Systems* (T. A. Skotheim, ed.), Marcel Dekker, New York, 1988, Chap. 2.
8. H. Naarmann and P. Strohriegel, in *Handbook of Polymer Synthesis* (H. R. Kricheldorf, ed.), Marcel Dekker, New York, 1992, Chap. 21.

9. T. A. Skotheim, ed., *Handbook of Conducting Polymers*, Vol. 1, Marcel Dekker, New York, 1986, p. 82.
10. A. Diaz, in *Organic Electrochemistry* (H. Lund and M. M. Baizer, eds.), Marcel Dekker, New York, Chap. 33.

Polythiophenes

11. G. Tourillon and F. Garnier, *J. Electroanal. Chem.*, *135*: 173 (1982).
12. J. Roncali, *Chem. Rev.*, *92*: 711 (1992).
13. E. M. Genies, G. Bidan, and A. Diaz, *J. Electroanal. Chem.*, *149*: 101 (1983).
14. A. F. Diaz and J. Bargon, in *Handbook of Conducting Polymers*, Vol. 1 (T. A. Skotheim, ed.), Marcel Dekker, New York, 1986, p. 82.
15. M. Ballauf, *Angew. Chem. Int. Ed. Engl.*, *101*: 261 (1989).
16. R. L. Elsenbaumer, K. Y. Jen, and R. Oboodi, *Synth. Metals*, *15*: 169 (1986).
17. M. A. Sato, S. Tanaka, and K. Kaeriyama, *J. Chem. Soc. Chem. Commun.*, 873 (1986).
18. S. Hotta, S. D. D. V. Rughooputh, A. J. Heeger, and F. Wudl, *Macromolecules*, *20*: 212 (1987).
19. J. Roncali, R. Garreau, A. Yassar, P. Marque, F. Garnier, and M. Lemaire, *J. Phys. Chem.*, *91*: 6706 (1987).
20. R. Sugimoto, S. Takeda, H. B. Gu, and K. Yoshino, *Chem. Express*, *1*: 635 (1986).
21. M. Leclerc, F. M. Diaz, and G. Wegner, *Makromol. Chem.*, *190*: 3105 (1989).
22. R. M. Souto Maior, K. Hinkelmann, H. Eckert, and F. Wudl, *Macromolecules*, *23*: 1268 (1990).
23. M. A. Sato and H. Morii, *Macromolecules*, *24*: 1196 (1991).
24. A. Bolognesi, M. Catellani, S. Destri, and W. Porzio, *Makromol. Chem. Rapid Commun.*, *12*: 9 (1991).
25. M. J. Winokur, P. Wamsley, J. Moulton, P. Smith, and A. J. Heeger, *Macromolecules*, *24*: 3812 (1991).
26. T. J. Prosa, M. J. Winokur, J. Moulton, P. Smith, and A. J. Heeger, *Macromolecules*, *25*: 4364 (1992).
27. J. Mardalen, E. J. Samuelsen, O. R. Gautun, and P. H. Carlsen, *Synth. Metals*, *48*: 363 (1992).
28. R. D. McCullough and R. D. Lowe, *J. Chem. Soc. Chem. Commun.*, *70*: (1992).
29. T. A. Chen and R. D. Rieke, *J. Am. Chem. Soc.*, *114*: 10087 (1992).
30. H. Masuda, K. Kaeriyama, H. Suezawa, and M. Hirota, *J. Polym. Sci. Polym. Chem. Ed.*, *30*: 945 (1992).
31. J. P. Ferraris and M. D. Newton, *Polymer*, *33*: 391 (1991).
32. M. C. Gallazi, L. Castellani, G. Zerbi, and P. Sozzani, *Synth. Metals*, *41*: 495 (1991).
33. A. O. Patil, Y. Ikenoue, F. Wudl, and A. J. Heeger, *J. Am. Chem. Soc.*, *109*: 1858 (1987).
34. Y. Ikenoue, J. Chiang, A. O. Patil, F. Wudl, and A. J. Heeger, *J. Am. Chem. Soc.*, *110*: 2983 (1988).
35. A. C. Chang, R. L. Blankespoor, and L. L. Miller, *J. Electroanal. Chem.*, *236*: 239 (1987).

36. M. Feldhues, G. Kampf, H. Litterer, and T. Mecklenburg, *Synth. Metals, 28*: C487 (1989).
37. M. Leclerc and G. Daoust, *J. Chem. Soc. Chem. Commun.*, 273 (1990).
38. G. Daoust and M. Leclerc, *Macromolecules, 24*: 455 (1991).
39. R. Cloutier and M. Leclerc, *J. Chem. Soc. Chem. Commun.*, 1194 (1991).
40. G. Heywang and F. Jonas, *Adv. Mater., 4*: 116 (1992).
41. W. Buchner, R. Garreau, M. Lemaire, and J. Roncali, *J. Electroanal. Chem., 277*: 355 (1990).
42. M. Lemaire, R. Garreau, D. Delabouglise, J. Roncali, H. K. Youssoufi, and F. Garnier, *New J. Chem., 14*: 359 (1990).
43. M. Ueda, Y. Miyaji, T. Ito, Y. Oba, and T. Sone, *Macromolecules, 24*: 2694 (1991).
44. Q. Pei, H. Jarvinen, J. E. Osterholm, O. Inganas, and J. Laakso, *Macromolecules, 25*: 4297 (1992).
45. J. Roncali, R. Garreau, D. Delabouglise, F. Garnier, and M. Lemaire, *Synth. Metals, 28*: C341 (1989).
46. B. Zinger, Y. Greenwald, and I. Rubinstein, *Synth. Metals, 41*: 583 (1991).
47. L. H. Shi, F. Garnier, and J. Roncali, *Macromolecules, 25*: 6425 (1992).
48. J. P. Ruiz, K. Nayak, D. S. Marynick, and J. R. Reynolds, *Macromolecules, 22*: 1231 (1989).
49. F. Andreani, P. Costa Bizzari, C. Della Casa, and E. Salatelli, *Polym. Bull., 27*: 117 (1991).
50. R. Mirrazaei, D. Parker, and H. S. Munro, *Synth. Metals, 30*: 265 (1989).
51. J. Grinshaw and S. D. Perara, *J. Electroanal. Chem., 278*: 287 (1990).
52. P. Bauerle and K. U. Gaudl, *Adv. Mater., 2*: 185 (1990).
53. M. Lemaire, D. Delabouglise, R. Garreau, A. Guy, and J. Roncali, *J. Chem. Soc. Chem. Commun.*, 658 (1988).
54. D. Kotkar, V. Joshi, and K. Gosh, *J. Chem. Soc. Chem. Commun.*, 917 (1988).

Polypyrroles

55. A. F. Diaz, K. K. Kanazawa, and G. P. Gardini, *J. Chem. Soc. Chem. Commun.*, 635 (1979).
56. G. B. Street, in *Handbook of Conducting Polymers*, Vol. 1 (T. A. Skotheim, ed.), Marcel Dekker, New York, 1986, p. 265.
57. A. Diaz, J. Castillo, K. K. Kanazawa, J. A. Logan, M. Salmon, and O. Fajardo, *J. Electroanal. Chem., 133*: 233 (1982).
58. M. Salmon, M. Aguilar, and M. Saloma, *J. Chem. Soc. Chem. Commun.*, 570 (1983).
59. S. Cosnier, A. Deronzier, and J. C. Moutet, *J. Electroanal. Chem., 193*: 193 (1985).
60. A. Deronzier and J. C. Moutet, *Acc. Chem. Res., 22*: 249 (1989), and references therein.
61. M. Salmon, M. Saloma, G. Bidan, and E. M. Genies, *Electrochim. Acta, 34*: 117 (1989).
62. J. C. Moutet, E. Saint-Aman, F. Tran-Van, P. Angibeaud, and J. P. Utille, *Adv. Mater., 4*: 511 (1992).

63. S. Martina, Thesis in Chemistry, University of Mainz, Germany, 1993.
64. S. Martina, V. Enkelmann, A.-D. Schluter, G. Wegner, G. Zotti, and G. Zerbi, *Synth. Metals*, *55*: 1096, (1993).
65. S. Martina, V. Enkelmann, G. Wegner, and A.-D. Schulter, *Synth. Metals*, *51*: 299 (1992).
66. G. Zotti, S. Martina, G. Wegner, and A.-D. Schluter, *Adv. Mater.*, *4*: 798 (1992).
67. G. Zerbi, M. Veronelli, S. Martina, A.-D. Schluter, and G. Wegner, *J. Chem. Phys.*, *100*: 978 (1994).
68. G. Zerbi, M. Veronelli, S. Martina, and A.-D. Schluter, *Adv. Mater.*, *6*: 385 (1994), submitted.
69. M. R. Bryce, A. Chissel, P. Kathirgamanathan, D. Parker, and N. M. R. Smith, *J. Chem. Soc. Chem. Commun.*, 466 (1987).
70. J. Ruhe, T. Ezquerra, and G. Wegner, *Makromol. Chem. Rapid Commun.*, *10*: 103 (1989).
71. H. Masuda, S. Tanaka, and K. Kaeriyama, *J. Chem. Soc. Chem. Commun.*, 25 (1989).
72. G. Zotti, G. Schiavon, A. Berlin, and G. Pagani, *Synth. Metals*, *28*: C183 (1989).
73. R. Casas, A. Dicko, J. M. Ribo, M. A. Valles, N. Ferrer-Anglada, R. Bonnett, N. Hanly, and D. Bloor, *Synth. Metals*, *39*: 275 (1990).
74. D. Delabouglise and F. Garnier, *Synth. Metals*, *39*: 117 (1990).
75. A. B. Kon, J. S. Foos, and T. L. Rose, *Chem. Mater.*, *4*: 416 (1992).
76. E. E. Havinga, W. ten Hoeve, E. W. Meijer, and H. Wynberg, *Chem. Mater.*, *1*: 650 (1989).
77. A. Berlin, G. Pagani, G. Zotti, and G. Schiavon, *Makromol. Chem.*, *193*: 399 (1992).

Polyanilines

78. H. Letheby, *J. Chem. Soc.*, *15*: 161 (1862).
79. A. G. Green and A. E. Woodhead, *J. Chem. Soc.*, *97*: 2388 (1910).
80. R. de Surville, M. Jozefowicz, L. T. Yu, J. Perichon, and R. Buvet, *Electrochim. Acta*, *13*: 1451 (1968).
81. A. F. Diaz and J. A. Logan, *J. Electroanal. Chem.*, *111*: 111 (1980).
82. R. Noufi, A. J. Nozik, J. White, and L. Warren, *J. Electrochem. Soc.*, *129*: 2261 (1982).
83. A. G. MacDiarmid, J. C. Chiang, M. Halpern, W. S. Huang, S. L. Mu, N. L. D. Somarisi, W. Wu, and S. I. Yaniger, *Mol. Cryst. Liq. Cryst.*, *121*: 173 (1985).
84. W. S. Huang, B. D. Humphrey, and A. G. MacDiarmid, *J. Chem. Soc. Faraday Trans. 1*, *82*: 2385 (1986).
85. E. M. Genies, C. Tsintavis, and A. A. Syed, *Mol. Cryst. Liq. Cryst.*, *121*: 181 (1985).
86. E. M. Genies and C. Tsintavis, *J. Electroanal. Chem.*, *195*: 109 (1985).
87. J. G. Masters, Y. Sun, A. G. MacDiarmid, and A. J. Epstein, *Synth. Metals*, *41*: 715 (1991).

88. A. J. Epstein, J. M. Ginder, F. Zuo, R. W. Bigelow, H. S. Woo, D. B. Tanner, A. F. Richter, W. S. Huang, and A. G. MacDiarmid, *Synth. Metals, 18*: 303 (1987).
89. W. W. Focke, G. E. Wnek, and Y. Wei, *J. Phys. Chem., 91*: 5813 (1987).
90. A. G. MacDiarmid, J. C. Chiang, A. F. Richter, N. L. D. Somarisi, and A. J. Epstein, in *Conducting Polymers* (L. Alcacer, ed.), D. Reidel, Dordrecht, The Netherlands, 1987, p. 105.
91. F. Wudl, R. O. Angus, F. L. Lu, P. M. Allemand, D. J. Vachon, M. Nowak, Z. X. Liu, and A. J. Heeger, *J. Am. Chem. Soc., 109*: 3677 (1987).
92. S. P. Armes and J. F. Miller, *Synth. Metals, 22*: 385 (1988).
93. Y. Cao, A. Andreatta, A. J. Heeger, and P. Smith, *Polymer, 30*: 2305 (1989).
94. Y. Sun, A. G. MacDiarmid, and A. J. Epstein, *J. Chem. Soc. Chem. Commun.*, 529 (1990).
95. Y. Cao, P. Smith, and A. J. Heeger, *Synth. Metals, 32*: 263 (1989).
96. G. D'Aprano, M. Leclerc, and G. Zotti, *Macromolecules, 25*: 2145 (1992).
97. M. Angelopoulos, G. E. Asturias, S. P. Ermer, A. Ray, E. M. Scherr, A. G. MacDiarmid, M. Akhtar, Z. Kiss, and A. J. Epstein, *Mol. Cryst. Liq. Cryst., 160*: 151 (1988).
98. A. Andreatta, Y. Cao, J. C. Chiang, A. J. Heeger, and P. Smith, *Synth. Metals, 26*: 383 (1988).
99. M. Leclerc, J. Guay, L. H. Dao, *Macromolecules, 22*: 649 (1989).
100. Y. Wei, W. W. Focke, G. E. Wnek, A. Ray, and A. G. MacDiarmid, *J. Phys. Chem., 93*: 495 (1989).
101. Y. Wei, R. Hariharan, and S. A. Patel, *Macromolecules, 23*: 758 (1990).
102. G. Bidan, E. M. Genies, and J. F. Penneau, *J. Electroanal. Chem., 271*: 59 (1989).
103. E. M. Genies and P. Noel, *J. Electroanal. Chem., 310*: 89 (1990).
104. D. MacInnes and B. L. Funt, *Synth. Metals, 25*: 235 (1988).
105. J. C. Lacroix, P. Garcia, J. P. Audiere, R. Clement, and O. Kahn, *Synth. Metals, 44*: 117 (1991).
106. G. Zotti, N. Comisso, G. D'Aprano, and M. Leclerc, *Adv. Mater., 4*: 749 (1992).
107. M. T. Nguyen and L. H. Dao, *J. Electroanal. Chem., 289*: 37 (1990).
108. M. T. Nguyen, R. Paynter, and L. H. Dao, *Polymer, 33*: 214 (1992).
109. J. Yue and A. J. Epstein, *J. Am. Chem. Soc., 112*: 2800 (1990).
110. J. Yue, Z. H. Wang, K. R. Cromack, A. J. Epstein, and A. G. MacDiarmid, *J. Am. Chem. Soc., 113*: 2665 (1991).
111. C. DeArmitt, C. P. Armes, J. Winter, F. A. Uribe, J. Gottesfeld, and C. Mombourquette, *Polymer, 34*: 158 (1993).
112. J.-Y. Bergeron and L. D. Dao, *Macromolecules, 25*: 3332 (1992).
113. M. T. Nguyen, P. Kasai, J. L. Miller, and A. Diaz, *Macromolecules, 27*: 3625 (1994) submitted.
114. A. Watanabe, K. Mori, A. Iwabuchi, Y. Iwasaki, Y. Nakamura, and O. Ito, *Macromolecules, 22*: 3521 (1989).
115. P. Hany, E. M. Genies, and C. Santier, *Synth. Metals, 31*: 369 (1989).
116. J. Y. Bergeron, J. W. Chevalier, and L. H. Dao, *J. Chem. Soc. Chem. Commun.*, 180 (1990).

117. J. W. Chevalier, J. Y. Bergeron, and L. H. Dao, *Macromolecules, 25*: 3325 (1992).

Polyphenylenes and Related Polymers

118. M. Rehahn, A. D. Schluter, G. Wegner, and W. J. Feast, *Polymer, 30*: 1060 (1989).
119. M. Rehahn, A. D. Schluter, and G. Wegner, *Makromol. Chem., 191*: 1991 (1990).
120. H. Eckhardt, L. W. Shacklette, K. Y. Jen, and R. L. Elsenbaumer, *J. Chem. Phys., 91*: 1303 (1989).
121. D. R. Gagnon, J. D. Capistran, F. E. Karasz, and R. W. Lenz, *Polym. Bull., 12*: 293 (1984).
122. I. Murase, T. Ohnishi, and M. Hirooka, *Polym. Commun., 25*: 327 (1984).
123. I. Murase, T. Ohnishi, T. Noguchi, and M. Hirooka, *Polym. Commun., 26*: 362 (1985).
124. S. H. Askari, S. D. Rughooputh, F. Wudl, and A. J. Heeger, *Polym. Prepr. (Am. Chem. Soc. Div. Polym. Chem).*, *30*: 157 (1989).
125. S. Shi and F. Wudl, *Macromolecules, 23*: 2119 (1990).
126. K. Y. Jen, T. R. Jow, and R. L. Elsenbaumer, *J. Chem. Soc. Chem. Commun.*, 1113 (1987).
127. K. Y. Jen, T. R. Eckhardt, T. R. Jow, L. W. Shacklette, and R. L. Elsenbaumer, *J. Chem. Soc. Chem. Commun.*, 215 (1988).
128. K. Y. Jen, R. Jow, L. W. Shacklette, M. Maxfield, H. Eckhardt, and R. L. Elsenbaumer, *Mol. Cryst. Liq. Cryst., 160*: 69 (1988).
129. P. C. van Dort, J. E. Pickett, and M. L. Blohm, *Synth. Metals, 41–43*: 2305 (1991).
130. J. R. Ruiz, J. R. Dharia, J. R. Reynolds, and L. J. Buckley, *Macromolecules, 25*: 849 (1992).

Solution and Melt Processing

131. A. Andreatta, Y. Cao, J. C. Chiang, A. J. Heeger, and P. Smith, *Synth. Metals, 26*: 383 (1988).
132. Y. Cao, P. Smith, and A. J. Heeger, *Synth. Metals, 48*: 91 (1992).
133. M. E. Jozefowicz, R. Laversanne, H. H. S. Javadi, A. J. Epstein, J. P. Pouget, X. Tang, and A. G. MacDiarmid, *Phys. Rev. B, 39*: 12958 (1989).
134. J. P. Pouget, M. E. Jozefowicz, A. J. Epstein, X. Tang, and A. G. MacDiarmid, *Macromolecules, 24*: 779 (1991).
135. A. P. Monkman and P. Adams, *Synth. Metals, 40*: 87 (1991).
136. A. G. MacDiarmid, S. K. Manohar, E. M. Scherr, X. Tang, M. A. Druy, P. J. Glatkowski, and A. J. Epstein, *Polym. Mater. Sci. Eng. (Am. Chem. Soc. Div. Polym. Mater. Sci. Eng.), 64*: 254 (1991).
137. J. M. Machado and F. E. Karasz, *Polym. Prepr. (Am. Chem. Soc. Div. Polym. Chem.), 30*: 154 (1989).
138. S. Tokito, P. Smith, and A. J. Heeger, *Polymer, 32*: 464 (1991).
139. K. Yoshino, S. Nakajima, M. Fujii, and R. Sugimoto, *Polym. Commun., 28*: 309 (1987).
140. S. Hotta, M. Soga, and N. Sonoda, *J. Phys. Chem., 93*: 4994 (1989).
141. J. Moulton and P. Smith, *Synth. Metals, 40*: 13 (1991).

142. R. M. Gregorius and F. E. Karasz, *Synth. Metals, 53*: 11 (1992).
143. J. Moulton and P. Smith, *J. Polym. Sci. Polym. Phys. Ed. Pt. B, 30*: 871 (1992).
144. K. Yoshino, K. Nakao, and M. Onoda, *Jpn. J. Appl. Phys., 28*: L1032 (1989).
145. S. P. Armes and M. Aldissi, *Synth. Metals, 37*: 137 (1990); ibid., *Prog. Org. Coatings, 19*: 21 (1991).
146. B. Vincent and J. Waterson, *J. Chem. Soc. Chem. Commun.*, 683 (1990).
147. G. Markham, T. M. Obey, and B. Vincent, *Colloids Surfaces, 51*: 239 (1990).

Blends and Composites

148. S. Hotta, S. D. D. V. Rughooputh, and A. J. Heeger, *Synth. Metals, 22*: 79 (1987).
149. J. Laakso, J. E. Osterholm, and P. Nyholm, *Synth. Metals, 28*: C467 (1989).
150. J. E. Osterholm, J. Laakso, P. Nyholm, H. Isotalo, H. Stubb, O. Inganas, and W. R. Salaneck, *Synth. Metals, 28*: C435 (1989).
151. M. Isotalo, H. Stubb, P. Yli-Lahti, P. Kuivalainen, J. E. Osterholm, and J. Laakso, *Synth. Metals, 28*: C461 (1989).
152. B. Wessling, in *Electronic Properties of Conjugated Polymers III* (H. Kuzmany, M. Mehring, and S. Roth, eds.), Springer-Verlag, Heidelberg, 1989, p. 447.
153. A. Andreatta, A. J. Heeger, and P. Smith, *Polym. Commun., 31*: 275 (1990).
154. Y. Cao and A. J. Heeger, *Synth. Metals, 52*: 193 (1992).
155. L. W. Shacklette and C. C. Han, *Synth. Metals*, in press.

Langmuir–Blodgett Films

156. A. K. M. Rahman, L. Samuelson, D. Minehan, S. Clough, S. Tripathy, T. Inagaki, X. Q. Yang, T. A. Skotheim, and Y. Okamoto, *Synth. Metals, 28*: C237 (1989).
157. M. Ando, Y. Watanabe, T. Syoda, K. Honda, and T. Shimidzu, *Thin Solid Films, 179*: 225 (1989).
158. C. L. Callender, C. A. Carere, G. Daoust, and M. Leclerc, *Thin Solid Films, 204*, 451 (1991).
159. I. Watanabe, K. Hong, and M. F. Rubner, *J. Chem. Soc. Chem. Commun.*, 123 (1989).
160. I. Watanabe, K. Hong, and M. F. Rubner, *Langmuir, 6*: 1164 (1990).
161. P. Yli-Lahti, E. Punkka, H. Stubb, P. Kuivalainen, and J. Laakso, *Thin Solid Films, 179*: 221 (1989).
162. M. A. Sato, S. Okada, H. Matsuda, H. Nakanishi, and M. Kato, *Thin Solid Films, 179*: 429 (1989).
163. A. T. Royappa and M. F. Rubner, *Langmuir, 8*: 3168 (1992).

Light-Emitting Diodes

164. J. H. Burroughes, D. D. C. Bradley, A. R. Brown, R. N. Marks, K. Mackay, R. H. Friend, P. L. Burns, and A. B. Holmes, *Nature, 347*: 539 (1990).
165. R. H. Friend, *Synth. Metals, 51*: 357 (1992).
166. D. Braun and A. J. Heeger, *Appl. Phys. Lett., 58*: 1982 (1991).
167. G. Grem, G. Leditzky, B. Ullrich, and G. Leising, *Adv. Mater., 4*: 36 (1992).
168. G. Grem, G. Leditzky, B. Ullrich, and G. Leising, *Synth. Metals, 51*: 383 (1992).

169. G. Gustafsson, Y. Cao, G. M. Treacy, F. Klavetter, N. Colaneri, and A. J. Heeger, *Nature*, *357*: 477 (1992).
170. Y. Ohmori, M. Uchida, K. Muro, and K. Yoshino, *Jpn. J. Appl. Phys.*, *30*: L1938 (1991).
171. T. Yamamoto, H. Wakayama, T. Fukuda, and T. Kanbara, *J. Phys. Chem.*, *96*: 8677 (1992).
172. P. L. Burn, A. B. Holmes, A. Kraft, D. D. C. Bradley, A. R. Brown, and R. H. Friend, *J. Chem. Soc. Chem. Commun.*, 32 (1992).
173. P. L. Burn, A. B. Holmes, A. Kraft, D. D. C. Bradley, A. R. Brown, R. H. Friend, and R. W. Gymer, *Nature*, *356*: 47 (1992).
174. D. A. Halliday, P. L. Burn, R. H. Friend, and A. B. Holmes, *J. Chem. Soc. Chem. Commun.*, 1685 (1992).
175. R. Friend, D. Bradley, and A. Holmes, *Phys. World*, 42 (1992).
176. N. C. Greenham, R. H. Friend, A. R. Brown, D. D. C. Bradley, and K. Pichler, *Proc. SPIE*, *1910*: 84 (1993).
177. C. Hosokawa, N. Kawasaki, S. Sakamoto, and T. Kusumoto, *Appl. Phys. Lett.*, *62*: 2503 (1992).
178. R. Dagani, *C&EN*, June 29, 1992, p. 27.

Electrochromic Windows

179. C. M. Lambert, *Solar Energy Mater.*, *11*: 1 (1984).
180. W. R. Shieh, S. C. Yang, C. Mazzacco, and J. H. Hwang, *Mater. Res. Soc. Symp. Proc.*, *173*: 329 (1990).
181. E. A. R. Duek, M. A. De Paoli, and M. Mastragostino, *Adv. Mater.*, *4*: 287 (1992).
182. M. Gazard, p. 673 in Ref. 9.
183. T. Kobayashi, H. Yoneyama, and H. Tamura, *J. Electroanal. Chem.*, *161*: 419 (1984).
184. M. Leclerc, J. Guay, and L. H. Dao, *J. Electroanal. Chem.*, *251*: 21 (1989).
185. M. T. Nguyen and L. H. Dao, *J. Electrochem. Soc.*, *136*: 2131 (1989).
186. L. H. Dao and M. T. Nguyen, in *Electrochromic Materials* (M. K. Carpenter and D. A. Corrigan, eds.), ECS 90-2, 246 (1990).
187. L. H. Dao and M. T. Nguyen, *Proc. 24th IEEE Intersociety Energy Conversion, Engineering Conf.*, *4*, 1737 (1989).

Batteries

188. M. G. Kanatzidis, *C&EN*, Dec. 3, 1990, p. 36.
189. T. Enomoto and D. P. Aller, Bridgestone Corp., Bridgestone New Release, Sept. 9, 1987, International Public Relations, Tokyo.

Sensors

190. S. E. Wolowacz, B. F. Y. Yon Hin, and C. R. Lowe, *Anal. Chem.*, *64*: 1541 (1992).
191. Y. Kajiya, H. Sugai, C. Iwakura, and H. Yoneyama, *Anal. Chem.*, *63*: 49 (1991).
192. M. Shaolin, X. Huaiguo, and Q. Bidong, *J. Electroanal. Chem.*, *304*: 7 (1991).

193. C. G. J. Koopal, M. C. Feiters, R. M. Nolte, D. de Ruiter, R. B. M. Schasfoort, R. Czajka, and H. Van Kempen, *Synth. Metals, 51*: 397 (1992).
194. B. Ballarin, C. J. Brumlik, D. R. Lawson, W. Liang, L. S. Van Dyke, and C. R. Martin, *Anal. Chem., 64*: 2647 (1992).
195. G. Zotti, *Synth. Metals, 51*: 373 (1992).
196. M. Nishizawa, T. Matsue, and I. Uchida, *Anal. Chem., 64*: 2642 (1992).
197. D. T. Hoa, T. N. S. Kumar, N. S. Punekar, R. S. Srinivasa, R. Lal, and A. Q. Contractor, *Anal. Chem., 64*: 2645 (1992).
198. T. Hanawa and H. Yoneyana, *Synth. Metals, 30*: 341 (1989).
199. C. Nylander, M. Armgarth, and I. Lundstrom, *Proc. International Meeting Chemical Sensors*, Fukuoka (T. Seyema, K. Fueki, J. Shiokawa, and S. Suzuki, eds.), 1983, p. 203.
200. J. J. Miasik, A. Hooper, and B. C. Tofield, *J. Chem. Soc. Faraday Trans. 1, 82*: 1117 (1983).
201. N. M. Ratcliffe, *Anal. Chim. Acta, 239*: 257 (1990).
202. M. Josowicz and J. Janata, *Anal. Chem., 58*: 514 (1986).
203. S. Chao and M. S. Wrighton, *J. Am. Chem. Soc., 109*: 6627 (1987).

Lithiography

204. M. Angelopoulos, J. M. Shaw, W. S. Huang, and R. D. Kaplan, *Mol. Cryst. Liq. Cryst., 189*: 221 (1990).
205. M. Angelopoulos, J. M. Shaw, W. S. Huang, M. A. Lecorre, and M. Tisser, *J. Vacuum Sci. Technol. B, 9*(6): 3428 (1991).
206. W. S. Huang, M. Angelopoulos, J. R. White, and J. M. Park, *Mol. Cryst. Liq. Cryst., 189*: 227 (1990).
207. M. Angelopoulos, M. Patel, J. M. Shaw, N. C. Labianca, and S. A. Rishton, *J. Vacuum Sci. Technol. B, 11*(6): 2794 (1993).
208. M. S. Abdou, M. I. Arroyo, G. Diaz-Quijada, and S. Holdcroft, *Chem. Mater., 3*: 1003 (1991).
209. M. S. Abdou, Z. W. Xie, A. M. Leung, and S. Holdcroft, *Synth. Metals, 52*: 159 (1992).
210. L. H. Dao and M. T. Nguyen, *Polym. Prepr., 33*: 408 (1992).

Others

211. G. Horowitz, X. Z. Peng, D. Fichou, and F. Garnier, *Synth. Metals, 51*: 419 (1992).
212. Y. Ohmori, H. Takahashi, K. Muro, M. Uchida, T. Kawai, and K. Yoshino, *Jpn. J. Appl. Phys., 30*: L610 (1991).
213. J. Lei, W. Liang, C. J. Brumlik, and C. Martin, *Synth. Metals, 47*: 351 (1992).
214. D. D. C. Bradley, *Chem. Britain*, 719 (1991).

Index